KU-693-049

SITY OF
GHAM

Environmental Microbiology

Environmental Microbiology
From genomes to biogeochemistry

Eugene L. Madsen

Cornell University, Department of Microbiology

Blackwell
Publishing

© 2008 by Eugene L. Madsen

BLACKWELL PUBLISHING
350 Main Street, Malden, MA 02148-5020, USA
9600 Garsington Road, Oxford OX4 2DQ, UK
550 Swanston Street, Carlton, Victoria 3053, Australia

The right of Eugene L. Madsen to be identified as the Author of this Work has been asserted in accordance with the UK Copyright, Designs, and Patents Act 1988.

All rights reserved. No part of this publication may be reproduced, stored in a retrieval system, or transmitted, in any form or by any means, electronic, mechanical, photocopying, recording or otherwise, except as permitted by the UK Copyright, Designs, and Patents Act 1988, without the prior permission of the publisher.

First published 2008 by Blackwell Publishing Ltd

2   2011

Library of Congress Cataloging-in-Publication Data

Madsen, Eugene L.
   Environmental microbiology / Eugene L. Madsen.
       p.   cm.
   Includes bibliographical references and index.
   ISBN-13: 978-1-4051-3647-1 (hardcover: alk. paper)   *26350238*
   ISBN-10: 1-4051-3647-2 (hardcover: alk. paper)
   1. Microbial ecology.   I. Title.   [DNLM: 1. Environmental Microbiology.   QW 55 M183e 2008]

   QR100.M33 2008
   579'.17—dc22

                                                                                  2007018804

A catalogue record for this title is available from the British Library.

Set in 10/12.5pt Meridien
by Graphicraft Limited, Hong Kong
Printed and bound in Singapore
by Markono Print Media Pte Ltd

The publisher's policy is to use permanent paper from mills that operate a sustainable forestry policy, and which has been manufactured from pulp processed using acid-free and elementary chlorine-free practices. Furthermore, the publisher ensures that the text paper and cover board used have met acceptable environmental accreditation standards.

For further information on
Blackwell Publishing, visit our website:
www.blackwellpublishing.com

Contents

Preface

Over the past 20 years, environmental microbiology has emerged from a rather obscure, applied niche within microbiology to become a prominent, ground-breaking area of biology. Environmental microbiology's rise in scholarly stature cannot be simply explained. But one factor was certainly pivotal in bringing environmental microbiology into the ranks of other key biological disciplines. That factor was molecular techniques. Thanks largely to Dr. Norman Pace (in conjunction with his many students) and Gary Olson and Carl Woese, nucleic acid analysis procedures began to flow into environmental microbiology in the mid-1980s. Subsequently, a long series of discoveries have flooded out of environmental microbiology. This two-way flow is constantly accelerating and the discoveries increasingly strengthen the links between environmental microbiology and core areas of biology that include evolution, taxonomy, physiology, genetics, environment, genomics, and ecology.

This textbook has grown from a decade of efforts aimed at presenting environmental microbiology as a coherent discipline to both undergraduate and graduate students at Cornell University. The undergraduate course was initially team-taught by Drs. Martin Alexander and William C. Ghiorse. Later, W. C. Ghiorse and I taught the course. Still later I was the sole instructor. Still later I became instructor of an advanced graduate version of the course. The intended audience for this text is upper-level undergraduates, graduate students, and established scientists seeking to expand their areas of expertise.

Environmental microbiology is inherently multidisciplinary. It provides license to learn many things. Students in university courses will rebel if the subject they are learning fails to develop into a coherent body of knowledge. Thus, presenting environmental microbiology to students in a classroom setting becomes a challenge. How can so many disparate areas of science (e.g., analytical chemistry, geochemistry, soil science, limnology, public health, environmental engineering, ecology, physiology, biogeochemistry, evolution, molecular biology, genomics) be presented as a unified body of information?

This textbook is my attempt to answer that question. Perfection is always evasive. But I have used five core concepts (see Section 1.1) that are reiterated throughout the text, as criteria for selecting and organizing the contents of this book.

The majority of figures presented in this book appear as they were prepared by their original authors in their original sources. This approach is designed to illustrate for the reader that advancements in environmental microbiology are a community effort.

A website with downloadable artwork and answers to study questions is available to instructors at www.blackwellpublishing.com/madsen

I hope this book will stimulate new inquiries into what I feel is one of the most fascinating current areas of science. I welcome comments, suggestions, and feedback from readers of this book. I thank the many individuals who provided both direct and indirect sources of information and inspiration. I am particularly grateful to P. D. Butler for assistance in manuscript preparation, to J. Yavitt who guided me to the right destinations in the biogeochemistry literature, and to W. C. Ghiorse for his unbounded enthusiasm for the art and science of microbiology. Constructive comments from several anonymous reviewers are acknowledged. I also apologize for inadvertently failing to include and/or acknowledge scientific contributions from fellow environmental microbiologist friends and colleagues.

Eugene Madsen

Significance, History, and Challenges of Environmental Microbiology

This chapter is designed to instill in the reader a sense of the goals, scope, and excitement that permeate the discipline of environmental microbiology. We begin with five core concepts that unify the field. These are strengthened and expanded throughout the book. Next, an overview of the significance of environmental microbiology is presented, followed by a synopsis of key scholarly events contributing to environmental microbiology's rich heritage. The chapter closes by reminding the reader of the complexity of Earth's biogeochemical systems and that strategies integrating information from many scientific disciplines can improve our understanding of biosphere function.

1.1 CORE CONCEPTS CAN UNIFY ENVIRONMENTAL MICROBIOLOGY

Environmental microbiology is inherently multidisciplinary. Its many disparate areas of science need to be presented coherently. To work toward that synthesis, this text uses five recurrent core concepts to bind and organize facts and ideas.

Core concept 1. Environmental microbiology is like a child's picture of a house – it has (at least) five sides (a floor, two vertical sides, and two sloping roof pieces). The floor is evolution. The walls are thermodynamics and habitat diversity. The roof pieces are ecology and physiology. To learn environmental microbiology we must master and unite all sides of the house.

Core concept 2. The prime directive for microbial life is survival, maintenance, generation of adenosine triphosphate (ATP), and sporadic growth (generation of new cells). To predict and understand microbial processes in real-world waters, soils, sediments, and other habitats, it is helpful to keep the prime directive in mind.

Core concept 3. There is a mechanistic series of linkages between our planet's habitat diversity and what is recorded in the genomes of microorganisms found in the world today. Diversity in habitats is synonymous with diversity in selective pressures and resources. When operated upon by forces of evolution, the result is molecular, metabolic, and physiological diversity found in extant microorganisms and recorded in their genomes.

Core concept 4. Advancements in environmental microbiology depend upon convergent lines of independent evidence using many measurement procedures. These include microscopy, biomarkers, model cultivated microorganisms, molecular biology, and genomic techniques applied to laboratory- and field-based investigations.

Core concept 5. Environmental microbiology is a dynamic, methods-limited discipline. Each methodology used by environmental microbiologists has its own set of strengths, weaknesses, and potential artifacts. As new methodologies deliver new types of information to environmental microbiology, practitioners need a sound foundation that affords interpretation of the meaning and place of the incoming discoveries.

1.2 SYNOPSIS OF THE SIGNIFICANCE OF ENVIRONMENTAL MICROBIOLOGY

With the formation of planet Earth 4.6×10^9 years ago, an uncharted series of physical, chemical, biochemical, and (later) biological events began to unfold. Many of these events were slow or random or improbable. Regardless of the precise details of how life developed on Earth, (see Sections 2.3–2.7), it is now clear that for ~70% of life's history, prokaryotes were the sole or dominant life forms. Prokaryotes (*Bacteria* and *Archaea*) were (and remain) not just witnesses of geologic, atmospheric, geochemical, and climatic changes that have occurred over the eons. Prokaryotes are also active participants and causative agents of many geochemical reactions found in the geologic record. Admittedly, modern eukaryotes (especially land plants) have been major biogeochemical and ecological players on planet Earth during the most recent 1.4×10^9 years. Nonetheless, today, as always, prokaryotes remain the "hosts" of the planet. Prokaryotes comprise ~60% of the total biomass (Whitman et al., 1998; see Chapter 4), account for as much as 60% of total respiration of some terrestrial habitats (Velvis, 1997; Hanson et al., 2000), and also colonize a variety of Earth's habitats devoid of eukaryotic life due to topographic, climatic and geochemical extremes of elevation, depth, pressure, pH, salinity, heat, or light.

The Earth's habitats present complex gradients of environmental conditions that include variations in temperature, light, pH, pressure, salinity, and both inorganic and organic compounds. The inorganic materials range from elemental sulfur to ammonia, hydrogen gas, and methane and

Table 1.1

Microorganisms' unique combination of traits and their broad impact on the biosphere

Traits of microorganisms	Ecological consequences of traits
Small size	Geochemical cycling of elements
Ubiquitous distribution throughout Earth's habitats	Detoxification of organic pollutants
High specific surface areas	Detoxification of inorganic pollutants
Potentially high rate of metabolic activity	Release of essential limiting nutrients from the biomass in one generation to the next
Physiological responsiveness	Maintaining the chemical composition of soil, sediment, water, and atmosphere required by other forms of life
Genetic malleability	
Potential rapid growth rate	
Unrivaled nutritional diversity	
Unrivaled enzymatic diversity	

the organic materials range from cellulose to lignin, fats, proteins, lipids, nucleic acid, and humic substances (see Chapter 7). Each geochemical setting (e.g., anaerobic peatlands, oceanic hydrothermal vents, soil humus, deep subsurface sediments) features its own set of resources that can be physiologically exploited by microorganisms. The thermodynamically governed interactions between these resources, their settings, microorganisms themselves, and 3.6×10^9 years of evolution are probably the source of metabolic diversity of the microbial world.

Microorganisms are the primary agents of geochemical change. Their unique combination of traits (Table 1.1) cast microorganisms in the role of recycling agents for the biosphere. Enzymes accelerate reaction rates between thermodynamically unstable substances. Perhaps the most ecologically important types of enzymatic reactions are those that catalyze oxidation/reduction reactions between electron donors and electron acceptors. These allow microorganisms to generate metabolic energy, survive, and grow. Microorganisms procreate by carrying out complex, genetically regulated sequences of biosynthetic and assimilative intracellular processes. Each daughter cell has essentially the same macromolecular and elemental composition as its parent. Thus, integrated metabolism of all nutrients (e.g., carbon, nitrogen, phosphorus, sulfur, oxygen, hydrogen, etc.) is implicit in microbial growth. This growth and survival of microorganisms drives the geochemical cycling of the elements, detoxifies many contaminant organic and inorganic compounds, makes essential nutrients present in the biomass of one generation available to the next, and maintains the conditions required by other inhabitants of the biosphere (Table 1.1). Processes carried out by microorganisms in soils, sediments, oceans, lakes, and groundwaters have a major impact on environmental quality, agriculture, and global climate change. These processes are also the basis for current and emerging biotechnologies with industrial and environmental applications (see Chapter 8). Table 1.2 presents

Table 1.2

Examples of nutrient cycling and physiological processes catalyzed by microorganisms in biosphere habitats (reproduced with permission from *Nature Reviews Microbiology* from Madsen, E.L. 2005. Identifying microorganisms responsible for ecologically significant biogeochemical processes. *Nature Rev. Microbiol.* **3**:439–446. Macmillan Magazines, www.nature.com/reviews)

Nutrient cycle	Process	Nature of process	Typical habitat	References
Carbon	Photosynthesis	Light-driven CO_2 fixation into biomass	FwS, Os, Ow	Pichard et al., 1997; Partensky et al., 1999; Ting et al., 2002
	Carbon respiration	Oxidation of organic C to CO_2	Sl	Heemsbergen, 2004
	Cellulose decomposition	Depolymerization, respiration	Sl	Jones et al., 1998
	Methanogenesis	Methane production	FwS, Os, Sw	Conrad, 1996; Schink, 1997
	Aerobic methane oxidation	Methane becomes CO_2	Fw, Ow, Sl	Segers, 1998; Bull et al., 2000
	Anaerobic methane oxidation	Methane becomes CO_2	Os	Boetius et al., 2000
Biodegradation	Synthetic organic compounds	Decomposition, CO_2 formation	All habitats	Alexander, 1999; Boxall et al., 2004
	Petroleum hydrocarbons	Decomposition, CO_2 formation	All habitats	Van Hamme et al., 2003
	Fuel additives (MTBE)	Decomposition, CO_2 formation	Gw, Sl, Sw	Deeb et al., 2003
	Nitroaromatics	Decomposition	Gw, Sl, Sw	Spain et al., 2000; Esteve-Núñez et al., 2001
	Pharmaceuticals, personal care products	Decomposition	Gw, Sl, Sw	Alexander, 1999; Ternes et al., 2004
	Chlorinated solvents	Compounds are dechlorinated via respiration in anaerobic habitats	Gw, Sl, Sw	Maymo-Gatell et al., 1997; Adrian et al., 2000

	Process	Description	Environment	Reference
Nitrogen	Nitrogen fixation	N_2 gas becomes ammonia	Ow, Sl	Karl et al., 2002
	Ammonium oxidation	Ammonia becomes nitrite and nitrate	Sl, Sw	Stark and Hart, 1997; Kowalchuk and Stephen, 2001
	Anaerobic ammonium oxidation	Nitrite and ammonia become N_2 gas	Os, Sw	Dalsgaard et al., 2003; van Niftrik et al., 2004
	Denitrification	Nitrate is used as an electron acceptor and converted to N_2 gas	Sl, Sw	Zumft, 1997; van Breemen et al., 2002
Sulfur	Sulfur oxidation	Sulfide and sulfur become sulfate	Os	Taylor and Wirsen, 1997
	Sulfate reduction	Sulfate is used as an electron acceptor and converted to sulfur and sulfide	Os	Habicht and Canfield, 1996
Other elements	Hydrogen oxidation	Hydrogen is oxidized to H^+, electrons reduce other substances	Sl, Os, Sw	Schink, 1997
	Mercury methylation and reduction	Organic mercury is formed and mercury ion is converted to metallic mercury	FwS, Os	Morel et al., 1998; Sigel et al., 2005
	(Per)chlorate reduction	Oxidants in rocket fuel and other sources are converted to chloride	Gw	Coates and Achenbach, 2004
	Uranium reduction	Uranium oxyanion is used as an electron acceptor; hence immobilized	Gw	Lovley, 2003
	Arsenate reduction	Arsenic oxyanion is used as an electron acceptor; hence toxicity is diminished	FwS, Gw	Oremland and Stolz, 2003
	Iron oxidation, acid mine drainage	Iron sulfide ores are oxidized, strong acidity is generated	FwS, Gw	Edwards et al., 2000

Fw, freshwater; FwS, freshwater sediment; Gw, groundwater; Os, ocean sediments; Ow, ocean waters; Sl, soil; Sw, sewage.

a sampling of the ecological and biogeochemical processes that microorganisms catalyze in aquatic or terrestrial habitats. Additional details of biogeochemical processes and ways to recognize and understand them are presented in Chapters 3 and 7.

1.3 A BRIEF HISTORY OF ENVIRONMENTAL MICROBIOLOGY

Early foundations of microbiology rest with microscopic observations of fungal sporulation (by Robert Hooke in 1665) and "wee animalcules" – true bacterial structures (by Antonie van Leeuwenhoek in 1684). In the latter half of the nineteenth century, Ferdinand Cohn, Louis Pasteur, and Robert Koch were responsible for methodological innovations in aseptic technique and isolation of microorganisms (Madigan and Martinko, 2006). These, in turn, allowed major advances pertinent to spontaneous generation, disease causation, and germ theory.

Environmental microbiology also experienced major advancements in the nineteenth century; these extend through to the present. Environmental microbiology's roots span many continents and countries (Russia, Japan, Europe, and England) and a complex tapestry of contributions has developed. To a large degree, the challenges and discoveries in environmental microbiology have been habitat-specific. Thus, one approach for grasping the history and traditions of environmental microbiology is to recognize subdisciplines such as marine microbiology, soil microbiology, rumen microbiology, sediment microbiology, geomicrobiology, and subsurface microbiology. In addition, the contributions from various centers of training can also sometimes be easily discerned. These necessarily revolved around various investigators and the institutions where they were based.

As early as 1838 in Germany, C. G. Ehrenberg was developing theories about the influence of the bacterium, *Gallionella ferruginea*, on the generation of iron deposits in bogs (Ehrlich, 2002). Furthermore, early forays into marine microbiology by A. Certes (in 1882), H. L. Russell, P. Regnard, B. Fischer, and P. and G. C. Frankland allowed the completion of preliminary surveys of microorganisms from far-ranging oceanic waters and sediments (Litchfield, 1976).

At the University of Delft (the Netherlands) near the end of the nineteenth century, M. W. Beijerinck (Figure 1.1) founded the Delft School traditions of elective enrichment techniques (see Section 6.2) that allowed Beijerinck's crucial discoveries including microbiological transformations of nitrogen and carbon, and also other elements such as manganese (van Niel, 1967; Atlas and Bartha, 1998; Madigan and Martinko, 2006). The helm of the Delft School changed hands from Beijerinck to A. J. Kluyver, and the traditions have been continued in the Netherlands, Germany, and other parts of Europe through to the present. After training in Delft with Beijerinck and Kluyver, C. B. van Niel was asked by L. G. M. Baas Becking to established a research program at Stanford University's Hopkins Marine Station (done in 1929), where R. Y. Stainer, R. Hungate, M. Doudoroff and many others were trained, later establishing their own research programs at other institutions in the United States (van Niel, 1967).

S. Winogradsky (Figure 1.2) is regarded by many as the founder of soil microbiology (Atlas and Bartha, 1998). Working in the latter part of the nineteenth and early decades of the twentieth centuries, Winogradsky's career contributed immensely to our knowledge of soil and environmental microbiology, especially regarding microbial metabolism of sulfur, iron, nitrogen, and manganese. In 1949, much of Winogradsky's work was published as a major treatise entitled, *Microbiologie du sol, problémes et methods: cinquante ans de recherches. Oeuvres Complétes* (Winogradsky, 1949).

Many of the marine microbiologists in the early twentieth century focused their attention on photoluminescent bacteria (E. Pluger, E. W. Harvey, H. Molisch, W. Beneche, G. H. Drew, and J. W. Hastings). Later, transformation by marine microorganisms of carbon and nitrogen were explored, as well as adaptation to low-temperature habitats (S. A. Waksman, C. E. ZoBell, S. J. Niskin, O. Holm-Hansen, and N. V. and V. S. Butkevich). The mid-twentieth century marine studies continued

Figure 1.1 Martinus Beijerinck (1851–1931). Founder of the Delft School of Microbiology, M. Beijerinck worked until the age of 70 at the University of Delft, the Netherlands. He made major discoveries in elective enrichment techniques and used them to advance the understanding of how microorganisms transform nitrogen, sulfur, and other elements. (Reproduced with permission from the American Society for Microbiology Archives, USA.)

exploration of the physiological and structural responses of microorganisms to salt, low temperature, and pressure (J. M. Shewan, H. W. Jannasch, R. Y. Morita, R. R. Colwell, E. Wada, A. Hattori, and N. Taga). Also, studies of nutrient uptake (J. E. Hobbie) and food chains constituting the "microbial loop" were conducted (L. R. Pomeroy).

Figure 1.2 Sergei Winogradsky (1856–1953). A major contributor to knowledge of soil microbiology, S. Winogradsy described microbial cycling of sulfur and nitrogen compounds. He developed the "Winogradsky column" for growing diverse physiological types of aerobic and anaerobic, heterotrophic and photosynthetic bacteria across gradients of oxygen, sulfur, and light. (Reproduced with permission from the Smith College Archives, Smith College.)

At Rutgers University, Selman A. Waksman was perhaps the foremost American scholar in the discipline of soil microbiology. Many of the Rutgers traditions in soil microbiology were initiated by J. Lipman, Waksman's predecessor (R. Bartha, personal communication; Waksman, 1952). Waksman produced numerous treatises that summarized the history, status, and frontiers of soil microbiology, often in collaboration with R. Starkey. Among the prominent works published by Waksman are "Soil microbiology in 1924: an attempt at an analysis and a synthesis" (Waksman, 1925), *Principles of Soil Microbiology* (Waksman, 1927), "Soil microbiology as a field of science" (Waksman, 1945), and *Soil Microbiology* (Waksman, 1952). A steady flow of Rutgers-based contributions to environmental microbiology continue to be published (e.g., Young and Cerniglia, 1995; Haggblom and Bossert, 2003).

In the 1920s and 1930s, E. B. Fred and collaborators, I. L. Baldwin and E. McCoy, comprised a unique cluster of investigators whose interests focused on the *Rhizobium*–legume symbiosis. Several decades later at the University of Wisconsin, T. D. Brock and his students made important contributions to microbial ecology, thermophily, and general microbiology. Another graduate of the University of Wisconsin, H. L. Ehrlich earned a Ph.D. in 1951 and, after moving to Rensselaer Polytechnic Institute, carried out studies on the bacteriology of manganese nodules, among other topics. Author of four comprehensive editions of *Geomicrobiology*, H. L. Ehrlich is, for many, the founder of this discipline.

Another University of Wisconsin graduate, M. Alexander, moved to Cornell University in 1955. For four decades prior to Alexander's arrival, soil microbiological research was conducted at Cornell by J. K. Wilson and F. Broadbent. From 1955 to the present, Alexander's contributions to soil microbiology have examined a broad diversity of phenomena, which include various transformations of nitrogen, predator–prey relations, microbial metabolism of pesticides and environmental pollutants, and advancements in environmental toxicology. Many environmental microbiologists have received training with M. Alexander and become prominent investigators, including J. M. Tiedje.

Other schools and individuals in Britain, Italy, France, Belgium and other parts of Europe, Japan, Russia and other parts of Asia, Africa, Australia, the United States and other parts of the Americas certainly have contributed in significant ways to advancements in environmental microbiology. An insightful review of the history of soil microbiology, with special emphasis on eastern European and Russian developments was written by Macura (1974).

The many historical milestones in the development of environmental microbiology (most of which are shared with broader fields of biology and microbiology) have been reviewed by Atlas and Bartha (1998), Brock (1961), Lechevalier and Solotorovsky (1965), Macura (1974), Madigan and Martinko (2006), van Niel (1967), Waksman (1925, 1927, 1952), and others. Some of the highlights are listed in Table 1.3.

Table 1.3
Selected landmark events in the history of environmental microbiology

- The first visualization of microscopic life by van Leeuwenhoek in 1684
- The role of microorganisms as causative agents of fermentations discovered by Pasteur in 1857
- The use of gelatin plates for enumeration of soil microorganisms by Koch in 1881
- Nitrogen fixation by nodules on the roots of legumes discovered by Hellriegel and Wilfarth in 1885
- The use of elective enrichment methods, by Beijerinck and Winogradsky, in the isolation of single organisms able to carry out ammonification, nitrification, and both symbiotic and nonsymbiotic nitrogen fixation
- Recognition of the diverse populations in soil (e.g., bacteria, fungi, algae, protozoa, nematodes, insect larvae)
- Documentation of anaerobic cellulose decomposition by Omelianskii in 1902
- The study of sulfur-utilizing phototrophic bacteria by van Niel and others
- The specificity of legume-nodulating bacteria (Fred et al., 1932)
- The discovery and development of antibiotics
- Direct microscopic methods of examining environmental microorganisms via staining and contact-slide procedures
- The development of radiotracer techniques
- A diversity of advancements in analytical chemistry for detecting and quantifying biochemically and environmentally relevant compounds
- Developments in molecular phylogeny (Woese, 1987, 1992; Pace, 1997)
- The application of molecular methods to environmental microbiology (Olsen et al., 1986; Pace et al., 1986; Amann et al., 1991, 1995; Ward et al., 1993; White, 1994; van Elsas et al., 1997; Madigan and Martinko, 2006)

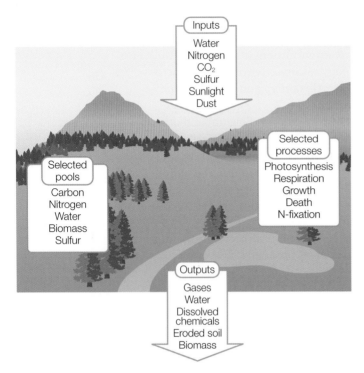

As this historical treatment reaches into the twenty-first century, the branches and traditions in environmental microbiology become so complex that patterns of individual contributions become difficult to discern. A complete list of schools, individual investigators, and their respective discoveries is beyond the scope of this section. The author apologizes for his biases, limited education, and any and all inadvertent omissions that readers may notice in this brief historical overview.

1.4 COMPLEXITY OF OUR WORLD

Although we humans are capable of developing ideas or concepts or models that partially describe the biosphere we live in, real-world complexity of ecological systems and subsystems remains generally beyond full scientific description. Figures 1.3 and 1.4 are designed to begin to develop for the reader a sense of the complexity of real-world ecosystems – in this case a temperate forested watershed. The watershed depicted in Figure 1.3 is open (energy and materials flow through it) and features dynamic

Figure 1.3 Watershed in a temperate forest ecosystem. Arrows show the inputs and outflows from the system. Reservoirs for carbon, nitrogen, and other nutrients include biomass, soil litter layer, soil mineral layer, subsoil, snow, streams, and lakes. Dominant physiological processes carried out by biota include photosynthesis, grazing, decomposition, respiration, nitrogen fixation, ammonification, and nitrification. Key abiotic processes include insolation (sunlight), transport, precipitation, runoff, infiltration, dissolution, and acid/base and oxidation/reduction reactions (see Table 1.4). Net budgets can be constructed for ecosystems; when inputs match outputs, the systems are said to be "steady state".

changes in time and space. The watershed system contains many components ranging from the site geology and soils to both small and large creatures, including microorganisms. Climate-related influences are major variables that, in turn, cause variations in how the creatures and their habitat interact. Biogeochemical processes are manifestations of such interactions. These processes include chemical and physical reactions, as well as the diverse physiological reactions and behavior (Table 1.4). The physical, chemical, nutritional, and ecological conditions for watershed inhabitants vary from the scale of micrometers to kilometers. Regarding temporal variability, in situ processes that directly and indirectly

Table 1.4
Types of biogeochemical processes that typically occur and interact in real-world habitats

Type	Processes
Physical	Insolation (sunlight), atmospheric precipitation, water infiltration, water evaporation, transport, erosion, runoff, dilution, advection, dispersion, volatilization, sorption
Chemical	Dissolution of minerals and organic compounds, precipitation, formation of secondary minerals, photolysis, acid/base reactions, reactions catalyzed by clay-mineral surfaces, reduction, oxidation, organic equilibria, inorganic equilibria
Biological	Growth, death, excretion, differentiation, food webs, grazing, migration, predation, competition, parasitism, symbiosis, decomposition of high molecular weight biopolymers to low molecular weight monomers, respiration, photosynthesis, nitrogen fixation, nitrification, denitrification, ammonification, sulfate reduction, sulfur oxidation, iron oxidation/reduction, manganese oxidation/reduction, anaerobic oxidation of methane, anaerobic oxidation of ammonia, acetogenesis, methanogenesis

influence fluxes of materials into, out of, and within the system are also dynamic.

At the scale of ~1 m, humans are able to survey habitats and map the occurrence of both abiotic (rocks, soils, gasses, water) and biotic (plants, animals) components of the watershed. At this scale, much progress has been made toward understanding ecosystems. Biogeochemical ecosystem ecologists have gained far-reaching insights into how such systems work by performing a variety of measurements in basins whose sealed bedrock foundations allow ecosystem budgets to be constructed (Figure 1.3). When integrated over time and space, the chemical constituents (water, carbon, nitrogen, sulfur, etc.) measured in incoming precipitation, in outflowing waters, and in storage reservoirs (lakes, soil, the biota) can provide a rigorous basis for understanding how watersheds work and how they respond to perturbations (Likens and Bormann, 1995). Understanding watershed (as well as global) biogeochemical cycles relies upon rigorous data sets and well-defined physical and conceptual boundaries. For a given system, regardless of its size, if it is in steady state, the inputs must equal the outputs (Figure 1.3). By the same token, if input and output terms for a given system are not in balance, key biogeochemical parameters of interest may be changing with time. Net loss or gain is dependent on relative rates of consumption and production. Biogeochemical data sets provide a means for answering crucial ecological questions such as: Is the system in steady state? Are carbon and nitrogen accruing or diminishing? Does input of atmospheric pollutants impact ecosystem function? What goods and services do intact watersheds provide in terms of water and soil quality? More details on measuring and modeling biogeochemical cycles are presented in Chapter 7.

Large-scale watershed data capture net changes in complex, open systems. Though profound and insightful, this approach leaves mechanistic microscale cause-and-effect linkages unaddressed. Measures of net change do not address dynamic controls on rates of processes that generate (versus those that consume) components of a given nutrient pool. Indeed, the intricate microscale interactions between biotic and abiotic field processes are often masked in data gathered in large-scale systems. Thus, ecosystem-level biogeochemical data may often fail to satisfy the scientific need for details of the processes of interest. An example of steps toward a mechanistic understanding of ecosystem process is shown in Figure 1.4. This model shows a partial synthesis of ecosystem processes that govern the fate of nitrogen in a watershed. Inputs, flows, nutrient pools, biological players, physiological reactions, and transport processes are depicted. Understanding and measuring the sizes of nitrogenous pools, their transformations, rates, fluxes, and the active biotic agents represents a major challenge for both biogeochemists and microbiologists. Yet Figure 1.4 considerably simplifies the processes that actually occur in real-world watersheds because many details are missing and comparably complex reactions and interactions apply simultaneously to other nutrient elements (C, S, P, O, H, etc.). Consider a data set in which concentrations of ammonium (a key form of nitrogen) are found to fluctuate in stream sediments. Interpreting such field measurements is very difficult because the ammonium pool at any given moment is controlled by processes of production (e.g., ammonification by microorganisms), consumption (e.g., aerobic and anaerobic ammonia-oxidizing microorganisms, nutrient uptake by all organisms) and transport (e.g., entrainment in flowing water, diffusion, dilution, physical disturbance of sediment). Clearly, the many compounded intricacies of nutrient cycling and trophic and biochemical interactions in a field habitat make biogeochemical processes, especially those catalyzed by microorganisms, difficult to decipher.

1.5 MANY DISCIPLINES AND THEIR INTEGRATION

- Given the complexity of real-world habitats that are home to microorganisms (see above), what is to be done?
- How can we contend with complexity?
- What approaches can productively yield clear information that enhances our understanding of the role of microorganisms in maintaining our world?
- How do microorganisms carry out specific transformations on specific compounds in soils, sediments, and waters?

Answer: The optimistic answer to these questions is simple: We use the many tools on hand to twenty-first century science.

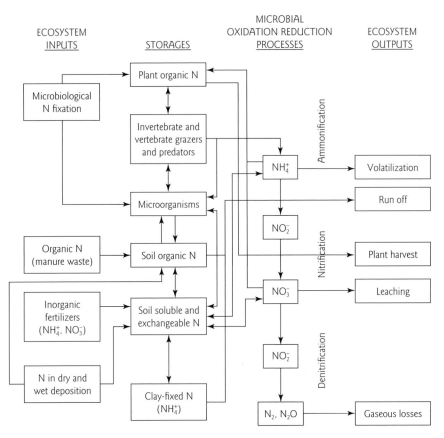

Figure 1.4 Flow model of nitrogen (N) cycling in terrestrial ecosystems. Shown are basic inputs, storages, microbial processes, outputs, and both biotic and abiotic interactions. (Reprinted and modified with permission from Madsen, E.L. 1998. Epistemology of environmental microbiology. *Environ. Sci. Technol.* **32**:429–439. Copyright 1998, American Chemical Society.)

The principles are sound, the insights are broad, and the sophisticated technologies are ever expanding. To counterbalance the challenges of ecosystem complexity, we can utilize: (i) robust, predictable rules of chemical thermodynamics, geochemical reactions, physiology, and biochemistry; (ii) measurement techniques from analytical chemistry, hydrogeology, physiology, microbiology, molecular biology; and (iii) compound-specific properties such as solubility, volatility, toxicity, and susceptibility to biotic and abiotic reactions. A partial listing of the many areas of science that contribute to advancements in environmental microbiology, with accompanying synopses and references, appears in Table 1.5.

Conceptually, environmental microbiology resides at the interface between two vigorously expanding disciplines: environmental science and

Table 1.5
Disciplines that contribute to environmental microbiology

Discipline	Subject matter and contribution to Environmental Microbiology	References
Environmental microbiology	The study of microorganisms that inhabit the Earth and their roles in carrying out processes in both natural and human-made systems; emphasis is on interfaces between environmental sciences and microbial diversity	Maier et al., 2000; Rochelle, 2001; Spencer and Ragout de Spencer, 2004
Microbial ecology	The study of interrelationships between microorganisms and their biotic and abiotic surroundings	Atlas and Bartha, 1998; Burlage et al., 1998; Staley and Reysenbach, 2002; Osborn, 2003; McArthur, 2006
Soil microbiology	Environmental microbiology and microbial ecology of the soil habitat; with emphasis on nutrient cycling, plant and animal life, and terrestrial ecosystems	van Elsas et al., 1997; Huang et al., 2002; Sylvia et al., 2005
Aquatic microbiology	Environmental microbiology and microbial ecology of aquatic habitats (oceans, lakes, streams, groundwaters)	Ford, 1993; Bitton, 1999; Kirchman, 2000
Microbiology	Holistic study of the function of microbial cells and their impact on medicine, industry, environment, and technology	Madigan and Martinko, 2006; Schaechter et al., 2006
Microbial physiology	Integrated mechanistic examination of bacterially mediated processes, especially growth and metabolism	Gottschalk, 1986; White, 1995; Lengeler et al., 1999
Microscopy	The use of optics, lenses, microscopes, imaging devices, and image analysis systems to visualize small structures	Murphy, 2001
Biochemistry	Molecular examination of the structure and function of subcellular processes, especially ATP generation, organelles, biopolymers, enzymes, and membranes	Devlin, 2001; Nicholls and Ferguson, 2002; Nelson and Cox, 2005
Biotechnology	The integrated use of biochemistry, molecular, biology, genetics, microbiology, plant and animal science, and chemical engineering to achieve industrial goods and services	Glick and Pasternack, 2003
Biogeochemisty	Systems approach to the chemical reactions between biological, geological, and atmospheric components of the Earth	Schlesinger, 1997; Fenchel et al., 1998
Microbial genetics	Molecular mechanistic basis of heredity, evolution, mutation in prokaryotes, and their biotechnological application	Snyder and Champness, 2003
Omics	Umbrella term that encompasses bioinfomatics-based systematic analysis of genes (genomics), proteins (proteomics), mRNA (transcriptomics), metabolites (metabolomics), etc.	Baxevanis and Ouellette, 2001; Sensen, 2002; Twyman, 2004; Zhou et al., 2004

Table 1.5 *Continued*

Discipline	Subject matter and contribution to Environmental Microbiology	References
Aquatic chemistry	Fundamental reactions of aqueous inorganic and organic chemistry and their quantification based on thermodynamics, equilibrium, and kinetics	Stumm and Morgan, 1996; Millero, 2001; Schwarzenbach et al., 2002
Geochemistry	Chemical basis for rock–water interactions involving thermodynamics, mineral equilibria, and solid-, liquid-, and vapor-phase reactions	Albaréde, 2003; Andrews, 2004
Soil science	Study of the intrinsic properties of soils and examination of physical, chemical, and biotic processes that lead to soil formation; the crucial role of soils in agriculture and ecosystems	Brady and Weil, 1999
Limnology	The study of freshwater ecosystems, especially lakes and streams	Wetzel and Likens, 2000; Wetzel, 2001
Oceanography	The study of saltwater ecosystems, especially oceans	Sverdrup et al., 2003
Hydrogeology	The study of the physical flow and migration of water in geological systems	Fetter, 1994
Analytical chemistry	Methods and technologies for detecting, separating, and identifying molecular structures of organic and inorganic compounds	Fifield, 2000; Christian, 2003
Civil and environmental engineering	Physical, chemical, hydraulic, and biological principles applied to the quantitative design of water supply, wastewater, and other engineering needs	Rittmann and McCarty, 2001; Tchobanoglous et al., 2002
Ecology	Integration of relationships between the biosphere and its inhabitants, with emphases on evolution, trophic dynamics, and emergent properties	Rickleffs and Miller, 2000; Krebs, 2001; Chapin et al., 2002
Environmental science	Multidisciplinary study of how the Earth functions, with emphasis on human influences on life support systems	Miller, 2004

microbial ecology (Figure 1.5). Both disciplines (spheres in Figure 1.5) seek to understand highly complex and underexplored systems. Each discipline currently consists of a significant body of facts and principles (green inner areas of spheres in Figure 1.5), with expanding zones of research (pink bands). But the chances are high that information awaiting discovery (blue areas) greatly exceeds current knowledge. For example, nearly all current information about prokaryotic microorganisms is based upon measurements performed on less than 6500 isolated species. These cultivated species represent approximately 0.1% of the total estimated diversity of microorganism in the biosphere (see Sections 5.1–5.7). The exciting new discoveries in environmental microbiology emerge by examining how

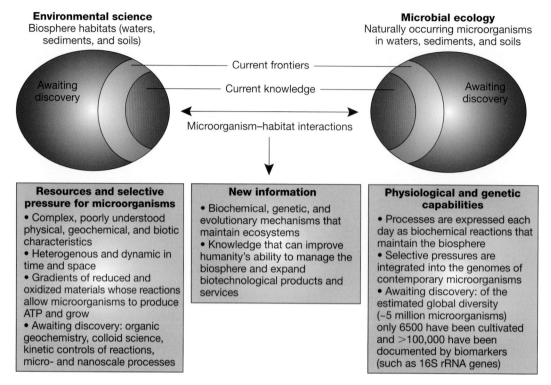

Figure 1.5 Conceptual representation of how the disciplines of environmental science (left sphere) and microbial ecology (right sphere) interact to allow new discoveries at the interface between microorganisms and their habitats. Information in each discipline is depicted as a combination of current knowledge, current frontiers, and knowledge awaiting discovery. Microbial Ecology and Environmental Microbiology have considerable disciplinary overlap (see Table 1.5); nonetheless, advancements in the latter are represented by the central, downward arrow. (Reproduced and modified with permission from *Nature Reviews Microbiology*, from Madsen, E.L. 2005. Identifying microorganisms responsible for ecologically significant biogeochemical processes. *Nature Rev. Microbiol.* **3**:439–446. Macmillan Magazines Ltd, www.nature.com/reviews.)

microorganisms interact with their habitats (central downward arrow in Figure 1.5).

Thus, the path toward progress in environmental microbiology involves multidisciplinary approaches, assembling convergent lines of independent evidence, and testing alternative hypotheses. Ongoing integration of new methodologies (e.g., from environmental science, microbial ecology and other disciplines listed in Table 1.5) into environmental microbiology ensures that the number of lines and the robustness of both their convergence and their tests will increase. A conceptual paradigm that graphically depicts the synergistic relationship between microbiological processes in field sites, reductionistic biological disciplines, and iterative methodological linkages between these disciplines is presented in Figure 1.6.

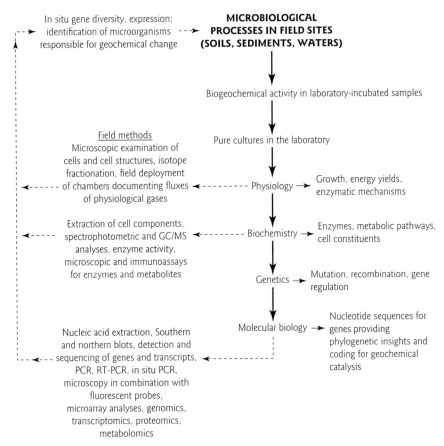

Figure 1.6 Paradigm for how the integration of disciplines and their respective methodologies can extend knowledge of environmental microbiology. Relationships between microorganisms responsible for field biogeochemical processes, reductionistic disciplines, and their application to microorganisms in field sites are depicted. The three different types of arrows indicate sequential refinements in biological disciplines (large downward-pointing solid arrows), resultant information (small arrows pointing to the right), and innovative methodological applications to naturally occurring microbial communities (dashed arrows). GC/MS, gas chromatography/mass spectrometry; PCR, polymerase chain reaction; RT, reverse transcriptase. (Reprinted and modified with permission from Madsen, E.L. 1998. Epistemology of environmental microbiology. *Environ. Sci. Technol.* **32**:429–439. Copyright 1998, American Chemical Society.)

Observations of microorganisms in natural settings instigate a series of procedures progressing through mixed cultures, pure cultures, and physiological, biochemical, genetic, and molecular biological inquiries that each stand alone scientifically. But appreciable new knowledge of naturally occurring microorganisms is gained when advancements from the pure biological sciences are directed back to microorganisms in their field habitats. These methodological advancements (shown as dashed arrows in Figure 1.6; see Chapter 6 for methodologies and their impacts) and the knowledge they generate accrue with each new cycle from field observations to molecular biology and back. Thus, integration of many disciplines is the path forward in environmental microbiology.

STUDY QUESTIONS

1 Core concept 1 presumes a two-dimensional house like that drawn on paper by school children. If you were to expand the concept to three dimensions, then two more walls would be required to keep the "house of environmental microbiology" from falling down. What two disciplines would you add and why? (Hint: for suggestions see Table 1.5.)

2 Core concept 3 uses the phrase "mechanistic series of linkages between our planet's habitat diversity and what is recorded in the genomes of microorganisms found in the world today". This is a hypothesis. If you wanted to test the hypothesis by completing measurements and assembling a data set, what would you do? Specifically, what experimental design would readily test the hypothesis? And what would you measure? What methodological barriers might hamper assembling a useful data set? How might these be overcome? (Hint: Sections 3.2 and 3.3 discusses genomic tools. Answer this question before *and* after reading Chapter 3.)

3 Many names of microorganisms are designed to recognize individual microbiologists who have contributed to the discipline. For instance, the genera *Pasteurella*, *Thauera*, and *Shewanella* are named after people. Similarly, the species designations in *Vibrio harveyii*, *Desulfomonile tiedjei*, *Thermotoga jannaschii*, *Nitrobacter winogradkyi*, and *Acetobacterium woodii* are also named for people. Use the world wide web or a resource like *Bergey's Manual of Systematic Bacteriology* or the *International Journal of Systematic and Evolutionary Microbiology* to discover the legacy of at least one person memorialized in the name of a microorganism.

4 Go for a walk outside to visit a forest, agricultural field, garden, or pond, stream or other body of water. Sit down and examine (literally, and aided by your imagination) the biotic and abiotic components of a cubic meter of water, sediment, or soil. This cubic meter defines a study system. What to you see? Divide a piece of paper into six columns with the headings "Materials and energy entering and leaving", "Inorganic materials", "Organic materials", "Organisms", "Interactions between system components", and "Biological processes". Add at least five entries under each column heading. Then imagine how each entry would change over the course of a year. Compare and contrast what you compiled in your listing with information in Figures 1.3–1.6 and Tables 1.2 and 1.4.

REFERENCES

Adrian, L., U. Szewyk, J. Wecke, and H. Görisch. 2000. Bacterial dehalorespiration with chlorinated benzenes. *Nature* **408**:580–583.

Albaréde, F. 2003. *Geochemistry: An introduction.* Cambridge University Press, New York.

Alexander, M. 1999. *Biodegradation and Bioremediation*, 2nd edn. Academic Press, San Diego, CA.

Amann, R.I., W. Ludwig, and K.-H. Schleifer. 1995. Phylogenetic identification and in situ detection of individual microbial cells without cultivation. *Microbiol. Rev.* **59**:143–169.

Amann, R., N. Springer, W. Ludwig, H.-D. Görtz, and K.-H. Schleifer. 1991. Identification in situ and phylogeny of uncultured bacterial endosymbionts. *Nature* **351**:161–164.

Andrews, J.E. 2004. *An Introduction to Environmental Chemistry*. Blackwell Science, Oxford.

Atlas, R.M. and R. Bartha. 1998. *Microbial Ecology: Fundamentals and applications*, 4th edn. Benjamin Cummings, Menlo Park, CA.

Baxevanis, A.D. and B.F.F. Ouellette. 2001. *Bioinformatics: A practical guide to the analysis of genes and proteins*, 2nd edn. Wiley Interscience, New York.

Bitton, G.E. 1999. *Wastewater Microbiology*, 2nd edn. Wiley Liss, New York.

Boetius, A., K. Ravenschlag, C.J. Schubert, et al. 2000. A marine microbial consortium apparently mediating anaerobic oxidation of methane. *Nature* **407**:623–626.

Boxall, A.B.A., C.J. Sinclair, K. Fenner, D. Kolpin, and S.J. Maund. 2004. When synthetic chemicals degrade in the environment. *Environ. Sci. Technol.* **38**:368A–375A.

Brady, N.C. and R.R. Weil. 1999. *The Nature and Properties of Soils*, 12th edn. Prentice Hall, Upper Saddle River, NJ.

Brock, T.D. 1961. *Milestones in Microbiology*. Prentice Hall, Englewood Cliffs, NJ.

Bull, I.D., N.R. Parekh, G.H. Hall, P. Ineson, and R.P. Evershed. 2000. Detection and classification of atmospheric methane oxidizing bacteria in soil. *Nature* **405**:175–178.

Burlage, R., R. Atlas, D. Stahl, G. Geesey, and G. Sayler (eds) 1998. *Techniques in Microbial Ecology*. Oxford University Press, New York.

Chapin, F.S., P.A. Matson, and H.A. Mooney. 2002. *Principles of Terrestrial Ecosystem Ecology*. Springer Verlag, New York.

Christian, G.D. 2003. *Analytical Chemistry*, 6th edn. Wiley and Sons, New York.

Coates, J.D. and L.A. Achenbach. 2004. Microbial perchlorate reduction: rocket-fuelled metabolism. *Nature Rev. Microbiol.* **2**:569–580.

Conrad, R. 1996. Soil microorganisms as controllers of atmospheric trace gases (H_2, CO, CH_4, OCS, N_2O, and NO). *Microbiol. Rev.* **60**:609–640.

Dalsgaard, T., D.E. Canfield, J. Petersen, B. Thamdrup and J. Acuña-González. 2003. N_2 production by the anammox reaction in the anoxic water column of Golfo Culce, Costa Rica. *Nature* **422**:606–608.

Deeb, R.A., K.-H. Chu, T. Shih, et al. 2003. MTBE and other oxygenates: Environmental sources, analysis, occurrence, and treatment. *Environ. Eng. Sci.* **20**:433–447.

Devlin, T.M. (ed.) 2001. *Textbook of Biochemistry with Clinical Correlations*, 5th edn. Wiley and Sons, New York.

Edwards, K.J., P.L. Bond, T.M. Gihrling, and J.F. Banfield. 2000. An archaeal iron-oxidizing extreme acidophile important in acid mine drainage. *Science* **287**:1796–1799.

Ehrlich, H.L. 2002. *Geomicrobiology*, 4th edn. Marcel Dekker, New York.

Esteve-Núñez, A., A. Caballero, and J.L. Ramos. 2001. Biological degradation of 2,4,6-trinitrotoluene. *Microbiol. Molec. Biol. Rev.* **65**:335–352.

Fenchel, T., G.M. King, and T.H. Blackburn. 1998. *Bacterial Biogeochemistry: The ecophysiology of mineral cycling*. Academic Press, San Diego, CA.

Fetter, G.W. 1994. *Applied Hydrogeology*. Prentice Hall, Upper Saddle River, NJ.

Fifield, F.W. 2000. *Principles and Practice of Analytical Chemistry*, 5th edn. Blackwell Science, Oxford.

Ford, T.E. (ed.) 1993. *Aquatic Microbiology: An ecological approach*. Blackwell Scientific Publications, Oxford.

Fred, E.B., I.L. Baldwin, and E. McCoy. 1932. *Root Nodule Bacteria and Leguminous Plants*. University of Wisconsin Studies of Science No. 5. University of Wisconsin, Madison, WI.

Glick, B.R. and J.J. Pasternack. 2003. *Molecular Biotechnology: Principles and application of recombinant DNA*, 3rd edn. American Society of Microbiology Press, Washington, DC.

Gottschalk, G. 1986. *Bacterial Metabolism*, 2nd edn. Springer-Verlag, New York.

Habicht, K.S. and D.E. Canfield. 1996. Sulphur isotope fractionation in modern microbial mats and the evolution of the sulphur cycle. *Nature* **25**:342–343.

Haggblom, M. and I.D. Bossert (eds) 2003. *Dehalogenation: Microbial processes and environmental applications*. Kluwer Academic Publications, Boston, MA.

Hanson, P.J., N.T. Edwards, C.T. Garten, and J.A. Andrews. 2000. Separating root and soil microbial contributions to soil respiration: A review of methods and observations. *Biogeochemistry* **48**:115–146.

Heemsbergen, D.A. 2004. Biodiversity effects on soil processes explained by interspecific functional dissimilarity. *Science* **306**:1019–1020.

Huang, P.M., J.-M. Bollag, and N. Senesi (eds) 2002. *Interactions between Soil Particles and Microorganisms: Impact on the terrestrial ecosystem.* Wiley and Sons, New York.

Jones, T.H., L.J. Thompson, J.H. Lawton, et al. 1998. Impacts of rising atmospheric carbon dioxide on model terrestrial ecosystems. *Science* **280**:441–443.

Karl, D., A. Michaels, B. Bergman, et al. 2002. Dinitrogen fixation in the world's oceans. *Biogeochemistry* **57/58**:47–98.

Kirchman, D.L. (ed.) 2000. *Microbial Ecology of the Oceans.* Wiley and Sons, New York.

Kowalchuk, G.A. and J.R. Stephen. 2001. Ammonia-oxidizing bacteria: A model for molecular microbial ecology. *Annu. Rev. Microbiol.* **55**:485–529.

Krebs, C.I. 2001. *Ecology: The experimental analysis of distribution and abundance.* Benjamin Cummings, San Francisco, CA.

Lechevalier, H.A. and M. Solotorovsky. 1965. *Three Centuries of Microbiology.* McGraw-Hill, New York.

Lengeler, J.W., G. Drews, and H.G. Schlegel (eds) 1999. *Biology of Prokaryotes.* Blackwell Science, Stuttgart.

Likens, G.E. and F.H. Bormann. 1995. *Biogeochemistry of a Forested Ecosystem,* 2nd edn. Springer-Verlag, New York.

Litchfield, C.D. (ed.) 1976. *Marine Microbiology.* Benchmark Papers in Microbiology No. 11. Dowden, Hutchinson and Ross Inc., Stroudsburg, PA.

Lovley, D.R. 2003. Cleaning up with genomics: Applying molecular biology to bioremediation. *Nature Rev. Microbiol.* **1**:35–44.

Macura, J. 1974. Trends and advances in soil microbiology from 1924 to 1974. *Geoderma* **12**:311–329.

Madigan, M.T. and J.M. Martinko. 2006. *Brock Biology of Microorganisms,* 11th edn. Prentice Hall, Englewood Cliffs, NJ.

Madsen, E.L. 1998. Epistemology of environmental microbiology. *Environ. Sci. Technol.* **32**:429–439.

Madsen, E.L. 2005. Identifying microorganisms responsible for ecologically significant biogeochemical processes. *Nature Rev. Microbiol.* **3**:439–446.

Maier, R.M., I.L. Pepper, and C.P. Gerba. 2000. *Environmental Microbiology.* Academic Press, San Diego, CA.

Maymo-Gatell, X., Y.T. Chien, J.M. Gossett, and S.H. Zinder. 1997. Isolation of a bacterium that reductively dechlorinates tetrachloroethene to ethene. *Science* **276**:1568–1571.

McArthur, J V. 2006. *Microbial Ecology: An evolutionary approach.* Elsevier Publishing, Amsterdam.

Miller, G.T., Jr. 2004. *Living in the Environment: Principles, connections, and solutions,* 13th edn. Brooks/Cole-Thomson Learning Inc., Pacific Grove, CA.

Millero, F.J. 2001. *The Physical Chemistry of Natural Waters.* Wiley Interscience, New York.

Morel, F.M., A.M.L. Kraepiel, and M. Amyiot. 1998. The chemical cycle and bioaccumulation of mercury. *Annu. Rev. Ecol. Systemat.* **29**:543–566.

Murphy, D.M. 2001. *Fundamentals of Light Microscopy and Electronic Imaging.* Wiley-Liss, New York.

Nelson, D.L. and M.M. Cox. 2005. *Lehninger Principles of Biochemistry,* 4th edn. W.H. Freeman, New York.

Nicholls, D.G. and S.J. Ferguson. 2002. *Bioenergetics 3.* Academic Press, London.

Olsen, G.J., D.J. Lane, S.J. Giovannoni, and N.R. Pace. 1986. Microbial ecology and evolution: A ribosomal RNA approach. *Annu. Rev. Microbiol.* **40**:337–365.

Oremland, R.S. and F.J. Stolz. 2003. The ecology of arsenic. *Science* **300**:939–944.

Osborn, A.M. and C.J. Smith (eds) 2005. *Molecular Microbial Ecology.* Taylor and Francis, New York.

Pace, N.R. 1997. A molecular view of microbial diversity and the biosphere. *Science* **276**:734–740.

Pace, N.R., D.A. Stahl, D.J. Lane, and G.J. Olsen. 1986. The analysis of natural microbial populations by ribosomal RNA sequences. *Adv. Microbial Ecol.* **9**:1–55.

Partensky, F., W.R. Hess, and D. Vaulot. 1999. *Prochlorococcus,* a marine photosynthetic prokaryote of global significance. *Microbiol. Mol. Biol. Rev.* **63**:106–127.

Pichard, S.L., L. Campbell, K. Carder, et al. 1997. Analysis of ribulose bisphosphate carboxylase gene expression in natural phytoplankton communities by group-specific gene probing. *Marine Ecol. Progr. Series* **149**:239–253.

Rickleffs, R.E. and G.L. Miller. 2000. *Ecology,* 4th edn. W.H. Freeman, New York.

Rittmann, B.E. and P.L. McCarty. 2001. *Environmental Biotechnology: Principles and applications*. McGraw-Hill, Boston, MA.

Rochelle, P.A. (ed.) 2001. *Environmental Molecular Microbiology: Protocols and applications*. Horizon Scientific, Wymondham, UK.

Schaechter, M., J.L. Ingraham, and F.C. Niedhardt. 2006. *Microbe*. American Society for Microbiology Press, Washington, DC.

Schink, B. 1997. Energetics of syntrophic cooperation in methanogenic degradation. *Microbiol. Mol. Biol. Rev.* **61**:262–280.

Schlesinger, W.H. 1997. *Biogeochemistry: An analysis of global change*, 2nd edn. Academic Press, San Diego, CA.

Schwarzenbach, R.P., P. Gschwend, and D.M. Imboden. 2002. *Environmental Organic Chemistry*, 2nd edn. Wiley and Sons, New York.

Segers, R. 1998. Methane production and methane consumption: A review of processes underlying wetland methane fluxes. *Biogeochemistry* **41**:23–51.

Sensen, C.W. (ed.) 2002. *Essentials of Genomics and Bioinformatics*. Wiley and Sons, New York.

Sigel, A., H. Sigel, and R. Sigel (eds) 2005. *Metal Ions in Biological Systems*, Vol. 43. *Biogeochemical Cycles of Elements*. Marcel Dekker, New York.

Snyder, L. and W. Champness. 2003. *Molecular Genetics of Bacteria*, 2nd edn. American Society for Microbiology, Washington, DC.

Spain, J.C., J.B. Hughes, and H.-J. Knackmuss (eds) 2000. *Biodegradation of Nitroaromatic Compounds and Explosives*. Lewis Publishing, Boca Raton, FL.

Spencer, J.F.T. and A.L. Ragout de Spencer (eds) 2004. *Environmental Microbiology: Methods and protocols*. Humana Press, Totowa, NJ.

Staley, J.T. and A.-L. Reysenbach (eds) 2002. *Biodiversity of Microbial Life: Foundation of Earth's biosphere*. Wiley and Sons, New York.

Stark, J.M. and S.C. Hart. 1997. High rates of nitrification and nitrate turnover in undisturbed coniferous forests. *Nature* **385**:61–64.

Stumm, W. and J.J. Morgan. 1996. *Aquatic Chemistry: Chemical equilibria and rates in natural waters*, 3rd edn. Wiley and Sons, New York.

Sverdrup, K.A., A.C. Duxbury, and A.B. Duxbury. 2003. *An Introduction to the World's Oceans*. McGraw-Hill, Boston, MA.

Sylvia, D.M., J.J. Fuhrmann, P.G. Hartel, and D.A. Zuberer. 2005. *Principles and Application of Soil Microbiology*, 2nd edn. Prentice Hall, Upper Saddle River, NJ.

Taylor, C.D. and C.O. Wirsen. 1997. Microbiology and ecology of filamentous sulfur formation. *Science* **277**:1483–1485.

Tchobanoglous, G., F.L. Burton, and H.D. Stensel. 2002. *Wastewater Engineering: Treatment and reuse*. McGraw-Hill, New York.

Ternes, T.A., A. Joss, and H. Seigrist. 2004. Scrutinizing personal care products. *Environ. Sci. Technol.* **38**:393A–399A.

Ting, C.S., G. Rocap, J. King, and S.W. Chisholm. 2002. Cyanobacterial photosynthesis in the oceans: the origins and significance of divergent light-harvesting strategies. *Trends Microbiol.* **10**:134–142.

Twyman, R.M. 2004. *Principles of Proteomics*. BIOS Scientific Publishing, New York.

van Breemen, N., E.W. Boyer, C.L. Goodale, et al. 2002. Where did all the nitrogen go? Fate of nitrogen inputs to large watersheds in the northeastern USA. *Biogeochemistry* **57/58**:267–293.

van Elsas, J.D., J.T. Trevors, and E.M.H. Wellington. 1997. *Modern Soil Microbiology*. Marcel Dekker, New York.

Van Hamme, J.C., A. Singh, and O.P. Ward. 2003. Recent advances in petroleum microbiology. *Microbiol. Mol. Biol. Rev.* **67**:503–549.

van Niel, C.B. 1967. The education of a microbiologist: some reflections. *Annu. Rev. Microbiol.* **21**:1–30.

van Niftrik, L.A., J.A. Fuerst, J.S.S. Damsté, J.G. Kuenen, M.S.M. Jetten, and M. Strous. 2004. The anammoxosome: an intracytoplasmic compartment in anammox bacteria. *FEMS Microbiol. Lett.* **233**:7–31.

Velvis, H. 1997. Evaluation of the selective respiratory inhibition method for measuring the ratio of fungal:bacterial activity in acid agricultural soils. *Biol. Fertil. Soils* **25**:354–360.

Waksman, S.A. 1925. Soil microbiology in 1924: an attempt at an analysis and a synthesis. *Soil Sci.* **19**:201–249.

Waksman, S.A. 1927. *Principles of Soil Microbiology*. Williams and Wilkins, Baltimore, MD.

Waksman, S.A. 1945. Soil microbiology as a field of science. *Science* **102**:339–344.

Waksman, S.A. 1952. *Soil Microbiology*. Wiley and Sons, New York.

Ward, D.M., M.M. Bateson, R. Weller, and A.L. Ruff-Roberts. 1993. Ribosomal RNA analysis of microorganisms as they occur in nature. *Adv. Microbial Ecol.* **12**:219–286.

Wetzel, R.G. 2001. *Limnology: Lake and river ecosystems*, 3rd edn. Academic Press, San Diego, CA.

Wetzel, R.G. and G.E. Likens. 2000. *Limnological Analyses*, 3rd edn. Springer-Verlag, New York.

White, D.C. 1994. Is there anything else you need to understand about the microbiota that cannot be derived from analysis of nucleic acids? *Microbial Ecol.* **28**:163–166.

White, D. 1995. *The Physiology and Biochemistry of Prokaryotes*. Oxford University Press, New York.

Whitman, W.B., D.C. Coleman, and W.J. Wiebe. 1998. Prokaryotes: the unseen majority. *Proc. Natl. Acad. Sci. USA* **95**:6578–6583.

Winogradsky, S. 1949. *Microbiologie du sol, problémes et methods: cinquante ans de recherches. Oeuvres completes*. Masson, Paris.

Woese, C.R. 1987. Bacterial evolution. *Microbiol. Rev.* **51**:221–271.

Woese, C.R. 1992. Prokaryotic systematics: the evolution of a science. *In*: A. Balows, H.G. Trüper, M. Dworkin, W. Harder, and K.-H. Schleifer (eds) *The Prokaryotes*, 2nd edn, pp. 3–18. Springer-Verlag, New York.

Young, L. and C. Cerniglia (eds) 1995. *Microbial Transformation and Degradation of Toxic Organic Chemicals*. Wiley-Liss, New York.

Zhou, J., D.K. Thompson, Y. Zu, and J.M. Tiedje (eds) 2004. *Microbial Functional Genomics*. Wiley-Liss, Hoboken, NJ.

Zumft, W.G. 1997. Cell biology and molecular basis of denitrification. *Microbiol. Mol. Biol. Rev.* **61**:533–616.

FURTHER READING

Carter, J.P. and J.M. Lynch. 1993. Immunological and molecular techniques for studying the dynamic of microbial populations and communities in soil. *In*: J.-M. Bollag and G. Stozky (eds) *Soil Bio-chemistry*, Vol. 8, pp. 239–272. Marcel Dekker, New York.

de Duve, C. 2002. *Life Evolving: Molecules, mind, and meaning*. Oxford University Press, Oxford.

2

Formation of the Biosphere: Key Biogeochemical and Evolutionary Events

This chapter provides an overview of the history of Earth and its forms of life. We review state-of-the-art tools, principles, and logic used to generate information addressing how our world progressed from its ancient prebiotic state to its contemporary biotic state. Key events included: planetary cooling, geochemical reactions at mineral surfaces on the floor of primordial seas, an "RNA world", development of primitive cells, the "last universal common ancestor", anoxygenic photosynthesis, oxygenic photosynthesis, the rise of oxygen in the atmosphere, the development of the ozone shield, and the evolution of higher forms of eukaryotes. The chapter closes by reviewing endosymbiotic theory and key biochemical and structural contrasts between prokaryotic and eukaryotic cells.

Chapter 2 Outline

2.1   Issues and methods in Earth's history and evolution
2.2   Formation of early planet Earth
2.3   Did life reach Earth from Mars?
2.4   Plausible stages in the development of early life
2.5   Mineral surfaces: the early iron/sulfur world could have driven biosynthesis
2.6   Encapsulation: a key to cellular life
2.7   A plausible definition of the tree of life's "last universal common ancestor"
2.8   The rise of oxygen
2.9   Evidence for oxygen and cellular life in the sedimentary record
2.10 The evolution of oxygenic photosynthesis
2.11 Consequences of oxygenic photosynthesis: molecular oxygen in the atmosphere and large pools of organic carbon
2.12 Eukaryotic evolution: endosymbiotic theory and the blending of traits from *Archaea* and *Bacteria*

2.1 ISSUES AND METHODS IN EARTH'S HISTORY AND EVOLUTION

- **How do we know what happened long ago?**
- **How old is the Earth?**
- **How did life begin?**
- **When did life begin?**
- **How have life and the Earth changed through the ages?**

These and related questions have likely been pondered by humans for thousands of years. In our quest for understanding extant micro-organisms that dwell in biosphere habitats, it is essential to place them in historical, metabolic, and evolutionary context. To achieve this, we would ideally be able to superimpose continuous, independent timelines derived from the geologic record, the fossil record, the climate record, the evolutionary record, and the molecular phylogenetic record. Conceivably this superimposition could allow cause-and-effect interactions to be documented, linking specific events such as changes in atmospheric composition, glaciation, tectonic movements, and the rise and fall of biotic adaptations. This ideal has not yet been achieved at high resolution. Instead, we have only glimpses here and there of our planet's complex, shrouded past (Table 2.1). However, recent advances have made progress toward achieving a synthesis that may solve the puzzles of Earth's history. The key tools used to discover and decipher planetary history are listed in Table 2.2 and further explained in Boxes 2.1 and 2.2. By knowing Earth's global distribution of land forms, rocks, and minerals, geologists have identified where to look for clues about ancient Earth and life's beginnings (Figure 2.1). Discovery of the clues and their assembly into a convincing, coherent body of knowledge is ongoing – reliant upon insights from geology, paleontology, nuclear chemistry, analytical chemistry, experimental biochemistry, as well as molecular phylogeny (Table 2.2).

2.2 FORMATION OF EARLY PLANET EARTH

Explosions from supernovae 4.6×10^9 years ago are thought to have instigated the formation of our solar system (Nisbet and Sleep, 2001). The inner planets (Earth, Mars, Venus, Mercury) were produced from collisions between planetisemels. Early Earth featured huge pools of surface magma which cooled rapidly (~2×10^6 years) to ~100°C. Later, water condensed, creating the oceans. Volcanism and bombardment by meteors were common. These collisions are thought to have repeatedly heated the oceans to >100°C, causing extensive vaporizing of water. Our moon was likely formed 4.5×10^9 years ago when molten mantle was ejected into orbit

Table 2.1

Key event and conditions of early Earth

Time (10^9 years before present)	Events	Conditions
4.6	<ul><li>Colliding planetisemals</li><li>Earth formed</li><li>Moon formed</li><li>Loss of water and hydrogen from atmosphere</li><li>Volcanism</li><li>Cooling of surface</li><li>Faint young Sun?</li><li>Glaciation?</li></ul>	<ul><li>Heat</li><li>Meteor bombardment and impacts</li><li>Lightning</li><li>UV radiation</li><li>Hot oceans followed each bombardment</li><li>Atmosphere: N_2, CO_2, CO, H_2, NH_3, CH_4, HCN</li><li>Ocean chemistry: H_2S, Fe^{2+}, heavy metals</li></ul>
4.2	<ul><li>Bombardment ceased (?)</li></ul>	
4.0	<ul><li>RNA world, iron/sulfur world</li><li>Last universal common ancestor</li></ul>	
3.8	<ul><li>Anoxygenic photosynthesis</li></ul>	
3.5–3.4	<ul><li>Fossils resembling bacterial filaments on stromatolite microbial mats</li></ul>	
2.7	<ul><li>Biomarker for cyanobacteria</li><li>Biomarker for primitive eukaryotes</li></ul>	<ul><li>Banded iron geologic formations</li></ul>
2.4	<ul><li>Signs of oxygen at low concentration in atmosphere</li></ul>	<ul><li>Red bed geologic formations</li><li>Oxygen in atmosphere ~1%</li></ul>
1.4	<ul><li>Nucleated eukaryotic algae</li></ul>	<ul><li>Ozone shield</li></ul>
0.6		<ul><li>Oxygen in atmosphere ~21%</li></ul>
0.4	<ul><li>Cambrian explosion of eukaryotic diversity</li></ul>	
0.1	<ul><li>Dinosaurs, higher plants, mammals</li></ul>	

after Earth was struck by another planet about the size of Mars. Bombardment diminished perhaps by $4.2–4.0 \times 10^9$ years ago. The scale of geologic time, from planet formation to the present, is shown in Figure 2.2.

The influence of ancient atmospheres upon surface conditions was critical. Abundances of greenhouse and other gases (especially CO_2, NH_3, H_2O, CO, CH_4, HCN, N_2) were probably highly dynamic. In combination with variations in solar radiation, atmospheric conditions may have contributed to periods of high surface temperatures (~100°C) that perhaps alternated with low-temperature (glaciated) periods. Clearly, conditions on prebiotic Earth were turbulent – characterized by fluctuating temperatures, aqueous reactions with magma, input of materials from meteorites (including organic carbon), electrical discharges from the atmosphere, and reduced (nonoxidizing) gases in the atmosphere.

Table 2.2

Scientific tools providing information about Earth history and evolution

Discipline	Tool	Insights
Geology	Global surveys of terrestrial and oceanic rocks	Sedimentary, igneous, and metamorphic formations reveal tectonic and other processes governing Earth's evolution
Nuclear chemistry	Radioisotopic dating	Ages of rocks, minerals, and their components are revealed
Paleontology	Fossil record	Organism structures preserved in stratified sediments provide records of evolution
Analytical chemistry of biomarkers	Analytical determination of biomolecules via chromatography and mass spectrometry	Molecular remnants of biomolecules (membranes, pigments, cell walls, etc.) document ancient biota
Analytical chemistry of isotopic ratios	Isotope ratio mass spectrometry	Enzyme reactions favor substrate molecules composed of lighter atoms. Biomass assimilates the lighter isotope and the remaining isotopic pool becomes "heavier" for a given process
Experimental biochemistry	Model systems that simulate ancient Earth	Discovery of precursors of cellular structures and their self-assembling properties
Molecular phylogeny	Sequencing and analysis of informational biomolecules	Alignment of sequences from DNA, proteins, and other molecules allow evolutionary inferences to be drawn, especially regarding the three domains of life
Mineralogy and geochemistry	X-ray diffraction and wet-chemical analysis of rocks	Chemical reactions and reactants of past ages can be inferred from the composition and oxidation/reduction status of ancient sediments
Biochemistry	Comparative biochemistry of cellular materials	Trends in evolutionary relatedness among and between members of *Bacteria*, *Archaea*, and *Eukarya*

2.3 DID LIFE REACH EARTH FROM MARS?

There is a general consensus that stable isotopic ratios (see Tables 2.1, 2.2 and Box 2.2) in graphite isolated from the Isua supracrustal belt (West Greenland; see Figure 2.1) prove that life, manifest as anoxygenic photosynthesis, was present 3.8×10^9 years ago (Nisbet and Sleep, 2001). Before focusing upon a plausible scenario of how life evolved on Earth, an alternative, perhaps equally plausible, hypothesis must briefly be considered: Panspermia. In the early history of our solar system, Earth, Venus, and Mars, close neighbors, were simultaneously undergoing planetary

Box 2.1

The age of the Earth and biota

Radioactive decay in rocks
Measurements performed on rock containing radioactive elements (nuclides) can reveal the age of the rock. For example, ^{238}U (half-life $= 4.5 \times 10^9$ years) decays to helium and ^{206}Pb. Each atom of ^{238}U that decomposes forms eight atoms of helium (with total mass 32) leaving one atom of ^{206}Pb. In 4.5×10^9 years, 1 g of ^{238}Pb becomes 0.5000 g of ^{238}U and 0.174 g of He and 0.326 g of ^{206}Pb. If analyses document nuclides present in the rock in the above ratios, the age would be 4.5×10^9 years. Other radioactive elements have their own characteristic half-lives and decay products; thus, ratios of $^{235}U/^{207}Pb$, $^{232}Th/^{208}Pb$, $^{40}K/^{40}Ar$, and $^{87}Rb/^{87}S$ are also insightful for determining the ages of rocks. The present estimate of the age of Earth and the other inner planets in our solar system is 4.6×10^9 years.

Carbon dating of life
About one in every 10^{12} carbon atoms on Earth is radioactive (^{14}C) and has a half-life of 5760 years. Carbon dioxide, radioactive and nonradioactive alike, is absorbed by plants and incorporated into the biota that consume plants. When a plant or animal dies, its ^{14}C atoms begin undergoing radioactive decay. After 11,520 years (two half-lives), only one-quarter of the original radioactivity is left. Accordingly, by determining the ^{14}C radioactivity of a sample of carbon from wood, flesh, charcoal, skin, horn, or other plant or animal remains, the number of years that have passed since the carbon was removed from atmospheric input of ^{14}C can be determined.

Box 2.2

Biomarkers and isotopic fractionation

Biomarkers
Biomarker compounds are molecules of known biosynthetic origin. As such, their detection in geologic samples (ancient buried soils, rocks, sediments) associates the biosynthetic pathway and/or its host organism with the source material. Biomarker geochemistry has been routinely applied to petroleum exploration and also has been insightful in analyzing rocks (e.g., 2.7×10^9-year-old shales from northwestern Australia; Brocks et al., 1999). Examples of biomarkers that have been extracted from rocks are 2-methyl-hopanes (derived from 2-methyl-bacteriohopane polyols, which are membrane lipids synthesized by cyanobacteria) and pyrrole molecules, essential building blocks of the photosynthetic (chlorophyll) and respiratory (cytochrome) apparatus.

Box 2.2 *Continued*

Isotopic fractionation

Many chemical elements on Earth occur as mixtures of atoms with differing numbers of neutrons in their nuclei. For instance, the natural abundance of stable (nonradioactive) carbon with six neutrons and six protons (^{12}C) is 98.9%, while ~1.1% of the total carbon pool has seven neutrons (^{13}C). Enzymes involved in photosynthesis show subtle selectivity in acting on their substrate (CO_2) when it is composed of the lighter (^{12}C) carbon isotope. Photosynthesis fixes atmospheric CO_2 into biomass; therefore, the biomass is enriched in ^{12}C – it is "light". Correspondingly, as ^{12}C-CO_2 is removed from the atmosphere, the remaining pool is enriched in ^{13}C-CO_2 – it becomes "heavy". Such shifts in isotopic ratios can be detected in carbon and other elements (especially sulfur) extracted from ancient rocks. Because enzymatic selectivity is the only known mechanism for such shifts, these constitute evidence for biological processes.

As listed in Table 2.2, carbon isotopic ratios are determined using an analytical technique known as isotope ratio mass spectrometry. The means of expressing the ratio uses a "del ^{13}C" value, which contrasts the $^{13}C/^{12}C$ ratio in a sample with that of a standard:

$$\delta^{13}C = \frac{(^{13}C/^{12}C \text{ sample}) - (^{13}C/^{12}C \text{ standard})}{^{13}C/^{12}C \text{ standard}} \times 1000$$

Note that the $\delta^{13}C$ value becomes negative ("light") if the sample is depleted in ^{13}C, relative to the standard. Extensive surveys of carbon pools found in nature have been cataloged. This compilation of characteristic values (Grossman, 2002) allows the origin of many carbon pools to be ascertained:

Pool of carbon	Range of $\delta^{13}C$
Marine carbonate	−5 to +8
Atmospheric CO_2	−8 to −5
Calvin cycle plants (C_3)	−27 to −21
C_4 plants	−17 to −9
Petroleum	−32 to −22
Thermogenic methane	−48 to −34
Microbial methane	−90 to −48
Cyanobacteria	−30 to −18
Purple sulfur bacteria	−35 to −20
Green sulfur bacteria	−20 to −9
Recent marine sediments	−35 to −10

development. Mars, in particular, featured an abundance of water and other geochemical conditions that may have been favorable for, and led to the development of, life. Meteor bombardment was rampant in the early solar system. Such collisions transferred materials between planets. Microbial life buried within the interstices of Martian rocks may have survived transit to Earth and landed in seas capable of supporting growth. Once seeded, Earth-specific evolutionary forces would have taken hold. The discussion below on the possible origin of life applies to Earth, as well as other planets.

Figure 2.1 Map of the world showing locations of rock formations that provide insights into the coevolution of life and the Earth.

2.4 PLAUSIBLE STAGES IN THE DEVELOPMENT OF EARLY LIFE

Among life's many attributes is the creation of order out of disorder. The Second Law of Thermodynamics mandates that order be created at the expense of energy and the production of entropy. Mechanistically, life is manifest as the synthesis of molecular structures that facilitate metabolic and genetic processes. Such structures are antientropic – requiring energy for synthesis and assembly. Fortunately, abundant physical energy sources prevailed in sterile, prebiotic Earth: these included heat, UV radiation, and electrical discharges (lightning). Investigations by S. Miller in the 1950s proved that amino acids can be chemically synthesized under conditions simulating ancient seas. This de novo synthesis of organic compounds, supplemented with ones borne on meteorites, leaves little doubt that an organic geochemical broth developed.

The transition from a soup of life's primitive potential building blocks to advanced cellular life is thought to have proceeded through many stages of increasing complexity (Figures 2.2, 2.3). The fundamental conceptual foundation in developing and testing theories about the origin of life is "getting here from there". We need to define "here", define "there", and do our best in devising feasible, continuous connections between the two. "Here" refers to the highly complex characteristics of modern cellular life: heredity (DNA), transcription (RNA), translation (ribosomes), catalysis (proteins), compartmentalization (membrane-enclosed cells and organelles), metabolic energy production [e.g., electron transport, adenosine triphosphate (ATP), and ATP synthase], and biosynthesis (using energy for

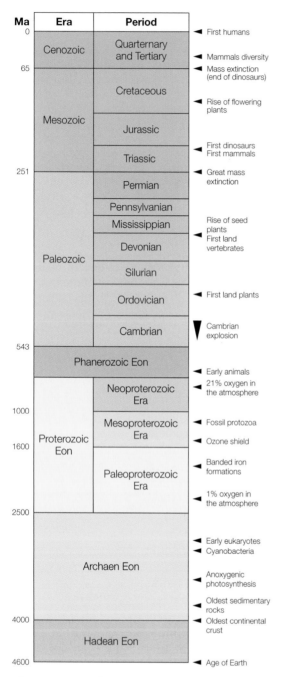

Figure 2.2 Geological timescales and evolutionary events. Note that the scale is not linear (Ma, 10^6 years ago). (Modified from Knoll, A.H. 2003. *Life on a Young Planet*. Copyright 2003, Princeton University Press. Reprinted by permission of Princeton University Press.)

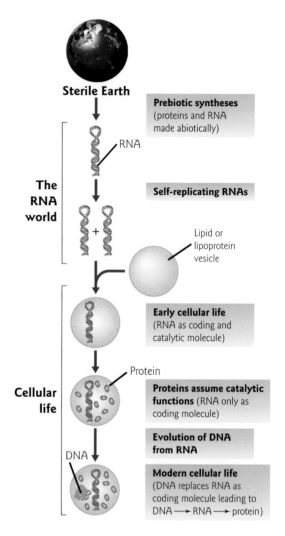

Figure 2.3 A general model of biochemical and biological evolution. Connections between sterile prebiotic and cellular stages of life rely on gradually increasing complexity. Key milestones were chemical synthesis, surface catalyzed reactions, the RNA world, the last universal common ancestral community, and compartmentalization to form free-living cells. (From MADIGAN, M. and J. MARTINKO. 2006. *Brock Biology of Microorganisms*, 11th edn, p. 304. Copyright 2006, reprinted by permission of Pearson Education, Inc., Upper Saddle River, NJ.)

Science and the citizen

War of the Worlds and martian life

Headline news from English literature and American radio

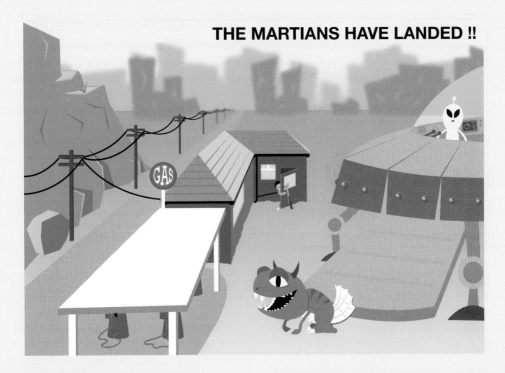

The 1898 novel by H. G. Wells depicted an invasion of England by aliens from Mars. Meteor-like, cylindrical spaceships landed throughout the countryside. Tentacled creatures assembled armed fighting machines that brought fear and destruction to humanity. In 1938, the American writer, director, and actor, Orson Wells (not related to H. G. Wells) broadcast a radio show based on *War of the Worlds*. The radio broadcast was so realistic that it caused widespread panic among radio listeners.

SCIENCE: Real martian life?

Contrary to H. G. Wells and Orson Wells' depiction of sophisticated martians with advanced technology, the real news about extraterrestrial life is microbial. On August 16, 1996, an article by McKay et al., entitled "Search for past life on Mars: Possible relic biogenic activity in martian meteorite ALH84001" appeared in the prestigious *Science* magazine. The authors' hypothesis was that microorganisms on Mars carried out metabolic activities that caused the formation of carbonate globules, magnetite mineral, iron sulfide mineral, and cell biomass. The latter was proposed to have been converted to polycyclic aromatic hydrocarbons (see Section 8.3 and Box 8.7) during transport from Mars to Earth.

The final paragraph of the article summarized the information presented that argued for past microbial life on Mars.

In examining the martian meteorite ALH84001 we have found that the following evidence is compatible with the existence of past life on Mars:

(i)   an igneous Mars rock (of unknown geologic context) that was penetrated by a fluid along fractures and pore spaces, which then became the sites of secondary mineral formation and possible biogenic activity;

(ii)   a formation age for the carbonate globules younger than the age of the igneous rock;

(iii) scanning electron micrograph and transmission electron micrograph images of carbonate globules and features resembling terrestrial microorganisms, terrestrial biogenic carbonate structures, or microfossils;

(iv) magnetite and iron sulfide particles that could have resulted from oxidation and reduction reactions known to be important in terrestrial microbial systems; and

(v)   the presence of polycyclic aromatic hydrocarbons associated with surfaces rich in carbonate globules.

None of these observations is in itself conclusive for the existence of past life. Although there are alternative explanations for each of these phenomena taken individually, when they are considered collectively, particularly in view of their spatial association, we conclude that they are evidence for primitive life on early Mars.

About 3 years subsequent to McKay and colleague's publication, each of the five arguments for ancient martian microbial life was challenged in the scientific literature (e.g., Anders, 1996; Borg et al., 1999). Alternative, largely chemical, mechanisms were found for the formation of what McKay et al. had argued to be biogenic structures. Now, the general consensus is that this hypothesis about ancient martian life has been disproven.

The intellectual and scientific exercise of seeking extraterrestrial life has, nonetheless, been beneficial to environmental microbiology. A new discipline has been born – astrobiology. Advances in the astrobiology scientific community have enabled it to be far better prepared to document new forms of microbial life.

Research essay assignment

The terms "extremophile", "exobiology", and "astrobiology" have been used extensively in both the scientific and nonscientific literature. After finding about six published works addressing these topics, write a 3–5 page essay that merges two aspects of astrobiology: (i) the human preoccupation with alien life forms; and (ii) the genesis and goals of the astrobiology field of science.

cellular replication). Figure 2.3 presents a possible scenario of events leading from sterile Earth through the "RNA world" to cellular life. RNA-based proto-life is a likely intermediary step because RNA has both self-replication and catalytic traits. However, proteins are superior to RNA as catalysts and DNA is superior to RNA as a stable reservoir of genetic information. In the scenario shown in Figure 2.3, RNA's role in metabolism gradually shifted to intermediary template between information-bearing DNA and substrate-specific protein catalysts. Several of the key steps thought crucial to life's development are discussed below.

2.5 MINERAL SURFACES: THE EARLY IRON/SULFUR WORLD COULD HAVE DRIVEN BIOSYNTHESIS

Although some organic chemicals can be synthesized from simple inorganic gases in the presence of electrical discharges (see above), the open waters of ancient seas are not the likely site of life's key early developmental stages. One reason for this is that water participates in hydrolytic cleavage reactions – which are not conducive to building complex organic molecules. In contrast, mineral surfaces [particularly iron monosulfide (FeS) minerals lining microporous rocks at the bottom of ancient seas] are currently thought to be the site where early life began (Martin and Russell, 2003). These rock formations, analogous to today's hydrothermal vents (Figure 2.4), offered three-dimensional compartments of

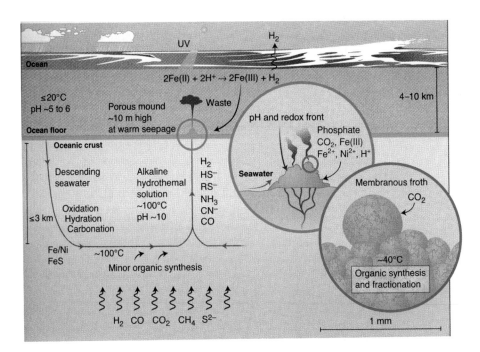

Figure 2.4 The submarine setting for the emergence of life. (From Russell, M.J. 2003. Geochemistry: the importance of being alkaline. *Science* **302**:580–581. Reprinted with permission of AAAS.)

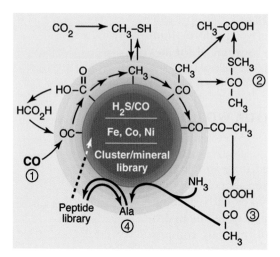

Figure 2.5 Reactions in the iron/sulfur world. Shown are the proposed reaction schemes for primordial molecules at catalytic surfaces in the iron/sulfur world. Reaction steps lead to the conversion of carbon monoxide (1) to a variety of key biomolecules, including: methyl thioacetate (2), pyruvate (3), and alanine (4). (From Wachtershauser, G. 2000. Origin of life: Life as we don't know it. *Science* **289**:1307–1308. Reprinted with permission of AAAS.)

diffusion-limited hydrophobic surfaces that could bind and concentrate organic compounds. In addition, FeS and nickel monosulfide (NiS) catalysts lining the porous cavities are capable of forging carbon–carbon bonds (Huber and Wachtershauser, 2006; see Section 7.1). The geochemical context of these porous mineral surfaces was along gradients of oxidation/reduction potential, pH, and temperature where sulfide-rich hydrothermal fluid mixed with Fe(II)-containing waters of the ocean floor. Under such conditions oxidative formation of pyrite (FeS_2) occurs spontaneously:

$$FeS + H_2S \rightarrow FeS_2 + H_2 \quad \Delta G = -38.4 \text{ kJ/mol}$$

This type of exothermic reaction produces reducing power (H_2), often essential in biosynthetic reactions. It also can drive the autocatalytic assembly of complex organic molecules (Figure 2.5; Wachtershauser, 1990, 1992). Current thought (Martin and Russell, 2003) holds that the chemistry of the RNA world (including reduction of CO and CO_2; peptide bond formation; synthesis of nucleotides; and formation of thioester precursors of ATP) occurred in these FeS cavities.

2.6 ENCAPSULATION: A KEY TO CELLULAR LIFE

As defined above, proto-life (organic catalysis and replication) was confined to the rocky pores where the RNA world began. The mobile, compartmentalized character of modern cells had yet to be invented. The active sites of modern enzymes still rely upon FeS- and NiS-type moieties; thus, it is likely that the early enzymes simply incorporated bits of their mineral heritage (see Section 7.1). Regarding encapsulation into membrane-bound compartments, experiments by D. Deamer in the 1970s showed that fatty acids have the capacity to self-assemble into membrane-like vesicles. More recently, Hanczyc et al. (2003) have demonstrated that conditions likely to prevail on an ancient seafloor (catalytic surfaces, hydrodynamic forces, alkaline conditions) have the potential to foster formation, growth, and division of fatty acid-based membranes that enclose biomolecules. Thus, the rudimentary mechanisms leading from surface-catalyzed proto-life to free-living cells seem to have been established (see Figure 2.3).

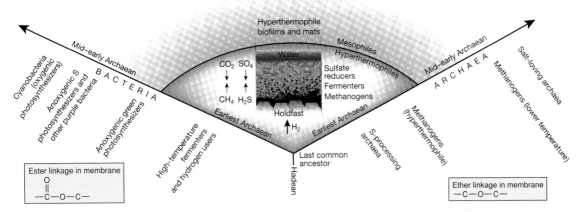

Figure 2.6 Divergence from the root of the tree of life: the last universal common ancestor. (Reprinted by permission from Macmillan Publishers Ltd: *Nature*, from Nisbet, E.G. and N.H. Sleep. 2001. The habitat and nature of early life. *Nature* **409**:1083–1091. Copyright 2001.)

2.7 A PLAUSIBLE DEFINITION OF THE TREE OF LIFE'S "LAST UNIVERSAL COMMON ANCESTOR"

Molecular phylogenetic analyses of genes encoding small subunit ribosomal RNA (shared by all existing cellular life forms) have led to a tree of life with three primary domains, *Bacteria*, *Archaea*, and *Eukarya* (see Section 5.5). Insights from molecular phylogeny into evolutionary relationships and diversity are far-reaching. What is most germane to the present chapter is the root and main trunks of the tree of life. Perusal of the tree's base (Figure 2.6) shows a single point of bifurcation where the bacterial trunk diverges from the trunk destined to be the precursor of the other two domains. That point of divergence is conceptually profound – it represents the last universal common ancestor. For the molecular phylogenist, the last universal common ancestor represents both an abstract idea, and a tangible entity. This precursor to all life that we know today must have carried metabolic and evolutionary traits reflecting conditions and resources on ancient Earth. Indeed, by knowing physiological traits (thermophily, autotrophy, etc.) of today's organisms that reside near the base of the phylogenetic tree, inferences can be drawn about selective pressures of ancient life. The last universal common ancestor, spawned by the RNA world, is the endpoint of prebiotic evolution, which blossomed into organismic biology.

The last universal common ancestor was unlikely to have been a single well-defined entity. Rather, it would have been a community of precellular genetic complexes (Woese, 1998; Woese et al., 2000) that readily mixed their genetic and metabolic traits. In the model advanced by

C. Woese, genuine organisms exhibiting distinctive lines of hereditary descent can only be found after the blurring effects of unrestricted gene transfer have largely ceased (Woese, 1998). Vertical gene transfer is heritage passed from parent to progeny. Lateral gene transfer is exchange of genetic material between forms of life that have become well-differentiated entities (e.g., *Bacteria* and *Archaea*). The tools of molecular phylogeny examine patterns of heredity; therefore, such tools can only be applied after distinctive lines of descent have been established. Thus, the root, or origin, of the tree of life cannot be determined by ribosomal RNA sequences.

In the scenario developed by Martin and Russell (2003), the last universal common ancestor was the sophisticated precellular offspring of the RNA world that flourished in hydrothermal FeS cavities. Fundamental biochemical differences in the membranes of *Bacteria* and *Archaea* argue for *evolutionary divergence and encapsulation to be one and the same events* (Figure 2.6). The two earliest evolutionary lineages (*Bacteria* and the line that was to become *Archaea*) exhibit highly distinctive membrane lipid architecture: isoprenoid *ether-type* membranes in *Archaea* and fatty-acid *ester-type* membranes in *Bacteria*. These are thought to have arisen in precellular life via repeated fissions and fusions of membranes of early cells, leading to proto-*Archaea* and proto-*Bacteria* with selective membrane biosynthetic capabilities. It is generally agreed that the *Archaea* and *Eukarya* arose from a second bifurcation in the trunk on the tree of life (see Sections 2.12 and 5.5). The presence of (seemingly a reversion to) ester-type membrane lipids in the *Eukarya* may have resulted from a merging of traits between an established bacterial line and a newly emerged eukaryal line – but the details of such developments are not yet clear (de Duve, 2007).

2.8 THE RISE OF OXYGEN

Regarding biological complexity, defining "here" (modern life), "there" (a sterile prebiotic Earth), and chronicling likely mechanistic evolutionary connections between the two is an ongoing challenge (see Sections 2.4–2.7). An analogous challenge confronts atmospheric chemists whose goal is to understand the transition between a highly reducing, abiotic Earth 4.6×10^9 years ago and the highly oxidizing, biotic Earth of today. The rise of biota and the rise of oxygen in Earth's atmosphere (from 0% to 21%) went hand in hand. Analogous to scenarios woven to explain the evolution of modern life, there are several key facts (milestones, benchmarks) that are critical for understanding the oxygenation of planet Earth. These benchmarks include: ancient evidence of oxygen as oxide minerals in the sedimentary record, the photosynthetic mechanism of oxygen production, and the consequences of oxygen production – including an atmospheric ozone shield, aerobic metabolism, and the appearance of higher eukaryotes late in evolution (see Table 2.1 and Figure 2.2).

2.9 EVIDENCE FOR OXYGEN AND CELLULAR LIFE IN THE SEDIMENTARY RECORD

In his eloquent portrayal of early Earth, A. Knoll (2003) describes P. Cloud's efforts to identify a demarcation line in the sedimentary record (between 2.4 and 2.2×10^9 years ago) for the appearance of oxygen on our planet. The iron sulfide mineral, pyrite (FeS_2), is thermodynamically unstable in the presence of molecular oxygen, particularly during cycles of erosion and deposition. Sedimentary rocks have been surveyed globally and those younger than 2.4×10^9 years generally do not contain pyrite while many older ones do. This argues for the absence of widespread atmospheric oxygen before 2.4×10^9 years ago. Evidence for the presence of atmospheric oxygen after 2.2×10^9 years ago resides in the formation of iron oxide minerals. These are manifest as "red beds" such as the vividly colored red rock formations in sedimentary successions visible today in Utah and Arizona, western United States (see Figure 2.1).

Another important line of evidence in deciphering the history of oxygen has been provided by J. Farquhar. Prior to being transformed and deposited in sediments, sulfur dioxide, an atmospheric gas, interacts with sunlight in the atmosphere leaving a characteristic signature in the sulfur isotopic ratios. If molecular oxygen is present, the characteristic isotopic ratios are erased. According to Kerr (2005), Farquhar and colleagues have found the sulfur signature in rocks older than 2.4×10^9 years, but not in younger rocks.

While the vast red bed formations and related facts argue convincingly for a widespread oxygenated atmosphere 2.4–2.2×10^9 years ago, this does not preclude the possibility of small-scale hot spots of oxygen production earlier. H. Ohmoto has reported the occurrence of oxidized minerals as old as 3×10^9 years. It is possible that these represented localized islands of oxygen-producing biota surrounded by confining sinks for oxygen. Molecular oxygen spontaneously reacts with (oxidizes) many reduced chemical compounds (e.g., H_2S, Fe^{2+}) that prevailed in the Proterozoic era. This means that appreciable atmospheric oxygen concentrations could only have developed after such oxygen sinks were depleted. Indeed, it has been argued that early oxygen in the atmosphere hovered between 1% and 10% for as long as 1.8×10^9 years after oxygenic photosynthesis evolved. At these low concentrations, oxygen would not be expected to penetrate the deep oceans. D. Canfield and A. Knoll have interpreted the sulfur isotopic record as indicating that all but the upper oceanic layers were anoxic long after oxygen began creating the red bed deposits. Isotopic ratios of molybdenum, also an indicator of oxygen concentration, have confirmed widespread hypoxia in the early oceans.

The era of the sulfur-dominated Canfied Ocean ended by 0.6×10^9 years ago when atmospheric oxygen reached its current concentration (21%). Several theories have been offered to explain why oxygen accumulated so significantly. Oxygenic photosynthesis is directly linked to conversion of CO_2 to organic carbon in biota. In a steady-state condition, oxygenic photosynthesis is balanced by the reverse reaction: aerobic respiration by heterotrophic life of fixed organic carbon. Oxygen's rise in the atmosphere hinged upon a major shift in the balance between the two processes. This resulted in the formation of massive amounts of plant biomass (organic carbon), that were subsequently buried in ocean sediments. As this organic carbon entered long-term storage in sedimentary basins (sequestered from the biosphere), oxygen concentrations climbed. One potential geochemical explanation for carbon sequestration is its reaction with seafloor clays that may have been capable of protecting organic matter from respiration. Another geochemical explanation favors plate tectonic formation of a supercontinent that relieved nutrient limitations in the ocean, thereby stimulating carbon fixation and burial. Potential biological explanations for increased photosynthetic activity and carbon burial include: (i) the evolution of land-based lichens that may have boosted fluxes of otherwise limiting nutrients to the oceans; and (ii) the evolution of zooplankton, whose carbon-rich fecal pellets sank rapidly to the ocean floor.

Table 2.2 lists "paleontology" and the "fossil record" as valuable sources of information about early Earth. We know that skeletons of many animals are well preserved in geologic strata. It seems prudent to seek corresponding preserved structures for prokaryotic life. Among the most robust evidence for fossil prokaryotic structures are stromatolites. Stromatolites are macroscopic layered microbial communities that grow today in shallow coastal waters such as those in northwestern Mexico and Western Australia (Figure 2.7a). Filamentous cyanobacteria contribute significantly to these microbial mat (stromatolite) ecosystems. Fossilized stromatolites have been found on many continents and are well represented in samples from South Africa's Barberton Mountains, dated 2.5×10^9 years old (see Figures 2.1, 2.7b). The Australian Pilbara chert deposits (3.4×10^9 years old) have revealed unusual forms of stromatolite-like structures (Allwood et al., 2006). Smaller scale bacteria-like structures have been reported in some ancient rocks, such as the 3.5×10^9-year-old pillow lavas in the Barberton Greenstone Belt in South Africa and in the Warrawoona Group, Western Australia.

2.10 THE EVOLUTION OF OXYGENIC PHOTOSYNTHESIS

D. Des Marais (2000) points out that one key index of life and its success is the accumulating mass of fixed carbon, otherwise known as biomass (see Section 2.9). Biogeochemically, it is reducing power that provides a means of converting the pool of inorganic oxidized carbon (CO_2) to the organic form that constitutes living organisms. When CO_2 is converted by autotrophic life forms to organic carbon (CH_2O), the carbon is reduced (accepts electrons) from an oxidation state of +4 to 0 (see

Figure 2.7 Examples of both modern (a) and ancient (b) microbial mats. (a) Contemporary stromatolites in Shark Bay, Western Australia (Wikipedia). (b) Stromatolite fossil aged 2.5×10^9 years from the Barberton Mountains, South Africa. Note the scale: the visible rock face in this image is 20 cm wide by 17 cm tall. (With permission from David J. Des Marais, NASA Ames Research Center.)

Section 3.6 and Box 3.4; see also Sections 7.3 and 7.4). Geochemical sources of reducing power in early Earth were limited largely to hydrothermal sources of H_2S, Fe^{2+}, Mn^{2+}, H_2, and CH_4. These only occurred at relatively low concentrations across the globe. However, another vast pool of potential reducing power was there *if it could be tapped*: water. The supply of water is virtually unlimited – it occurs at a concentration of 55 M. If a sun-driven biochemical mechanism evolved to use the atoms of oxygen in water as a source of electrons ($H_2O \rightarrow {}^1/_2O_2 + 2H^+ + 2e^-$), biology would soar.

Where did oxygenic photosynthesis come from? It features a complex, membrane-bound electron-transport assembly that includes chlorophyll, two light reaction centers, a manganese-containing water-splitting enzyme, a cytochrome proton pump, a ferredoxin NADP (nicotinamide adenine dinucleotide phosphate) reductase, ATP synthase, and light-harvesting pigments. Overall, light is harvested and used to steal electrons from water molecules and then electrons are funneled through transport chains to generate ATP and to reduce CO_2. Perhaps predictably, the elegant contemporary oxygenic photosynthesis apparatus (Figure 2.8) is resolvable into a series of components that each has its own origin and evolution (Knoll, 2003; Olson and Blankenship, 2004). Remarkably, a primitive version of this apparatus was present in representatives of cyanobacteria 2.7×10^9 years ago (see Table 2.1, Figure 2.2, and Box 2.2).

The fundamental functional unit of photosynthesis is the "reaction center" (RC) (Olson and Blankenship, 2004). Biochemically, RCs are membrane-bound associations of porphyrin-containing chlorophyll molecules, proteins, and light-harvesting pigments. The deep roots of RCs are unclear – they may have had an early precellular (last universal common ancestor) phase, based on the aqueous chemistry of porphyrins. Alternatively, the earliest RCs may have developed from early membrane-bound respiratory cytochrome proteins.

The oxygenic photosynthetic apparatus of modern cyanobacteria and higher plants is depicted in Figure 2.8. In the classic "Z scheme" of electron flow (a "Z" on its side), light energy activates a manganese-containing RC (photosystem II) which liberates electrons from water, generating molecular oxygen and protons. Electrons are then carried from a reduced (activated) chlorophyll molecule through a series of quinones, and cytochromes to a protein that donates the electrons to another RC (photosystem I), whose chlorophyll catalytic site contains magnesium. The electrons are accepted by the RC chlorophyll of photosystem I, which initiates light-activated transfer of electrons through several carriers, leading to the production of reducing power ($NADP^+$). Thus, modern oxygenic photosynthesis is accomplished by two different photosystems linked together. During flow of an electron from the RC in photosystem II to the RC in photosystem I, the energy state of the electron is boosted twice by light quanta, and after each boost electron transport flows in a thermodynamically favorable (negative to positive) direction. This

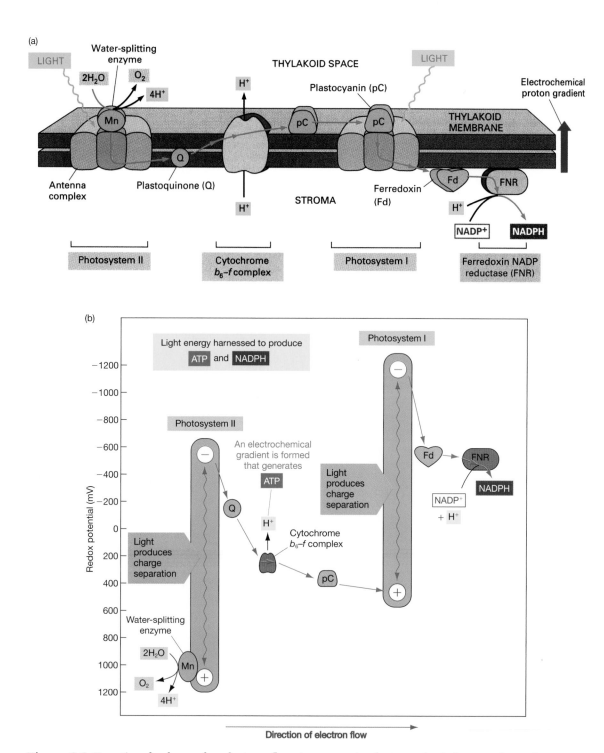

Figure 2.8 Functional scheme for electron flow in oxygenic photosynthesis in cyanobacteria and in higher plants. (a) Three-dimensional arrangement of photosystem components in their supporting membrane structure. (b) Diagram of energy and electron flow through photosystem II (left) to photosystem I (right). For a further explanation, see the text. (From Alberts, B., A. Johnson, J. Lewis, M. Raff, K. Roberts, and P. Walter. 2002. *Molecular Biology of the Cell,* 4th edn. Garland Science Publications, Taylor and Francis Group, New York.)

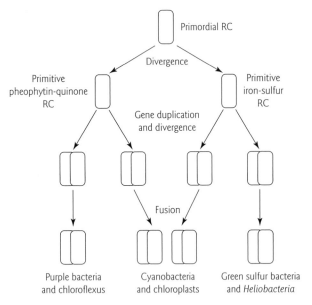

Figure 2.9 Evolution of photosynthetic reaction centers (RCs) that are the basis for photosystems I and II. Contemporary agents of oxygenic photosynthesis (cyanobacteria and chloroplasts) host photosystems I and II. (From Olson, J.M. and R.E. Blankenship. 2004. Thinking about the evolution of photosynthesis. *Photosynthesis Res.* **80**:373–386, fig. 4, p. 379. With kind permission of Springer Science and Business Media.)

generates both reducing power and a proton motive force (see Section 3.6) that allows ATP synthase to form ATP. The reducing power and ATP can then fuel the reduction of CO_2 and formation of biomass, often through the Calvin cycle via the enzyme ribulose bisphosphate carboxylase (see Section 7.4). Because CO_2 fixation is independent of light, this process is often referred to as "dark reactions".

The modular nature of oxygenic photosynthesis is obvious: two separate distinctive photosystems (with their respective RCs) complement one another. In Olson and Blankenship's (2004) essay summarizing current thought on the evolution of photosynthesis, the authors admit that certainty about ancient events is unlikely to ever be achieved. But they have prepared a feasible scenario drawing evidence from geology, biogeochemistry, comparative biochemistry, and molecular evolution. This scenario presents a model showing how the genotype (and phenotype) of a primordial RC may have developed into the three RC types represented in contemporary photosynthetic prokaryotes: (i) the (anoxygenic) purple nonsulfur and green nonsulfur bacteria; (ii) the (anoxygenic) green sulfur bacteria and *Heliobacteria*; and (iii) the oxygenic cyanobacteria and chloroplasts (Figure 2.9). In the proposed scenario, the ancestral RC diverged twice, creating two pairs of RCs – each featuring different electron-accepting proteins: pheophytin-quinone or iron-sulfur. A sequence of gene duplication and protein fusion events is thought to have combined the pheophytin-quinone RC with manganese-dependent catalase to create photosystem II, which used light energy to oxidize water. An early cyanobacterial cell was host to both types of RC. In this context, oxygenic, "Z scheme" photosynthesis may have been invented. One clear inference from Figure 2.9 is that photosystem II developed late – it was preceded by many prior RC combinations in many hosts.

As mentioned above, typical anoxygenic photosynthetic bacteria of today include purple bacteria, green sulfur bacteria, and *Heliobacteria*, whose electron donors range from H_2S to $S_2O_3^{2-}$, S^0, or Fe^{2+}. Owing to the relatively large pool of Fe^{2+} in ancient oceans, it is widely accepted that the ancient anoxygenic photosynthesis was responsible for oxidizing Fe^{2+} to Fe^{3+}.

Insoluble Fe^{3+} precipitated and was deposited on seafloor beds, perhaps creating the banded iron formations (BIFs) prevalent in the global sedimentary record – approximately 2×10^9 years ago.

2.11 CONSEQUENCES OF OXYGENIC PHOTOSYNTHESIS: MOLECULAR OXYGEN IN THE ATMOSPHERE AND LARGE POOLS OF ORGANIC CARBON

Figure 2.10 uses a clock-like metaphor to depict the chronology of oxygen-related and other events on life's development on Earth. The great inventors, oxygen-producing cyanobacteria, began their transformation of the biosphere about 2.7×10^9 years ago. The light-driven photosynthetic apparatus created atmospheric oxygen – simultaneously providing ATP and reducing power to fix CO_2 into organic carbon. The manifestation of oxygenic photosysnthesis,

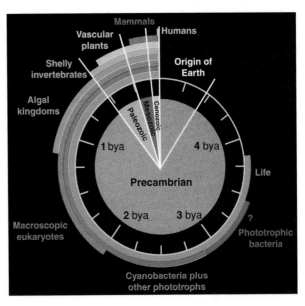

Figure 2.10 Earth's biological clock. bya, billions of years ago. (From Des Marais, D.J. 2000. When did photosynthesis emerge on Earth? *Science* **289**:1703–1705, with permission.)

light reactions making oxygen plus ATP, and dark reactions making biomass, impacted the Earth and its inhabitants via at least five different pathways (Figure 2.11). In the physiological roles of an electron donor and carbon source (see Section 3.3), the reduced organic carbon fueled biochemical innovation for both prokaryotic and eukaryotic heterotrophic life. Moreover, the molecular oxygen had drastic geochemical effects displayed in the sedimentary record and also fostered biochemical innovation in prokaryotic and eukaryotic physiology. Large ATP yields became available to organisms utilizing oxygen-dependent respiratory chains (see Sections 3.8 and 3.9). This allowed a "Cambrian explosion" in biological diversity – especially among eukaryotes (see Figure 2.2). The molecular oxygen also formed an ozone shield that curtailed damage to biota by UV radiation and allowed life's transition to new ecological niches on land.

There are physiological drawbacks to oxygen, however. Its metabolic byproducts can be toxic. Figure 2.12 provides a summary of the many chemical transformations that molecular oxygen (known as "triplet" or "ground state" oxygen) can routinely undergo during the reduction of O_2 to H_2O in aerobic respiration. Toxic, transient forms of oxygen include singlet oxygen, superoxide radical, and hydrogen peroxide. Without protection from these reactive oxygen species (ROS), the structure of essential cellular components, (e.g., DNA, lipids, proteins, carbohydrates) can

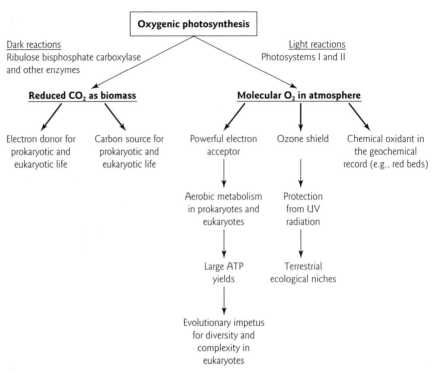

Figure 2.11 Five major chemical, physiological, and evolutionary impacts of oxygenic photosynthesis.

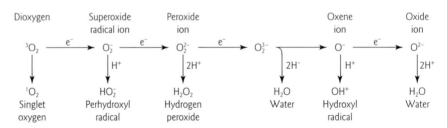

Figure 2.12 Oxygen's toxic downside: generation of reactive oxygen species (ROS) by common energy transfer reactions in cells. (From Rodriguez, R. and R. Redman 2005. Balancing the generation and elimination of reactive oxygen species. *Proc. Natl. Acad. Sci. USA* **102**:3175–3176. Copyright 2005, National Academy of Sciences, USA.)

be severely altered. The result can be failure in membrane and reproductive function. The many innovations of aerobic life could never have developed without the evolution, in parallel, of both enzymatic and nonenzymatic metabolic systems that eliminated ROS. Well-known enzymatic ROS-elimination systems include catalase (converting H_2O_2 to H_2O and

O_2), peroxidase (converting H_2O_2 to H_2O), superoxide dismutase (converting O_2^- to H_2O_2 and O_2), and superoxide reductase (converting O_2^- to H_2O_2). Nonenzymatic ROS scavengers include glutathione and proline. ROS are unavoidable byproducts of biochemical pathways (such as respiration, glycolysis, and photosynthesis) that are central to energy production and storage strategies of aerobic microorganisms, animals, and plants. Thus, the growth and reproduction of all aerobic life is a balancing act between the generation of ROS and the capacity of antioxidation systems to eliminate them.

As the Earth's pool of molecular oxygen gradually grew, the biosphere's original prokaryotic life forms that flourished in the absence of oxygen were forced to adapt. They did so either by acquiring oxygen-defense mechanisms (via direct evolution or lateral gene transfer from other organisms) or by receding into refuges protected from toxic oxygen. Earth features many habitats (e.g., waters, soils, sediments, gastrointestinal tracts of animals) where oxygen fails to penetrate. Oxygen-free locales reflect the dynamic balance between the rate of oxygen influx by diffusion/convection and the rate of consumption by oxygen-respiring microbial communities. Respiratory demand by the microorganisms dwelling in a thin layer of mud at the sediment–water interface in lakes, streams, and the ocean can easily outpace the rate of oxygen influx. This is because oxygen has limited solubility in water (~9 mg/L at 20°C when oxygen is 21% of the atmosphere) and oxygen diffuses into water from the atmosphere quite slowly.

2.12 EUKARYOTIC EVOLUTION: ENDOSYMBIOTIC THEORY AND THE BLENDING OF TRAITS FROM *ARCHAEA* AND *BACTERIA*

The tree of life is a portrait of evolution based on molecular phylogeny of genes encoding small subunit ribosomal RNA (see Sections 2.7 and 5.5). The three-domain concept provides a framework for refining hypotheses about how life developed. If developments were based solely upon linear evolutionary trajectories, once a new positive trait arose, it might be expected to remain exclusively within its original line of descent (see Section 5.9). In practice, however, phylogenetic trajectories often depart from linearity. This departure probably is caused by two major factors: (i) genes being transferred laterally between lines of descent; and (ii) fluctuating environmental conditions (hence selective pressures) which over the eons foster hereditary discontinuities, via loss of genes and/or their host via extinction.

Insights into the robustness of the tree of life can be obtained by examining other traits (morphological, genetic, biochemical) carried by current members of the *Bacteria*, *Archaea*, and *Eukarya*. Table 2.3 provides a comparison of such traits across the three domains of life. The traits listed

Table 2.3

A comparison of morphological and biochemical traits of *Bacteria*, *Archaea*, and *Eukarya* (modified from MADIGAN, M. and MARTINKO, J. 2006. *Brock Biology of Microorganisms*, 11th edn, p. 309. Prentice Hall, Upper Saddle River, NJ. Reprinted by permission of Pearson Education, Inc., Upper Saddle River, NJ)

Trait	Bacteria	Archaea	Eukarya
Prokaryotic cell structure	+	+	–
Chromosomal DNA in closed circle	+	+	–
Histone proteins with DNA	–	+	+
Nucleus	–	–	+
Mitochondria and/or chloroplast organelles	–	–	+
Cell wall with muramic acid	+	–	–
Membrane lipids	Ester-linked	Ether-linked	Ester-linked
Ribosome mass	70S	70S	80S
Genes with introns	–	–	+
Genes as operons	+	+	–
mRNA tailed with polyA	–	–	+
Sensitivity to antibiotics (chlorophenicol, streptomycin, kanamycin) that increase errors during protein synthesis	+	–	–
Growth above 80°C	+	+	–
Chemolithotrophy	+	+	–
Nitrogen fixation	+	+	–
Denitrification	+	+	–
Dissimilatory reduction of SO_4^{2-}, Fe^{3+}, Mn^{4+}	+	+	–
Methanogenesis	–	+	–

in Table 2.3 include overall cell structure, chromosome structure, the presence of a nucleus, the presence of organelles, membrane structure, the structure of ribosomes, gene organization, mechanisms of protein synthesis, sensitivity to high temperature, and metabolic diversity. The pattern of pluses and minuses in the table clearly supports the broad message from the tree of life: *Bacteria*, *Eukarya*, and *Archaea* are distinctive, though *Archaea* have commonalities with the other two domains. Refined phylogenetic results and biochemical correlations have shown that the genetic lines of *Eukarya* and *Archaea* have a common ancestral branch that is independent of the one that gave rise to *Bacteria*. Thus *Eukarya* and *Archaea* are more closely related to one another than either is to *Bacteria*.

The hypothetical route from simple precellular life to the advanced multicellular eukaryotes of today is presented in Figure 2.13. After *Bacteria* and the predecessor of the other two domains diverged from the last universal common ancestor, another division in the main trunk occurred that created *Archaea* and *Eukarya*. The *Eukarya* developed a membrane-bound nucleus to accommodate an expanded genome, which contributed to increased cell size. The eukaryotic nuclear membrane and other membrane systems may be the result of early, poorly regulated expression of bacterial membrane replication: vesicles may have formed and fortuitously accumulated in the cytosol near their site of synthesis.

Endosymbiosis, as advanced by L. Margulis in the 1960s, is likely to have played a crucial role in eukaryotic development (see Section 8.1). Mitochondria, the ATP-generating organelles carried almost universally by *Eukarya*, feature cell structures (membranes, 16S RNA not 18S rRNA) that unquestionably are of bacterial origin. In fact, the 16S rRNA gene sequences of mitochondria are closely related to β-Proteobacteria. This suggests that symbiotic uptake by a unicellular eukaryote allowed an originally free-living aerobic bacterium to evolve into mitochondria by specializing in ATP generation (Figure 2.13). Similarly, the photosynthetic organelles of green plants (chloroplasts) feature undeniable bacterial characteristics and a 16S rRNA gene sequence closely related to that of cyanobacteria. Endosymbiotic cyanobacteria became chloroplasts. Accordingly, algae and higher plants of today appear to be the product of two sequential endosymbiotic events (Figure 2.13).

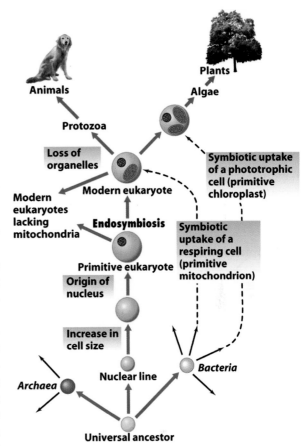

Figure 2.13 Hypothetical evolutionary events leading from the last universal common ancestor through the three domains to the endosymbiotic production of modern eukaryotes and their developmental achievements. (From MADIGAN, M. and J. MARTINKO, 2006. *Brock Biology of Microorganisms*, 11th edn, p. 307. Copyright 2006, reprinted by permission of Pearson Education, Inc., Upper Saddle River, NJ.)

To close this section and chapter, it seems prudent to develop a rudimentary sketch of the key characteristics of microorganisms. The six contemporary types of microorganisms are *Bacteria*, *Archaea*, fungi, protozoa, algae, and viruses. Information in Box 2.3 solidifies major themes about the structural distinctions between prokaryotic and eukaryotic forms of life and how microbiologists do their work. Additional details on the diverse cellular and physiological traits of microorganisms appear in Chapters 3, 7, and 8, and particularly in Chapter 5.

Box 2.3

A primer on what microorganisms are and how they are studied: small size scale makes a big difference for microbiologists

When cataloging the characteristics of creatures, all biologists seek and find unifying themes and fascinating contrasts in the structure, behavior, and ecology of their objects of study. Thus, there are many commonalities between how microbiologists study microorganisms and how plant and animal biologists study plants and animals. But unlike plants and animals, individual microorganisms (*Bacteria*, *Archaea*, protozoa, fungi, algae, and viruses) can be seen only with a microscope. The size of prokaryotic cells (*Bacteria* and *Archaea*) is about 1 μm.

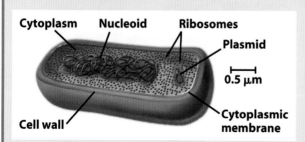

(a)

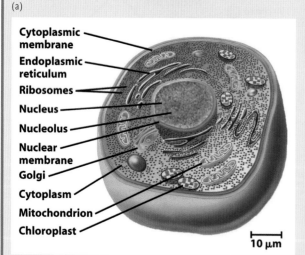

(b)

Figure 1 Comparison of key structural features of (a) prokaryotic and (b) eukaryotic cells. (From MADIGAN, M. and J. MARTINKO. 2006. *Brock Biology of Microorganisms*, 11th edn, p. 22. Copyright 2006, reprinted by permission of Pearson Education, Inc., Upper Saddle River, NJ.]

Shapes of individuals include spheres (cocci), rods, or spiral forms; these may be in clusters, chains, or long filaments.

Except for viruses (which are intracellular parasites; see Section 5.8), within each unicellular microorganism are all of the genetic and biochemical structures that allow metabolism and self-replication to occur. Growth of microbial cells relies upon the coordination of approximately 2000 chemical reactions that both generate energy (especially as ATP) and utilize that energy in biosynthetic production of all the materials (membranes, cell walls, proteins, DNA, RNA, ribosomes, etc.; see also Sections 3.3 and 3.9) needed to build new organisms. Figure 1 provides a generic overview comparing prokaryotic and eukaryotic cells.

As is clear from Figure 1 in this box and information in Table 2.3, major contrasts between prokaryotic microorganisms (*Bacteria* and *Archaea*) and eukaryotic microorganisms (fungi, algae, protozoa) include size (~1 μm versus >10 μm, respectively), the organization of genetic material (as nucleoid and plasmid DNA versus a membrane-enclosed nucleus, respectively), and the presence of other specialized membrane-enclosed organelles (e.g., mitochondria and chloroplasts) only in eukaryotes.

While "macrobiologists" can inspect an individual eagle or flower or aardvark to gain deep insight into the structure and function of each creature, microbiologists often take a "collective" approach to studying microbial biology and taxonomy. Microbiologists routinely rely upon growing and isolating colonies of a single (purified) microorganism on solid media (e.g., agar) in the laboratory. Colonies of microorganisms are composed of thousands to millions of cells derived from a single cell that grew exponentially, via binary fission, on nutrients provided by the solid medium. During binary fission, all cell components (DNA, ribosomes, protein complexes, inorganic ions, etc.) essentially double within an elongating mother cell. After partitioning of the intracellular materials, a septum forms in the middle of the elongated cell. Next, completion of new cell wall synthesis at the septum allows the two daughter cells to separate and be released.

Operationally, a pure culture of a given microorganism is one that exhibits consistent phenotype and genotype after sequential subculturing (it must "breed true"). Subcultures of purified microorganisms can be inoculated into many types of liquid or solid media and the abilities of ~10^9 individual cells to grow and/or produce particular metabolic byproducts (acids, gases, metabolites) can be assessed. For microbiologists, these abilities [to grow (or not grow) and produce (or not produce) metabolites] constitute key phenotypic and genotypic information. Such information is analogous to leaf shape and flower morphology – used by botanists to classify plants. One widely used phenotypic trait for classifying *Bacteria* is the Gram stain. In 1884, Christian Gram discovered a fundamental structural distinction between two main types of bacterial cells. When stained with crystal violet and iodine, Gram-positive cells (with a thick multilayered peptidoglycan cell wall) retain crystal violet and its purple color after being rinsed with alcohol. However, Gram-negative cells (with thin peptidoglycan walls and lipid-rich outer membranes) are decolorized by alcohol.

Chapter 5 conveys a substantial amount of additional information about microorganisms and microbial diversity. Within Chapter 5, Section 5.6 provides a sampling of the traits of many prokaryotic and eukaryotic microorganisms and their phylogenetic relationships.

STUDY QUESTIONS

1 Devise a way to test the "panspermia hypothesis". To do this, presume that you have access to samples from many planets near Earth and access to all tools listed in Table 2.2. Begin by defining the logic that you would use. Then state what measurements would be performed on which samples and how you would interpret the data.

2 Table 2.1 indicates that the early surface of Earth was subjected to extreme conditions. What habitat might have been spared these extremes? Might these habitats have been preferred sites for life's early stages?

3 ^{14}C *dating of a peat core*. You are studying the microbiology of peat bogs. A core of peat is retrieved from a depth of 6 m and you want to know the age of the preserved plant material. Presume that, since being originally deposited, there has been no replenishment of ^{14}C from the atmosphere. Thus, the amount of ^{14}C there is the residual since radioactive decay began.

The measurements you complete reveal that the ^{14}C content is 0.03125 times that of surface material. How many half-lives have passed? How old is the material? Why is the limit for dating based on radioactive decay limited to about 10 half-lives? (Hint: use information in Box 2.1 to answer this question.)

4 Wachtershauser's theory of early metabolism relies on the notion that mineral surfaces at the seafloor–water interface catalyzed chemical reactions. Are there mineral surfaces today that catalyze chemical reactions? If so, please provide some examples. (To answer this, search library sources and/or the world wide web.)

5 The last universal common ancestor is an intriguing concept. The genetic heritage that you share with your siblings (plus parents, grandparents, and great grandparents) establishes a "line of descent". Name one major factor that is absolutely necessary for lines of descent to be traced. Name another factor that would readily "blur" the lines. Briefly explain both factors.

6 Why are cyanobacteria (and their ancestors) the "great innovators" of evolution? List and explain six major impacts (direct or indirect) that cyanobacteria had on evolution.

REFERENCES

Alberts, B., A. Johnson, J. Lewis, M. Raff, K. Roberts, and P. Walter. 2002. *Molecular Biology of the Cell*, 4th edn. Garland Science Publications, Taylor and Francis Group, New York.

Allwood, A.C., M.R. Walker, B.S. Kamber, C.P. Marshall and I.W. Burch. 2006. Stromatolite reef from the early Archaean era of Australia. *Nature* **441**:714–718.

Anders, E. 1996. Evaluating the evidence for past life on Mars. *Science* **274**:2119–2121.

Borg, L.E., S.N. Connelly, L.E. Nyquist et al. 1999. The age of the carbonates in martian meteor ALH84001. *Science* **286**:90–94.

Brocks, J.J., G.A. Logan, R. Buick and R.E. Summons. 1999. Archaen molecular fossils and the early rise of eukaryotes. *Science* **285**:1033–1036.

de Duve, C. 2007. The origin of Eukaryotes: a reappraisal. *Nature Rev. Genetics* **8**:395–403.

Grossman, E.L. 2002. Stable carbon isotopes as indicators of microbial activities in aquifers. *In*: C.J. Hurst, R.L. Crawford. G.R. Knudsen, M.I. McInerney, and L.D. Stetzenbach (eds) *Manual of Environmental Microbiology*, 2nd edn, pp. 728–742. American Society for Microbiology, Washington, DC.

Hanczyc, M.M., S.M. Fujikawa, and J.W. Szostak. 2003. Experimental models of primitive cellular compartments: Encapsulation, growth, and division. *Science* **302**:618–622.

Huber, C. and G. Wachtershauser. 2006. α-Hydroxy and α-amino acids under possible Hadean, volcanic origin-of-life conditions. *Science* **314**:630–632.

Kerr, R.A. 2005. The story of O$_2$. *Science* **308**: 1730–1732.

Knoll, A.H. 2003. *Life on a Young Planet: The first three billion years of life on Earth*. Princeton University Press, Princeton, NJ.

Madigan, M.T. and J.M. Martinko. 2006. *Brock Biology of Microorganisms*, 11th edn. Prentice Hall, Upper Saddle River, NJ.

Martin, W. and M. J. Russell. 2003. On the origins of cells: a hypothesis for the evolutionary transitions from abiotic geochemistry to chemoautotrophic prokaryotes, and from prokaryotes to nucleated cells. *Phil. Trans. Royal Soc. London Ser. B. Biol. Sci.* **358**:59–83.

McKay, D.S., E.K. Gibson, Jr., K.L. Thomas-Keprta, et al. 1996. Search for past life on Mars: Possible relic biogenic activity in martian meteorite ALH84001. *Science* **273**:924–930.

Nisbet, E.G. and N.H. Sleep. 2001. The habitat and nature of early life. *Nature* **409**:1083–1091.

Olson, J.M. and R.E. Blankenship. 2004. Thinking about the evolution of photosynthesis. *Photosynthesis Res.* **80**:373–386.

Rodriguez, R. and R. Redman. 2005. Balancing the generation and elimination of reactive oxygen species. *Proc. Natl. Acad. Sci. USA* **102**:3175–3176.

Russell, M.J. 2003. Geochemistry: the importance of being alkaline. *Science* **302**:580–581.

Wachtershauser, G. 1990. Evolution of the first metabolic cycles. *Proc. Natl. Acad. Sci. USA* **87**: 200–204.

Wachtershauser, G. 1992. Groundworks for an evolutionary biochemistry: The iron-sulphur world. *Prog. Biophys. Molec. Biol.* **58**:85–201.

Wachtershauser, G. 2000. Origin of life: Life as we don't know it. *Science* **289**:1307–1308.

Woese, C. 1998. The universal ancestor. *Proc. Natl. Acad. Sci. USA* **95**:6854–6859.

Woese, C.R., G.J. Olsen, M. Ibba, and D. Soll. 2000. Aminoacyl-tRNA synthases, the genetic code, and the evolutionary process. *Microbiol. Molec. Biol. Rev.* **64**:202–236.

FURTHER READING

Anbar, D. and A.H. Knoll. 2002. Proterozoic ocean chemistry and evolution: a bioinorganic bridge? *Science* **297**:1137–1142.

Atlas, R.M. and R. Bartha. 1997. *Microbial Ecology*, 4th edn. Benjamin/Cummings, Menlo Park, CA.

Banfield, J.F. and C.R. Marshall. 2000. Genomics and the geoscience. *Science* **287**:605–606.

Baross, J.A. 1998. Do the geological and geochemical records of the early Earth support the prediction from global phylogenetic models of a thermophilic cenancestor? *In*: J. Wiegel and M. Adams (eds) *Thermophiles: The key to molecular evolution and the origin of life?*, pp. 3–18. Taylor and Francis, London.

Brack, A. (ed.) 1998. *The Molecular Origins of Life: Assembling pieces of the puzzle*. Cambridge University Press, London.

Castresana, J. and D. Moreira. 1999. Respiratory chains in the last common ancestor of living organisms. *J. Molec. Evol.* **49**:453–460.

Davis, B.K. 2002. Molecular evolution before the origin of species. *Prog. Biophys. Molec. Biol.* **79**:77–133.

de Duve, C. 2002. *Life Evolving: Molecules, mind, and meaning*. Oxford University Press, Oxford.

Des Marais, D.J. 2000. When did photosynthesis emerge on Earth? *Science* **289**:1703–1705.

DiGiub, M. 2003. The universal ancestor and the ancestor of bacteria were hyperthermophiles. *J. Molec. Evol.* **57**:721–730.

Ehrlich, H.L. 2002. *Geomicrobiology*, 4th edn. Marcel Dekker, New York.

Franchi, M. and E. Gallori. 2005. A surface-mediated origin of the RNA world: Biogenic activities of clay-absorbed RNA molecules. *Gene* **346**:205–214.

Furnes, H., N.R. Banarjee, K. Muehlenbachs, H. Stadigel and M. deWitt. 2004. Early life recorded in pillow lavas. *Science* **304**:578–581.

Lahav, N., S. Nir, and A.C. Elitzur. 2001. The emergence of life on earth. *Prog. Biophys. Molec. Biol.* **75**:75–120.

Lazcano, A. and S.L. Miller. 1999. On the origin of metabolic pathways. *J. Molec. Evol.* **49**:424–431.

Macalady, J. and J. Banfield. 2003. Molecular geomicrobiology: Genes and geochemical cycling. *Earth Plant. Sci. Lett.* **209**:1–17.

Schopf, J.W. 1988. Tracing the roots of the universal tree of life. *In*: Brack, A. (ed.) *The Molecular Origins of Life: Assembling pieces of the puzzle*, pp. 337–362. Cambridge University Press, London.

Schopf, J.W. and B.M. Packer. 1987. Early Archaean (3.3 billion year old) microfossils from The Warawoona Group, Australia. *Science* **237**:70–73.

VanZullen, M.A., A. Lepland, and G. Arrhenius. 2002. Reassessing the evidence for the earliest traces of life. *Nature* **418**:627–630.

Wachtershauser, G. 2006. Origins of life: RNA world versus autocatalytic anabolism. *In*: M. Dworkin, S. Falknow, E. Rosenberg, K.-H. Schleifer and E. Stackebrandt (eds) *The Prokaryotes*, Vol. 1, 3rd edn, pp. 275–283. Springer-Verlag, New York.

Westall, F. 1999. Fossil bacteria. *In*: J. Seckbach (ed.) *Enigmatic Microorganisms and Life in Extreme Environments*, pp. 73–88. Kluwer Publications, Dordrecht.

Zhaxybyeva, J. and J.P. Gogarten. 2004. Cladogenesis, coalescence and the origin of the 3 domains of life. *Trends Genetics* **20**:182–187.

3

Physiological Ecology: Resource Exploitation by Microorganisms

This chapter begins by presenting an important concept: Habitats provide selective pressures over evolutionary time. *Thus, facts and principles in Chapters 3 and 4 are intimately linked. Next, we develop a genome-based definition of microorganisms and review universal functional categories of cellular processes. This is followed by discussion of habitat-specific factors that are likely to have shaped microbial evolution: nutrient availability and a need for dormancy. Conditions for life on Earth are then presented as "mixtures of materials in chemical disequilibrium". Thermodynamics provides a way to systematically organize and quantify the many biogeochemical reactions that microorganisms catalyze. The chapter ends by discussing metabolism, the logic of electron transport reactions for adenosine triphosphate (ATP) production, and the diversity of lithotrophic metabolic reactions.*

Chapter 3 Outline

3.1 THE CAUSE OF PHYSIOLOGICAL DIVERSITY: DIVERSE HABITATS PROVIDE SELECTIVE PRESSURES OVER EVOLUTIONARY TIME

Chapter 2 presented the conditions of early Earth and the transition from its abiotic past to its biotic present. Metabolism, replication, and heredity are critical traits of life. These three processes are inseparable from the environmental context that provides resources for metabolism and selective pressures for both replication and heredity. Figure 3.1 conceptually depicts the long (3.8×10^9 years) and dynamic dialog between Earth's habitats and microorganisms. The dialog is framed by thermodynamics because, by systematically examining thermodynamically favored geochemical reactions, we can understand and predict selective pressures that act on microorganisms. There is a mechanistic series of linkages between our planet's habitat diversity and what is recorded in the genomes of microorganisms found in the world today. Diversity in habitats is synonymous with diversity in selective pressures and resources. When operated upon by forces of evolution, the result is molecular, metabolic, and physiological diversity found in extant microorganisms. This chapter will focus on the microbiological component (nutritional, genomic, biochemical, and physiological; upper right-hand sphere of Figure 3.1) and Chapter 4 will focus on the broad catalog of Earth habitats (upper left-hand sphere of Figure 3.1).

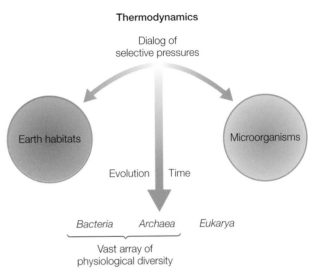

Figure 3.1 Conceptualized mechanism for the development of past and current microbial metabolic diversity: diverse habitats provide selective pressures over evolutionary time.

3.2 BIOLOGICAL AND EVOLUTIONARY INSIGHTS FROM GENOMICS

Genome size

To begin to discern what prokaryotes are and the evolutionary pressures that have shaped them, we need only turn to the ongoing revolution in biology: genomics (e.g., Fraser et al., 2004; Konstantinidis and Tiedje, 2004a; Zhou et al., 2004). The sequence of an organism's genome reveals the blueprint of its current physiological capabilities and the integrated history of its heritage. In the last decade, genome-sequencing efforts have provided a remarkable amount of fundamental information about microbial life. As of 2007, ~450 bacterial genomes, 35 archaeal genomes, and 15 eukaryotic microbial (fungi, protozoa) genomes were either in completed or draft forms. Hundreds more were in progress. Table 3.1 provides a representative list of genomes of prokaryotes isolated from ocean water, soils, sediments, sewage, or hot springs. These are cultured microorganisms (see Section 5.1) whose physiological properties make them valuable for understanding environmentally significant biogeochemical

Table 3.1

A sampling of completed prokaryotic genomes (modified from US Department of Environment Joint Genome Institute)

Organism	Genome size (Mb)	ORFs	Phenotype/habitat
Bacteria			
Acinetobacter sp. ADP1	3.59	3425	Aerobic chemoheterotroph and human pathogen/water, soil, human skin
Agrobacterium tumefaciens C58 Dupont	5.67	5467	Aerobic chemoheterotroph/soil, plant pathogen
Anabaena variabilis ATCC29413	7.10	5720	Oxygenic photosynthesis, nitrogen fixation/water, soil
Bacillus cereus ATCC 14579	5.42	5397	Aerobic spore-forming heterotroph/soil
Bradyrhizobium japonicum USDA 110	9.10	8371	Aerobic heterotroph, nitrogen-fixing symbiont on soybean roots/soil
Buchnera aphidicola APS	0.655	609	Insect endosymbiont
Burkholderia xenovorans LB400	9.73	8784	Aerobic heterotroph, metabolizes polychlorinated biphenyl/soil
Carsonella rudii	0.160	182	Obligate endosymbiont of a psyllid, a plant sap-feeding insect
Clostridium acetobutylicum ATCC 824	4.13	3955	Obligate anaerobic spore-forming chemoorganotroph/soil
Dechloromonas aromatica RCB	4.50	4247	Facultative benzene degrader/water, soil
Deinococcus radiodurans R1	3.28	3239	Chemoheterotroph, aerobe, highly radiation resistant/soil
Desulfovibrio desulfuricans G20	3.73	3853	Strict anaerobe, chemoorganotroph using substrate as electron acceptors/sediment
Dehalococcoides ethenogenes strain 195	1.47	1629	Strict anaerobe using chlorinated solvents as final electron acceptor/sewage, groundwater
Escherichia coli K12	4.64	4359	Facultative chemoheterotroph/human intestine
Escherichia coli O157:H7 EDL933	5.62	55434	Facultative chemoautotroph, human pathogen/intestine, food
Geobacter metallireducens GS-15	4.01	3587	Anaerobic chemoheterotroph using metal anions as electron acceptors/water, subsurface sediment

Table 3.1 *Continued*

Organism	Genome size (Mb)	ORFs	Phenotype/habitat
Methylococcus capsulatus Bath	3.30	3012	Aerobic methanotroph/soil water
Mycoplasma genitalium	0.580	516	Intracellular human parasite of urogenital tract
Nitrobacter winogradskyi Nb-255	3.40	3174	Facultative autotroph, CO_2 fixation, nitrite oxidizer/soil
Polaromonas naphthalenivorans CJ2	5.34	5022	Chemoorgano- and lithotrophic aerobe/ terrestrial sediment
Prochlorococcus marinus MIT 9312	1.70	1852	Photosynthetic bacterioplankton/ocean
Prochlorococcus marinus MIT 9313	2.41	2321	Photosynthetic bacterioplankton/ocean
Pseudomonas fluorescens Pf-5	7.07	6223	Aerobic chemoheterotroph/soil
Pseudomonas putida KT2440	6.18	5446	Aerobic chemoautotroph/soil
Rhodopseudomonas palustris CGA009	5.86	4897	Physiologically versatile-facultative photosynthetic organism/water, soil
Shewanella oneidensis MR-1	5.13	4601	Metabolically versatile chemoheterotroph, metal reduction/lake sediment
Silicibacter pomeroyi DSS-3	4.60	4314	Aerobic heterotroph important in the sulfur cycle/seawater
Streptomyces avermitilis MA-4680	9.11	7759	Aerobic chemoheterotroph, filamentous spore former/soil
Synechococcus elongatus PCC 7942	2.74	2712	Photosynthetic bacterioplankton/ocean
Thermotoga maritima MSB8	1.86	1907	Hyperthermophilic anaerobe/geothermal marine sediment
Trichodesmium erythraeum ImS101	7.75	4494	Photosynthetic nitrogen-fixing filamentous cyanobacteria/ocean water
Archaea			
Archaeoglobus fulgidus DSM 4304	2.18	2517	Strict anaerobe, hyperthermophilic, sulfate reduction/hot springs
Halobacterium salinarum NRC-1	2.57	2726	Chemoorganotrophic aerobe/highly saline ponds and lakes
Methanopyrus kandleri AV19	1.69	1765	Strict anaerobe hyperthermophilic, methanogenesis/hot springs
Methanosarcina barkeri Fusaro	4.87	3695	Strict anaerobe, methanogenesis, cellulose metabolism/marine mud, sludge
Nanoarchaeum equitans Kin4-M	0.49	573	Intracellular parasite, anaerobic hyperthermophile/hot springs
Pyrococcus furiosus DSM 3638	1.90	2171	Strict anaerobe hyperthermophilic, radiation resistant, sulfur respiration/hot springs
Sulfolobus solfataricus P2	2.99	3104	Strict aerobe hyperthermophilic, acidophile, sulfur oxidizer/hot springs
Thermoplasma volcanium GSS1	1.58	1610	Facultative acidophilic thermophile/hydrothermal vent

Mb, mega bases, 10^6 base pairs; ORF, open reading frame.

processes such as the cycling of carbon, nitrogen, sulfur, metals, and organic environmental pollutants. Also represented in Table 3.1 are several well-characterized parasitic and pathogenic prokaryotes because these reveal several key features of prokaryotic biology.

One lesson from the data in Table 3.1 is the rough correspondence between 10^3 base pairs (bp) of DNA and a single open reading frame (ORF) or gene. Although there is significant variation in size (number of amino acids) among proteins, they average somewhat more than 300 amino acids each. At three codons per amino acid, each ORF consists of about 10^3 bases. Bioinformatic software divides the long stretches of DNA bases into ORFs based on recognition of start and stop sites for DNA replication and protein translation on the encoded mRNA. It is important to realize that the initial judgments on ORFs are theoretical; their accuracy must be confirmed subsequently by genetic and physiological tests. Another lesson from the genome size comparison shown in Table 3.1 is that two strains of the same species can possess strikingly different genotypes and phenotypes. This is illustrated by the *Escherichia coli* and *Prochlorococcus* entries. Acquisition of DNA by an ancestor of *E. coli* strain K12 appears to have converted the benign intestinal tract resident to an often lethal human pathogen (*E. coli* strain O157:H7; see Chapter 5, Science and the citizen box). Similarly, the *Prochlorococcus* strain with the larger genome has been shown to occupy a distinctive ecological niche characterized by low light intensity (Rocap et al., 2003).

Another clear trend in the entries in Table 3.1 is the wide range in sizes of prokaryotic genomes. The smallest genomes are those of intracellular parasites: *Carsonella rudii* (featuring 182 ORFs encoded by a DNA sequence of 159,662 bp; Nakabachi et al., 2006), *Buchnera aphidicola* APS (featuring 609 ORFs encoded by a DNA sequence of 655,725 bp), *Mycoplasma genitalium* (featuring 516 ORFs encoded by a DNA sequence of 580,074 bp), and *Nanoarchaeum equitans* Kin4-M (featuring 573 ORFs encoded by a DNA sequence of 490,885 bp). In contrast, the largest prokaryotic genomes are free-living heterotrophic *Bacteria* from the soil habitat: *Bradyrhizobium japonicum* (featuring 8371 ORFs, encoded by a DNA sequence of 9,105,828 bp), *Burkholderia xenovorans* LB400 (featuring 8784 ORFs encoded by a DNA sequence of 9,703,676 bp), and *Streptomyces avermililis* (featuring 7759 ORFs encoded by a DNA sequence of 9,119,895 bp). After comparing the genomic composition of 115 prokaryotes, Konstantinidis and Tiedje (2004b) argued persuasively for two hypotheses: (i) the small genomes of intracellular parasites are streamlined because they are relieved of selective pressures for maintaining elaborate metabolic pathways required for a free-living lifestyle; and (ii) the large genomes of soil-dwelling microorganisms accommodate many genes (both regulatory and structural) that confer traits such as slow growth and exploitation of resources that are both diverse and scarce.

Gene content within microbial genomes

After the bioinformatics software applies recognition algorithms that delineate individual genes (ORFs), then the ORFs are aligned with sequences of homologous genes of known function. These two steps allow amino acid similarities in proteins encoded by new and previously characterized genes to be computed so that the biochemical function of the new genes can be tentatively assigned. Several different levels of confidence for assigning gene function have been proposed (Kolker et al., 2005) and these vary, from high to low, depending upon the degree of similarity with known genes and proteins.

The template used by the US DOE Joint Genome Institute for categorizing and organizing genome sequence data is shown in Table 3.2. There are 23 functional categories of expected genes, known as COGs (clusters of orthologous groups) which span three well-established necessities of cellular life: "information storage and processing", "cellular processes", and "metabolism". The remainder of genes in a genome fall into the "poorly characterized" category ("function unknown" and "general function prediction only"). Results of COG analysis for the 6.18 Mb genome (5452 ORFs) of *Pseudomonas putida* KT2440 are presented in Table 3.2. Note the high apportionment of this soil bacterium's genome to energy production (5.4%) and amino acid uptake and metabolism (9.5%); this issue will be revisited in Section 3.3.

The "poorly characterized" or "unknown hypothetical" category of genes is a barometer for the completeness of our understanding of prokaryotic genomes and (by inference) prokaryotic biology. The proportion of unknown hypothetical genes in all genome-sequencing projects hovers at roughly one-third (e.g., Kolker et al., 2005). This proportion of unknown hypothetical genes even applies to the thoroughly studied bacterium, *E. coli*. While it is possible that poorly characterized genes have no function (they may be noncoding or "junk" DNA), it is equally plausible that these genes are blatant reminders of how little we truly understand about real-world ecological pressures and the physiological processes expressed by microorganisms in their native habitats. *Because laboratory cultivation of microorganisms likely fails to mimic the physiological and ecological conditions experienced by microorganisms in field settings, many genetic traits may go unexpressed by the organism; hence, unobserved by the microbiologist.* Understanding the function of unknown hypothetical genes represents one of the major frontiers in biology. The lack of knowledge of a significant proportion of each microbial genome limits our ability to advance biology to a more predictive science (Kolker et al., 2005). Thus, although genomics has provided genetic blueprints for individual microorganisms, a substantial proportion of the blueprints remains "illegible" and mysterious.

Table 3.2

Framework for categorizing and organizing open reading frames (ORFs) found during genome-sequencing projects. Twenty five gene categories (cluster of orthologous groups: COGs) and their respective contributions to the sequenced genome of *Pseudomonas putida* KT2240 are listed (from US Department of Environment Joint Genome Institute)

Functional group of genes (COG)	Example genome (*Pseudomonas putida* KT2440): number of genes in each functional group (% of total)
Information storage and processing	
J Translation, ribosomal structure and biogenesis	187 (3.4)
A RNA processing and modification	3 (0.05)
K Transcription	440 (8.1)
L DNA replication, recombination and repair	221 (4.0)
B Chromatin structure and dynamics	2 (0.04)
Cellular processes	
D Cell division and chromosome partitioning	42 (0.77)
Y Nuclear structure	–
V Defense mechanisms	63 (1.2)
T Signal transduction mechanisms	330 (6.1)
M Cell envelope biogenesis, outer membrane	266 (4.9)
N Cell motility and secretion	127 (2.3)
Z Cytoskeleton	–
W Extracellular structures	–
U Intracellular trafficking, secretion, and vesicular transport	123 (2.3)
O Posttranslational modification, protein turnover, chaperones	174 (3.2)
Metabolism	
C Energy production and conversion	293 (5.4)
G Carbohydrate transport and metabolism	228 (4.2)
E Amino acid transport and metabolism	516 (9.5)
F Nucleotide transport and metabolism	91 (1.7)
H Coenzyme metabolism	186 (3.4)
I Lipid metabolism	181 (3.3)
P Inorganic ion transport and metabolism	275 (5.1)
Q Secondary metabolites biosynthesis, transport and catabolism	128 (2.3)
Poorly characterized	
R General function prediction only	564 (10.4)
S Function unknown	1014 (18.6)

Integrating genome data

Even with roughly one-third of each microbial genome obscured (see above) the insights from the remaining (legible) portion of each genome are tremendously powerful. For each sequenced genome, a template for cell function is used to organize the genes from the 23 functional gene categories of Table 3.2. Figure 3.2 shows the metabolic template for a model cell. The generalized conceptual cell model exhibits an interior (cytoplasm) and an exterior bounded by the cytoplasmic membrane and the cell wall. The periphery of the cell is lined with membrane-bound transport proteins that regulate cytoplasmic composition by acting on inorganic cations, inorganic anions, carbohydrates, amino acids, peptides, purines, pyrimidines, other nitrogenous compounds, carboxylates, aromatic compounds, other carbon compounds, and water. Many of the transport mechanisms are energy-coupled (requiring ATP) and show a recognizable ATP binding cassette (ABC) motif. Channel proteins transport materials both into and out of the cell, as do P-type ATPase transport systems for uptake and efflux.

The interior of the model cell (Figure 3.2) contains a template for organizing products (proteins) of recognized genes involved in metabolism. Similar templates apply to genes functioning in information storage and other cellular processing. Structural and regulatory genes encoding metabolic processes (respiration, the tricarboxylic acid cycle, other biochemical pathways, etc.) appear as networks within the cell's interior. Figure 3.3 extends the general model to a portion of the functional gene data presented in Table 3.2. Shown in Figure 3.3 is an overview of the integrated genetic information for metabolism and transport in *Pseudomonas putida* KT2440. Analysis of this 6.18 Mb genome provides a paradigm for the physiological and ecological function of pseudomonads – a broad class of opportunistic and versatile bacteria that are widespread in terrestrial and aquatic environments. The genome data are the basis for a wealth of hypotheses about genetic determinants for transporters, oxygenase enzymes, electron transport chains, sulfur metabolism proteins, and microbial mechanisms for protection from toxic pollutants and their metabolites (Nelson et al., 2002).

Genomic representations such as that shown in Figure 3.3 provide unrivaled and comprehensive views of what microorganisms are, what they can do, and what each of their evolutionary histories have been. Genome-sequencing projects are not an end in themselves. Rather, such blueprints are the gateway to a variety of tools that extend and refine our understanding of prokaryotic biology. To test hypotheses that arise from genome-sequencing data, other essential tools include biochemical assays, physiological assays, transcriptome assays, proteome assays, site-directed mutagenesis creating knockout and other mutations, genetic complementation, computational modeling, systems biology (see Secion 9.1), and tests of ecological relevance (the latter are described in Sections 6.10 and 9.2).

Figure 3.2
Conceptual cell model with bioinformatic template for organizing recognized metabolic genes from each microbial genome-sequencing project into a model prokaryotic cell. (Modified from M. Kanehisa, Kyoto University and KEGG, with permission.)

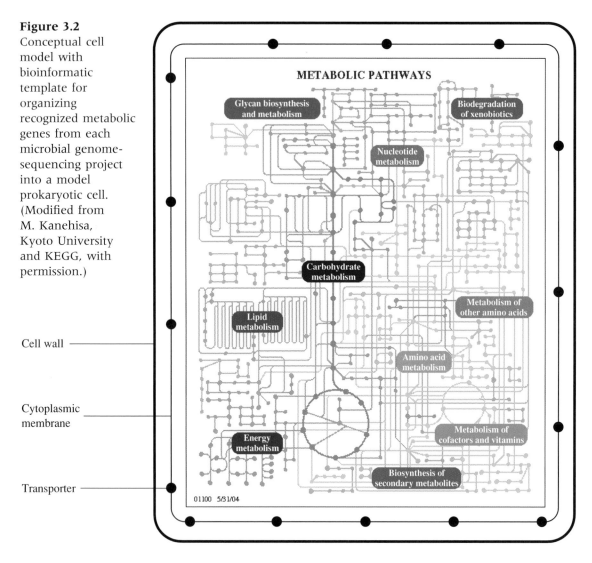

METABOLIC PATHWAYS

Glycan biosynthesis and metabolism

Biodegradation of xenobiotics

Nucleotide metabolism

Carbohydrate metabolism

Metabolism of other amino acids

Lipid metabolism

Amino acid metabolism

Metabolism of cofactors and vitamins

Energy metabolism

Biosynthesis of secondary metabolites

01100 5/31/04

Cell wall

Cytoplasmic membrane

Transporter

Figure 3.3 (*opposite*) A modeled genome for *Pseudomonas putida* KT2440. Shown is an overview of metabolism and transport based on the 6.18 Mb completed genome. Predicted pathways for energy production and metabolism of organic compounds are shown. Predicted transporters are grouped by substrate specificity: inorganic cations (light green), inorganic anions (pink), carbohydrates (yellow), amino acids, peptides, amines, purines, pyrimidines, and other nitrogenous compounds (red), carbohydrates, aromatic compounds, and other carbon sources (dark green), water (blue), drug efflux and other (dark gray). Question marks indicate uncertainty about the substrate transported. Export or import of solutes is designated by the direction of the arrow through the transporter. The energy-coupling mechanisms of the transporters are also shown: solutes transported by channel proteins are shown with a double-headed arrow; secondary transporters are shown with two arrowed lines indicating both the solute and the coupling ion; ATP-driven transporters are indicated by the ATP hydrolysis reaction; transporters with an unknown energy-coupling mechanism are shown with only a single arrow. The P-type ATPases are shown with a double-headed arrow to indicate that they include both uptake and efflux systems. Where multiple homologous transporters with similar predicted substrate exist, the number of that type of transporter is indicated in parentheses. The outer and inner membrane are sketched in gray, the periplasmic space is indicated in light turquoise, and the cytosol in turquoise. (From Nelson, K.E., C. Weinel, I.T. Paulsen, et al., 2002. Complete genome sequence and comparative analysis of the metabolically versatile *Pseudomonas putida* KT2440. *Environ. Microbiol.* **4**:799–808. With permission from Blackwell Publishing, Oxford, UK.)

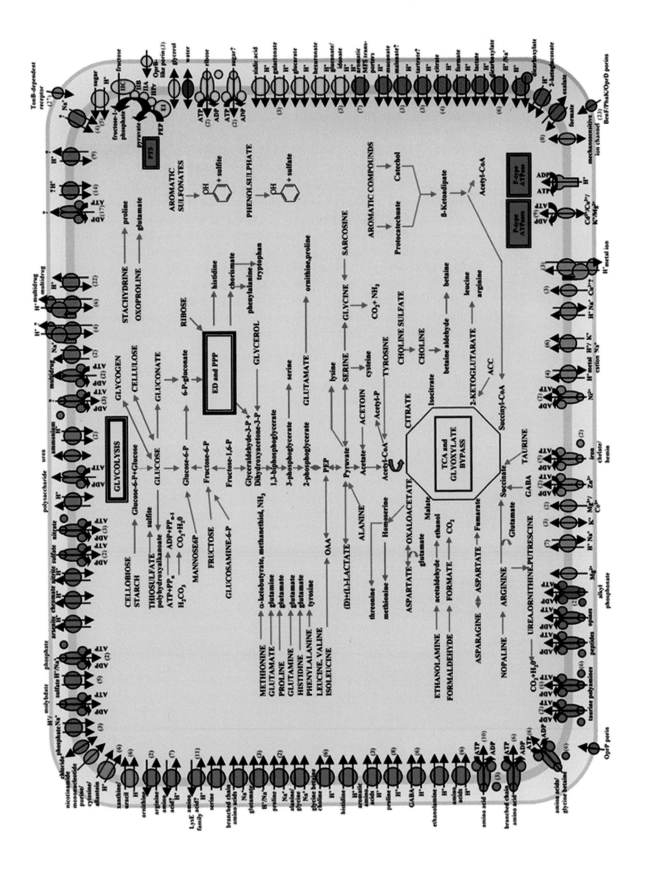

3.3 FUNDAMENTALS OF NUTRITION: CARBON- AND ENERGY-SOURCE UTILIZATION PROVIDE A FOUNDATION FOR PHYSIOLOGICAL ECOLOGY

All forms of life (from prokaryotes to humans) need to make a living. To achieve this, we all require at least two physiological resources: (i) an energy source for generating ATP; and (ii) a carbon source for assembling the cellular building blocks during maintenance of existing cells and/or creation of new cells (growth). No life form can exist without abilities to successfully exploit well-defined energy and carbon sources. Therefore, when microbial physiologists and microbial ecologists approach a new organism or a new habitat (from deep subsurface sediments, to ocean water, to insect guts, to Mars), the first questions to ask are: "What drives metabolism here? What are the energy and carbon sources?" The answers provide fundamental nutritional bases for subsequent hypotheses and inquiries about past and ongoing biogeochemical and ecological processes.

Table 3.3 provides a useful framework for classifying the nutritional needs of individual microorganism and ecosystems alike. The matrix shows

Table 3.3

Physiological classification of life forms based on energy source and carbon source. The five categories assist in understanding both individual microorganisms and biogeochemical systems

Carbon source	Energy source		
	Chemical, organic	Chemical, inorganic	Light
Fixed organic	**Chemosynthetic organoheterotroph** (Example: humans, fungi)	**Chemosynthetic lithoheterotroph** (Example: *Thiobacillus* sp.)	**Photosynthetic heterotroph** (Example: purple and green bacteria)
Gaseous CO_2		**Chemosynthetic lithoautotroph** (Example: hydrogen- and sulfur-oxidizing bacteria)	**Photosynthetic autotroph** (Example: plants, algae)

Terminology:
* Autotroph: carbon from CO_2 fixation
* Heterotroph: carbon assimilated from (fixed) organic compounds
* Photosynthetic: energy from light
* Chemosynthetic: energy by oxidizing reduced chemicals
* Chemolitho: energy by oxidizing inorganic reduced chemicals
* Chemoorgano: energy by oxidizing organic reduced chemicals.

the energy source along the top and carbon source along the left side. Energy sources fall into two main categories: chemical and light. The chemical sources of energy are further divided into two: inorganic (such as hydogen gas, ammonia, methane, and elemental sulfur) or organic compounds (generally containing multiple C-C bonds). The carbon sources also fall into two categories: gaseous CO_2 or fixed organic carbon. The matrix in Table 3.3 provides a means for nutritionally classifying virtually all forms of life on Earth into five groups. The two that are perhaps easiest to recognize are plants and animals (including humans). Plants are *photosynthetic autotrophs* – deriving energy from light and carbon from gaseous CO_2. Humans, fungi, and microorganisms that cycle plant-derived (and other) organic substrates are *chemosynthetic organoheterotrophs*. Microorganisms that oxidize inorganic compounds (e.g., hydrogen gas, ammonia, elemental sulfur) for energy and assimilate CO_2 into their biomass are *chemosynthetic lithoautotrophs*. Microorganisms that use light as an energy source and assimilate fixed carbon into their biomass are *photosynthetic heterotrophs* (this lifestyle is exclusively prokaryotic and rare). Microorganisms that use inorganic compounds as energy sources and assimilate fixed carbon into their biomass are *chemosynthetic lithoheterotrophs* (this lifestyle is exclusively prokaryotic and extremely rare). It should be recognized that the boundaries between the five lifestyle categories in Table 3.3 may sometimes be blurred. For instance, some microorganisms are genetically and physiologically versatile – adopting a heterotrophic lifestyle when fixed organic carbon is available, then resorting to CO_2 fixation in the absence of organic carbon. Regardless of occasional minor ambiguities, the above-described nutritional concepts and terminology are insightful and will be used throughout this text.

Figure 3.4 shows a bar graph revealing general trends in the proportions of genes represented in genomes in four major prokaryotic lifestyles. COG analysis (see Section 3.2) of three organisms in each nutritional class suggests that the transport and metabolism of both amino acid and carbohydrates have been emphasized during the evolution of chemosynthetic organoheterotrophs (especially soil-dwelling bacteria). For photosynthetic heterotrophs, a relatively large portion of the genome has been relegated to energy production and conservation. Further, there seems to be a trend in photoautotroph evolution that de-emphasizes a need for inorganic ion transport and metabolism. The trends suggested in Figure 3.4 may change as additional genomes are completed and/or the members of each nutritional class are expanded. Nonetheless, the notion of mechanistic links between evolutionary pressures and gene content is a compelling one that is likely to be extensively pursued in the future.

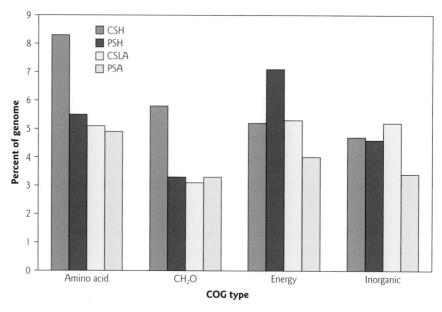

Figure 3.4 Apportionment of metabolic genes (as percent of total genome) among representatives of the four major nutritional categories in prokaryotes: chemosynthetic organoheterotrophs (CSH), photosynthetic heterorophs (PSH), chemosynthetic lithoautotrophs (CSLA), and photosynthetic autotrophs (PSA). Bar graph shows the percent of the genome devoted to amino acid transport + metabolism, carbohydrate (CH_2O) transport + metabolism, energy production + conversion, and inorganic ion transport + metabolism. Data are based on COG analysis of 12 genomes. Representatives of chemosynthetic organoheterotrophs were *Pseudomonas putida* KT2440, *Bacillus cereus* ATCC 14579, and *E. coli* K12. Representatives of photosynthetic heterotrophs were *Chlorobium tepidum* TLS, *Chlorochromatium aggregatum*, and *Chlorobium lumicola* CSMZ 245 (T). Representatives of chemosynthetic lithoautotrophs were *Nitrosomonas europea* ATCC 19718, *Nitrobacter winogradskyi* Nb-255, and *Thiobacillus denitrificans* sp. ATCC 25259. Representatives of photosynthetic autotrophs were *Prochlorocuccus marinus* MIT 9313, *Trichodesmium erythraeum* IMS101, and *Anabaena* sp. PCC7120. (Data from US Department of Energy Joint Genome Institute.)

3.4 SELECTIVE PRESSURES: ECOSYSTEM NUTRIENT fIUXES REGULATE THE PHYSIOLOGICAL STATUS AND COMPOSITION OF MICROBIAL COMMUNITIES

In keeping with the concepts displayed in Figure 3.1, it is the dialog between Earth's habitats and their microbial inhabitants that has directed the course of prokaryotic evolution and the development of microbial diversity. Thus, knowing the resources, the nutritional status, and the geochemical composition of our biosphere can help us understand pressures for genetic selection and adaptation. Consider the following questions.

- **What was it like to be an early microorganism in Earth's primordial seas?**
- **What physiological and ecological pressures do contemporary microorganisms face in aquatic and terrestrial habitats?**
- **What are their carbon and energy sources?**
- **How important is nutrient uptake? Starvation? Competition? Predation? Parasitism?**
- **To what degrees do issues such as acid stress, oxidative stress, temperature extremes, desiccation, or UV radiation influence physiology?**

The unifying answer to questions such as those above is based on Darwinian evolutionary theory: *the prime directive for prokaryotic life is survival, cell maintenance, ATP generation, and growth.* The answer applies to ancient and modern microorganisms under all environmental conditions. Without growth, replication of genes cannot occur and evolution is thwarted.

- **But what is the nature of microbial growth in real-world habitats (waters, sediments, soils)?**
- **Is it rapid or slow? Constant or sporadic?**
- **And what habitat conditions control microbial growth?**

To emphasize the rarity of extended periods of rapid microbial growth in natural habitats, Stanier et al. (1986) developed the following scenario:

In 48 h, a single bacterium (weighing 10^{-12} g) exponentially doubling its biomass every 20 min would produce progeny weighing 2.2×10^{31} g, or roughly 4000 times the weight of the Earth!

Clearly then, extended rapid exponential growth is not the status quo for microorganisms in nature. The alternatives to rapid exponential growth are: (i) sporadic growth (rapid growth when resources are available followed by a quiescent stage); (ii) slow growth; and/or (iii) dormancy. Figure 3.5 displays the key types of physiological and ecological pressures that confront microorganisms in nature. The illustration depicts a planktonic microorganism in the water column of a lake, sea, or ocean. The forces of nature confronted by this cell and fellow members of its population are complex. Physical, biological, and chemical properties of habitats are dynamic – varying in space and in time. The cell has three potential fates resulting from its interactions with its habitat: (i) ecological success; (ii) ecological failure; or (iii) survival/maintenance.

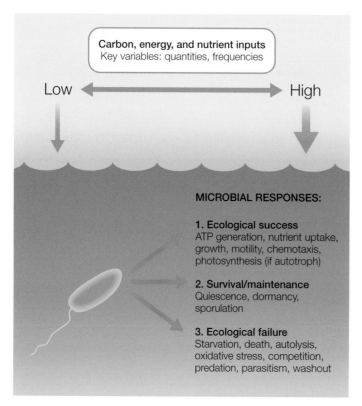

Figure 3.5 Selective pressures influencing a typical microbial cell dwelling in the water column of a lake, sea, or ocean. Three outcomes of cell responses are ecological success, ecological failure, or survival/maintenance.

As shown in Figure 3.5, ecological success would be manifest as growth and increased population size. Ecological success occurs when energy and carbon courses are exploited, other nutrients are taken up, and potentially adverse physiological and ecological obstacles are overcome. Ecological failure is manifest by population decline, death, and/or elimination from the habitat. Ecological failure occurs when detrimental environmental, physiological, and/or ecological factors (starvation, competition, predation, parasitism, washout, oxidative stress, toxins, UV light, etc.) overwhelm a population's ability to survive and grow. The survival/maintenance state is the "neutral, wait and see" middle ground between ecological failure and success. The survival/maintenance state relies upon a suspension of active growth – substituting a quiescent, dormant status in which cellular activities have drastically slowed or stopped (for further discussion of dormancy, see Section 3.5).

In describing microbial habitats, a crucial feature of Figure 3.5 is the input of carbon and energy (and other nutrient) sources. These input fluxes provide adaptive pressures and exert major influences on system productivity, standing biomass, metabolic rates, physiological status, and the composition of microbial communities. Poindexter (1981) suggested that it is appropriate to define a given habitat in terms of the average flux of nutrients. A nutrient-rich lake and areas of ocean shores are *eutrophic* when they have a flux of organic carbon of at least 5 mg C/L/day (Poindexter, 1981). In contrast, a nutrient-poor lake and areas of the open ocean offering *oligotrophic conditions* are defined as having a flux of organic carbon that does not exceed 0.1 mg C/L/day (see Box 3.1 for definitions of terms pertinent to nutrients, habitats, and microorganisms).

Types of microorganisms adapted to ecological success in the above two habitat types have been termed eutrophs (alternatively copiotrophs; Poindexter, 1981) and oligotrophs, respectively. According to Poindexter

Box 3.1

Terms, concepts, and definition relevant to the nutritional status of habitats and their microbial inhabitants

Term	Definition/use
Oligotrophic	Adjective that describes a habitat or microorganism. Low nutrient status. For heterotrophs in lakes, the carbon (C) flux is <0.1 mg C/L/day
Oligotroph and olilgocarbophile	Class of microorganism adapted to life in nutrient-poor, oligotrophic habitats
Eutrophic	Adjective that describes a habitat or microorganism. High nutrient status. For heterotrophs in lakes, the C flux is >5 mg C/L/day
Eutroph and copiotroph	Class of microorganism adapted to life in nutrient-rich, eutrophic habitats
r-selection, *r*-selected	Ecological terms referring to lifestyle strategy that features brief bursts of rapid growth in response to sporadic nutrient input
K-selection, *K*-selected	Ecological terms referring to lifestyle strategy that features slow constant growth
Maintenance energy	Nongrowth-related cellular energy demand used: to maintain intracellular pH and potential across the cytoplasmic membrane; for osmotic regulation; to transport solutes; to resynthesize macromolecules; for motility; and to counteract energy dissipation by proton leakage and ATP hydrolysis
Stenoheterotrophic	Ecological term referring to a habitat with a narrow fluctuation in nutrient status (Horowitz et al., 1983)
Euryheterotrophic	Ecological term referring to a habitat with wide fluctuations in nutrient status
Dormancy	Quiescent, resting state assumed by vegetative cells, especially those unable to develop specialized resting stages. In the dormant state, energy expenditure counteracts macromolecular damage (depurination of DNA and racemization of amino acids (Morita, 2000; Price and Sowers, 2004)
Sporulation	A complex series of developmental steps in many prokaryotes and fungi that leads to the formation of a resistant resting structure, the spore. Spore formation is typically triggered by nutrient deprivation

(1981), oligotrophic bacteria are conceived to be those whose survival in nature depends on their ability to multiply in habitats of low nutrient flux. Proteins involved in membrane transport, substrate binding, and catalysis should be distinctive for oligotrophs, when compared to proteins that in eutrophic organisms operate at substrate concentrations at least 50-fold higher. Morita (1997) has argued that the term "oligotrophic" should be applied only to habitats and not to microorganisms. Morita's (1997) rationale is that the scientific literature does not provide experimental evidence supporting the existence of "oligotrophic bacteria" and "copiotrophic bacteria"; therefore the terms should be dropped from the literature. In the same year, Schut et al. (1997) suggested that the "oligotrophic way of life" is a widespread transient physiological characteristic – especially of marine bacteria. Thus, controversy surrounds ideas and terminology related to oligotrophic microorganisms.

The frequency at which pulses of nutrients are delivered to a given habitat (Figure 3.5) is likely to be as influential as the sizes of a given pulse. Some habitats (especially guts of carnivorous animals, soil litter layers beneath vegetation, and even the ocean floor which occasionally receives whale carcasses; Smith and Baco, 2003) can experience long periods with little or no nutrient inputs. These can be followed by a substantive input pulse. The frequency and quantities of nutrients delivered to waters, sediments, and soils has obvious implications for adaptive pressures of the microbial inhabitants and their chances of ecological success. For example, a "feast and famine" existence is very different from one that has low but constant nutrient inputs.

In recognition of the dynamic and widely varying nutritional characteristic of aquatic and terrestrial habitats (and their respective selective pressures), ecologists have developed the concept of r-selected and K-selected growth strategies (Box 3.1). r-Selected species are ones adapted to high rates of reproduction – they exploit nutrient inputs rapidly, exhibiting high rates of growth in an uncrowded habitat. In contrast, K-selected species are adapted to conserve resources – they exhibit slow, constant growth rates appropriate for habitats featuring crowded, high-density populations (Andrews and Harris, 1986; Atlas and Bartha, 1998). Using a foot-race metaphor for microbial growth strategies, r-selected species are good sprinters, while K-selected species are slow, steady endurance runners.

As used above, the terms oligotrophic and eutrophic apply to organic carbon-dependent chemoorganoheterotrophic nutrition. Table 3.4 extends the approach of habitat classification to two additional major physiological types of microorganisms: photoautotophs and chemolithotrophs. Each habitat type listed in Table 3.4 (ocean water, lake water, sediment, soil, and subsurface sediment) has its own set of physical, chemical, and biological characteristics that govern, enhance, and/or constrain microbial activity. For phototrophic life, clearly diurnal variations in sunlight represent a major habitat-imposed constraint. However, minerals may also govern the growth of photosynthetic microorganisms (see Section 7.1) – the open ocean is typically limited by iron availability while freshwaters are often limited by phosphorus. Regarding chemolithotrophic life, the fluxes and types of energy-limiting, reduced inorganic compounds vary with habitat types. Typically systems at the biosphere–geosphere boundary are dependent upon fluxes of volcanic or geothermal resources (especially H_2, CH_4, H_2S), while soil and freshwater sediment-type habitats rely on inorganic inputs (such as H_2, NH_3, and CH_4) produced by the processing of deceased biomass by anaerobic microbial food chains (see Sections 3.10 and 7.3).

Table 3.4

Predominant ecological limitations for energy and growth for three physiological classes of microorganisms in a variety of habitats

Habitat characteristics and nutrient limitations faced by three physiological classes of microorganisms			
Habitat type	**Photoautotroph**	**Chemolithotroph**	**Chemoorganoheterotroph**
Ocean water	Daily light cycle, light penetration depth; scarce iron	Flux of reduced inorganic compounds, especially NH_3, H_2S, H_2, or CH_4 from nutrient turnover and hydrothermal vents	Carbon flux from phototrophs, dead biomass, and influent waters
Lake water	Daily light cycle, light penetration depth; scarce phosphorus	Flux of reduced inorganic materials, especially NH_3, H_2, and CH_4 from nutrient turnover	Carbon flux from phototrophs, dead biomass and influent waters
Sediment (freshwater and oceanic)	Daily light cycle, light pentration depth	Flux of reduced inorganic materials, especially NH_3 and H_2 from nutrient turnover or H_2, H_2S, or CH_4 from hydrothermal vents	Flux of organic carbon from phototrophs and dead biomass; flux of final electron acceptors to carbon-rich anaerobic strata
Soil	Daily light cycle, light penetration depth	Flux of reduced gaseous substrates, especially methane from nutrient turnover by anaerobes	Slow turnover of soil humus, dead biomass, plant root exudates; leaf fall from vegetation
Subsurface sediment	No light	Flux of reduced inorganic materials, especially H_2 and CH_4 from geothermal origin	Carbon flux from nutrient turnover

3.5 CELLULAR RESPONSES TO STARVATION: RESTING STAGES, ENVIRONMENTAL SENSING CIRCUITS, GENE REGULATION, DORMANCY, AND SLOW GROWTH

The life cycle of some specialized prokaryotes, fungi, and protozoa includes a resistant, quiescent stage, variously termed endospore, myxospore, cyst, conidium, etc. Such developmental stages in microbial life cycles are triggered by environmental cues, especially starvation, received by the microorganism. The resulting resting stages typically surround vital cytoplasmic constituents with a thick-walled structure that confers resistance not only to starvation, but also to extreme environmental conditions ranging from heat, to desiccation, to acidity, to γ-irradiation, to salinity, to UV light. In many prokaryotes, the ultrastructure, physiology, and genetics of endospore formation have been well characterized (Box 3.2; e.g., Nicholson et al., 2000; Hilbert and Piggot, 2004).

Box 3.2

Stages of *Bacillus* endospore formation and an electron micrograph of an endospore of *Sporosarcina ureae*

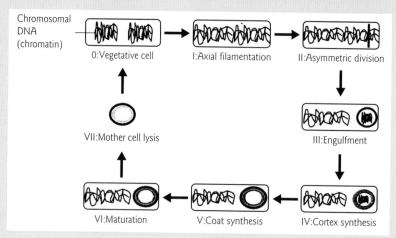

Figure 1 Schematic representation of the stages of spore formation. A vegetatively growing cell is defined as stage 0. It is shown as having completed DNA replication and containing two complete chromosomes (represented as disordered lines within the cells), although replication is not completed at the start of spore formation. Formation of an axial filament of chromatin, where both chromosomes (or a partially replicated chromosome) form a continuous structure that stretches across the long axis of the cell, is defined as stage I. Asymmetric division occurs at stage II, dividing the cell into the larger mother cell and smaller prespore; for clarity, the septum is indicated as a single line. At the time of division, only approximately 30% of a chromosome is trapped in the prespore, but the DNA translocase SpoIIIE will rapidly pump in the remaining 70%. Stage III is defined as completion of engulfment, and the prespore now exists as a free-floating protoplast within the mother cell enveloped by two membranes, represented by a single ellipse. Synthesis of the primordial germ cell wall and cortex, a distinctive form of peptidoglycan, between the membranes surrounding the prespore is defined as stage IV and is represented as thickening and graying of the ellipse. Deposition of the spore coat, protective layers of proteins around the prespore, is defined as stage V. The coat is represented as the black layer surrounding the engulfed prespore. Coincident with coat and cortex formation, the engulfed prespore is dehydrated, giving it a phase-bright appearance in a microscope, represented here as a light gray shading. Stage VI is maturation, when the spore acquires its full resistance properties, although no obvious morphological changes occur. Stage VII represents lysis of the mother cell, which releases the mature spore into the environment. (From Hilbert, D.W. and P.J. Piggot, 2004. Compartmentalization of gene expression during *Bacillus subtilis* spore formation. *Microbiol. Molec. Biol. Rev.* **68**:234–263. With permission from the American Society for Microbiology, Washington, DC.)

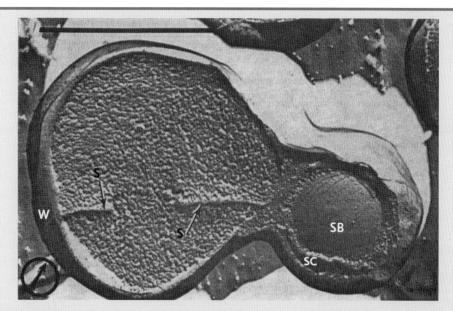

Figure 2 A freeze-etch prepared electron micrograph of an endosphere within an *Sporosarcina ureae* cell. SB, spore body; SC, spore coat; W, cell wall. Scale bar, 0.5 μm. (From Holt, S.C. and E.R. Leadbetter, 1969. Comparative ultrastucture of selected aeorobic spore-forming bacteria: a freeze-etching study. *Bacteriol. Rev.* **33**:346–378. With permission from the American Society for Microbiology, Washington. DC.)

Specialized morphological adaptations to starvation, however, are not the rule in the microbial world. Recent studies have examined the physiological and genetic responses of microorganisms to starvation using pure cultures (such as *E. coli*) that have exhausted nutrient supplies in laboratory growth media. In these model systems, genomic and biochemical assays have been applied to prokaryotes experiencing oxidative stress, senescence, programmed cell death, arrested growth, and mutations in long-term stationary phase culture (e.g., Nystrom, 2004; Finkle, 2006). Results have reinforced the view that responsiveness to environmental conditions, especially those that cause physiological stress, is one of the hallmarks of prokaryotic behavior. Information about the chemical, physical, and nutritional status of a microorganism's habitat must be transmitted to the genetic regulatory networks within. Thus, sensing the environment is a well-honed ability in prokaryotes. Figure 3.6 shows a generalized scheme of environmental sensing systems used by prokaryotes. Commonly the sensor is a protein embedded within, but extending out from, the cytoplasmic membrane of the cell. The environmental change

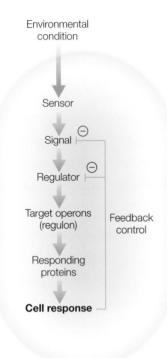

Environmental
condition

Sensor

Signal

Regulator

Target operons
(regulon)

Feedback
control

Responding
proteins

Cell response

Figure 3.6
Environmental
sensing by bacteria.
Diagram shows a
simplified sensor
circuit that transmits
signals (such as low
nutrient status) to
regulatory genes.
The eventual
protein-mediated cell
response completes
the negative
feedback loop for the
circuit. (Modified
from Niedhardt, F.C.,
J.L. Ingraham, and
M. Schaechter. 1990.
*Physiology of the
Bacterial Cell: A
molecular approach.*
With permission
from Sinauer
Associates,
Sunderland, MA.)

causes an allosteric (structural) alteration in protein conformation which leads to its self-catalyzed binding to a phosphate molecule. The term "sensor kinase" applies. The
phosphorylation triggers a subsequent series of phosphorylation events that influence the activity of one or more regulatory proteins. These, in turn, control gene transcription by
binding to the promoter or attenuator regions of one or
more operons. Following translation of the transcribed
genes, protein-catalyzed metabolic changes in the cell eventually deliver negative feedback to the regulatory circuit
(Figure 3.6). Overall, the cell's response is matched to the sever-
ity of the nutritional stress. Furthermore, when the stress is
relieved, the cytoplasmic sensor resumes its previous non-
phosphorylated state. Sensor kinase-based control of gene
expression is an effective way for cells to respond to environmental cues. The cues may lead to genetic control of large
constellations of operons in regulatory networks (regulons).
Table 3.5 provides a list of 11 well-characterized regulatory
response systems that allow prokaryotes to contend with
nutrient-related environmental conditions.

It is virtually certain that all microorganisms have evolved
environmental response systems as sophisticated as those
described for model prokaryotes listed in Table 3.5. However, the science
of microbiology has not yet progressed far enough to know how widely
these specific genetic adaptations have reached through the prokaryotic
world. To gain insight into adaptations to starvation found broadly
among prokaryotes, we turn to a well-established literature examining
the physiological response of microorganisms to starvation. Some of the
key scholars addressing the issue of survival of microorganisms in
nature, especially under conditions of nutrient limitation and slow
growth, include J. Poindexter (1981), H. Jannasch (1969, 1979), D.
Roszak and R. Colwell (1987), A. Koch (1971, 1997), D. Button (1998),
A. Matin et al. (1989), Y. Henis (1987), S. Pirt (1982), D. Tempest et al.
(1983), E. Dawes (1989), R. Kolter et al. (1993), R. Morita (1982, 1997,
2000), and P. Price and T. Sowers (2004). In reviewing the problems of
survival of heterotrophic prokaryotes in the marine environment, Morita
(1982) proposed five processes that render populations "fit" for starvation survival:

1 All metabolic processes are reduced to a dormant or near-dormant state.
2 When starved, many species will increase in cell number – resulting
   in reduced cell size (see point 5, below).
3 In the starvation/survival process, any cellular energy reserve material
   is used to prepare the cell for survival.
4 All metabolic mechanisms are directed to the formation of specific proteins, ATP, and RNA so that the cell, when it encounters a substrate,

Table 3.5

Prominent gene regulation systems that allow bacteria to sense and coordinate metabolism according to environmental conditions, especially starvation-related stresses (from Schaechter, M., J.L. Ingram, and F.C. Niedhardt. 2006. *Microbe*. American Society for Microbiology Press, Washington, DC. With permission from the American Society for Microbiology.)

Stimulus/conditions	System	Organism(s)	Regulatory genes (and their products)	Regulated genes (and their products)	Type of regulation
Nutrient utilization					
Carbon limitation	Catabolite repression	Enteric bacteria	*crp* (transcription activator CAP); *cya* (adenylate cyclase)	Genes encoding catabolic enzymes (*lac, mal, gal, ara, tna, dsd, hut*, etc.)	Activation by CAP protein complexed with cAMP as a signal of carbon source limitations
Amino acid or energy limitation	Stringent response	Enteric bacteria and many others	*relA* and *spoT* (enzymes of (p)ppGpp metabolism)	Genes (>200) for ribosomes, other proteins involved in translation and biosynthetic enzymes	(p)ppGpp thought to modify promoter recognition by RNA polymerase
Ammonia limitation	Ntr system (enhances ability to acquire nitrogen from organic sources and from low ammonia concentrations)	Some enteric bacteria	*glnB, glnD, glnG, glnL* (transcriptional regulators and enzyme modifiers)	*glnA* (glutamine synthetase), *hut*, and others encoding deaminases	Complex
Ammonia limitation	Nif system (nitrogen fixation)	*Klebsiella aerogenes* and many others	Multiple genes, including those controlling ammonia assimilation	Multiple genes encoding nitrogenase (for nitrogen fixation)	Complex; ammonia represses activity of NtrC; under low ammonia status, NtrC is active and promotes transcription of NifA, the activator protein for *nif* transcription
Phosphate limitation	Pho system (acquisition of inorganic phosphate)	Enteric bacteria	*phoB* (PhoB, response regulator), *phoR* (PhoR, sensor kinase), *phoU*, and *pstA, -B, -C, -S* (facilitate PhoR function)	*PhoA* (alkaline phosphatase) and ~40 other genes involved in utilizing organophosphates	Two-component regulation; transcriptional activation by PhoB upon signal of low phosphate from the sensor kinase, PhoR

Table 3.5 Continued

Stimulus/conditions	System	Organism(s)	Regulatory genes (and their products)	Regulated genes (and their products)	Type of regulation
Energy metabolism					
Presence of oxygen	Arc system (aerobic respiration)	E. coli	ArcA (ArcA, repressor) and arcB (ArcB, modulator)	Many genes (>30) for aerobic enzymes	Repression of genes of aerobic enzymes by ArcA upon signal from ArcB of low oxygen
Presence of electron acceptors other than oxygen	Anaerobic respiration	E. coli	fnr (Fnr)	Genes for nitrate reductase and other enzymes of anaerobic respiration	Transcriptional activation by Fnr
Absence of usable electron acceptors	Fermentation	E. coli and other facultative bacteria	Unknown	Genes (>20) for enzymes of fermentation pathways	Unknown
Miscellaneous global systems					
Growth-supporting property of environment	Growth rate control	All bacteria	fis (Fis), hns (H-NS), relA (RelA), and spoT (SpoT)	Hundreds of genes, many involved in macromolecule synthesis	Complex; involves availability of RNA polymerase/sigma-70 haloenzyme influenced by passive control
Starvation or inhibition	Stationary phase	All bacteria	rpoS (sigma-S), lrp (Lrp), crp (CAP), dsrA, rprA, and oxyS (regulatory sRNA molecules), and many other regulatory genes	Hundreds of genes affecting structure and metabolism	Multiple modes of regulation in a complex network; involves several global regulatory systems, in addition to selection of genes with promoters recognized by σ^s
Starvation	Sporulation	Bacillus subtilis and other spore formers	SpoOA (activator), spoOF (modulator), and many other regulatory genes	Many (>100) genes for spore formation	Complex; cells respond to nutrient deprivation with SpoOA phosphorelay; a cluster of sigma factors assist in controlling seven stages of differentiation leading to endospore formation and release

cAMP, cyclic adenosine monophosphate; CAP, catabolite activator protein.

is equipped to use it immediately without a delay that otherwise would occur if initial amounts of energy had to be expended for the synthesis of RNA and protein. Both RNA and protein synthesis are high-energy-consuming processes, and the high ATP level per viable cell is thus available and used primarily for active transport of substrates across the membrane.

5 The change to a smaller cell size on starvation (miniaturization) permits greater efficiency in scavenging what little energy-yielding substrates there are in the environment and also enhances survival prospects against other adverse environmental factors.

These five fitness traits have been refined and extended (see text below).

In a more recent essay on bacterial starvation, B. Schink (Lengeler et al., 1999) points out that starvation for a carbon (energy) source is very different from starvation for a source of nitrogen, phosphorus, or sulfur because scarcity of mineral nutrients, while preventing an increase in cell mass, may still allow cell maintenance. Should nutrient concentrations diminish substantially, one well-documented cell response is to enhance substrate uptake – either by increasing the number of transport systems (e.g., permeases, ABC-type transporters; see Section 3.2) or shifting to alternate high-affinity systems such as those known for glycerol, phosphate, sugars, and amino acids. Efficient utilization of cellular reserve materials is another clear survival strategy. Many microorganisms accumulate intracellular storage bodies when nutrients are transiently abundant in the cell's immediate surroundings. Three main classes of cellular reserves are recognized: carbohydrates (polyglucans, glycogen), lipids (poly-β-hydroxy alkanoates, especially poly-β-hydroxybutyrate), and polyphosphates. A fourth type of storage body, cyanophycin, is a nitrogen source (composed of the amino acids arginine and aspartic acid) found in some cyanobacteria. Clearly, when energy, phosphorus, and nitrogen sources are scarce in the external environment, the ability to draw on intracellular reserves offers an advantage in the struggle for survival. The relative importance of uptake versus cellular reserve survival strategies is undoubtedly dependent upon habitat-specific and organism-specific factors. Nonetheless, Poindexter (1981) perceptively remarked that "over extended periods of time, conservative utilization of nutrients once they are in the cell . . . may be more important than high-affinity uptake systems".

It is important to develop an appreciation for the range of growth rates for microorganisms and the degree to which their survival strategies have been successful. Globally, the significance and ecological success of prokaryotic life in the biosphere are undisputed (see Section 1.2). Detailed insights into this success can be gained by examining habitat-specific trends. Table 3.6 provides estimates of growth rates and survival of microorganisms in nature. The habitats surveyed for growth rate include freshwaters, the ocean, mammals, soil, and the deep subsurface.

Table 3.6

Estimates of microbial growth rate, dormancy, and duration of dormancy and survival in nature

Habitat	Organism	Doubling time (DT) or survival time (ST)	References
Growth rate			
Laboratory medium	*E. coli*	20 min DT	Koch, 1971
Human intestine	*E. coli*	12 h DT	Koch, 1971
Mouse	*Salmonella typhimurium*	10–24 h DT	Brock, 1971
Rumen	Heterotrophic bacteria	~12 h DT	Brock, 1971
Pond	Heterotrophic bacteria	2–10 h DT	Brock, 1971
Lake water	Heterotrophic bacteria	10–280 h DT	Brock, 1971
Ocean	Heterotrophic bacteria	20–200 h DT	Jannasch, 1969
Ocean	Autotroph, *Prochlorococcus*	~24 h DT	Vaulot et al., 1995
Soil	Heterotrophs: α Proteobacteria, rhizobia	100 days DT	Gray and Williams, 1971
Shallow groundwater	Heterotrophs: *Acidovorax, Commamonas*	15 days DT	Mailloux and Fuller, 2003
Deep subsurface	Heterotrophs	100 years DT	Phelps et al., 1994; Fredrickson and Onstott, 2001
Duration of dormancy or survival			
Laboratory test tube	*Clostridium aceticum* endospore	34 years ST	Braun et al., 1981
Lake Vostok beneath the Antarctic ice sheet	Dormant nitrifying prokaryotes	$>1.4 \times 10^5$ years ST	Sowers, 2001; Price and Sowers, 2004
Gut of extinct bee trapped in amber	Heterotroph, spore-forming *Bacillus*	$25–40 \times 10^6$ years ST	Cano and Borucki, 1995
Deeply buried clay and shale	Dormant heterotrophs	100×10^6 years ST	Phelps et al., 1994; Price and Sowers, 2004
Precambrian salt crystals	Heterotroph, endospore	250×10^6 years ST	Vreeland et al., 2000

Despite the fact that the data shown in Table 3.6 reflect approaches and methodologies spanning several decades, several clear patterns emerge. The first is that (compare the first two entries), despite an innate genetic potential to grow with extreme rapidity in laboratory media, *E. coli*'s doubling time in its native habitat (the human intestine) is slowed about 35-fold. This undoubtedly reflects habitat-imposed nutrient limitations. Another contrast in the data shown in Table 3.6 is the immense range of doubling times. Among the 10 nonlaboratory growth rate entries, the highest (2–10 h) was based on microscopic examination of microcolonies on glass slides immersed in pond water. An extremely slow doubling time of several centuries was estimated for inhabitants of the deep subsurface where low-permeability sediment and nearly constant geochemical conditions place severe limitations on potential growth rates. In the lower portion of Table 3.6 are three entries attesting to the effectiveness of endospores in conferring longevity to microorganisms. Survival estimates range from a minimum of several decades (in a controlled laboratory setting) to 250×10^6 years, based on the recovery of cells preserved in fluids trapped within an ancient salt crystal. There are also two entries describing dormancy. Sowers (2001) reported that nitrifying prokaryotes residing in liquid veins deep beneath the Antarctic ice sheet had remained viable for $>1.4 \times 10^5$ years. Similarly, deep terrestrial sediments as old as 10^8 years have been found to harbor viable heterotrophic prokaryotes.

Price and Sowers (2004) recently gained insights into starvation/survival strategies of prokaryotes by reviewing results from more than 30 high-quality studies that emphasized very cold habitats (e.g., ice cores) and deep subsurface sediments (both terrestrial and oceanic). The integrated data set summarized rates of processes carried out by microbial communities in laboratory simulations of nature and from geochemical gradients found in field sites. By plotting rates of microbial metabolism versus temperature, Price and Sowers (2004) were able to discern three distinctive metabolic regimes: growth, maintenance, and survival (dormancy). As expected, energy demand during cellular maintenance is several orders of magnitude lower than for growth; while survival energy demand is orders of magnitude lower than that of maintenance. One shocking finding was that there seems to be no evidence of a minimum temperature for metabolism: even at a temperature of −40°C in ice, about one turnover of cellular carbon is expected every 10^8 years. Another major conclusion reinforced and extended prior ideas by Morita (2000): extremely slow rates of metabolism characteristic of dormancy work to counteract chemical instability of amino acids (subject to racemization) and nucleic acids (subject to depurination). The extremely low energy requirements for DNA and protein repair during dormancy may be provided to microorganisms in deep, sometimes cold, habitats via slow diffusion of hydrogen gas from the adjacent geologic strata (Morita, 2000; Price and Sowers, 2004).

3.6 A PLANET OF COMPLEX MIXTURES IN CHEMICAL DISEQUILIBRIUM

A theme of this chapter is the generation of ATP, the energy currency of the cell. Without ATP, cell maintenance, cell motility, biosynthetic reactions, replication, heredity, and cell growth would be impossible. The biochemical mechanism of ATP generation relies upon either substrate-level phosphorylation (very significant for many anaerobic microorganisms; Schmitz et al., 2006) or membrane-bound electron transport chains. Electron transport chains create the proton motive force which drives ATP synthases embedded in cytoplasmic membranes (White, 1995; Nichols and Ferguson, 2002; Nelson and Cox, 2005; Devlin, 2006; Madigan and Martinko, 2006; Schaechter et al., 2006). But the ultimate driver of ATP synthesis is the variety of thermodynamically unstable materials that are commingled in the waters, sediments, and soils of the biosphere.

Consider the material present in a liter of seawater, in a handful of soil, or in a scoop of freshwater sediment:

- **What is commingled here? (Answer:** Gases, solids, water, minerals, organic compounds, inorganic compounds, soluble materials, microorganisms, and other life forms – some alive, some perished.)
- **What is the chemical composition of these environmental samples?**
- **What chemical and biochemical reactions are occurring?**
- **Are these mixtures in a state of chemical equilibrium?**

The Earth, its habitats, subhabitats and microenvironments, have always been in this complex state consisting of commingled materials. Figure 3.7 provides a global view of our planet. Its status can be summarized as follows: *a heterogeneous mixture of rock, water, gases, and other materials bathed in sunlight.* This is the place where life evolved and where we and microorganisms make a living each day, hour by hour, minute by minute. How do we, as scientists, make sense of biosphere complexity? (For a partial answer, see Section 1.4.) A question more germane to the matter at hand is "How do microorganisms make *physiological* sense of this complexity?" One way to achieve a unified and orderly view of our planet is with thermodynamics. Thermodynamics is the branch of chemistry that rigorously predicts the chemical reactions that are energetically favorable and ones that are not. Thermodynamics can systematically arrange the type of chemical reactions that occur in complex, hetero-

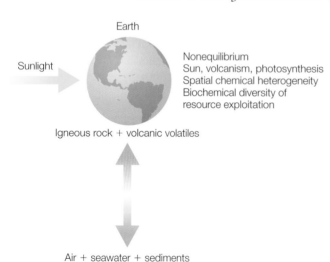

Figure 3.7 Global view of the Earth: a heterogeneous mixture of complex materials (rock, water, gases, others) is maintained in a nonequilibrium state by sunlight and volcanism.

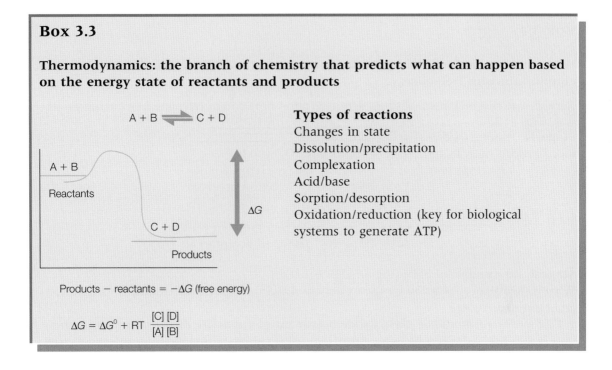

Box 3.3

Thermodynamics: the branch of chemistry that predicts what can happen based on the energy state of reactants and products

A + B ⇌ C + D

A + B

Reactants

C + D

Products

ΔG

Types of reactions
Changes in state
Dissolution/precipitation
Complexation
Acid/base
Sorption/desorption
Oxidation/reduction (key for biological systems to generate ATP)

Products − reactants = −ΔG (free energy)

$$\Delta G = \Delta G^0 + RT \frac{[C]\,[D]}{[A]\,[B]}$$

geneous mixtures typical of biosphere habitats – we can predict and cat-alog the energetically favorable reactions. It is these that are the resources which microorganisms (and humans) exploit as energy sources that generate ATP. It is the thermodynamically unstable resources that have provided selective pressure for microbial energy-production strate-gies throughout evolution. Box 3.3 displays several key thermodynamic principles for predicting chemical and physiological reactions. Box 3.4 elab-orates on the principles from Box 3.3 – providing specific examples of half reactions and how the thermodynamics of redox reactions can be used to calculate free energy change.

One of the major features of the information conveyed in Figure 3.7 is that Earth is in a dynamic state of constant disequilibrium. Key inputs of energy to the biosphere are light from the Sun and a slow release of heat and materials from the Earth's core via volcanism. These global energy inputs are imposed upon a spatially varied tapestry of elevation (from the deepest ocean trench to the summit of Mt. Everest), latitude (from equator to the poles), climate, and chemical conditions (pH, salinity, oxy-gen concentration, other gases, etc.) spanning both aquatic and terrest-rial habitats around the globe. Details of habitat diversity are presented in Chapter 4.

Box 3.4

Four stages to understanding oxidation/reduction reactions, chemical equilibrium, and free energy

Here we use rust and metabolism to define electron flow and to calculate free energy yields.

Stage 1. Defining oxidation-reduction reactions

Rust formation (iron metal) $4Fe^0 = 3O_2 \rightarrow 2Fe_2^{3+}O_3^{2-}$ (rust)

Electron flow: $4Fe^0 \rightarrow 4Fe^{3+} + 12e^-$ (electron donor)

$12e^- + 3O_2^0 \rightarrow 6O^{2-}$ (electron acceptor)

Combustion and respiration $CH_2O + O_2 \rightarrow CO_2 + H_2O$

Electron flow: $C^0 \rightarrow C^{4+} + 4e^-$ (electron donor)

$4e^- + O_2 \rightarrow 2O^{2-}$ (electron acceptor)

Stage 2. An example of balancing a redox equation

What is the balanced reaction for the oxidation of H_2S to SO_4^2 by O_2 (modified from Madigan and Martinko, 2006)?

1 *Electron donor half reaction.* First, decide how many electrons are involved in the oxidation of H_2S to SO_4^{2-}. This can be easily calculated using simple arithmetic from rules of charge balance and the fixed and/or limited oxidation states of the common atoms. Because H has an oxidation state of +1, the oxidation state of S in H_2S is −2. Because the reduced form of O has an oxidation state of −2, the oxidation state of S in SO_4^{2-} is +6. Thus, the oxidation of S^{2-} to SO_4^{2-} involves an eight-electron transfer (S changes from −2 to +6):

$S^{2-} \rightarrow SO_4^{2-} + 8e^-$

2 *Electron-acceptor half reaction.* Because each O atom in O_2 can accept two electrons (the oxidation state of O in O_2 is zero, but in H_2O is −2), this means that two molecules of molecular oxygen, O_2, are required to provide sufficient electron-accepting capacity to accommodate the eight electrons from S^{2-}:

$8e^- + 2O_2 \rightarrow 2O_2^{2-}$

Thus, at this point, we know that the reaction requires $1H_2S$ and $2O_2$ on the left side of the equation, and $1SO_4^{2-}$ on the right side. To achieve an ionic balance, we must have two positive charges on the right side of the equation to balance the two negative charges of S^{2-} and SO_4^{2-}. Thus $2H^+$ must be added to the both sides of the equation making the overall reaction:

$H_2S + 2O_2 \rightarrow SO_4^{2-} + 2H^+$

By inspection, it can be seen that this equation is balanced in terms of the total number of atoms of each kind on each side of the equation.

In general for microbiological reactions, the first step is quantitatively balancing electron flux between donor and acceptor. Next, mass balances of O and H can be achieved by adding H_2O to the side of the reaction with an O deficit and then H^+ to the other side to compensate for an H deficit. Because all reactions take place in an aqueous medium, H^+ production shows that a reaction generates acidity while H^+ consumption indicates generation of alkalinity.

Stage 3. Examples of calculating free energy yields

Chemical reactions under standard conditions (for chemists)
The procedures to calculate ΔG for chemical reactions are well established (e.g., Dolfing, 2003; Madigan and Martinko, 2006). Briefly, calculations of the changes in Gibbs free energy of a system are made according to the equation:

$$\Delta G° = \Sigma G_f° \text{ (products)} - \Sigma G_f° \text{ (reactants)}$$

The naut sign (°) indicates that the calculations are made for standard conditions, i.e., concentrations of 1 M, temperature of 25°C, and a partial pressure of 1 atm for gases. This approach is appropriate for the reaction between H_2S and O_2 shown above.

Once an equation has been balanced, the free energy yield can be calculated by inserting the values for the free energy of formation of each reactant and product from standard tables (Stumm and Morgan, 1996; Dolfing, 2003; Madigan and Martinko, 2006). For instance, for the equation:

$$H_2S + 2O_2 \rightarrow SO_4^{2-} + 2H^+$$
$\rightarrow G_f'$ values $\rightarrow$ (−27.87) + (0)   (−744.6) + 2(−39.83) (assuming pH 7)
$\Delta G_f° = -796.39$ kJ/reaction

The G_f' values for the products (right side of the equation) are summed and the G_f' values for the reactants (left side of the equation) are subtracted, taking care to ensure that the arithmetic signs are correct.

Biochemical reactions under physiological conditions (for microbiologists)
For evaluations of changes of free energy under actual physiological conditions, many assumptions made by chemists are not really relevant. Under environmentally relevant conditions, the concentrations of substrates and products are not 1 M and the partial pressures are not 1 atm. One way to compensate for deviations from standard conditions is to adjust calculations according to the Nernst equation, because the change in Gibbs free energy values of a reaction are directly linked to its equilibrium constant. This is reflected in ΔG values. For a hypothetical reaction:

$$aA + bB \rightarrow cC + dD$$

ΔG values are calculated by using the mass equation:

$$\Delta G = \Delta G° + RT\ln [C]^c[D]^d/[A]^a[B]^b$$

where R is a constant (8.29 J/mol/°K) and T is temperature in °K.

Under dynamic biological conditions that feature very low concentrations of some reactants and products, free energy yields can shift drastically from those predicted under standard conditions.

Box 3.4 *Continued*

Stage 4. A preferred approach to calculating biochemical free energy yield using electron potential and Figure 3.9

Reduction potentials of many redox half-reaction pairs are shown graphically in Figure 3.9. The amount of energy that can be released from two half reactions can be calculated from the differences in reduction potentials of the two reactions and from the number of electrons transferred. The further apart the two half reactions are, and the greater the number of electrons, the more energy is released. The conversion of potential difference to free energy is given by the formula $\Delta G^{\circ\prime} = -nF\Delta E_0^\prime$, where n is the number of electrons, F is the Faraday constant (96.48 kJ/V), and $\Delta E_0^\prime$ is the difference in potentials (electron acceptor minus electron donor). Energetically favorable oxidations are those in which the flow of electrons is from (reduced) electron donors (lower right in Figure 3.9) up to (oxidized) electron acceptors (upper left in Figure 3.9) (see Section 3.8).

Example: hydrogen gas oxidation by microorganisms using molecular oxygen and nitrate as terminal electron acceptors

The couple $2H^+/H_2$ has the potential of -0.41 V and the $^1/_2O_2/H_2O$ pair has a potential of $+0.82$ V. The potential difference is 1.23 V, which (because two electrons are involved) is *equivalent to a free energy yield (ΔG°) of -237.34 kJ* [= $(-2)(96.48$ kJ/V$)(1.23$ V$)$].

By contrast, if oxygen is unavailable and anaerobic nitrate respirers (producing nitrite) are active, the potential difference between the $2H^+/H_2$ and the NO_3^-/NO_2^- reactions is 0.84 V, which is *equivalent to a free energy yield of -162.08 kJ* [= $(-2)(96.48$ kJ/V$)(0.84$ V$)$].

Because many biochemical reactions are two-electron transfers, it is often useful to give energy yields for two-electron reactions, even if more electrons are involved. Thus, the SO_4^{2-}/H_2 redox pair involves eight electrons, and complete reduction of SO_4^{2-} with H_2 requires $4H_2$ (equivalent to eight electrons). From the reduction potential difference between $2H^+/H_2$ and SO_4^{2-}/H_2S (0.19 V), a free energy yield of -146.64 kJ is calculated, or -36.66 kV per two electrons.

$4H_2 + SO_4^{2-} \rightarrow 4H_2O$ (as written, eight electrons are transferred)

$\Delta G^{\circ\prime} = -nF\Delta E_0^\prime$: $\Delta G^{\circ\prime} = (-8e)$ (96.48 kJ/V) (0.19 V)$ = -146.64$ kJ

Normalized to 2e–: $\Delta G^{\circ\prime} = (-2e)$ (96.48 kJ/V) (0.19 V)$ = -36.66$ kJ

3.7 A THERMODYNAMIC HIERARCHY DESCRIBING BIOSPHERE SELECTIVE PRESSURES, ENERGY SOURCES, AND BIOGEOCHEMICAL REACTIONS

In Chapter 2, we emphasized that the development of photosynthesis and the photosynthetic apparatus was a major event in evolution – it allowed life to harvest light energy from the Sun, led to the accumulation of fixed carbon as biomass, and created a global pool of gaseous oxygen (that itself

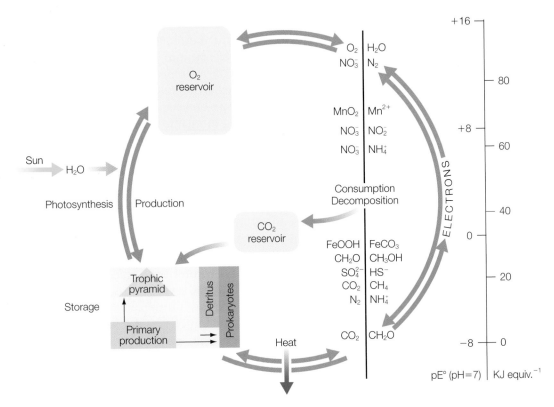

Figure 3.8 A global thermodynamic model of the biosphere. Sunlight and ecosystem processes are arranged along a vertical scale of oxidation–reduction potential. (From Zehnder, A.J.B. and W. Stumm. 1988. Geochemistry and biogeochemistry of anaerobic habitats. *In*: A.J.B. Zehnder (ed.) *Biology of Anaerobic Microorganisms*, pp. 1–38. Wiley and Sons, New York. Reprinted with permission from John Wiley and Sons, Inc., New York.)

had vast metabolic and evolutionary repercussions). Thermodynamics helps to systematically define the far-ranging influence of photosynthesis upon biogeochemical processes. Figure 3.8, devised by Zehnder and Stumm (1988), presents an insightful portrait of the Earth's biogeochemistry. In a single figure these biogeochemists were able to create a conceptual masterpiece that captures the essence of mechanistic relationships between oxidation–reduction processes and ecosystem function. On the left-hand side of Figure 3.8 is an arrow representing sunlight impinging on the Earth. As sunlight drives photosynthesis, water is split into oxygen gas (arrow up to the O_2 reservoir) and reducing power (arrow down leading to reduced carbon in the box labeled storage, which surrounds primary production, trophic pyramid, detritus, and prokaryotes). By capturing sunlight and simultaneously creating pools of both oxygen and reduced carbon in biomass, photosynthesis assures that our planet is in a state of disequilibrium. Photosynthesis effectively (very effectively) reverses the universal drift toward thermodynamic equilibrium on our planet (Zehnder and Stumm, 1988; Stumm and Morgan, 1996). *It is the commingling of reduced biomass with oxygen and other oxidized compounds* (such as nitrate, Fe oxides, Mn oxides, sulfate, and CO_2) *that sets the stage for metabolism by chemosynthetic organotrophic life* (see Section 3.3). Likewise the commingling of reduced inorganic compounds (e.g., hydrogen gas, NH_3, H_2S) with oxygen and other oxidized materials sets the stage for metabolism by chemosynthetic lithotrophic life. A nonequilibrium state is created by co-occurring pairs of reduced and oxidized materials in waters, sediments, and soil of the biosphere. These very same pairs of reduced and oxidized materials constitute the selective pressures that have driven evolution of ATP generation in nonphotosynthetic organisms since life began. To the right of Figure 3.8 is a vertical scale of oxidation/reduction half reactions arranged with O_2/H_2O at the top and CO_2/CH_2O at the bottom. An expanded view of this vertical scale (hierarchy) is discussed further below. The figure also features a vertical scale whose units are pE and kilojoules of energy per electron transferred in a reaction. The pE values range from a low of -8 to a high of $+16$. As described in Box 3.5, pE is to electrons what pH is to protons. pE is an index of how oxidized or how reduced a given setting may be. Just as a high pH value (e.g., 12) indicates an exceedingly low concentration of protons in solution, a high pE indicates rareness of electrons. Rareness of electrons is synonymous with oxidizing conditions and the converse is synonymous with reducing conditions.

In any habitat with a significant partial pressure of oxygen gas, the pE is poised at a value of about $+14$. For this reason, both the oxygen reservoir and the O_2/H_2O half reaction are placed at the top of Figure 3.8. If

---

Box 3.5

Understanding pE: an analogy to pH that is the vertical scale of the oxidation/reduction hierarchy (Figures 3.8, 3.9) and describes the oxidation/reduction status of biosphere habitats

$$pH = -\log[H^+]$$
$$HA = H + A^-$$

$$K = \frac{[H^+][A^-]}{[HA]}$$

$$pH = pK + \log\frac{[A]}{[HA]}$$

pH = measure of a solution's tendency to donate or accept protons

$$pE = -\log[e^-]$$
$$Red = Ox + e^-$$

$$K = \frac{[Ox][e^-]}{[Red]}$$

$$pE = pK + \log\frac{[Ox]}{[Red]}$$

pE = measure of a solution's tendency to donate or accept electrons

oxygen is absent from a system (e.g., the prebiotic Earth or contemporary anaerobic freshwater sediments) the pE drops to the next highest redox half reaction that happens to predominate in the habitat of interest. Depending on local geochemical conditions and ecological processes, the dominant oxidation–reduction half reaction may be NO_3^-/N_2, $FeOOH/FeCO_3$, or CO_2/CH_2O. Note that in Figure 3.8, the "storage box" of reduced organic carbon and the CO_2/CH_2O redox couple occur at the same height on the pE redox scale. From a thermodynamic viewpoint, the life forms that dwell in the biosphere are simply reservoirs of reduced organic carbon, waiting to be oxidized.

3.8 USING THE THERMODYNAMIC HIERARCHY OF HALF REACTIONS TO PREDICT BIOGEOCHEMICAL REACTIONS IN TIME AND SPACE

Figure 3.9 shows an expanded view of the scale of half reactions in Figure 3.8. This diagram can be considered a general guide for predicting and interpreting a large portion of the microbially mediated ATP-generating biochemical reactions in the biosphere. The information in Figure 3.9 is a tool that can help us maneuver through the physiological maze of reactions – in this sense, the figure is a "compass for environmental microbiologists".

  Figure 3.9 graphically depicts the relationship between reduced and oxidized substrates as a vertically arranged hierarchy of oxidation–reduction half reactions. The vertical axes are electron potential (as E_O) and pE. Compounds on the left of the half reaction hierarchy are in an oxidized state (electron acceptors), while those on the right are in the reduced form (electron donors). Furthermore, the transition from oxidized to reduced forms is governed by the redox status of the system of interest and by catalytic mechanisms of microbially produced enzyme systems. Highly oxidizing conditions appear in the upper portion of the hierarchy in Figure 3.9, while highly reducing conditions are listed in the lower portion. This figure can be used to predict which combinations of half reaction pairs are thermodynamically possible because, under standard conditions, the lower reaction proceeds leftward (electron producing) and the upper reaction proceeds rightward (electron accepting). Graphically, pairs of thermodynamically favorable half reactions can be linked simply by drawing arrows diagonally from the lower right to the upper left portion of the hierarchy. Fundamental reactions of the carbon cycle tie the oxidation of photosynthetically produced organic carbon (e.g., CH_2O; lower right of the hierarchy in Figure 3.9) to the variety of final electron acceptors that may be present in natural habitats (O_2, NO_3^-, Mn^{4+}, Fe^{3+}, SO_4^{2-}, CO_2). Each of these coupled half reactions is mediated by chemosynthetic organotrophic microorganisms. Moreover, when diagonal arrows directing

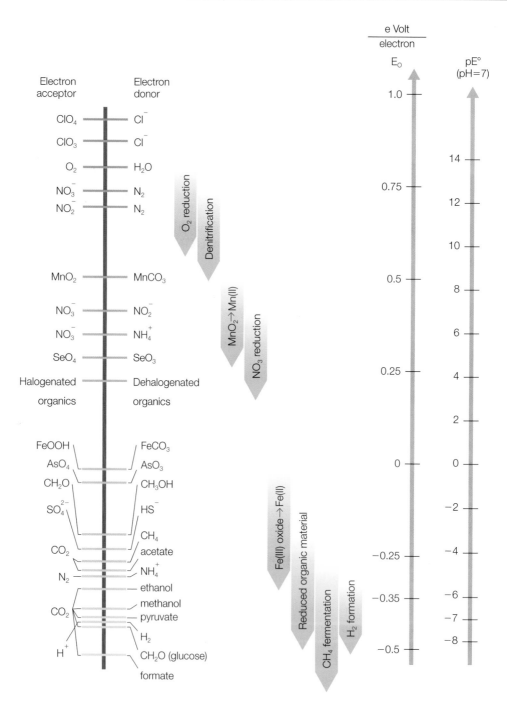

Figure 3.9 The hierarchy of half reactions, between electron donors and electron acceptors, defines biogeochemical reactions carried out by microorganisms in waters, soils, and sediments. (Modified from Zehnder, A.J.B. and W. Stumm. 1988. Geochemistry and biogeochemistry of anaerobic habitats. *In*: A.J.B. Zehnder (ed.) *Biology of Anaerobic Microorganisms*, pp. 1–38. Wiley and Sons, New York. Reprinted with permission from John Wiley and Sons, Inc., New York.)

Science and the citizen

Microorganisms can breath using chlorinated solvents

Headline news: groundwater pollution by chlorinated solvents, PCE, and TCE, and reductive dechlorination by microorganisms

Cl2C=CCl2 is the chemical formula for tetrachloroethene (also known as perchloroethene, perchloroethylene, and PCE). HClC=CCl2 is the chemical formula for trichlorothene (also known as trichloroethylene and TCE). For many decades, these synthetic compounds have been used widely in the dry cleaning, metal machining, and electronics industries. PCE and TCE are effective at leaving surfaces very clean – a critical need in many manufacturing processes. But both PCE and TCE, suspected carcinogens, have been improperly handled. These compounds are among the most ubiquitous groundwater pollutants in the world. They have a density greater than 1 g/cm³. This means that, after being spilled, they penetrate the soil and sediment and reach groundwater. They then continue to sink and become nearly intractable in geologic formations below. Once released to a subsurface habitat, pools of TCE and PCE can seldom be retrieved and their slow dissolution can contaminate huge volumes of adjacent flowing groundwater.

SCIENCE: Microorganisms to the rescue?

In standard *aerobic* biodegradation tests, PCE and TCE are generally found to be non-biodegradable. The carbon atoms in PCE and TCE are not good electron donors for aerobic microorganisms. However, under *anaerobic* conditions it has been discovered that PCE and TCE are microbiologically useful, physiological electron acceptors (McCarty, 1997). This discovery was made both in the laboratory and in field sites. In laboratory-incubated bottles, anaerobic microorganisms from groundwater sediment and sewage sludge were found to consume PCE and TCE when the microorganisms were supplied with electron donors like methanol and hydrogen gas. As PCE was consumed, TCE appeared. Then, sequentially, new compounds appeared: dichloroethene (DCE), monochloroethene (vinyl chloride or VC) and ethene. This process has become known as sequential "reductive dechlorination" or "halorespiration". As hydrogen atoms and electrons are added to the two-carbon ethene backbone of the pollutant molecules (as reduction occurs), the molecules are gradually stripped of their chlorine atoms.

In field sites where PCE and TCE were spilled along with electron donors (like methanol), the same series of metabolites can be found. Initially, spilled chemicals and groundwaters are free of DCE, VC, and ethene. Yet as time passes, DCE, VC, and ethene appear as groundwater constituents.

This is biogeochemistry in action. Microbially mediated, reductive dechlorination has the potential to counteract groundwater pollution by chlorinated solvents.

- **Is microbially mediated reductive dechlorination of PCE and TCE always reliable and 100% beneficial?**
  Answer: Unfortunately, the answer is "not always".

The figure below (from McCarty, 1997) provides an overview of many of the physiological and ecological factors that govern if and when reductive dechlorination will be complete. As is show in Figure 3.11, anaerobic processes are often carried out by many cooperating populations that constitute cooperative food chains. Organic materials such as those derived from plant biomass (see Section 7.3) are fermented in anaerobic habitats leading to transient extracellular pools of hydrogen and acetate. The hydrogen and acetate are used directly as electron donors by at least four key functional groups of microorganisms: sulfate reducers, iron reducers, methanogens, and reductive dechlorinators. Understanding how the four physiological groups of anaerobic populations compete for electron donors is an important goal in managing contaminated sites in ways that optimize reductive dechlorination.

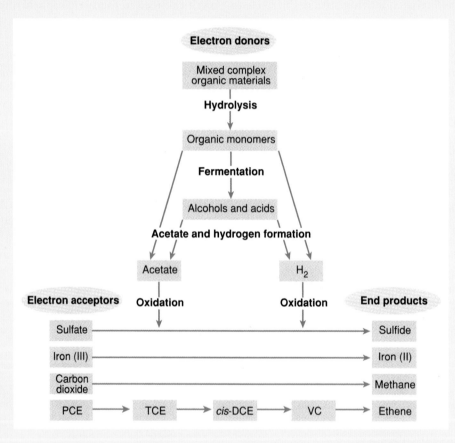

Figure 1 Detoxification: a competitive situation, showing the electron flow from electron donors to electron acceptors in the anaerobic oxidation of mixed and complex organic materials. Microorganisms that can use chlorinated compounds (PCE, TCE, *cis*-DCE, and VC) as electron acceptors in reductive dechlorination compete for the electrons in the acetate and hydrogen intermediates with microorganisms that can use sulfate, iron (III), and carbon dioxide. (From McCarty, P.L. 1997, Microbiology: breathing with chlorinated solvents. *Science* **276**:1521–1522. Reprinted with permission from AAAS.)

Note that along the bottom of the figure, the endproduct is ethene – a nontoxic, naturally occurring compound. But the immediate precursor of ethene is VC, a proven carcinogen. Thus, if the reductive dechlorination process stops short, it may convert suspected carcinogenic compounds (PCE and TCE) to a known carcinogen (VC). Clearly, then, there are some risks in applying reductive dechlorination-based strategies in technologies aimed at environmental cleanup of contaminated sites (for more information on biodegradation and bioremediation, see Section 8.3). Ongoing research continues to address scientific issues pertinent to the biogeochemistry, the molecular microbial ecology, and the genomics of microorganisms capable of metabolizing PCE, TCE, and their daughter products (e.g., Maymo-Gatell et al., 1997; He et al., 2003; Seshadri et al., 2005).

Research essay assignment
PCE and TCE were used industrially for many years before the environmental and health hazards they pose were recognized. Prepare an essay that documents the sequence of events linking improper disposal of chlorinated solvents to legislation recognizing their environmental threats. Next, tie in the discovery of microbial cleanup technologies. Base the essay on a search of the scientific and news literature.

carbohydrate oxidation to the reduction of these electron acceptors are drawn, the length of each arrow is proportional to the free energy gained by the microorganisms. Thus, microorganisms metabolizing carbohydrates with O_2 as a final acceptor are able to generate more ATP than those carrying out nitrate respiration. These microorganisms, in turn, gain more energy than those using Mn^{4+} and Fe^{3+} as final electron acceptors. This pattern continues down the hierarchy of electron-accepting regimes until methanogenesis (CO_2 as the final electron acceptor) is reached. There is a three-way convergence between the thermodynamics of half reactions, the physiology of microorganisms, and the presence of geochemical constituents actually found in field sites. It is notable that synthetic halogenated compounds (such as tetrachloroethene and polychlorinated biphenyls) are also present in the hierarchy depicted in Figure 3.9. Halogenated compounds can be utilized as final electron acceptors by microorganisms (see also Section 8.3). Other oxidation–reduction half reactions of important inorganic environmental pollutants (e.g., arsenate, selenium, and perchlorate) also appear in Figure 3.9.

The beauty of the scheme presented in Figure 3.9 is that it makes sense of what might be perceived as the overwhelming complexity of real-world conditions that prevail in aquatic and terrestrial environments. The predictions work for two reasons: (i) the power of thermodynamics; and (ii) the fact that only a few forms of a few elements (e.g., C, O, N, H_2, Fe, Mn) are prevalent in biosphere habitats. Table 3.7 formally defines the eight common processes that are recognized to occur in carbon-rich

Table 3.7

Hierarchy of oxidation–reduction processes typical of carbon-rich environments. When carbonaceous materials (CH_2O) are electron donors, individual microorganisms or consortia of populations can mediate electron transfer reactions. See also Figure 3.11. (Modified from Stumm, W. and J.J. Morgan. 1996. *Aquatic Chemistry: Chemical equilibria and rates in natural waters*, 3rd edn. Wiley and Sons, New York. Reprinted with permission from John Wiley and Sons, New York)

Process	PE regimes ($PE° \approx \log K$)	Heterotrophic reactions	$\Delta G°$ (kJ/eq.)
Aerobic respiration	$\frac{1}{4}O_2\,(g) + H^+ + e = \frac{1}{2}H_2O$	+13.75	$CH_2O + O_2 \rightarrow CO_2 + H_2O$	−125
Denitrification	$\frac{1}{5}NO_3^- + \frac{6}{5}H^+ + e = \frac{1}{10}N_2 + \frac{3}{5}H_2O$	+12.65	$5CH_2O + 4NO_3^- + 4H^+ \rightarrow 5CO_2 + 2N_2 + 7H_2O$	−119
Manganese reduction	$\frac{1}{2}MnO2(s) + \frac{1}{2}HCO_3^- + \frac{3}{2}H^+ + e = \frac{1}{2}MnCO_3(s) + H_2O$	+8.9	$CH_2O + 2MnO_2 + 4H^+ \rightarrow CO_2 + 2Mn^{2+} + 3H_2O$	−98
Iron reduction	$FeOOH(S) + HCO_3^- + 2H^+ + e = FeCO_3(s) + 2H_2O$	−0.8	$CH_2O + 4FeOOH + 8H^+ \rightarrow CO_2 + 4Fe^{2+} + 7H_2O$	−42
Fermentation	$\frac{1}{2}CH_2O + H^+ + e = \frac{1}{2}CH_3OH$	−3.01	$3CH_2O \rightarrow CO_2 + CH_3CH_2OH$	−27
Sulfate reduction	$\frac{1}{8}SO_4^{2-} = \frac{9}{8}H^+ + e = \frac{1}{8}H_2S(g) + \frac{1}{2}H_2O$	−3.75	$2CH_2O + SO_4 + 2H^+ \rightarrow 2CO_2 + H_2S + 2H_2O$	−25
Methanogenesis	$\frac{1}{8}CO_2(g) + H^+ + e = \frac{1}{8}CH_4(g) + \frac{1}{4}H_2O$	−4.13	$2CH_2O \rightarrow CO_2 + CH_4$	−23
Acetogencsis	$\frac{1}{4}CO_2(g) + H^+ + e = \frac{1}{8}CH_3COOH + \frac{1}{4}H_2O$	−4.2	$2CH_2O \rightarrow CO_3COOH$	−22

habitats. These coupled biogeochemical reactions are: aerobic respiration, denitrification, Mn reduction, Fe reduction, fermentation, sulfate reduction, methanogenesis, and acetogenesis. The hierarchy of the reactions is clear: aerobic respiration occurs at a pE of 13.75 and its free energy yield exceeds that of the other reactions. Thus, in a given aquatic or terrestrial habitat rich in carbon, microorganisms endowed with the physiological capacity to carry out aerobic respiration will have an advantage – their ATP-generating ability exceeds that of the other microbial residents. As long as the supply of oxygen is adequate, aerobic respiration will predominate. Once the supply of oxygen is exhausted, nitrate-respiring organisms will become predominant . . . and so on down the hierarchy. This predictable sequence of physiological processes is reinforced at the level of gene regulation within individual microbial cells. Generally speaking, the final electron acceptor allowing the highest free energy yield inhibits expression of genes required for utilization of final electron acceptors residing lower in the hierarchy (Saffarini et al., 2003; Gralnick et al., 2005). The reader should be aware that kinetic considerations, such as substrate uptake affinities by microorganisms and biochemical reaction rates, can also be important in determining which electron-accepting process will predominate in a given habitat. Vital for anaerobic physiology and food chains (see Section 3.10), hydrogen gas production occurs via reduction of protons. This process (mediated by hydrogenase enzymes) may be energetically unfavorable, but allows cells to dispose of excess reducing power, thereby maintaining redox balance (Hedderich and Forzi, 2005).

The information in Table 3.7 accurately represents the energetics of key biogeochemical processes. However, the information overlooks both the biochemical and ecological manifestations of the hierarchy. Examples of these, especially the involvement of anaerobic food chains are presented in Sections 3.10 and 7.3–7.5. The power of the information embodied by the thermodynamic "compass" (Figure 3.9) can be illustrated by its ability to successfully predict, in time and space, what biogeochemical reactions will occur. A laboratory experiment, illustrating the sequence of physiological processes (also known as terminal electron acceptor processes, TEAPs) is described in Box 3.6. The rationale of the experiment is to provide the heterotrophic soil community of microorganisms with an excess supply of electron donor (potato starch) but limited supplies of common electron acceptors (O_2, NO_3^-, $Mn(OH)_4$, $Fe(OH)_3$, SO_4^{2-}, CO_2). If water samples are removed and analyzed over time, the thermodynamically predicted patterns shown in Box 3.6 emerge. The most energetically favorable reactions occur first and the least favorable last. The sequence of depleted electron acceptors is O_2, NO_3^-, Mn^{4+}, Fe^{3+}, and SO_4^{2-}. Each electron acceptor is converted to its reduced form. In an aqueous medium, newly formed H_2O would not be detectable (unless the original O_2 were labeled with the stable isotope ^{18}O, and analyses were performed with a

Box 3.6

Laboratory demonstration of the sequence of thermodynamically-predicted physiological processes

- *Experimental design*: closed vessels containing water (1 L), 1 g potato starch, dissolved O_2, NO_3^-, SO_4^{2-}, $Fe(OH)_3$, $Mn(OH)_4$, CO_2, and soil (10 g).
- *Rationale*: provide carbon and energy for heterotrophs. Potato starch is the electron donor. Oxygen and mineral salts are electron acceptors.
- *Measure*: time course of geochemical change.

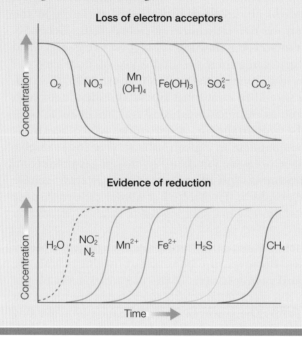

mass spectrometer). But if the TEAP experiment is carefully implemented, reduced forms of the other acceptors can be measured. Nitrate is converted by denitrifying microorganisms first to nitrite, then nitrogen gas (an alternative endpoint for some microorganisms is ammonia; see Sections 7.3 and 7.4). $Mn(OH)_4$, a solid mineral, is converted to soluble Mn^{2+} by manganese reducers. $Fe(OH)_3$, also a solid mineral, is converted to soluble Fe^{2+} by iron reducers. Sulfate-reducing bacteria convert sulfate anions to H_2S, a material that smells like rotten eggs and that readily reacts with many metal ions to form a black precipitate. Finally, methanogenic bacteria convert CO_2 to methane. Carbon dioxide is universally present in heterotrophic microbial systems as a result of its formation during respiration (though there may be exceptions: for example, in high pH environments, CO_2 availability may be drastically reduced). This endogenous source of CO_2 accounts for the constancy of CO_2 in the top panel of Box 3.6, and also assures the potential utilization of this often energetically least favorable major electron acceptor in all anaerobic habitats.

Acetogenesis is a process that coexists with methanogenesis in many anaerobic environments. Many acetogenic bacteria are chemolitho-autotrophs that carry out reactions between the following electron donor/acceptor pairs: H_2/CO_2, CO/CO_2, and H_2/CO. But acetogens are metabolically versatile – able to utilize a variety of electron donors and acceptors and to utilize fixed carbon sources when available (Drake et al., 2006). A key trait of acetogens is production of acetic acid via the Wood–Ljungdahl pathway that relies upon the enzyme acetyl-CoA synthase for CO_2 fixation and in terminal electron-accepting energy conservation (Drake et al., 2006). Factors that regulate and control the relative importance of methanogenesis and acetogenesis in anaerobic habitats are not fully understood. Based on simple energetic consideration, methanogenesis from H_2 is a more favorable process than acetogenesis (-136 kJ/mol versus -105 kJ/mol, respectively; Table 3.7). However, in many environments (e.g., termite gut, microaerophilic zones), acetogens can compete successfully with methanogens by positioning themselves closer to the H_2 source or by supplementing their nutrition with fixed organic compounds or by tolerating exposure to oxygen. At low temperatures, low pH (e.g., habitats such as tundra wetland soils) acetogenic populations can flourish. Furthermore, in high-temperature habitats featuring high acetate concentrations (e.g., anaerobic digesters), acetogens and methanogens can form a syntrophic relationship that converts acetate to methane (Schink, 1997). Regardless of which is the dominant process, the metabolic activities of both methanogens and acetogens are clearly a very important part of the global carbon cycling in virtually all ecosystems.

The laboratory demonstration of redox reaction chronology (Box 3.6) has been manifest innumerable times in real-world contaminated field sites. Box 3.7 provides an example of a groundwater contamination scenario that leads to spatially distinctive zones of oxidation/reduction reactions developing down-gradient from a gasoline spill. Envision a deep, sandy sediment housing an underground storage tank used to dispense automobile fuel. The tank corrodes and gasoline spills into the subsurface – reaching the water table and dissolving in the aqueous phase. The prespill conditions of the groundwater show significant concentrations of O_2, NO_3^-, and sulfate. Furthermore, the sand grains are coated with both iron and manganese oxides. Gasoline is a complex carbon source (alkanes, aromatics, and other compounds; see Sections 7.3 and 8.3) whose components can be metabolized by a subset of populations residing in the subsurface microbial community. Immediately adjacent to the spill, microorganisms flourish using gasoline components as electron donors

Box 3.7

Field demonstration of the sequence of thermodynamically predicted physiological processes

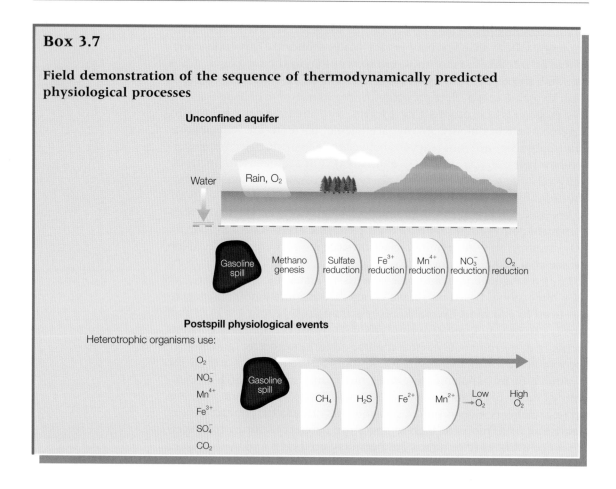

and oxygen as the electron acceptor. This zone of aerobic respiratory activity spreads down-gradient with the gasoline moving in groundwater, leaving behind a zone of oxygen depletion. Then, nitrate-respiring microorganisms flourish in an expanding front until nitrate is exhausted. This process of electron-acceptor exhaustion continues through the thermodynamic hierarchy of free energy-yielding reactions shown in Figure 3.9 and Table 3.7 until CO_2 is reached. The bottom panel of Box 3.7 depicts the electron-accepting processes that characteristically occur down-gradient in a mature, contaminated groundwater site. Closest to the contamination, methane can be found. Further down-gradient, samples of water and/or sediment routinely reveal elevated concentrations of sulfide. Further along are zones rich in Fe^{2+} and Mn^{2+} and nitrite. Finally, far from the contamination, an unaffected aerobic zone can be found. The spatially distinctive zones of microbial physiological processes shown in Box 3.7 are striking evidence for the responsiveness of microorganisms to environmental perturbation (in this case, gasoline pollution) and for the geochemical impact of those responses. Simply by making a living as chemosynthetic organoheterotrophs, naturally occurring subsurface microorganisms consume (biodegrade) gasoline and drastically alter their geochemical setting. Most important: these alterations are predictable based on thermodynamic relationships shown in Figure 3.9.

3.9 OVERVIEW OF METABOLISM AND THE "LOGIC OF ELECTRON TRANSPORT"

It is beneficial to place the concepts of "resource exploitation" and "oxidation/reduction reactions" within the overall physiological function of the microbial cell. We have emphasized in Sections 3.3–3.8 that ATP generation assists cellular processes by fueling cell maintenance and cell growth. As shown in Figure 3.10, cellular metabolism broadly consists of catabolism and anabolism. Catabolism is the cell's network of reactions for ATP generation. Anabolism is the cell's network of reactions, fueled by ATP, that assemble molecular building blocks (inorganic nutrients and low molecular weight organic compounds transported to the cell's interior) into new cell constituents or into progeny cells. "Habitat resources" in Figure 3.10 are the wide variety of organic and inorganic compounds that occur in water, sediments, and soils. These include electron donors that are thermodynamically unstable in the presence of electron acceptors. During catabolism by microorganisms endowed with electron transport chains, the electron acceptor terminates the series of electron transport reactions within the cytoplasmic membrane. For this reason, the term, *terminal electron acceptor*, is frequently used. After the electrons have been accepted, the reduced materials (e.g., H_2O, NO_2^-, N_2, NH_3, Mn^{2+}, Fe^{2+}, CH_4) are waste materials, discarded extracellularly; they are not assimilated into cell material. During anabolic reactions (Figure 3.10) biosynthetic reactions (driven by ATP) forge new bonds, assemble macromolecules, and ultimately lead to the assembly of new (progeny) cells. As cellular components are assembled, oxidized inorganic compounds (e.g., NO_3^-, SO_4^{2-}, CO_2) may be reduced – as required for their assimilation into new biomass. This anabolically driven reduction of inorganic compounds, termed "assimilatory reduction", consumes ATP and is distinctive from dissimilatory reduction carried out during catabolism.

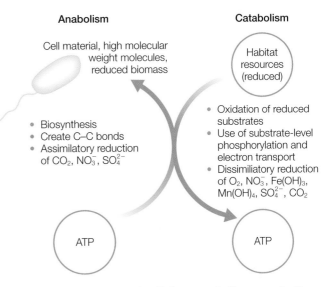

Figure 3.10 Overview of cellular metabolism: catabolism creates ATP which fuels the biosynthetic reactions of anabolism.

Anabolism

Cell material, high molecular weight molecules, reduced biomass

- Biosynthesis
- Create C–C bonds
- Assimilatory reduction of CO_2, NO_3^-, SO_4^{2-}

ATP

Catabolism

Habitat resources (reduced)

- Oxidation of reduced substrates
- Use of substrate-level phosphorylation and electron transport
- Dissimiliatory reduction of O_2, NO_3^-, $Fe(OH)_3$, $Mn(OH)_4$, SO_4^{2-}, CO_2

ATP

Box 3.8

The miracle of electron transport for ATP production: comparing a burning match to aerobic respiration

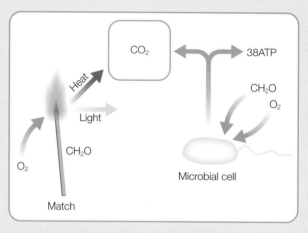

Summary of glycolysis, the citric acid cycle, and aerobic respiration

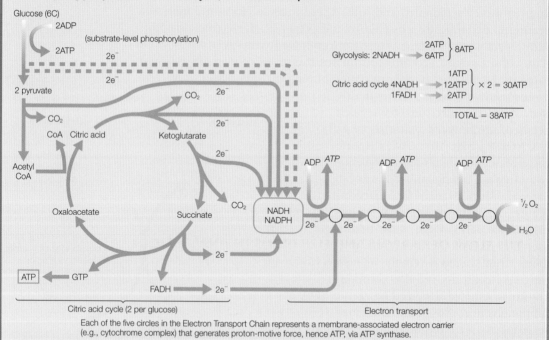

Each of the five circles in the Electron Transport Chain represents a membrane-associated electron carrier (e.g., cytochrome complex) that generates proton-motive force, hence ATP, via ATP synthase.

The lower panel in Box 3.8 provides an example of the biochemical manifestation of catabolism. During glycolysis (or the Embden–Meyerho–Parnas pathway), enzyme-mediated transformation of glucose (a six-carbon molecule) delivers two molecules of pyruvate (three carbons each) to the citric acid cycle. As electrons are removed from intermediates, CO_2 is produced and the electrons [as NADH (nicotinamide adenine dinucleotide phosphate) or NADPH] are delivered to an electron transport chain driven by oxygen as the terminal electron acceptor. The result is 38 ATP molecules per molecule of glucose. Glucose, $C_6(H_2O)_6$, is thermodynamically unstable in the presence of oxygen. If glucose were to burn via a direct chemical reaction with oxygen, it would produce light and heat, much like a matchstick (upper panel, Box 3.8). The energy released (heat and light) when a match burns is the same energy released when an equivalent mass of glucose is oxidized by a microbial cell. The membrane architecture and electron transport systems that convert the free energy of combustion to biochemically useful ATP are one of the many miracles of life. A key feature of the lower portion of Box 3.8 is exploitation of the electron transport chain. As long as there is an extracellular terminal electron acceptor available to consume electrons, and hence drive the flow of electrons through the electron transport chain, ATP production is assured. In Box 3.8, oxygen is the driver – but other compounds in the hierachy in Figure 3.9 function similarly. In general, microorganisms carrying out catabolism use fine-tuned enzymatic pathways (analogous to the citric acid cycle) to oxidize reduced substrates (inorganic compounds if chemolithotrophs; organic compounds if chemoorganotrophs) to deliver electrons to the respiratory chains of electron transport systems. Aerobic microorganisms utilizing glycolysis and the citric acid cycle have a high yield of ATP. If oxygen is unavailable, another biochemical mechanism (encoded by a corresponding set of genes) may be expressed and enable the cell to use other terminal electron acceptors (Figure 3.11a). The thermodynamic hierarchy (see Table 3.7 and Figure 3.9) shows the energetics of the reactions.

3.10 THE FLOW OF CARBON AND ELECTRONS IN ANAEROBIC FOOD CHAINS: SYNTROPHY IS THE RULE

In Section 3.9 and Box 3.8, we saw that glycolysis, the citric acid cycle, and the electron transport chain allow aerobic microorganisms to exploit a prevalent resource in the biosphere: glucose. The energetics of aerobic respiration (see Table 3.7 and Figure 3.9) predict that it will happen. But only the tools of physiology, biochemistry, and genetics applied to a particular model bacterium can reveal details of the mechanisms. After such tools have been applied to many modes of catabolism carried out by many model microorganisms, clear themes and patterns emerge (see

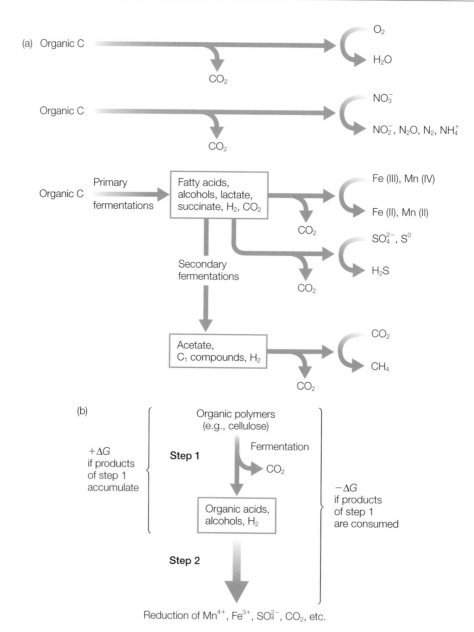

Figure 3.11 Flow of carbon and electrons in anaerobic food chains. (a) Contrasts between oxygen and nitrate respiration (carried out by individual microorganisms) and the cooperation of multiple populations of anaerobes that occur during iron reduction, manganese reduction, sulfate reduction, and methanogenesis. (b) How metabolic scavenging (step 2) allows anaerobic processes to overcome potential thermodynamic barriers. ((a) From Lengeler, J.W., G. Drews, H.G. Schlegel (eds) 1999. *Biology of Prokaryotes*. Blackwell Science, Stuttgart. With permission from Blackwell Science, Stuttgart.)

Section 7.5). These themes constitute an established body of physiological knowledge codified as mechanisms often named for their discoverers, such as the Embden–Meyerhoff–Parnas pathway, the Entner–Douderoff pathway, the Wood–Ljungdahl pathway, the Krebs tricarboxylic acid cycle, and the Calvin cycle (Kluyver and van Niel, 1954; Gottschalk, 1986; Zehnder and Stumm, 1988; Niedhardt et al., 1990; Ferry, 1993; Drake, 1994; White, 1995; Lengeler et al., 1999; Wackett and Hershberger, 2001; Drake et al., 2006).

Many of the processes, especially anaerobic ones, depicted by simple equations in Table 3.7 cannot be catalyzed by pure cultures of single microorganisms. Instead, they are only carried out by cooperating populations of physiologically distinct microorganisms. These *syntrophic associations of microorganisms* are often critical for organic carbon-driven iron reduction, manganese reduction, sulfate reduction, acetogenesis, and methanogenesis (lower group of entries in Table 3.7). In syntrophic associations, an anaerobic food web is established in which metabolic byproducts of one group of microorganisms are essential substrates for another. Figure 3.11 features the flow of carbon electrons and energy among cooperating populations of anaerobic microorganisms. Typical organic carbon materials reaching soils, lakes, sediments, and sewage treatment plants are high molecular weight polymers such as cellulose. Fermentative microorganisms hydrolyze the polymers into low molecular weight constituents that undergo fermentation reactions. Fermentation, by definition, represents no net change in oxidation–reduction status of the substrate. In chemical terms, fermentations are disproportionation reactions in which a portion of the substrate pool is oxidized to CO_2 while the remainder is reduced as an electron acceptor. The reduced waste products of fermentation include fatty acids (e.g., lactate, succinate, butyrate, acetate), alcohols, and hydrogen gas. These pools of fermentation waste products are valuable metabolic resources for the various metabolic groups know as iron reducers, manganese reducers, sulfate reducers, acetogens, and methanogens. Hydrogen (H_2) plays a particularly critical role in these interacting populations because it is an important electron donor used by most of these groups (see Section 3.8). Figure 3.11b depicts anaerobic food webs as two-step processes driven by populations responsible for the second step. Many of the fermentation reactions that occur in nature are not energetically favorable ($+\Delta G$) if the endproducts are allowed to accumulate. However, because populations involved in step two are effective at scavenging the fermentation products, especially H_2, the overall process becomes thermodynamically favorable. Thus, syntrophy (the metabolic cooperation between distinctive populations) is crucial for the success of many anaerobic processes. The term *interspecies hydrogen transfer* has been coined in recognition of hydrogen's role in this cooperative metabolism.

Table 3.8

Well-characterized chemolithotrophic reactions and their respective energy and growth yields (modified from Kelly, D.P. and A.P. Wood. 2006. The chemolithotrophic prokaryotes. *In*: M.W. Dworkin, S. Falkow, E. Rosenberg, K.-H. Schleifer, and E. Stackebrandt (eds) *The Prokaryotes*, Vol. 2, 3rd edn, pp. 441–456. Springer-Verlag, New York. With kind permission of Springer Science and Business Media.)

Reaction	Substrate oxidized	$\Delta G°$ (kJ/mol substrate)	Estimated number of mol ATP synthesized/ mol substrate
$H_2 + 0.5O_2 \rightarrow H_2O$	H_2	−237	2–3
$5H_2 + 2NO_3^- + 2H^+ \rightarrow N_2 + 6H_2O$	H_2	−241	
$4H_2 + CO_2 \rightarrow CH_4 + 2H_2O$	H_2	−35	<0.25?
$NH_4^+ + 1.5O_2 \rightarrow NO_2^- + H_2O + 2H^+$	NH_4^+	−272	1 or 2
$NH_4^+ + 0.75O_2 \rightarrow N_2 + 1.5\ H_2O$	NH_4^+	−315	
$NH_4^+ + NO_2^- \rightarrow N_2 + 2\ H_2O$	NH_4^+	−361	
$NH_2OH + O_2 \rightarrow NO_2^- + H_2O + H^+$	NH_2OH	−288	2
$NO_2^- + 0.5O_2 \rightarrow NO_3^-$	NO_2^-	−73	1
$H_2S + 0.5O_2 \rightarrow S^0 + H_2O$	H_2S	−209	1?
$S^0 + 1.5O_2 + H_2O \rightarrow H_2SO_4$	S^0	−505	1–3?
$HS^- + 2O_2 \rightarrow SO_4^{2-} + H^+$	HS^-	−733	1.5–4?
$S_2O_3^{2-} + 2O_2 + H_2O \rightarrow 2SO_4^{2-} + 2H^+$	$S_2O_3^{2-}$	−739	2.3
$5S_2O_3^{2-} + 8NO_3^- + H_2O \rightarrow 10SO_4^{2-} + 2H^+ + 4N_2$	$S_2O_3^{2-}$	−751	4–5
$S_4O_6^{2-} + 3.5O_2 + 3H_2O \rightarrow 4SO_4^{2-} + 6H^+$	$S_4O_6^{2-}$	−1245	5
$5S_4O_6^{2-} + 14NO_3^- + 8H_2O \rightarrow 20SO_4^{2-} + 16H^+ + 7N_2$	$S_4O_6^{2-}$	−1266	8–10
$2Fe^{2+} + 2H^+ + 0.5O_2 \rightarrow 2Fe^{3+} + H_2O$	Fe^{2+}	−47	0.5
$4FeS_2 + 15O_2 + 2H_2O \rightarrow 2Fe_2(SO_4)_3 + 2H_2SO_4$	FeS_2	−1210	
$Cu_2S + 0.5O_2 + H_2SO_4 \rightarrow CuS + CuSO_4 + H_2O$ (oxidation of Cu^+ to Cu^{2+})	Cu_2S	−120	1?
$CuSe + 0.5O_2 + H_2SO_4 \rightarrow CuSO_4 + Se^0 + H_2O$ (oxidation of selenide to selenium)	$CuSe$	−124	1?
$CH_4 + 2O_2 \rightarrow CO_2 + 2H_2O$	CH_4	−871	
$CH_4 + 8/3NO_2^- + 8/3H^+ \rightarrow CO_2 + 4/3N_2 + 10/3H_2O$	CH_4	−309	
$CH_4 + 8/5NO_3^- + 8/5H^+ \rightarrow CO_2 + 4/5N_2 + 14/5H_2O$	CH_4	−153	
$CH_4 + SO_4^{2-} \rightarrow HCO_3^- + H_2S^- + H_2O$	CH_4	−20 to −40	0.5?

3.11 THE DIVERSITY OF LITHOTROPHIC REACTIONS

The perceptive reader may have realized that the biogeochemical compass depicted in Figure 3.9 applies to organic and inorganic compounds equally well. To chart potential biogeochemical reactions between inorganic electron donors in Figure 3.9 (e.g., H_2S, H_2, NH_3, CH_4), one only needs to draw arrows connecting reduced substances on the right of the half-reaction scale to oxidized substances on the left (see also Box 3.4). For example, hydrogen-metabolizing microorganisms can link hydrogen oxidation to oxygen reduction (a long arrow), nitrate reduction (a slightly shorter arrow), sulfate reduction (a short arrow), or methanogenesis (the shortest arrow). Any reduced substrate (lower right on the half-reaction scale) can be linked to any oxidized substrate (upper left on the half-reaction scale). All combinations of electron-donor–electron-acceptor reactions predicted by thermodynamics may not yet have been documented as genuine biochemical processes in microorganisms. Two physiological processes, anaerobic oxidation of methane and ammonium, were suspected to be important biogeochemical process in aquatic habitats for decades – but have only been well documented, microbiologically, since the 1990s (see Sections 7.4 and 7.5). Another related physiological process, the use of methane as an electron donor and nitrate/nitrite as electron acceptors, was discovered in 2006 (Raghoebarsing et al., 2006). Table 3.8 provides a summary of 23 established chemolithotrophic reactions used by microorganisms. As illustrated for chemosynthetic organotrophs (see Sections 3.9 and 3.10), the thermodynamic compass reveals energetic impetus for catabolic reactions, not the biochemical mechanisms nor the genes that underlie them. Details of the biochemistry and genetics of chemolithoautotrophy are current areas of active research.

STUDY QUESTIONS

1 Regarding trends in genomics, how would you interpret discovery of an intracellular parasite with a large genome?

2 Unknown hypothetical genes constitute roughly one-third of all known genomes. As a curious microbiologist, you want to discover if a portion of these genes are still useful to their hosts. Describe experiments aimed at assessing when and if unknown hypothetical genes become active in their hosts in nature.

3 In the carbon–energy matrix of Table 3.3, one combination is clearly absent: chemosynthetic heteroautotrophs. Can you suggest a reason why this theoretically possible class of physiology does not seem to exist?

4 Table 3.5 briefly describes a variety of genetic systems that govern microbial responses to nutrient stress. The fourth entry lists how ammonia limitation activates nitrogen fixation (*nif*) genes. Explain the physiological benefit of this regulatory circuit.

5 In Section 3.5, a quote appears from J. Poindexter, "over extended periods of time, conservative utilization of nutrients once they are in the cell . . . may be more important than high-affinity uptake systems."

(A) Do you agree with this statement? Why or Why not?

(B) Under what circumstances is the statement likely to be true? Under what circumstances is it likely to be false?

(Hint: in preparing your answers, consider data in Table 3.6.)

6 Calculate the free energy yield available to two types of nitrate-reducing microorganisms – those converting nitrate to nitrogen gas (N_2; e.g., *Pseudomonas*) and those converting nitrate to ammonia (*E. coli*). Assume that the electron donor for each reaction is carbohydrate (CH_2O). To carry out the calculation, use Box 3.4 (see final preferred approach for calculating biochemical free energy yield) and Figure 3.9. Assume E_0' values for the electron donor and two electron acceptors are as follow: CO_2/CH_2O, -0.43 V; NO_3^-/NH_3, $+0.36$ V; NO_3^-/N_2, $+0.75$ V.

7 Regarding your answer to question 6, given the significantly smaller free energy yield when nitrate is reduced to ammonia, why would *E. coli* use dissimilatory reduction of nitrate to ammonia? (Hint: consider the stoichiometry, especially the number of electrons accepted. For additional, background, see Section 7.4.)

8 Note that the stoichiometric redox equations on the right-hand side of Table 3.9 are written with carbohydrate (CH_2O) as the electron donor.

  (A) Prepare balanced redox equations for metabolism of the fuel component, toluene (C_7H_8), under aerobic, denitrifying, iron-reducing, sulfate-reducing, and methanogenic conditions. Be sure to account first for the electrons transferred in the redox reactions, then use mass, charge, water, and protons to complete the balancing (see Box 3.4).

  (B) In both sets of equations (glucose and toluene) some of these metabolic processes consume H^+. What geochemical impact would this have on field sites where these processes occur?

9 A deep-sea hydrothermal vent emits dissolved concentrations of hydrogen gas, methane, hydrogen sulfide, and Fe^{2+} into aerobic waters.

  (A) Write balanced stoichiometric reactions between oxygen (electron acceptor) and each of the four electron donors.

  (B) Using the thermodynamic tools explained in Box 3.4 and Figure 3.9, calculate the free energy yield for each reaction to rank the four potential electron donors from most to least physiologically beneficial.

  (C) Regarding the microbiology of oceanic hydrothermal vent sites, what major factors other than those treated in (B) above are likely to determine processes and populations likely to flourish in vent communities?

  (D) How would you prove that each of these potentially useful energy sources was actually used by microorganisms in situ?

REFERENCES

Andrews, J.A. and R.F. Harris. 1986. *r*- and *K*-selection and microbial ecology. *Adv. Microb. Ecol.* **9**:99–147.

Atlas, R.M. and R. Bartha. 1998. *Microbial Ecology: Fundamentals and applications*, 4th edn. Benjamin Cummings, Menlo Park, CA.

Braun. M., F. Mayer, and G. Gottschalk. 1981. *Clostridium aceticum* (Wieringa): a microorganism producing acetic acid from molecular hydrogen and carbon dioxide. *Arch. Microbiol.* **128**:288–293.

Brock, T.D. 1971. Microbial growth rates in nature. *Bacteriol. Rev.* **35**:39–58.

Button, D.K. 1998. Nutrient uptake by microorganisms according to kinetic parameters from theory as related to cytoarchitecture. *Microbiol. Molec. Biol. Rev.* **62**:636–645.

Cano, R.J. and M.K. Borucki. 1995. Revival and identification of bacterial spores in 25- to 40-million-year-old Dominican amber. *Science* **268**: 1060–1064.

Dawes, E.A. 1989. Growth and survival of bacteria. *In*: J.S. Poindexter, and E.R. Leadbetter (eds) *Bacteria in Nature*, Vol. 3. pp. 67–187. Plenum Press, New York.

Devlin, T.M. (ed.) 2006. *Textbook of Biochemistry*, 6th edn. Wiley and Sons, New York.

Dolfing, J. 2003. Thermodynamic considerations for dehalogenation. *In*: M.M. Häggblom and I.D. Bossert (eds) *Dehalogenation: Microbial processes and environmental applications*, pp. 89–114. Kluwer Academic Publishers, Boston.

Drake, H.L. (ed.) 1994. *Acetogenesis*. Chapman and Hall, New York.

Drake, H.L., K. Küsel, and C. Mathies. 2006. Acetogenic prokaryotes. *In*: M. Dworkin, S. Falkow, E. Rosenberg, K.-H. Schleifer, and E. Stackebrandt (eds) *The Prokaryotes*, Vol. 2, 3rd edn, pp. 354–420. Springer-Verlag, New York.

Ferry, J.G. (ed.) 1993. *Methanogenesis: Ecology, physiology, biochemistry, and genetics*. Chapman and Hall, New York.

Finkle, S.E. 2006. Long-term survival during stationary phase: evolution and the GASP phenotype. *Nature Rev. Microbiol.* **4**:113–120.

Fraser, C.M., T.D. Read, and K.E. Nelson. 2004. *Microbial Genomes*. Humana Press, Totowa, NJ.

Fredrickson, J.K. and T.C. Onstott. 2001. Biogeocheimical and geological significance of subsurface microbiology. *In*: J.K. Fredrickson and M. Fletcher (eds) *Subsurface Microbiology and Biogeochemistry*, pp. 3–37. Wiley and Sons, New York.

Gottschalk, G. 1986. *Bacterial Metabolism*, 2nd edn. Springer-Verlag, New York.

Gralnick, J.A., C. Titus-Brown, and D.K. Newman. 2005. Anaerobic regulation by an atypical *Arc* system in *Shewanella oneidensis*. *Molec. Microbiol.* **56**:1347–1357.

Gray, T.R.G. and S.T. Williams. 1971. Microbial productivity in soil. *Symp. Soc. Gen. Microbiol.* **21**:256–286.

He, J., K.M. Ritalahti, K.-L. Yang, S.S. Koenigsberg, and F.E. Loeffler. 2003. Detoxification of vinyl chloride to ethene coupled to growth of an anaerobic bacterium. *Nature* **424**:62–65.

Hedderich, R. and L. Forzi. 2005. Energy-converting [NiFe] hydrogenases: more than just H_2 activation. *J. Molec. Microbiol. Biotechnol.* **10**:92–104.

Henis, Y. 1987. Survival and dormancy of bacteria. *In*: Y. Henis (ed.) *Survival and Dormancy of Microorganisms*, pp. 1–108. Wiley and Sons, New York.

Hilbert, D.W. and P.J. Piggot. 2004. Compartmentalization of gene expression during *Bacillus subtilis* spore formation. *Microbiol. Molec. Biol. Rev.* **68**:234–263.

Holt, S.C. and E.R. Leadbetter. 1969. Comparative ultrastucture of selected aeorobic spore-forming bacteria: a freeze-etching study. *Bacteriol. Rev.* **33**:346–378.

Horowitz, A., M.I. Krichevsky, and R.M. Atlas. 1983. Characteristics and diversity of subarctic marine oligotrophic, stenoheterotrophic, and euryheterotrophic bacterial populations. *Can. J. Microbiol.* **29**:527–535.

Jannasch, H.W. 1969. Estimation of bacterial growth rates in natural waters. *J. Bacteriol.* **99**:156–160.

Jannasch, H.W. 1979. Microbial ecology of aquatic low nutrient habitats. *In*: M. Shilo (ed.) *Extreme Environments*, pp. 243–260. Dahlem Konferenzen Life Sciences Research Report No. 13. Verlag Chemie, Weinheim.

Kelly, D.P. and A.P. Wood. 2006. The chemolithotrophic prokaryotes. *In*: M.W. Dworkin, S. Falkow, E. Rosenberg, K.-H. Schleifer, and E. Stackebrandt (eds) *The Prokaryotes*, Vol. 2, 3rd edn, pp. 441–456. Springer-Verlag, New York.

Kluyver, A.J. and C.B. van Niel. 1954. *The Microbe's Contributions to Biology*. Harvard University Press. Cambridge, MA.

Koch, A.L. 1971. The adaptive response of *Escherichia coli* to feast and famine existence. *Adv. Microbial Physiol.* **6**:147–217.

Koch, A.L. 1997. Microbial physiology and ecology of slow growth. *Microbiol. Molec. Biol. Rev.* **61**:305–318.

Kolker, E., A.F. Picone, M.Y. Galperin, et al. 2005. Global profiling of *Shewanella oneidensis* MR-1: expression of hypothetical genes and improved functional annotations. *Proc. Natl. Acad. Sci. USA* **102**:2099–2104.

Kolter, R., A. Siegele, and A. Tormo. 1993. The stationary phase of the bacterial life cycle. *Annu. Rev. Microbiol.* **47**:855–874.

Konstantinidis, K.T. and J.M. Tiedje. 2004a. Microbial diversity and genomics. *In*: J. Zhou, D.K. Thompson, J.M. Tiedje, and Y. Xu (eds) *Microbial Functional Genomics*, pp. 21–40. Wiley and Sons, New York.

Konstantinidis, K.T. and J.M. Tiedje. 2004b. Trends between gene content and genome size in prokaryotic species with larger genomes. *Proc. Natl. Acad. Sci. USA* **101**:3160–3165.

Lengeler, J.W., G. Drews, and H.G. Schlegel (eds) 1999. *Biology of Prokaryotes*. Blackwell Science, Stuttgart.

Madigan, M.T. and J.M. Martinko. 2006. *Brock Biology of Microorganisms*, 11th edn. Prentice Hall, Upper Saddle River, NJ.

Mailloux, B.J. and M.E. Fuller. 2003. Determination of in situ bacterial growth rates in aquifers and aquifer sediments. *Appl. Environ. Microbiol.* **69**: 3798–3808.

Matin, A., E.A. Auger, P.H. Blum, and J.E. Schultz. 1989. Genetic basis of starvation survival in non-differentiating bacteria. *Annu. Rev. Microbiol.* **43**: 293–314.

Maymo-Gatell, X., Y.-T. Chien, J.M. Gossett, and S.H. Zinder. 1997. Isolation of a bacterium that reductively dechlorinates tetrachloroethene to ethene. *Science* **276**:1568–1571.

McCarty, P.L. 1997. Microbiology: breathing with chlorinated solvents. *Science* **276**:1521–1522.

Morita, R.Y. 1982. Starvation–survival of heterotrophs in the marine environment. *Adv. Microbial Ecol.* **6**:171–198.

Morita, R.Y. 1997. *Bacteria in Oligotrophic Environments: Starvation–survival lifestyle*. Chapman and Hall. New York.

Morita, R.Y. 2000. Is H_2 the universal energy source for long-term survival? *Microbial Ecol.* **38**:307–320.

Nakabachi, A., A. Yamashita, H. Toh, et al. 2006. The 160-kilobase genome of bacterial endosymbiont *Carsonella*. *Science* **314**:267.

Nelson, D.L. and M.M. Cox. 2005. *Lehninger Principles of Biochemistry*, 4th edn. W.H. Freeman, New York.

Nelson, K.E., C. Weinel, I.T. Paulsen, et al. 2002. Complete genome sequence and comparative analysis of the metabolically versatile *Pseudomonas putida* KT2440. *Environ. Microbiol.* **4**:799–808.

Nichols, D.G. and S.J. Ferguson. 2002. *Bioenergetics 3*. Academic Press, San Diego, CA.

Nicholson, W.L., N. Munakata, G. Horneck, H.J. Melosh and P. Setlow. 2000. Resistance of *Bacillus* endospores to extreme terrestrial and extraterrestrial environments. *Microbiol. Molec. Biol. Rev.* **64**:548–572.

Niedhardt, F.C., J.L. Ingraham, and M. Schaechter. 1990. *Physiology of the Bacterial Cell: A molecular approach*. Sinauer Associates, Sunderland, MA.

Nystrom, T. 2004. Stationary-phase biology. *Annu. Rev. Microbiol.* **58**:161–181.

Phelps, T.J., E.M. Murphy, S.M. Pfifner, and D.C. White. 1994. Comparison between geochemical and biological estimates of subsurface microbial activities. *Microbial Ecol.* **28**:335–349.

Pirt, S.J. 1982. Maintenance energy: a general model for energy-limited and energy-sufficient growth. *Arch. Microbiol.* **133**:300–302.

Poindexter, J.S. 1981. Oligotrophy: fast and famine existence. *In*: M. Alexander (ed.) *Advances in Microbial Ecology*, Vol. 5, pp. 63–89. Plenum, New York.

Price, P.B. and T. Sowers. 2004. Temperature dependence of metabolic rates for microbial growth, maintenance and survival. *Proc. Natl. Acad. Sci. USA* **101**:4631–4636.

Raghoebarsing, A., A. Pol, K.T. van de Pas-Schoonen, et al. 2006. A microbial consortium couples anaerobic methane oxidation to denitrification. *Nature* **440**:918–921.

Rocap, G., F.W. Larimer, J. Lamerdin, et al. 2003. Genome divergence in two *Prochlorococcus* ecotypes reflects oceanic niche differentiation. *Nature* **424**:1042–1047.

Roszak, D.B. and R.R. Colwell. 1987. Survival strategies of bacteria in the natural environment. *Microbiol. Rev.* **51**:365–379.

Saffarini D.A., R. Schultz, and A. Beliaev. 2003. Involvement of cylic AMP (cAMP) and cAMP receptor protein in anaerobic respiration of *Shewanella oneidensis*. *J. Bacteriol.* **185**:3668–3671.

Schaechter, M., J.L. Ingraham, and F.C. Niedhardt. 2006. *Microbe*. American Society for Microbiology Press, Washington, DC.

Schink, B. 1997. Energetics of syntrophic cooperation in methanogenic degradation. *Microbiol. Mol. Biol. Rev.* **61**:262–280.

Schmitz, R.A., R. Daniel, U. Deppenmeir, and G. Gottschalk. 2006. The anaerobic way of life. *In*: M.W. Dworkin, S. Falkow, E. Rosenberg, K.-H. Schleifer, and E. Stackebrandt (eds) *The Prokaryotes*, Vol. 2, 3rd edn, pp. 86–101. Springer-Verlag, New York.

Schut, F., R.A. Prins, and J.C. Gottschal. 1997. Oligotrophy and pelagic marine bacteria: facts and fiction. *Aquat. Microb. Ecol.* **12**:177–202.

Seshadri, R., L. Adrian, D.E. Fouts, et al. 2005. Genome sequence of the PCE-dechlorinating bacterium *Dehalococcoides ethenogenes*. *Science* **307**: 105–108.

Smith, C.R. and A.R. Baco. 2003. Ecology of whale falls at the deep-sea floor. *Oceanogr. Marine Biol.* **41**:311–354.

Sowers, T. 2001. N$_2$O record spanning the penultimate deglaciation from the Vostok ice core. *J. Geophys. Res. Atmos.* **106**:31903–31914.

Stumm, W. and J.J. Morgan. 1996. *Aquatic Chemistry: Chemical equilibria and rates in natural waters*, 3rd edn. Wiley and Sons, New York.

Tempest, D.W., O.M. Neijssel, and W. Zevenboom. 1983. Properties and performance of microorganisms in laboratory culture; their relevance to growth in natural ecosystems. *In*: J.H. Slater, R. Whittenbury, and J.W.T. Wimpenny (eds) *Microbes in their Natural Environments*, pp. 119–152. Cambridge University Press, London.

Vaulot, D., D. Marie, R.J. Olson, and S.W. Chishholm. 1995. Growth of *Prochlorococcus*, a heterotrophic prokaryote, in the equatorial Pacific Ocean. *Science* **268**:1480–1482.

Vreeland, R.H., W.D. Rosenzweig, and D.W. Powers. 2000. Isolation of a 250-million-year-old halotolerant bacterium from a primary salt crystal. *Nature* **407**:897–900.

Wackett, L.P. and D.C. Hershberger. 2001. *Biocatalysis and Biodegradation*. American Society for Microbiology Press, Washington, DC.

White, D. 1995. *The Physiology and Biochemistry of Prokaryotes*. Oxford University Press, New York.

Zehnder, A.J.B. and W. Stumm. 1988. Geochemistry and biogeochemistry of anaerobic habitats. *In*: A.J.B. Zehnder (ed.) *Biology of Anaerobic Microorganisms*, pp. 1–38. Wiley and Sons, New York.

Zhou, J., D.K. Thompson, J.M. Tiedje, and Y. Xu. 2004. *Microbial Functional Genomics*. Wiley and Sons, New York.

FURTHER READING

Boetius, A., K. Ravenschlag, C.J. Schubert, et al. 2002. A marine microbial consortium apparently mediating anaerobic oxidation of methane. *Nature* **407**:623–626.

Gil, R., F.J. Silva, J. Peretó, and A. Moya. 2004. Determination of the core of a minimal bacterial gene set. *Microbiol. Molec. Biol. Rev.* **68**:518–537.

Karl, D.M. 1986. Determination of *in situ* microbial biomass, viability, metabolism, and growth. *In*: J.S. Poindexter and E.R. Leadbetter (eds) *Bacteria in Nature*, Vol. 2, pp. 85–176. Plenum, New York.

Kieft, T.L., and T.J. Phelps. 1997. Life in the slow lane: activities of microorganisms in the subsurface. *In*: P.S. Amy and D.L. Haldeman (eds) *The Microbiology of the Terrestrial Deep Subsurface*, pp. 137–163. Lewis Publishers, New York.

Konstantinidis, K.T. and J.M. Tiedje. 2005. Towards a genome-based taxonomy for prokaryotes. *J. Bacteriol.* **187**:6258–6264.

Kreth, J., J. Merritt, W. Shi, and F. Qi. 2005. Competition and coexistence between *Streptococcus mutans* and *Streptococcus sanguinis* in the dental biofilm. *J. Bacteriol.* **187**:7193–7203.

Madsen, E.L. 2002. Methods for determining biodegradability. *In*: C.J. Hurst, R.L. Crawford, G.R. Knudsen, M.I. McInerney, and L.D. Stetzenbach (eds) *Manual of Environmental Microbiology*, 2nd edn. American Society for Microbiology Press, Washington, DC.

Methé, B.A., K.E. Nelson, J.A. Eisen, et al. 2003. Genome of *Geobacter sulfurreducens*: metal reduction in subsurface environments. *Science* **302**: 1967–1968.

Postgate, J.R. 1976. Death in macrobes and microbes. *In*: T.R.G. Gray and J.R. Postgate (eds) *The Survival of Vegetative Microbes*, pp. 1–18. Cambridge University Press, London.

Stanier, R.Y., J.L. Ingraham, M.L. Wheelis, and P.R. Pantera. 1986. *The Microbial World*, 5th edn. Prentice Hall, Englewood Cliffs, NJ.

Van Verseveld, H.W. and R.K. Thauer. 1987. Energetics of C1-compound metabolism. *Antonie van Leeuwenhoek* **53**:37–45.

Wren, B.W. 2006. Prokaryotic genomics. *In*: M. Dworkin, S. Falkow, E. Rosenberg, K.-H. Schliefer, and E. Stackebrandt (eds) *The Prokaryotes*, Vol. 1, 3rd edn, pp. 246–260. Springer-Verlag, New York.

4

A Survey of the Earth's Microbial Habitats

Chapter 3 began with the concept that Earth habitats and life have been coevolving for 3.8×10^9 years. Furthermore, the coevolution can be metaphorically viewed as a dialog depicted as a double-headed arrow extending between the two spheres in Figure 3.1. The figure's left-hand sphere, labeled "Earth habitats", is the subject of this chapter. To set the stage for this chapter, you will need to recall the following points:

- *Some of the key physical events in planetary history include cooling from >100°C, alterations in the concentration of many atmospheric gases, meteor impacts, glaciation, volcanic activity, and plate tectonics (see Section 2.1 and Table 2.1).*
- *The histories of these physical planetary changes have been revealed by details in the geologic record (see Section 2.1).*
- *Key clues that have partially unlocked the history of the planetary biological record include fossils, ancient biomarkers, and stable isotopic ratios (see Section 2.1 and Table 2.2).*
- *The notion (the hypothesis) that genomes of extant microorganisms contain a genetic record of evolutionarily important selective pressures was presented in Section 3.2.*
- *In order to implement the prime directive (survive, maintain, grow: see Sections 1.1 and 3.4), all life forms must exploit planetary resources – especially those manifest as carbon and energy sources (see Section 3.3).*

Chapter 4 Outline

4.1 Terrestrial biomes

4.2 Soils: geographic features relevant to both vegetation and microorganisms

4.3 Aquatic habitats

4.4 Subsurface habitats: oceanic and terrestrial

4.5 Defining the prokaryotic biosphere: where do prokaryotes occur on Earth?

4.6 Life at the micron scale: an excursion into the microhabitat of soil microorganisms

4.7 Extreme habitats for life and microbiological adaptations

But the diverse habitats of Earth present many selective pressures in addition to carbon and energy sources. This chapter will take the reader on a tour of terrestrial and aquatic habitats that currently exist on Earth. After a survey of habitat characteristics (including prokaryotic biomass, itself), we will focus on extremes of environmental conditions that have confronted microorganisms and explore the adaptive biochemical mechanisms that have resulted.

4.1 TERRESTRIAL BIOMES

Major climatic determinants for world biomes are temperature (a function of both elevation and latitude) and atmospheric precipitation. Figure 4.1 depicts the present-day distribution of vegetation-based biomes across the globe. The 12 major biome types are tundra, boreal forest, temperate deciduous forest, temperate grassland, dry woodland/ shrubland (chaparral), desert, tropical rain forest/evergreen forest, tropical deciduous forest, tropical scrub forest, tropical savanna thorn forest, semidesert arid grassland, and mountains (complex biome zonation). Several striking patterns appear in Figure 4.1. Tundra predominates in

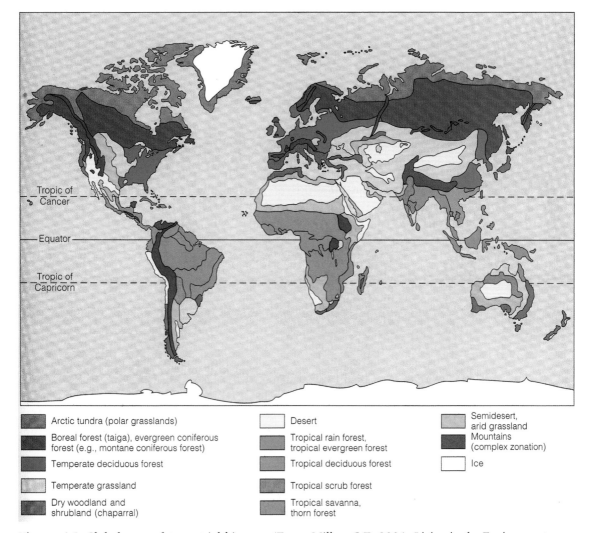

Figure 4.1 Global map of terrestrial biomes. (From Miller, G.T. 2004. *Living in the Environment*, 13th edn. Reprinted with permission of Brooks/Cole, a division of Thomson Learning, www.thomsonrights.com, fax 800 730-2215.)

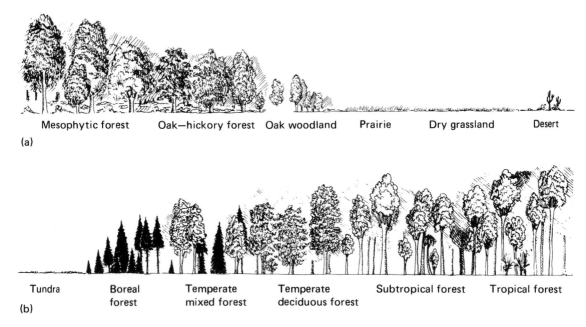

Mesophytic forest   Oak–hickory forest Oak woodland   Prairie   Dry grassland   Desert

(a)

Tundra   Boreal   Temperate   Temperate   Subtropical forest   Tropical forest
              forest   mixed forest   deciduous forest
(b)

Figure 4.2 Examples of regional vegetation gradients in North America from east to west and north to south. (a) The east–west gradient runs from the mixed mesophytic forest of the Appalachian Mountains through oak–hickory forests of the central states, to bur oak and grasslands, to the prairies, short-grass plains, and desert. The transect does not cut across the Rocky Mountains. This gradient reflects precipitation. (b) The north–south gradient reflects temperatures. The transect cuts across tundra, boreal coniferous forest, mixed mesophytic forests of the Appalachian Mountains, subtropical forests of Florida and Mexico, and tropical forests of southern Mexico. (From Smith, R.L. 1990. *Ecology and Field Biology*, 4th edn. Harper and Row, New York. Reprinted with permission of Pearson Education, Inc.)

circumpolar regions of the northern hemisphere. There is a broad band of boreal forest south of the tundra in the northern hemisphere and this blends into temperate deciduous forest and/or temperate grassland steppe at mid-latitudes in both hemispheres. Tropical savanna–thorn forest biomes occupy significant portions of northern Australia, Southeast Asia, India, the Arabian Peninsula, Africa, and South America. Tropical forests occupy broad equatorial swaths of the globe – across Central and South America, Central Africa, Southeast Asia, and the Malay Archipelago. Semidesert arid grasslands occur in subequatorial South America, equatorial Africa, and Australia. Deserts occur globally at mid-latitudes in southwestern North America, South Africa, Central Australia, and from northern Africa, eastward beyond the Arabian Peninsula.

Examples of regional vegetation-based biome gradients are depicted in Figure 4.2. Holding latitude roughly constant but varying precipitation, a hiker moving west from the forest in the Appalachian Mountains in North America will pass through mixed mesophytic forest (moist broad-leaved species including maple, beech, birch, ash, and rhododendron), oak–hickory forest, bur oak, and grasslands to prairie, then short-grass plains and on to the desert (Figure 4.2a). A north–south transect in North America (Figure 4.2b) extends from tundra through boreal coniferous forest, the mixed mesophytic forests of the Appalachians, to subtropical forests of Florida and Mexico, on to the tropical forests of southern Mexico.

4.2 SOILS: GEOGRAPHIC FEATURES RELEVANT TO BOTH VEGETATION AND MICROORGANISMS

Soils are vital for the biosphere – serving as the "skin of the Earth" in terrestrial habitats. Soils are the medium for root growth and nutrient uptake by plants. Soils are a major site for nutrient cycling (see Sections 7.3 and 7.4) – acting as reservoir for both plant-derived nutrients and for vast microbial diversity. At any given location on the Earth's terrestrial surface, the type of soil that develops reflects a combination of five factors: geologically derived parent material, climate, vegetation, time, and topography (Brady and Weil, 1999; Gardiner and Miller, 2004). The soil matrix is a three-dimensional porous array consisting of inorganic solids (sand, silt, and clay) intermingled with deceased biomass, humic materials, and organic and inorganic chemical coatings (humus and amorphous oxides), as well as viable organisms (micro- and meso-flora and fauna; especially fungi, protozoa, insects, nematodes, and burrowing animals). The pore spaces are shared in variable proportions by gases (whose composition reflects a balance between atmospheric diffusion and biotic activity) and aqueous soil solution (whose composition reflects complex equilibria between inorganic, organic, and biotic reactions; Madsen, 1996).

The global geography of specific soil types and the processes that form them have been well studied. Table 4.1 extends the relationships between climate and biome type to the associated soil types. In ice-free polar zones (mean temperature of the warmest month <10°C) where tundra predominates and many soils remain frozen within the top 200 cm, gelisols ("gel" for gel-like) occur. In these polar (and some montane and some recently eroded) regions, the underlying geologic parent material has undergone minimal soil development; these soils are known as entisols ("ent" for recent) and inceptisols ("incept" for inception, indicating a young embryonic stage with little or no horizon development). Table 4.2 provides a summary of the 12 global soil orders developed for the United States' soil taxonomy system. In subarctic polar zones (mean temperature of summer is 10°C; of winter, −3°C), histosols and spodosols are expected. Histosols ("hist" for tissue as in histology) occur in peat lands and bogs and are >20% organic matter. Spodosols typically develop beneath coniferous forests where the combination of acidity, humus, and infiltrating water form a gray-colored spodic soil horizon rich in aluminum

Table 4.1

Classification of major global biomes based on climate and vegetation, with their associated soil types (from Smith, R.L. 1990. *Ecology and Field Biology*, 4th edn. Harper and Row Publishers, New York. Reprinted with permission of Pearson Educational, Inc.)

Domain	Division	Temperature	Rainfall	Vegetation	Soil
Polar	Tundra	Mean temperature of warmest month <10°C	Water deficient during the cold session	Moss, grasses, and small shrubs	Tundra soils, gelisols, entisols, inceptisols, and associated histosols
	Subarctic	Mean temperature of summer, 10°C; of winter, −3°C	Rain even throughout the year	Forest, parklands	Podosol (spodosols and associated histosols)
Humid temperate	Warm continental	Coldest month below 0°C, warmest month <22°C	Adequate throughout the year	Seasonal forests, mixed coniferous–deciduous forests	Gray-brown podosols (spodosols, alfisols)
	Hot continental	Coldest month below 0°C, warmest month >22°C	Summer maximum	Deciduous forests	Gray-brown podosols (alfisols)
	Subtropical	Coldest month between +18 and −3°C, warmest month >22°C	Adequate throughout the year	Coniferous and mixed coniferous–deciduous forests	Red and yellow podzolic (ultisols)
	Marine	Coldest month between +18 and −3°C, warmest month >22°C	Maximum in winter	Coniferous forests	Brown forest and gray-brown podosols (alfisols)
	Prairie	Variable	Adequate all year, excepting dry years; maximum in summer	Tall grass, parklands	Prairie soils, chermozenes (mollisols)
	Mediterranean	Coldest month between +18 and −3°C, warmest month >22°C	Dry summers, rainy winters	Evergreen woodlands and shrubs	Mostly immature soils
Dry	Steppe	Variable, winters cold	Rain <50 cm/year	Short grass, shrubs	Chestnut, brown soils, and seriozems (mollisols, aridisols)
	Desert	High summer temperature, mild winters	Very dry in all seasons	Shrubs or sparse grasses	Desert (aridisols)
Humid tropic	Savanna	Coldest month >18°C, annual variation <12°C	Dry season with <6 cm/year	Open grassland, scattered trees	Latosols (oxisols)
	Rain forest	Coldest month >18°C, annual variation <3°C	Heavy rain, minimum 6 cm/month	Dense forest, heavy undergrowth	Latosols (oxisols)

Names in parentheses in the soil columns are soil taxonomy orders (USDA Soil System).

Table 4.2

Names of the 12 soil orders of the world, as specified in the US soil taxonomy, with their major characteristics (from GARDINER, D.T. and R.W. MILLER 2004. *Soils in our Environment*, 10th edn, p. 205. Pearson/Prentice Hall, Englewood Cliffs, NJ. Copyright 2004, reprinted by permission of Pearson Education, Inc., Upper Saddle River, NJ)

Soil order*	General features
Gelisols	Gelisols have permafrost within 200 cm of the surface
Entisols	Entisols have no profile development except perhaps a shallow marginal A horizon. Many recent river flood plains, volcanic ash deposits, unconsolidated deposits with horizons eroded away, and sands are entisols
Inceptisols	Inceptisols, especially in humid regions, have weak to moderate horizon development. Horizon development is minimal because of cold climates, waterlogged soils, or lack of time for stronger development
Andisols	Andisols are soils with more than 60% volcanic ejecta (ash, cinders, pumice, basalt) with bulk densities below 900 kg/m^3. They have enough weathering to produce dark A horizons and early-stage amorphous clays. Andisols have high adsorption and immobilization of phosphorus and very high cation exchange capacities
Histosols	Histosols are organic soils (peats and mucks) consisting of variable depths of accumulated plant remains in bogs, marshes, and swamps
Aridisols	Aridisols exist in dry climates, and some have developed horizons of lime or gypsum accumulations, salty layers and/or A and clay-rich Bt horizons
Mollisols	Mollisols are mostly in grasslands below some broadleaf forest-covered soils with relatively deep, dark A horizons; they often have B horizons with lime accumulations
Vertisols	Vertisols have high contents of clays that swell when wetted. A vertisol requires distinct wet and dry seasons to develop because deep wide cracks when the soil is dry are a necessary feature. Usually, vertisols have deep self-mixed A horizons (topsoil falls into cracks seasonally, gradually mixing the soil to the depth of the cracking). These soils exist most in temperate to tropical climates with distinct wet and dry seasons
Alfisols	Alfisols develop in humid and subhumid climates, have precipitation of 500–1300 mm, and are frequently under forest vegetation. Clay accumulation in a Bt horizon and available water much of the growing season are characteristic features. A thick E horizon is also common. They are slightly to moderately acidic
Spodosols	Spodosols are typically the sandy, leached soils of coniferous forests. Usually organic surface O horizons, strongly acidic profiles, and well-leached E horizons are expected. The most characteristic feature is a Bh or Bs horizon with accumulated organic material plus iron and aluminum oxides
Ultisols	Ultisols are strongly acidic, extensively weathered soils of tropical and subtropical climates. A thick E horizon and clay accumulation in a clay-rich Bt horizon are the most characteristic features
Oxisols	Oxisols are excessively weathered; few original minerals are left unweathered. Often, oxisols are more than 3 m deep, have low fertility, have dominantly iron and aluminum oxide clays, and are acidic. Oxisols develop only in tropical and subtropical climates

* Orders are arranged in approximate sequence from undeveloped soil to increased extent of profile development or increased extent of mineral weathering.

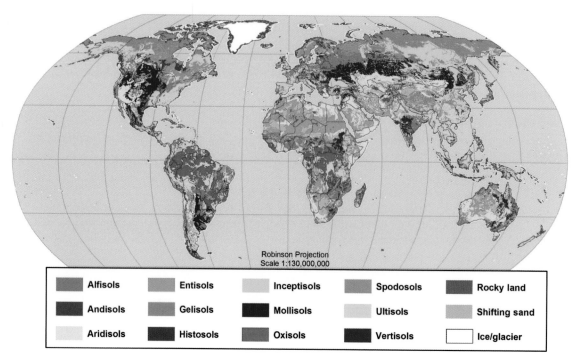

Figure 4.3 Distribution of 12 major soil orders throughout the world. (Reprinted with permission from USDA Natural Resources Conservation Service, Soil Survey Division, World Soil Resources, Washington, DC.)

and iron oxides. In humid temperate zones beneath mixed conifer–deciduous forests, a zone high in clay (known as an argillic horizon) often develops: this is a key characteristic of alfisols. Beneath prairie vegetation in humid temperate climate, the soil horizons are often dark and deep – fertile for agriculture and easy to work with a plow. These, known as mollisols ("moll" for mollify or soft), sometimes extend into dryer, short grass prairies, known as steppes. In the dry, warm conditions found in desert climates, only shrubs and sparse grasses contribute to the geochemical reactions governing soil formation. In this context aridisols develop. In the humid tropics, extremely weathered geologic strata occur at the Earth's surface. The heat, leaching from rainfall, and decomposing vegetation lead to the formation of highly oxidized minerals of iron and aluminum that are characteristic of oxisols. Figure 4.3 shows the distribution of the major members of the 12 major soil orders throughout the continents. Compare Figures 4.3 and 4.1 and note how vegetation biomes and soil types coincide.

One of the key characteristics of soil habitats is that they feature an astonishing physical, chemical, and biological heterogeneity – in space and time, and both locally and globally (see Section 4.6). All soils serve as depositories for deceased plant, animal, and microbial biomass. Unlike many aquatic habitats, there are few physical mechanisms for flushing materials through soil. What is deposited on the soil surface (e.g., plant material, deceased biomass, or microbial cells) generally remains there – serving as fodder for nutrient cycling and decay processes mediated by microorganisms (see Sections 7.3–7.6).

4.3 AQUATIC HABITATS

The Earth is the "blue planet": 70.8% of the surface is covered with water. Of the estimated 1.403×10^9 km^3 of water on the globe, 2% is ice pack (or glaciers), 97.61% is seawater, and only 0.3% is freshwater (in lakes, streams, and groundwater aquifers). Thus, despite their crucial role for maintaining plants, animals, and humans on land, freshwaters comprise only a small proportion of the global water resources (Wetzel, 2001). Table 4.3 compares several key characteristics of marine, freshwater, and groundwater reservoirs on Earth. Because of its huge volume and relatively small fluxes of incoming and outgoing waters, the residence time for water in the oceans is long (~4000 years). In contrast, the residence time of water in lakes, ponds, and rivers is relatively brief (~2 weeks to 10 years). After infiltrating through subsoil and entering the subsurface, the incoming water may re-emerge in surface habitats within ~2 weeks or may be trapped in deep aquifers for as long as 10,000 years. As shown in Table 4.3, subsurface habitats are quite distinctive, chemically and biologically, from open waters – largely due to the opaque, interstitial nature of rock and sediment matrices that influence geochemical reactions and exclude both large biota and light.

Freshwaters

Major freshwater resources of the globe include glaciers in polar and/or high elevation zones (North America, South America, Europe, Asia, Antarctica, Greenland), aquifers, and lakes. According to Wetzel (2001), inland waters cover less than 2% of the Earth's surface and about 20 lakes are extremely deep – in excess of 400 m. Lake Baikal in Siberian Russia (with a surface area of 31,500 km^2, an average depth of 740 m, and a maximum depth of 1620 m) is by far the largest freshwater body in terms of volume (23,000 km^3). Wetzel (2001) has presented a diagram comparing the world's largest lakes on an areal basis (Figure 4.4). The Mediterranean Sea has a connection to the Black Sea; thus, it is not truly an inland lake. The Caspian Sea (436,400 km^2), though saline, is the largest lake. The Laurentian Great Lakes of North America (Lakes Superior, Huron, Michigan, Ontario, and Eerie) constitute the greatest continuous mass of freshwater on Earth (245,240 km^2, with a collective volume of 24,620 km^3).

Table 4.3

Comparison between characteristics of three different aquatic habitats: marine, freshwater, and groundwater (from Madsen, E.L. and W.C. Ghiorse. 1993. Ground water microbiology: subsurface ecosystem processes. *In*: T. Ford (ed.) *Aquatic Microbiology: An ecological approach*, pp. 167–213. With permission from Blackwell Publishing, Oxford, UK)

Characteristics	Marine habitat	Freshwater habitat	Groundwater habitat
Physical			
Global location	Ocean basins	Continental depressions, valleys, basins	Beneath continental subsoils
Global surface area (%)	70.8	0.2	29.0
Global water volume (%)	94	<0.01	4
Residence time for water	~4000 years	2 weeks to ~10 years	2 weeks to 10,000 years
Hydrologic regime	Relatively deep water; very low percentage solids; little, if any, unsaturated zone; hydrologic stratification	Relatively shallow water; low percentage solids; little, if any, unsaturated zone, though some streams are ephemeral; hydrologic stratification	Interstitial water in solid matrix of variable porosity and variable degree of saturation; unsaturated zone may be substantial; hydrologic and geologic stratification
Biological			
Biota	Multicellular and unicellular algae; animals, protists, fungi, and prokaryotes	Multicellular and unicellular algae; animals, protists, fungi, and prokaryotes	Primarily prokaryotes; protists, rare algae, and cave-dwelling animals
Food chain	Photosynthesis, rare chemosynthesis, heterotrophy	Photosynthesis, rare chemosynthesis, heterotrophy	Heterotrophy, chemosynthesis at depth
Nutrient status	Nutrient-poor regions; productivity in up-welling zones	Broad range of oligotrophic, mesotrophic, and eutrophic conditions	Low levels of DOC and other nutrients common, but many nutrient-rich waters, i.e., beneath landfills
Water flow	Flow paths well defined	Flow paths well defined	Flow paths difficult to define

DOC, dissolved organic carbon.

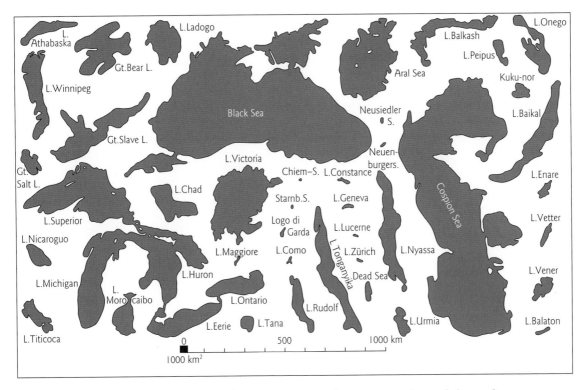

Figure 4.4 Major lakes of the world showing an approximate comparison of the surface areas of many of the larger inland waters, all drawn to the same scale. The Aral Sea has experienced catastrophic reductions in area (more than half of that depicted here) because of diversion of water for agriculture. (Reprinted from Wetzel, R.G. 2001. *Limnology: Lake and river ecosystems*, 3rd edn. Academic Press, San Diego, CA. Copyright 2001, with permission from Elsevier.)

In Africa, the largest freshwater body is Lake Victoria (68,870 km², with a volume of 2760 km³) – the source of the White Nile. In South America, the largest freshwater body is Lake Titicaca (8372 km², with a volume of 893 km³).

Lakes are termed *lentic* environments because their waters are calm and slow. Rivers and streams are *lotic* habitats because they feature water moving in response to gravity. Though only 0.0001% of the water on the Earth occurs in river channels, running waters are of enormous ecological and biogeochemical significance. They play key roles in the hydrologic cycle, deliver water across continents, and serve as critical habitats for many important aquatic species. Table 4.4 provides a summary of the catchment sizes and mean annual flows for major rivers in North America, South America, Europe, Africa, and Asia. Immediately derived from atmospheric precipitation, freshwaters generally have low dissolved concentrations of salts. After contacting fallen water, rock and soil generally contribute inorganic constituents to concentrations in the

Table 4.4

Catchment size and drainage for selected major river basins of the Earth (reprinted from Wetzel, R.G. 2001. *Limnology: Lake and river ecosystems*, 3rd edn. Academic Press, San Diego, CA. Copyright 2001, with permission from Elsevier)

Rivers by continents	Drainage area (10^3 km^2)	Mean annual flow (m^3/s)
North America	20,700	191,000
Colorado	624	580
Mississippi	3,222	17,300
Rio Grande	352	120
Yukon	932	9,100
South America	17,800	336,000
Amazon	5,578	212,000
Magdalena	241	7,500
Orinoco	881	17,000
Parana	2,305	14,900
San Francisco	673	2,800
Tocantins	907	10,000
Europe	9,800	1,000,000
Danube	817	6,200
Po	70	1,400
Rhine	145	2,200
Rhone	96	1,700
Vistula	197	1,100
Africa	30,300	136,000
Congo	4,015	40,000
Niger	1,114	6,100
Nile	2,980	2,800
Orange	640	350
Senegal	338	700
Zambezi	1,295	7,000
Asia	45,000	435,000
Bramahputra	935	20,000
Ganges	1,060	19,000
Indus	927	5,600
Irrawaddy	430	13,600
Mekong	803	11,000
Oh-Irtysh	2,430	12,000
Tigris-Euphrates	541	1,500
Yangtze	1,943	22,000
Yellow River (Huang Ho)	673	3,300

millimolar range (Stumm and Morgan, 1996). The major common ions in freshwaters are SO_4^{2-} (3×10^{-4} M), Cl^- (2.5×10^{-4} M), Ca^{2+} (4×10^{-4} M), Mg^{2+} (3×10^{-4} M), and Na^+ (2.5×10^{-4} M). Obviously, localized conditions throughout the globe contribute other dissolved and suspended aqueous constituents that determine major water quality characteristics such as pH, buffering capacity, dissolved organic matter, particulate organic matter, alkalinity, color, and turbidity.

Ocean waters

Each major oceanic basin (especially the Pacific, Atlantic, Indian, Antarctic, and North Atlantic) offers a unique set of geologic, physical, and biotic features. Physical oceanographers have defined key circulatory patterns of ocean waters that have major climatic and biotic implications. Figure 4.5 is a map in plan view of major ocean currents that result from a combination of the Earth's rotation and atmospheric forces. Vertical migration of ocean water also plays a critical role in biosphere function. Figure 4.6 displays a vertical cross-section of ocean circulation patterns centering on Antarctic waters. The three major circulation loops into the Atlantic, Pacific and Indian Oceans, respectively, are driven by density gradients in which cold, saline waters descend to the ocean bottom. These zones of down-welling are balanced by zones of up-welling, often along continental margins. The horizontal and vertical circulation patterns lead to heterogeneous nutrient distribution patterns (especially for N, P, and Fe) that directly govern the productivity of photosynthetic phytoplankton in ocean waters. Phytoplankton productivity indirectly governs fish productivity and harvest. Figure 4.7 provides an example of ocean productivity – represented by photosynthetic carbon dioxide fixation. Clearly the oceans are nutritionally and biologically heterogeneous.

An accurate perception of the vertical extent of the ocean basins is crucial for understanding these habitats. The vast majority of the oceans are deep, dark, and cold. Ninety percent to the volume of ocean waters remains at the stable temperature of approximately 3°C. The rest may be influenced by sunlight at the surface, by hydrothermal vents that mark the boundaries of tectonic plates along mid-ocean ridges, and/or by the continental margins. Figure 4.8 displays key categories of zonation at the ocean–continent interface. The *littoral zone* is the extremely shallow periphery influenced by waves and tidal action. The *neritic zone* is a slightly deeper boundary along the continental shelf. Deep ocean waters beyond the continental slope are known as the *pelagic zone*. Sunlight penetrates the upper pelagic (epipelagic) zone to about 200 m. Beneath this is a zone of very dim light (mesopelagic) which extends an additional 800 m. Deeper still are the bathypelagic and abyssopelagic zones (Figure 4.8). The hadal pelagic extreme is the Mariana Trench (in the Pacific Ocean northeast of Indonesia), 12,000 m below the ocean surface.

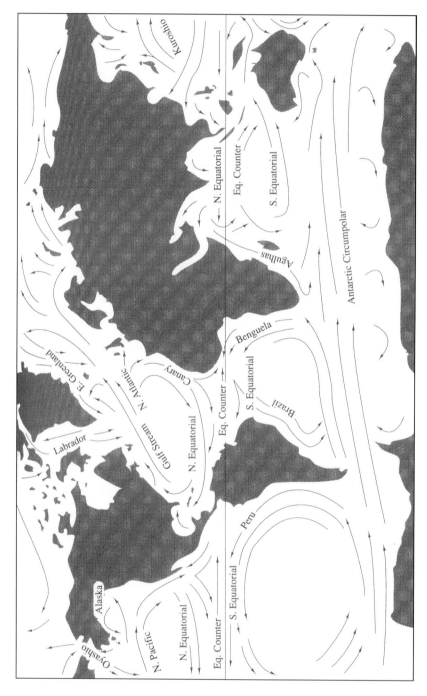

Figure 4.5 Global ocean currents. (Reprinted from Schlesinger, W.H. 1997. *Biogeochemistry: An analysis of global change*, 2nd edn. Academic Press, New York. Copyright 1997, with permission from Elsevier.)

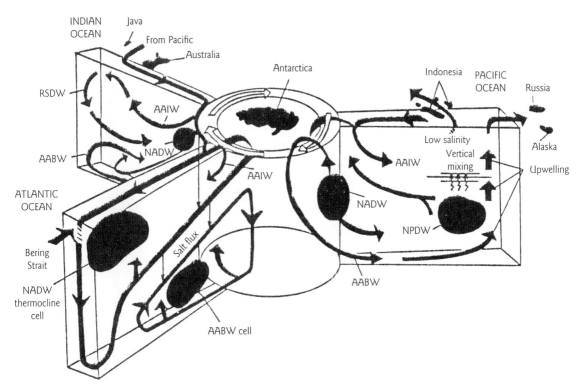

Figure 4.6 Vertical water circulation patterns of the ocean. Thermohaline circulation begins with the sinking cold, salty surface waters in the North Atlantic. The North Atlantic deep water (NADW) thus formed flows toward the south, where it wells up in the Antarctic, cools, mixes with other water types, and sinks again as Antarctic bottom water (AABW). AABW flows north along the ocean floor in all three ocean basins: the Atlantic, Pacific, and Indian. As it flows, its density is gradually reduced by mixing with waters from above. It thus moves upward, to become part of the intermediate-depth, southward-flowing "deep" bodies of water called NADW in the Atlantic, North Pacific deep water (NPDW) in the Pacific, and Red Sea deep water (RSDW) in the Indian Ocean. The combination of AABW flowing to the north and the "deep" water flowing to the south above it forms a deep "conveyor belt" circulation. There is also shallow circulation involving the formation of intermediate-depth water at lower latitudes of the Antarctic (AAIW). This water returns to the Antarctic by a variety of pathways, many not well understood. Also necessary to close the thermohaline circulation is a shallow northward flow to supply the NADW. This probably involves transport around South Africa as well as South America. (From Stumm, J.J. and W. Morgan. 1996. *Aquatic Chemistry*, 3rd edn. Wiley and Sons, New York. Reprinted with permission from John Wiley and Sons, Inc., New York.)

Seawater is salty. The salinity is, on average, 35 ppt (parts per thousand) or 35 g of salt per liter – as Na^+ (10.7 g/L), Mg^{2+} (1.29 g/L), Ca^{2+} (0.41 g/L), K^+ (0.4 g/L), Cl^- (19.4 g/L), SO_4^{2-} (2.7 g/L), and HCO_3^- (0.14 g/L) with moderate contributions from Sr^{2+}, Br^-, and B (Stumm and Morgan, 1996). Although seawater is derived from river waters exiting the continents, the chemistries of the two waters are quite distinctive.

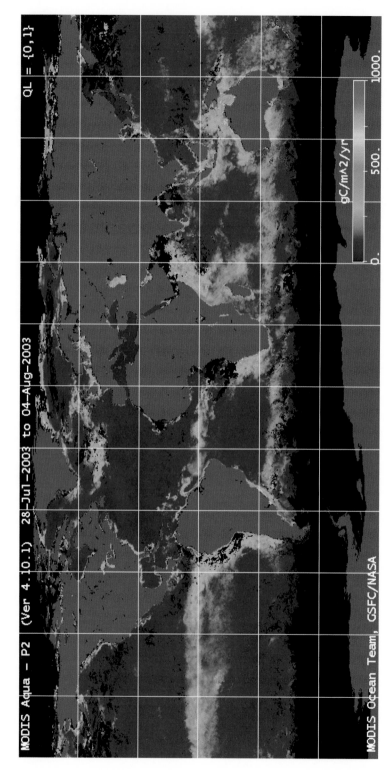

Figure 4.7 World map showing a snapshot of phytoplankton productivity in the oceans, August 2003. (Courtesy of NASA, MODIS Ocean Primary Productivity, with permission.)

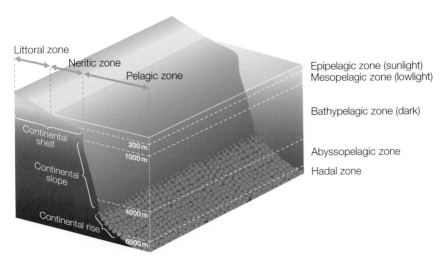

Figure 4.8 Vertical cross-section of the continental shelf.

Evaporated river water does not resemble the composition of the ocean. Furthermore, the dissolved materials currently present in the oceans are only a small fraction of those that have been delivered to the oceans by rivers over geologic time. Some constituents from incoming river waters are removed from the ocean, as mineral or other precipitates, approximately as fast as they are supplied; this avoids accrual (Stumm and Morgan, 1996). Overall, the composition of seawater is regulated by two key complementary mechanisms: (i) control by chemical equilibria between seawater and oceanic sediments; and (ii) kinetic regulation by three interacting rates: supply of individual chemical components, biological processes, and mixing processes.

4.4 SUBSURFACE HABITATS: OCEANIC AND TERRESTRIAL

Subsurface habitats are the deep layers of sediment and rock that extend far beneath soils (on the continents) and the ocean floor. The small pores and harsh conditions that prevail typically support only prokaryotic life. Conceptually, subsurface habitats exist as a spherical shell at the biosphere–geosphere interface. The upper edge of subsurface habitats has been defined in various ways: one definition focuses on zones below 8 m on continents and below 10 cm in the oceans (Whitman et al., 1998). The lower boundary of subsurface habitats is a depth of ~4 km, where average temperatures reach ~125°C, which is likely the upper limit for prokaryotic life (Amend and Teske, 2005).

A facile way to appreciate the extent and boundaries of subsurface habitats relies upon a plate tectonic map of the globe (Figure 4.9). On the

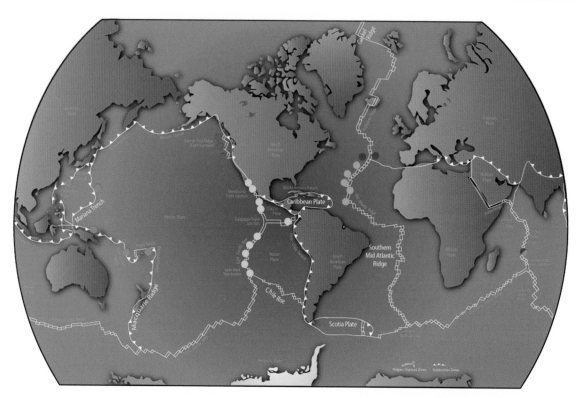

Figure 4.9 Plate tectonic map of the globe showing major plate boundaries, mid-ocean ridges and distribution of hydrothermal vent sites. Colored circles show vents with similar animal communities. (Courtesy of E. Paul Oberlander, with permission from Woods Hole Oceanographic Institute.)

continents, the Earth's crust (granitic in overall composition) is 20–80 km thick with an average age of 3×10^9 years. In contrast, the oceanic crust is of basalt, averaging only 10 km in thickness. Due to constant creation of new oceanic crust (at spreading centers) and consumption of oceanic crust (in subduction zones), its average age is $70–100 \times 10^6$ years. The mid-ocean ridge axis extends globally ~60,000 km. Known locations of deep ocean hydrothermal vents areas are shown in Figure 4.9. The basic unit of hydrothermal activity along the ridge axis is termed a *vent field*. Approximately 30 vent fields have been described to date (this number is sure to grow) and their areal extent is typically 60×100 m (Seyfried and Mottl, 1995).

Oceanic subsurface

The seafloor features obvious variations in topography (Figure 4.9). Perhaps less obviously, fluid circulation through the seafloor plays a critical role in geologic, geochemical, and microbiological processes (Seyfried and Mottl, 1995). Fluid circulation is responsible for large-scale cooling of magma, formation of the oceanic crust, geochemical cycling of the elements, and formation of polymetallic sulfide mineral deposits. Four fluid circulation regions have been identified (Seyfried and Mottl, 1995): (i) the seafloor spreading axis, in which the heat source is magma and sediment cover is generally lacking; (ii) mid-ocean ridge flanks, in which the crust is still young and hot and sediment cover is thin and patchy; (iii) ocean basins, in which the crust is cooler and sediment cover is thick and continuous; and (iv) subduction zones, where both the seafloor and accumulated sediments are consumed (Lisitzin, 1996).

As suggested above, the mature portions of the basaltic ocean floor are overlain by sedimentary deposits. These originate from the overlying water column and/or from the continents. By definition, ocean basins receive water- and wind-borne materials from higher ground. Streams and rivers are major conveyors of dissolved and particulate materials from the continents into the oceans. The settled particulate materials, collectively known as *sediments*, typically have both organic and inorganic components derived from eroding rock and soil. Lisitzin (1996) has mapped and categorized the sediments occurring in the ocean basins. Table 4.5 provides estimates of the thickness, areal extent, and volume of ocean sediments for each of the three major ocean basins. Column 1 of Table 4.5 divides the sediments into seven thickness categories (from <0.1 to >4.0 km). Reading across the table, the data display the areal extent of each thickness category and the corresponding sediment volumes for the Atlantic, Indian, and Pacific basins plus the "world ocean" average. Note that the Atlantic Ocean basin, though less than half the size of the Pacific, has a total sediment volume nearly 50% larger than that of the Pacific. This clearly reflects the relatively large number of high-volume rivers (e.g., Amazon, Mississippi, Congo, Rhine) delivering sediment to the Atlantic Ocean. The figures for average sediment thickness (bottom row of Table 4.5) reinforce the notion that sediments are thin (280 m) in the vast Pacific Ocean basin, relative to the other basins.

Regarding the composition of the ocean sediments, Lisitzin (1996) has created a world map estimating the proportion derived from the continents (terrigenous) versus that generated within the ocean by indigenous biota (phytoplankton, diatoms, other biomass) (Figure 4.10). The proportion from each source reflects a balance between proximity to major water- or wind-borne particulates and the biotic productivity of the water column. As expected, ocean basins near the mouths of major rivers are highly terrigenous. Surprisingly large areas of the central Pacific Ocean, seemingly out of range for continental impacts, also have been found to be highly terrigenous. Evidently, autochthonous inputs from biota are very low in these locales.

Table 4.5

Thickness, volume, and areal extent of sediments at the ocean floor of three major oceans basins (from Lisitzin, A.P. 1996. *Oceanic Sedimentation*, p. 20. Copyright 1996, American Geophysical Union. Reprinted by permission of American Geophysical Union)

Thickness, km	Atlantic		Indian		Pacific		World oceans	
	Area, km² × 10⁶	Volume, km³ × 10⁶	Area, km² × 10⁶	Volume, km³ × 10⁶	Area, km² × 10⁶	Volume, km³ × 10⁶	Area, km² × 10⁶	Volume, km³ × 10⁶
<0.1	11.46	0.573	18.21	0.910	53.22	2.661	82.89	4.144
01–0.3	7.99	1.198	14.17	2.834	51.80	10.360	73.96	14.392
0.3–0.5	17.47	6.115	11.92	4.768	28.19	11.276	57.58	22.159
0.5–1.0	13.65	10.238	7.58	5.685	27.29	20.468	48.52	36.391
1.0–2.0	13.52	20.208	3.39	5.085	0.14	0.210	17.05	25.575
2.0–4.0	7.83	23.460	1.89	5.670	–	–	9.71	29.130
>4.0	–	–	0.15	1.200	–	–	0.15	1.200
Total	71.92	61.864 (46.5)*	57.31	26.152 (19.7)*	160.64	44.975 (33.8)*	289.86	132.991 (100)*†
Total oceanic area without seas	82.48 (87.2)†		73.35 (78.1)†		164.8 (97.5)†		325.69 (89)†	
Average sediment thickness, m	860		456		280		459	

* Percentage of total volume of runoff.

† Considered or calculated area, percentage of total within each basin; the balance is continental shelf and slope.

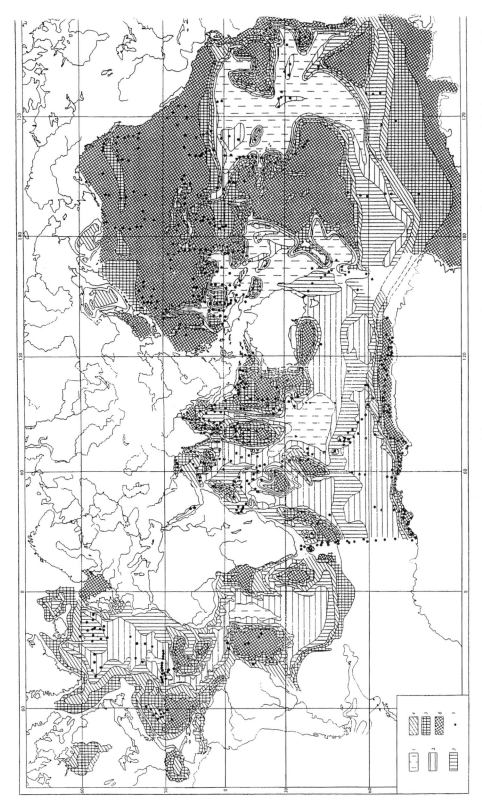

Figure 4.10 Composition of ocean sediments: percent derived from continents (terrigenous) versus allochothonous biotic sources. The six boxes in the key (numbered 1–6) show values from <10% (1, lightest shading) to >90% (6, darkest shading) terrigenous sediment. (From Lisitzin, A.P. 1996. *Oceanic Sedimentation*, pp. 28–29. American Geophysical Union, Washington, DC. Copyright 1996, American Geophysical Union. Reproduced by permission of American Geophysical Union.)

Science and the citizen

The Lost City – Atlantis' myth and reality

Headline news from Plato and Greek mythology

According to the original writings of Plato (Stewart, 1960; Jones, 2002), the mythical civilization of Atlantis was founded by Poseidon (the Greek god of the sea). The center of the civilization was the island of Atlantis, situated in the Atlantic Ocean west of the Straights of Gibraltar (the Pillars of Hercules). The island's mountains surrounded a central plain. By some accounts, the island "was larger than Libya and Asia altogether". The Atlantean royalty, led by Poseidon's son, Atlas (and his descendants), created an extensive Mediterranean dynasty. The royal city included intricate networks of canals, bridges, temples, and palaces. Atlantis' downfall came from three forces: (i) corruption and decadence within Atlantean society; (ii) defeat of the Atlantean army in battles against Athens to the east; and (iii) an earthquake and tidal wave (triggered by judgments of the gods) that caused the island to sink beneath the ocean waves forever.

SCIENCE: A genuine "lost city" beneath the waters of the Atlantic Ocean

Beneath the Atlantic Ocean, midway between North America and North Africa is a plate tectonic boundary known as the mid-Atlantic ridge (MAR; see Figure 4.9). At latitude 30°N, approximately 15 km to the west of the MAR, is a topographic feature on the seafloor known as the Atlantis massif. On a shelf below the summit of the massif at a depth of ~800 m below sea level is a remarkable geologic formation that was discovered in 2000 (Figure 1)

Figure 1 (*opposite*) A beehive of activity: microbial niches in serpentinization-influenced environments at the Lost City hydrothermal field. (A) Exothermic serpentinization reactions within the subsurface produce fluids of high pH enriched in methane and hydrogen, as well as some hydrocarbons. (B) Environments within the warm interior of carbonate chimneys in contact with end-member hydrothermal fluids host biofilms of *Methanosarcina*-like *Archaea* (green circles). These organisms may play a dominant role in methane production and methane oxidation within the diverse environments present in the chimneys. Bacterial communities within these sites are related to *Firmicutes* (purple rod-like cells). These organisms may be important for sulfate reduction at high temperature and high pH. (C) Moderate-temperature (40–70°C) endolithic environments with areas of sustained mixing of hydrothermal fluids and seawater support a diverse microbial community containing *Methanosarcina*-like *Archaea*, ANME-1 (a methane-oxidizing group of *Archaea*; blue rectangular cells), and bacteria that include ε- and γ-Proteobacteria (yellow filaments and red circles). The oxidation and reduction of sulfur compounds, the consumption and production of methane, and the oxidation of hydrogen most likely dictate the biogeochemistry of these environments. (D) In cooler environments (<40°C) associated with carbonate-filled fractures in serpentinized basement rocks, ANME-1 is the predominant archaeal group. The bacterial populations contain aerobic methanotrophs and sulfur oxidizers. (Copyright Taina Litwak 2005, p. 1421 in Boetius, A. 2005. Lost city life. *Science* **307**:1410–1422, with permission.)

(Boetius, 2005; Kelley et al., 2005). This dramatic "lost city" has resulted from a hydrothermal vent field that is distinctive from the customary "black smoker" hydrothermal vents of the MAR. Beneath this formation, the igneous rock known as peridotite predominates. Unlike the MAR igneous rock (basalt), peridotide is rich in the mineral olivine and is free of the mineral feldspar. Geochemical reactions between hydrothermal fluids and the peridotite form the secondary mineral, serpentine, and give rise to waters that are alkaline and rich in $CaCO_3$, H_2, and CH_4 – with small amounts of hydrocarbons (Figure 1). Spectacular towers, spires,

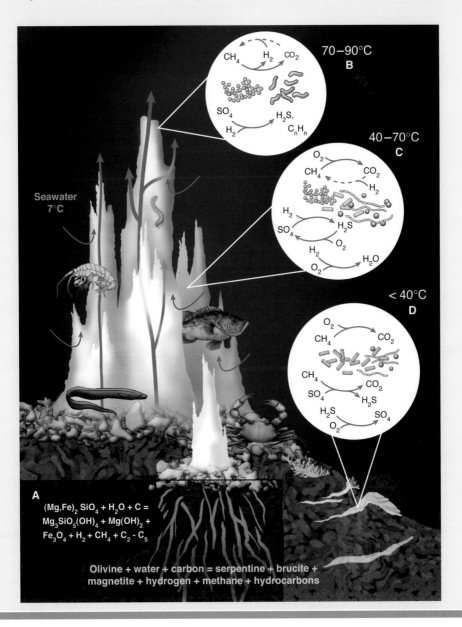

and beehive-like deposits of $CaCO_3$ dominate the seascape. The porous deposits have been colonized by microorganisms (especially *Bacteria* and *Archaea*) that catalyze approximately a dozen biogeochemical reactions. The electron donors for ATP production are H_2, CH_4, H_2S, and hydrocarbons. The electron acceptors are sulfate, CO_2, and O_2.

Research essay assignment

Environmental microbiology can boast of a long history of research events in which the exploration of a new habitat has led to the discovery of unique microbial adaptations. Among these explorations are: the cold, high-pressure deep sea; hydrothermal vents; caves; hot springs; Lake Vostock; the Lost City; and anoxic salty basins beneath the Mediterranean Sea. Search the scientific literature for two such discoveries and prepare an essay comparing them.

Terrestrial subsurface

Continental areas of the Earth are typically composed of materials in the following vertical sequence: the A and B soil horizons; the C soil horizon (from which the other soil horizons may have been derived: Madsen and Ghiorse, 1993; Brady and Weil, 1999); an unsaturated (or vadose) zone (that begins with the C soil horizon and ends at the water table); and a capillary fringe zone residing directly above a saturated zone which may extend through many different geologic strata (Figure 4.11).

• Where does the soil habitat end and the subsurface terrestrial habitat begin?

In some definitions, the subsurface terrestrial habitat begins immediately below the B soil horizon where soil scientists traditionally have felt that photosynthesis-based biological activity ceases (Madsen and Ghiorse, 1993; Brady and Weil, 1999). It is important to acknowledge, however, that the transition between soil and subsurface terrestrial habitats is not delineated by soil horizons per se. Indeed, plant roots may penetrate the C soil horizon, thereby supplementing the subsurface with photosynthetic carbon compounds that may stimulate microbial activity (Madsen, 1995).

Beneath the soil, which, by definition, is the zone of *pedogenesis* (soil formation), lie the unsaturated and saturated subsurface zones. This view of the subsurface habitat as being delineated in terms of the degree to which water occupies voids in a porous matrix (if air has been completely displaced by water, the system is "saturated"; if not, the system is "unsaturated") is satisfying, but it is also simplistic. For superimposed upon the degree of water saturation are the geologic, geographic, and climatic characteristics. At a given location on the Earth's surface, the stratigraphy beneath reflects a unique and complex history of geologic hydrologic, and chemical events (e.g., sedimentation, erosion, volcanism, tectonic activity, dissolution, precipitation, and biogeochemical activity). The result often is a heterogeneous geologic profile whose complexity may be compounded by variations in pore-water chemistry that may stem from localized aberrations in mineral phases or inorganic or organic solute concentrations. As mentioned in Section 4.3, the large surface area provided by rocks and sediments in the porous matrix may strongly influence the physical and chemical conditions of the groundwater habitat by altering concentrations of dissolved aqueous constituents at the surfaces and by adsorbing microbial cells (van Loosdrecht et al., 1990; Madsen and Ghiorse, 1993). Adsorption and aqueous equilibrium reactions are most likely to be influential in the saturated zone. But many subsurface habitats are dominated by unsaturated zones. In arid climates, the unsaturated zone may be hundreds of meters deep. Rainfall in such desert climates may be insufficient to allow saturated infiltration of soil to reach to the water table, except in restricted low-lying areas (Davis and DeWiest, 1966). Therefore, rather large portions of deserts may have unsaturated zones beneath them with little or no saturated water flux. Under such circumstances, vapor-phase reactions may be the prevalent form of geochemical change. Such conditions have important implications for microbial physiology and activity (Madsen, 1995).

The terrestrial subsurface is an important component of the landscape through which water passes as it cycles among the atmosphere, soil, lakes, streams, and oceans (Figure 4.11). Once water has infiltrated below the surface layer of soil, it has several possible fates. It may: (i) return to soil via capillary, gaseous, or saturated transport; (ii) be intercepted by plant roots; (iii) reach streams, lakes, or ponds via saturated flow; (iv) reverse its saturated flow direction from streams or lakes back into subsurface strata when levels of surface water are high; (v) directly reach the ocean via saturated flow; (vi) become mixed with seawater when groundwater withdrawal in coastal areas causes seawater to intrude inland; or (vii) enter a closed deep continental basin (Figure 4.11; Domenico and Schwartz, 1990). Regardless of the flow path taken through the subsurface, groundwater remains in the biosphere. However, the residence time before water exits the subsurface is highly variable. As mentioned in Section 4.3 and Table 4.3, return of the subsurface water to the soil may occur within a few days or weeks, though return from a deep continental basin may require thousands of years (Freeze and Cherry, 1979; Madsen and Ghiorse, 1993).

In conceptualizing the routes taken by water through the terrestrial segment of the hydrologic cycle, Chapelle's (2003) presentation of local, intermediate, and regional flow system is insightful. Chapelle provides the following definitions for these three flow systems based on relationships among surface topography, large-scale geologic structures, and the depth of water penetration along its path from recharge to discharge areas:

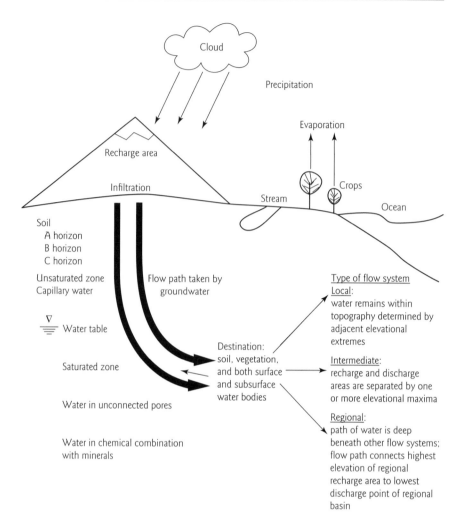

Figure 4.11 Conceptual flow system for understanding the role of soil and subsurface habitats in the hydrologic cycle. (Reprinted from Madsen, E.L. 1995. Impacts of agricultural practices on subsurface microbial ecology. *Adv. Agron.* **54**:1–67. Copyright 1995, with permission from Elsevier.)

1 A local (shallow flow) system has its recharge area at a topographic high and its discharge area at a topographic low which are located adjacent to each other.

2 An intermediate system occurs when recharge and discharge areas are separated by one or more topographic highs.

3 In a regional system, deep water flow bypasses local surface topography: the recharge area occupies the regional water divide and the discharge area occurs at the bottom of the basin.

Figure 4.11, which incorporates Chapelle's flow systems, illustrates the spatial and functional relationships between the geologic setting of the subsurface and its most dynamic component, water.

Freeze and Cherry (1979) have presented the idea of "chemical evolution" of groundwater as it passes from the atmosphere in recharge zones along the variety of flow paths such as those depicted in Figure 4.11. As precipitation, water begins as pure distillate containing only atmospheric gaseous and atmospheric particulate materials. After contact with soil and deeper subsurface sediments, the chemical composition of the water changes substantially. Not only do components in surface and subsurface matrixes dissolve, volatilize, and precipitate, but as the water reaches zones that are more remote from the atmosphere, complexation and oxidation/reduction reactions also occur. Many of the reactions are strictly geochemical (Domenico and Schwartz, 1990; Morel and Hering, 1993; Stumm and Morgan, 1996; Schwarzenbach et al., 2002; Chapelle, 2003), but many are also microbiologically mediated (see, for example, Sections 3.8 and 7.2–7.5).

The chemical composition of a given sample of groundwater reflects the integrated history of chemical and biochemical reactions that occur along a given flow path through soil and geologic strata. Because of the diversity of flow paths and biogeochemical reactions, the composition of groundwater is quite variable. Nonetheless, some generalizations can be made. In aquifers used for drinking water supplies that are not influenced significantly by human activity, major chemical constituents (>5 mg/L) typically include calcium, magnesium, silica, sodium, bicarbonate, chloride, and sulfate; while minor constituents (0.01–5 mg/L) include iron, potassium, boron, fluoride, and nitrate; with trace amounts (<0.1 mg/L) of many inorganics and organics (including humic acids, fulvic acids, carbohydrates, amino acids, tannins, lignins, hydrocarbons, acetate, and propionate) (Domenico and Schwartz, 1990). However, human activities (including septic systems, landfills, other types of waste disposal, and agricultural practices) may alter the chemistry of groundwater substantially by adding high concentrations of solutes such as both toxic and nontoxic organic carbon compounds and inorganic nutrients.

4.5 DEFINING THE PROKARYOTIC BIOSPHERE: WHERE DO PROKARYOTES OCCUR ON EARTH?

Whitman et al. (1998) completed a global survey of prokaryotic biomass in aquatic habitats, soils, and subsurface sediments, and in the intestinal tracts of selected animals. The approach required the investigators to scrutinize reports of the abundances of prokaryotic cells (number per unit volume) in representative habitats and multiply these cell densities by estimated global volumes of each of the habitats. Table 4.6 shows the global

Table 4.6

Number of prokaryotes in global aquatic habitats (from Whitman, W.B., D.C. Coleman, and W.J. Wiebe. 1998. Prokaryotes: the unseen majority. *Proc. Natl. Acad. Sci. USA* **95**:6578–6583. Copyright 1998, National Academy of Sciences, USA)

Habitat	Volume, cm^3	Cells, ml $\times 10^5$	Total number of cells, $\times 10^{26}$
Marine			
Continental shelf	2.03×10^{20}	5	1.0
Open ocean			
Water, upper 200 m	7.2×10^{22}	5	360
Water, below 200 m	1.3×10^{24}	0.5	650
Sediment, 0–10 cm	3.6×10^{19}	4600	170
Freshwater			
Lakes	1.25×10^{20}	10	1.3
Rivers	1.2×10^{18}	10	0.012
Saline lakes	1.04×10^{20}	10	1.0
Total			1180

Table 4.7

Number of prokaryotes in soils of global terrestrial biomes (from Whitman, W.B., D.C. Coleman, and W.J. Wiebe. 1998. Prokaryotes: the unseen majority. *Proc. Natl. Acad. Sci. USA* **95**:6578–6583. Copyright 1998, National Academy of Sciences, USA)

Ecosystem type	Area, $\times 10^{12}$ m^2	Number of cells, $\times 10^{27}$*
Tropical rain forest	17.0	1.0
Tropical seasonal forest	7.5	0.5
Temperate evergreen forest	5.0	0.3
Temperate deciduous forest	7.0	0.4
Boreal forest	12.0	0.6
Woodland and shrubland	8.0	28.1
Savanna	15.0	52.7
Temperate grassland	9.0	31.6
Desert scrub	18.0	63.2
Cultivated land	14.0	49.1
Tundra and alpine	8.0	20.8
Swamps and marsh	2.0	7.3
Total	123.0	255.6

* For forest soils, the number of prokaryotes in the top 1 m was 4×10^7 cells per gram of soil, and in the top 1–8 m, it was 10^6 cells per gram of soil. For other soils, the number of prokaryotes in the top 1 m was 2×10^9 cells per gram of soil, and in the top 1–8 m, it was 10^8 cells per gram of soil. The boreal forest, tundra, and alpine soils were only 1 m deep. A cubic meter of soil was taken as 1.3×10^6 g.

Table 4.8

Number of prokaryotes in global unconsolidated subsurface sediments (from Whitman, W.B., D.C. Coleman, and W.J. Wiebe. 1998. Prokaryotes: the unseen majority. *Proc. Natl. Acad. Sci. USA* **95**:6578–6583. Copyright 1998, National Academy of Sciences, USA)

Depth interval, m*	Cells, $cm^3 \times 10^6$	Number of cells $\times 10^{28}$		
		Deep oceans	Continents	Coastal plains
0.1	220.0	66.0	14.5	4.4
10	45.0	121.5	26.6	8.1
100	6.2	18.6	4.1	1.2
200	19.0	57.0	12.5	3.8
300	4.0	12.0	2.6	0.8
400	7.8		10.1	3.2
600	0.95		3.7	1.2
1200	0.61		3.2	1.0
2000	0.44		2.6	0.9
3000	0.34			0.7
Total		275.1	79.9	25.3
Grand total	$380 \times 10^{28} = 3.8 \times 10^{30}$			

* Depth intervals are designated by the upper boundary. Thus, "0.1" represents 0.1–10 m and "3000" represents 3000–4000 m.

survey of prokaryotes in aquatic habitats, exclusive of groundwater. Cell densities in freshwaters ($\sim10^6$/ml) average approximately 20 times those of deep ocean water and twice that of the upper ocean and continental-shelf waters. Surface sediment layers of the ocean typically support high numbers of microbial cells ($4.6 \times 10^8 \ cm^3$). Despite the low cell density in deep ocean water, this high-volume habitat supports more than half of the global aquatic prokaryotic biomass (Table 4.6). The estimated distribution of prokaryotic biomass in the soils of 12 terrestrial biomes is presented in Table 4.7. Note that the estimated density of soil microorganisms (footnote in Table 4.7) is generally very high (2×10^9 per gram of soil) for most surface soil types; though the density declines significantly with depth. Owing to the large global area and the abundance of bacteria in soil profiles of desert scrub ecosystems, these habitats harbor approximately 25% of the total soil microorganisms. Table 4.8 provides a depth profile and tabulation of prokaryotic biomass in sediments beneath the oceans and continents. Whitman et al. (1998) presumed that 20% of the continental subsurface was unconsolidated sediment – the remaining 80% (rock) was not counted. Clearly the upper 100 m of sediment (Table 4.8, first two rows) contain the majority of subsurface biomass. As will be discussed in Chapter 8 (Sections 8.1 and 8.2), higher life forms are, themselves, habitats available for colonization by microorganisms.

Table 4.9

Total number of prokaryotes in some representative animals (from Whitman, W.B., D.C. Coleman, and W.J. Wiebe. 1998. Prokaryotes: the unseen majority. *Proc. Natl. Acad. Sci. USA* **95**:6578–6583. Copyright 1998, National Academy of Sciences, USA)

Animal	Organ	Cells/ml or cells/g	Organ contents*	Number of animals	Number of cells, × 10²³
Human	Colon	3.2×10^{11}	220 g	5.6×10^9	3.9
Cattle	Rumen	2.1×10^{10}	106 L	1.3×10^9	29.0
Sheep and goats	Rumen	4.4×10^{10}	12 L	1.7×10^9	9.0
Pigs	Colon	5.4×10^{10}†	9 L	8.8×10^8	4.3
	Cecum	2.8×10^{10}	1 L	8.8×10^8	0.3
Domestic birds‡	Cecum	9.5×10^{10}	2 g	1.3×10^{10}	0.024
Termites	Hindgut	2.7×10^6§		2.4×10^{17}	6.5

* Organ contents in volume or grams of wet weight. For comparison, the volume of the human colon is 0.5 L. For domestic birds, wet weight was calculated from a volume of 2 ml, assuming that 1 ml = 1 g wet weight.
† The direct count was assumed to be 2.7 × viable count.
‡ Includes chickens, ducks, and turkeys.
§ Per termite.

Table 4.10

Number and biomass of prokaryotes in the world (from Whitman, W.B., D.C. Coleman, and W.J. Wiebe. 1998. Prokaryotes: the unseen majority. *Proc. Natl. Acad. Sci. USA* **95**:6578–6583. Copyright 1998, National Academy of Sciences, USA)

Environment	Number of prokaryotic cells, × 10²⁸	10¹⁵ g of C in prokaryotes
Aquatic habitats	12	2.2
Oceanic subsurface	355	303
Soil	26	26
Terrestrial subsurface	25–250	22–215
Total	415–640	353–546

Table 4.9 shows the tally of prokaryotic biomass that dwells in the digestive tracts of six types of higher organisms – cell densities can be extremely high (3.2×10^{11}/ml in humans).

When Whitman et al. (1998) computed the total global prokaryotic biomass (Table 4.10) the results were astounding: oceanic subsurface prokaryotes constitute well over half of the estimated global total (353–546×10^{15} g of C). The prokaryotic carbon pool is approximately 60–100% of the total carbon found globally in plants. Because prokaryotic biomass is relatively rich in nitrogen and phosphorus, the mass of each of these two essential nutrients in global prokaryotic biomass exceeds that in plants by an order of magnitude.

4.6 LIFE AT THE MICRON SCALE: AN EXCURSION INTO THE MICROHABITAT OF SOIL MICROORGANISMS

- What is it like to be a cell ~1 µm in size in a world whose diameter is 1.2×10^4 km, whose ocean depths can be 12,000 m, and whose soil aggregates are typically ~1 cm?

We have learned in Section 4.5 that the density of microorganisms in many soils is ~10^9 per gram. Is this a "crowded" state, or a "lonely" one? One way to answer this question is to focus on the microbial habitat at the microscale. In this section, we examine the soil habitat because many insights gained from soil are equally applicable to the microbial ecology of sediments and waters.

Appreciating habitat complexity in soil

Scholarly inquiry into the intrinsic properties of soil (pedology), separate from their impacts on plant growth (edaphology), developed significantly in Europe, Russia, and the United States in the nineteenth century (Brady and Weil, 1999). At least two complementary approaches to soil science have progressed simultaneously since then: field approaches to natural history and soil genesis; and laboratory approaches (chemical, biological, mineralogical, and physical determinations) applied to *soil samples*. Despite advancements in both approaches throughout the twentieth century, McBride (1994) has written, "much of soil science is empirical rather than theoretical in practice. This fact is a result of the extreme complexity and heterogeneity of soils, which are impossible to fully describe or quantify by simple chemical or physical models."

Soils are natural bodies, whose lateral and vertical boundaries usually occur as gradients between mixtures of materials of atmospheric, geologic, aquatic, and/or biotic origin. Soils are open systems subject to fluxes in energy (e.g., sunlight, wind) and materials (e.g., aqueous precipitation, erosion, deposition, inputs of organic compounds from activities of plants, human beings, and other animals). Furthermore, the intrinsic complexity of soil stems from its nature as an assemblage of solid, liquid, gaseous, organic, inorganic, and biological constituents whose chemical composition and random three-dimensional structure have not been completely characterized. In addition to physical complexity, the microbial (*Bacteria, Archaea*, fungi, algae, protozoa, viruses; see Chapter 5) physiological processes in soil and their multitude of interactions are dauntingly complicated. Compounding the challenge of understanding soil processes is the fact that abiotic reactions (e.g., precipitation, dilution, hydrolysis) must also be considered when attempting to understand soil biogeo-chemistry. In a field setting, plants, animals, and microorganisms effect geochemical change.

Attention must also be paid to the fact that the soil properties des-cribed above are subject to dynamic changes in time and space. No field setting is homogeneous or static. Regarding spatial inhomogeneity, the physical, chemical, nutritional, and ecological conditions for soil biota undoubtedly vary from the scale of micrometers to kilometers. Regard-ing temporal variability, in situ processes that directly and indirectly influence fluxes of material into, out of, and within soil are dynamic. Climate-related influences (such as temperature, sunlight, evaporation, and precipitation) are probably major variables that cause temporal vari-ations in biogeochemical processes in soil (Madsen, 1996).

A thermodynamic overview of inorganic soil reactions

Lindsay (1979) provided a unifying thermodynamic overview of soil in which dissolved substances in soil solution are in constant dynamic equi-libria with six independent chemical influences: (i) solid mineral phases; (ii) exchangeable ions and surface adsorption; (iii) nutrient uptake by plants; (iv) soil air; (v) organic matter and microorganisms; and (vi) water flux. The mineral phases of soil (typically 90% of the solid matter) have been described as "rock on its way to the ocean" (Lindsay, 1979). Primary min-erals (the parent material from which soils are derived) were often formed under conditions of high pressure and temperature. At the Earth's surface, subject to oxidative and hydrolytic weathering, the prim-ary minerals become secondary minerals as ionic species in solution are leached away and the remaining mineral structures seek lower free energy levels in their atomic arrangements. Soils contain numerous min-erals, some of which are crystalline, while others are amorphous or metastable. These minerals both respond to and control the dynamic pool of dissolved constituents in soil solution. A detailed discussion of soil min-eralogy and equilibria is beyond the scope of this chapter (for this see Lindsay, 1979; Dixon and Weed, 1989; Sposito, 1989; McBride, 1994); but it is critical to appreciate that soil and sediment habitats are in con-stant chemical transition, albeit at rates that are slow in human terms. Many of the mineral components are thermodynamically unstable and this instability is compounded by additional reaction pathways imposed by (micro)biological processes – especially those driven by plant-derived carbonaceous materials added via photosynthesis.

On size and microscale characterization of microbes in soil

Ladd et al. (1996) have reviewed relationships between soil components and the biological activity occurring therein. These authors emphasized the vastness in the range of scale of soil constituents (nine orders of magnitude, from atoms to rocks) and the hierarchical features of soil aggregates that form the three-dimensional fabric of soil. As shown in Figure 4.12, six size-based categories of aggregation were described (Ladd et al., 1996): (i) amorphous minerals develop at the nanometer to angstrom scale; (ii) clay microstructure colloids form at 10^{-7} m diameter; (iii) quasicrystals, domains, and assemblages (10^{-7}–10^{-5} m diameter) form between clay, silt, and smaller particles; (iv) microaggregates (0.1–250 µm diameter) occur between sand, silt, and smaller particles; (v) macroaggregates (250 µm to 25 mm diameter) occur between gravel, sand, and smaller particles; and (vi) clods (>25 mm diameter) occur between rocks, gravel, and smaller particles.

It is the aggregation, the *aggregate behavior*, of soil that contributes to its complexity. Tisdall and Oades (1982) insightfully presented a schematic model of the aggregate organization of soil (Figure 4.13). Emphasized in Figure 4.13 are both the hierarchical scales of soil aggregates and the mechanistically crucial binding agents responsible for aggregate formation. It is clear from Figure 4.13 that soil components of biological origin (typically referred to as soil organic matter – microorganisms plus humic substances; see Section 7.3) play a major role in creating soil structure, hence the habitat of soil microorganisms. Roots, fungal hyphae, plant debris, fungal debris, bacteria, and humic materials are specifically mentioned in Figure 4.13 because of their structural and nutritional contributions to the soil matrix.

Documentation of soil micromorphology (or "soil fabric"; Ringrose-Voase and Humphreys, 1994) has played a major role in establishing and reinforcing the type of model of soil aggregate organization shown in Figure 4.13. Microscopic procedures that have been applied, whenever possible, to intact soil samples (Foster, 1993; Ladd et al., 1996; Nunan et al., 2002, 2003; Gregory et al., 2003; Thieme et al., 2003; Young and Crawford, 2004) include: epifluorescent analysis of soil thin sections, transmission electron microscopy (TEM), scanning electron microscopy (SEM), electron microprobe analysis (EMP), environmental SEM, and X-ray tomography. These approaches have provided direct observations of intimate associations in soil aggregates of solid surfaces, root hairs, fungi, bacteria, extracellular polysaccharides, clay films (cutans), humic substances, and cellular debris. Such "ultrastructural" studies have revealed that soil microorganisms may occupy only 10^{-3} to 10^{-6}% of the soil surface area; this is because the clay fraction of soils can feature specific surface areas of up to ~10^3 m²/g. Microorganisms, though present in large numbers (~10^9 cells/g) are neither uniformly nor randomly distributed but, as revealed by TEM of soil sections, have been found clumped near or within cellular residues or in micropores (Ladd et al., 1996). Recent geostatistical analyses of bacteria captured in thin sections of undisturbed soils confirm a patchy, mosaic-like distribution of microorganisms in soil pores (Nunan

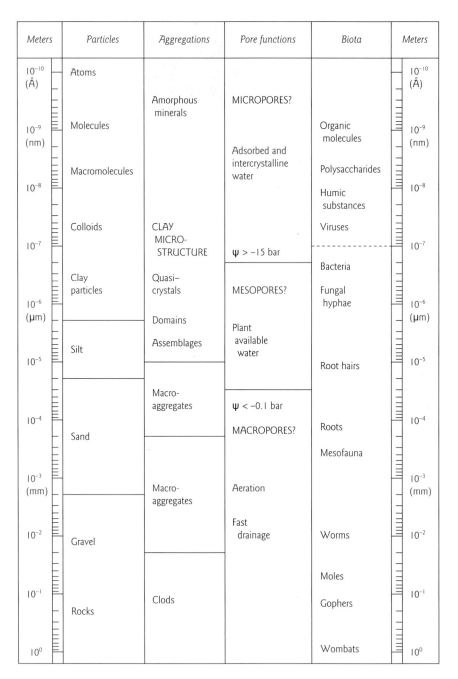

Figure 4.12 The vast range in scale in soil structure and the habitat of soil microorganisms. (From Ladd, J.N., R.C. Foster, P. Nannipieri, and J.M. Oades. 1996. Soil structure and biological activity. *In*: G. Stotzky and J.-M. Bollag (eds) *Soil Biochemistry*, Vol. 9, pp. 23–78. Copyright 1996, reproduced by permission of Taylor and Francis Group, a division of Informa plc.)

Major binding agent

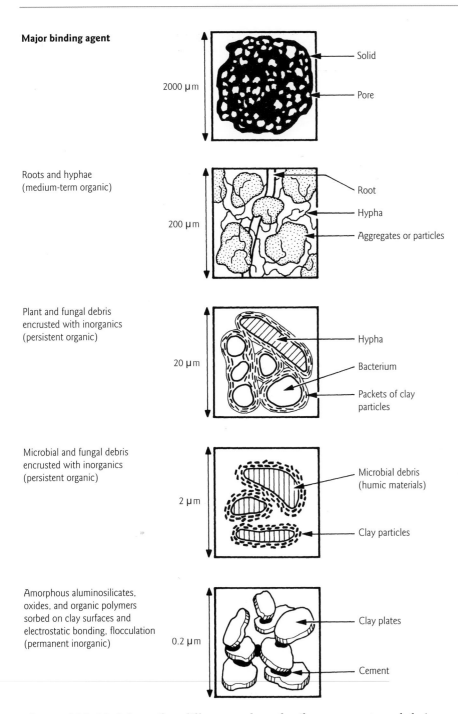

Roots and hyphae
(medium-term organic)

Plant and fungal debris
encrusted with inorganics
(persistent organic)

Microbial and fungal debris
encrusted with inorganics
(persistent organic)

Amorphous aluminosilicates,
oxides, and organic polymers
sorbed on clay surfaces and
electrostatic bonding, flocculation
(permanent inorganic)

Figure 4.13 Models, at five different scales, of soil components and their contribution to soil structure. (From Tisdall, J.M. and J.M. Oades. 1982. Organic matter and water-stable aggregates in soil. *J. Soil Sci.* **33**:141–163. With permission from Blackwell Publishing, Oxford, UK.)

Figure 4.14 A three-dimensional visualization of the soil matrix. (From Young, I.M. and J.W. Crawford. 2004. Interactions and self-organization in the soil–microbe complex. *Science* **304**:1634–1637. Reprinted with permission from AAAS.)

et al., 2003). Although such microscale imaging efforts are revealing, one well-recognized limitation of high-resolution soil microscopy is that each image surveys such a small soil volume that accruing information that is truly representative of bulk soil remains a challenge (Foster, 1993). Fortunately, three-dimensional tomography has recently begun to be used to assemble multiple cross-sectional images (Figure 4.14), potentially achieving an integrated picture of the micromorphology of the soil habitat.

4.7 EXTREME HABITATS FOR LIFE AND MICROBIOLOGICAL ADAPTATIONS

"Extreme" habitats are ones that, from a human point of view, seem inhospitable for life. For humans it may be inconceivable that life may survive and proliferate in hot springs, acid mine drainage, or at the bottom of glaciers. Yet prokaryotes often do (see Section 3.5). These prokaryotic "extremophiles" can be endemic (restricted) to extreme habitats and feature unusual biochemical adaptations not exhibited by prokaryotes dwelling in less extreme habitats. Conditions that prevail in extreme habitats, warranting the extremophile label, include broad ranges in temperature (from approximately −40 to >130°C), pH (from approximately −3.6

to 13), salinity (from rainwater to 5 M NaCl or $MgCl_2$), desiccation (from wet to extremely dry), pressure (from ~0.3 atm on Mt. Everest to 1200 atm in the Mariana Trench), and radiation (e.g., UV light, gamma rays). As discussed in Sections 4.1–4.5, contemporary global geography offers wide ranges in altitude, latitude, water availability, temperature, and light intensity. When these conditions are integrated over evolutionary time (see Sections 1.5 and 3.1), it is perhaps predictable that evolution will allow the development of adaptive traits.

Table 4.11 provides a sampling of information derived from microbiological studies of extreme habitats and the traits of microorganisms derived therefrom. Thermophilic (high-temperature loving) microorganisms have been discovered in terrestrial hot springs and in submarine hydrothermal vents where hydrostatic pressure allows superheating of the water. Kashefi and Lovley (2003) recently isolated an organism, strain 121 (related to *Pyrodictium*), capable of growth at 121°C! This combination of pressure and temperature is that of an autoclave. Regarding cold-tolerant microorganisms, snow- and ice-covered habitats have yielded cultures able to grow at temperatures as low as −17°C. Furthermore, Price and Sowers (2004) have used metabolism–temperature relationships (see Section 3.5) to predict prokaryotic metabolic activity at temperatures as low as −40°C. Regarding acid adaptation, acid-loving microorganisms have been isolated from acid mine drainage (with pH readings of ~0.5) and solfataras (hot, sulfur-rich, terrestrial volcanic vents with pH readings of ~0). High pH and salty habitats such as saline lakes, evaporation ponds, and hypersaline basins have yielded alkaliphilic and halophilic microorganisms able to grow, respectively, at pH >10 and salt concentrations over 5 M. Regarding radiation resistance, *Deinococcus radiodurans* is able to survive gamma-ray exposure 5000 times the dose that is lethal to humans (Schaechter et al., 2006). Other traits of prokaryotes featured in Table 4.11 include abilities to tolerate toxic organic compounds, high pressures, low nutrients, desiccation (low water activity), and conditions that prevail in geologic strata.

Comparative physiological and genetic studies of extremophilic prokaryotes have led to mechanistic explanations of how growth and survival occur under the extreme conditions described in Table 4.11. In general, vital molecular components of the cell (especially membranes, proteins, and nucleic acids) need to be modified to achieve functional stability. Table 4.12 provides a summary of adaptive biochemical mechanisms used by prokaryotes exposed to six types of environmental stress. To maintain metabolic function under extreme cold, cells must resist freezing and maintain flexibility in both their membranes and proteins. Membranes of cold-tolerant microorganisms are enriched in unsaturated fatty acids, which confer flexibility; while cold-tolerant enzymes are enriched in polar (not hydrophobic) amino acids and often feature a relatively high proportion of flexible α-helix-type protein tertiary structure (Lengeler et al., 1999; Madigan and Martinko, 2006; Schaechter et al., 2006).

Table 4.11

Examples of environmentally extreme habitats and extremophiles (modified from Cavicchioli, R. 2002. Extremophiles and the search for extraterrestrial life. *Astrobiology* **2**:281–292. With permission from Mary Ann Liebert Publications, Inc.)

Types of extreme condition	Environment	Organism*	Defining growth condition	References
High-temperature growth (hyperthermophile)	Hydrothermal vent	*Pyrodictium*-related strain 121 (A)	121°C	Kashefi and Lovley, 2003
High-temperature growth (hyperthermophile)	Submarine vent, terrestrial, hot spring	*Pyrodictium fumarii* (A)	T_{max} 113°C	Blochl et al., 1997
High-temperature survival	Soil, growth media contaminant	*Moorella thermoacetica* (spore) (B)	2 h, 121°C, 15 psi	Bryer et al., 2000
Cold temperature (psychrophile)	Lake Vostock beneath Antarctic ice sheet	Psychrophilic microbial community	−40°C	Price and Sowers, 2004
Cold temperature (psychrophile)	Snow, lakewater, sediment, ice	Numerous, e.g., *Vibrio*, *Arthrobacter*, *Pseudomonas* (B), and *Methanogenium* (A) spp.	−17°C	Carpenter et al., 2000; Cavicchioli and Thomas, 2000
High acid (acidophile)	Acid mine drainage (Iron Mountain, CA)	*Ferroplasma acidarmanis*	pH 0.5	Edwards et al., 2000
High acid (acidophile)	Dry, sulfur-rich, acid soil (solfatara)	*Picrophilus oshimae/ torridus* (A)	pH_{opt} 0.7 (1.2 M H_2SO_4)	Schleper et al., 1995; Johnson, 1998
High salt (halophile)	Saline lakes, evaporation ponds, salted foods	Mainly archaeal halophiles, e.g., *Halobacterium* and *Halorubrum* spp.	Saturated salt (up to 5.2 M)	Grant et al., 1998
High salt	Hypersaline basin in eastern Mediterranean Sea	Physiologically active, anaerobic microbial community	5 M $MgCl_2$	Van der Wielen, 2005
High alkaline (alkaliphile)	Soda lakes (hypersaline lakes rich in Na^+, low in Ca^{2+} and Mg^{2+})	*Bacillus* spp., *Clostridium paradoxum* (B), and *Halorubrum* spp. (A)	pH_{opt} >10	Jones et al., 1998

Radiation (radiation tolerant)	Soil, nuclear reactor water core, submarine vent	*Deinococcus radiodurans, Rubrobacter* spp., *Kineococcus* sp. (B), and *Pyrococcus furiosus* (A)	High γ-, UV, and X-ray radiation (e.g. >5000 Gy γ-radiation and >400 J/m² UV)	DiRuggerio et al., 1997; Ferreira et al., 1999; Battista, 2000
Toxicity (toxitolerant)	Toxic waste sites; industrial sites; organic solution and heavy metals	Numerous, e.g., *Rhodococcus* sp. (B)	Substance-specific (e.g., benzene-saturated water)	Isken and de Bont, 1996
High pressure (barophile or piezophiles)	Deep sea	Various, e.g., *Photobacterium* sp. (B) and *Pyrococcus* sp. (A)	Deep open ocean or submarine vent (e.g., pressure in Mariana Trench is >1000 atm)	Horikoshi, 1998
Low nutrients (oligotroph)	Pelagic and deep ocean, alpine and Antarctic lakes, various soils	*Sphingopyxis alaskensis* and *Caulobacter* spp. (B)	Growth with low concentration of nutrients (e.g., <1 mg/L^{-1} dissolved organic carbon) and inhibited by high concentrations	Schut et al., 1997
Low water activity (xerophile)	Rock surfaces (poikilohydrous), hypersaline, organic fluids (e.g., oils)	Particularly fungi (e.g., *Xeromyces bisporus*) and Archaea (e.g., *Halobacterium* sp.)	Water activity (a_w) <0.96 (e.g., *X. bisporus* 0.6 and *Halobacterium* 0.75)	Atlas and Bartha, 1998
Rock-dwelling (endolith)	Upper subsurface to deep subterranean	Various, e.g., *Methanobacterium subterranean* (A) and *Pseudomonas* sp. (B)	Resident in rock	Atlas and Bartha, 1998

* A, *Archaea*; B, *Bacteria*; Gy, Gray, a unit of ionizing radiation (1 Gy = 100 rad).

Table 4.12

Biochemical adaptations by microorganisms to extreme environmental stresses (compiled from Lengeler et al., 1999; Madigan and Martinko, 2006; Schaechter et al., 2006)

Environmental extreme	Adaptation
Cold	Enzymes are "cold active": • greater α-helix, polar amino acids • lesser β-sheet (rigid), hydrophobic amino acids Membranes have more unsaturated fatty acids (stay in fluid state) Cryoprotectants (?)
Heat	Enzymes are "heat stable": • key amino acid substitutions improve folding stability • salt bridges (ionic bonds between charged amino acids) • hydrophobic cores • high content of chaperonin molecules that maintain protein structure Membranes are rich in saturated fatty acids or lack fatty acids entirely (*Archaea*) Special DNA-stabilizing proteins
pH extremes	H^+ stabilizes membranes of acidophiles Intracellular pH kept moderate by membrane transport systems
High salt	Maintain intracellular solutes (pump inorganic ions into cells or concentrate organic solutes) to prevent water loss
High pressure	Enzymes fold so pressure does not alter substrate-binding sites Membranes have more unsaturated fatty acids Membrane composition changes to adjust permeability Pressure-controlled gene expression
Radiation/desiccation	Powerful DNA repair machinery. Many copies of DNA repair genes; multiple copies of chromosomes in novel ring-like structure

Thermophilic microorganisms feature a variety of key adaptations (Table 4.12). Not only do thermophilic enzymes resist denaturation and coagulation at high temperatures, some enzymes actually function optimally at elevated temperatures. By comparing characteristics of isofunctional enzymes in thermophiles and nonthermophiles, researchers have attributed thermostability to changes in a few key amino acids whose intramolecular hydrogen bonding and salt bridges stabilize tertiary structure. In addition, chaperonin proteins (that facilitate both folding and assembly of catalytic proteins) can play a large role in ensuring enzymatic functionality of thermophiles. Stabilization of DNA in thermophiles is thought to be the result of DNA supercoiling and association with both DNA-binding proteins and Mg^{2+}. Thermostability of cytoplasmic membranes is imparted by increased proportions of long-chain saturated fatty acids and (in *Archaea*) isoprene-like molecules linked by ether bonds to glycerol phosphate (Lengeler et al., 1999; Madigan and Martinko, 2006; Schaechter et al., 2006).

To accommodate extracellular extremes in pH, microorganisms have developed mechanisms that generally maintain moderate intracellular pH values – often via intracellular cytoplasmic buffering and membrane transport systems that can either pump out protons or pump in the counter ions, K^+ and Na^+. Remarkably, some acidophiles actually require high proton concentrations to maintain membrane stability (van de Vossenburg et al., 1998).

The challenge of existence in a high salt environment is maintaining the turgor pressure essential for cell wall growth and adequate water activity in the cytoplasm to maintain routine metabolic function. Halophilic (salt-loving) bacteria and related physiological types (osmophiles, high sugar; and xerophiles, lack of water) counterbalance the osmotic flow of water out of the cytoplasm by increasing the internal solute concentrations. Intracellular materials that retain water without disrupting cell physiology are termed "compatible solutes". These can be inorganic salts (e.g., KCl pumped into the cell from the external habitat). But more often, compatible solutes are organic molecules common to many cellular components (e.g., proteins, carbohydrates, lipids) that can be hydrolyzed intracellularly to yield sugar- and amino acid-building blocks that include ectoine, trehalose, glycerol, sucrose, L-proline, D-mannitol, and glycine betaine.

We learned earlier in this chapter about abyssal ocean depths. Hydrostatic pressure adds 1 atm for each 10 m depth of water; thus, the pressure in the Mariana Trench is 1200 atm. Microorganisms capable of tolerating (barotolerant) and those requiring (barophilic) high pressure have been described. Adaptations to high pressure include: enzymes whose conformation minimizes pressure-related changes in polypeptide folding, a high proportion of unsaturated fatty acids in membranes (to maintain flexibility), and adjustments in the expression of membrane transport proteins.

The astonishing degree of resistance to gamma-irradiation in *Deinococcus radiodurans* is considered the result of several factors that render the organism virtually immune to the ordinarily lethal impact of multiple breaks in DNA expected under conditions of extreme radiation and desiccation. Multiple DNA repair enzyme systems (including RecA) are active in *D. radiodurans*. The repair mechanisms require the close proximity of an undamaged DNA template. It is thought that undamaged templates are available for this microbe because in each cell there are 4–10 copies of the genome arranged in a dense, ring-like structure (a torroid). Furthermore, the cells occur in tightly linked clusters of four cells (tetrads), which are able to exchange DNA (Lengeler et al., 1999; Madigan and Martinko, 2006; Schaechter et al., 2006).

STUDY QUESTIONS

1 Based on the categories of biomes shown in Figure 4.1, what biome type do you live in?
   (A) Are there virgin tracts of this biome type set aside as park land near you? How have human land-use changes altered local biogeochemistry?
   (B) How have human land-use changes altered biodiversity and ecology?
   (C) How many additional biome types are within 100 km of your home? Within 1000 km? What are they?

2 Do you know the dominant soil order where you live (see Figure 4.3 and Tables 4.1 and 4.2)? How far must you travel to find another soil order? Briefly explain how each of the five factors mentioned in Section 4.2 influences soil development.

3 Use information in Section 4.3 to answer the following:
   (A) If the single outlet (the Angara River) exiting Lake Baikal flows at 60 km^3/year, what is the turnover time of the water in the lake? (Hint: turnover time is the volume of the reservoir divided by the influx or efflux of water.)
   (B) If the river flux out of Lake Titicaca was also 60 km^3/year, what would the turnover time be?
   (C) Consider the implications of turnover time for water pollution events in each lake system. From a management perspective, what are the pros and cons of living near a large water body?

4 Compare the general compositions of freshwaters and ocean waters given in Section 4.3. Identify the major anions in each.
   (A) In light of information in Chapter 3, which of the major anions is physiologically significant for microorganisms? Why?
   (B) Consider that 0.1 g of plant biomass (CH_2O) per liter may dissolve from aquatic macrophytes and be respired by heterotrophic microorganisms native to both freshwater and ocean-water habitats. The CH_2O is the major electron donor. What percent of the dissolved oxygen (assume the initial concentration to be 9 mg/L) would be consumed? After oxygen consumption, how much of the major anion (from part A) would also be respired? (Please show stoichiometries of the reactions and convert all units to millimoles.)
   (C) Given the answer to part B, would you expect distinctive physiological classes of microorganisms in freshwater versus saltwater habitats? If so, why? If not, why not?
   (D) What is the geochemical impact of the heterotrophic activity in fresh- and saltwater habitats? Name the dominant endproduct of electron flow in each habitat.

5 Use Chapelle's terminology (see Section 4.4) to describe the hydrogeologic regime where you live. Where does your drinking water come from? And where does it go?

6 Regarding Section 4.6, are microorganisms "crowded" or "lonely"? Compare and contrast their proximity to one another for a high clay soil and for your own intestines.

7 In acidic environments, the cellular adenosine triphosphate (ATP, energy) demand for maintaining an acceptable intracellular pH may be considerable. For a single class of autotrophic or heterotrophic microorganisms of your choice (see Table 3.3), can you suggest how acidophilic populations are compensated for this additional ATP expense? Please speculate about the evolutionary trade-offs for life in acidic versus neutral habitats.

REFERENCES

Amend, J.P. and A. Teske. 2005. Expanding frontiers in deep subsurface microbiology. *Paleogeog. Paleo-climat. Paleoecol.* **219**:131–155.

Atlas, R.M. and R. Bartha. 1998. *Microbial Ecology: Fundaments and applications.* Benjamin/Cummings, Menlo Park, CA.

Battista, J.R. 2000. Radiation resistance: the fragments that remain. *Curr. Biol.* **10**:R204–R205.

Blochl, E., R. Rachel, S. Burggraf, D. Hafenbradl, H.W. Jannasch, and K.O. Stetter. 1997. *Pyrolobus fumarii*, gen. and sp. nov., represents a novel group of Archea, extending the upper temperature limit for life to 113°C. *Extremophiles* **1**:14–21.

Boetius, A. 2005. Lost city life. *Science* **307**:1420–1422.

Brady, N.C. and R.R. Weil. 1999. *The Nature and Properties of Soils.* Prentice Hall, Upper Saddle River, NJ.

Bryer, D.E., F.R. Rainey, and J. Wiegel. 2000. Novel strains of *Moorella thermoacetica* from unusually heat resistant spores. *Arch. Microbiol.* **174**:334–339.

Carpenter, E.J., S.J. Lin, and D.G. Capone. 2000. Bacterial activity in South Pole snow. *Appl. Environ. Microbiol.* **66**:4514–4517.

Cavicchioli, R. 2002. Extremophiles and the search for extraterrestrial life. *Astrobiology* **2**:281–292.

Cavicchioli, R. and T. Thomas. 2000. Extremophiles, *In*: J. Lederberg, M. Alexander, B.R. Bloom, et al. (eds) *Encyclopedia of Microbiology*, 2nd edn, pp. 317–337. Academic Press, San Diego, CA.

Chapelle, F.H. 2003. *Groundwater Microbiology and Geochemistry*, 2nd edn. Wiley and Sons, New York.

Davis, S.N. and R.J.M. DeWiest. 1966. *Hydrology.* Wiley and Sons, New York.

DiRuggiero, J., N. Santangelo, Z. Nackerdien, J. Ravel, and F.T. Robb. 1997. Repair of extensive ionizing-radiation DNA damage at 95°C in the hyperthermophilic archaon *Pyrococcus furiosus*. *J. Bacteriol.* **179**:4643–4645.

Dixon, J.B. and S.B. Weed (eds) 1989. *Minerals in Soil Environments*, 2nd edn. Soil Science Society of America, Madison, WI.

Domenico, P.A. and F.W. Schwarz. 1990. *Physical and Chemical Hydrology*. Wiley and Sons, New York.

Edwards, K.J., P.L. Bond, T.M. Gihring, and J.F. Banfield. 2000. An archael iron-oxidizing extreme acidophile important in acid mine drainage. *Science* **287**:1796–1799.

Ferreira, A.C., M. Fernanda-Nobre, E. Moore, F.A. Rainey, J.R. Battista, and M.S. Da Costa. 1999. Characterization and radiation resistance of new isolates of *Rubrobacter radiotolerans* and *Rubrobacter xylamophilus*. *Extremophiles* **3**:235–238.

Foster, R.C. 1993. The ultramicromorphology of soil biota *in situ* in natural soils: a review. *In*: A.J. Ringrose-Voase and G.S. Humphreys (eds) *Soil Micromorphology: Studies in management and genesis*, pp. 381–393. Elsevier Science, New York.

Freeze, R.A. and J.A. Cherry. 1979. *Groundwater.* Prentice Hall, Englewood Cliffs, NJ.

Gardiner, D.T. and R.W. Miller. 2004. *Soils in our Environment*, 10th edn. Pearson/Prentice Hall Englewood Cliffs, NJ.

Grant, W.D., R.T. Gemmell, and T.J. McGenity. 1998. Halophiles. *In*: K. Horikoshi and W.C. Grant (eds) *Extremophiles: Microbial life in extreme environments*, pp. 99–132. Wiley-Liss, New York.

Gregory, P.J., D.J. Hutchinson, D.B. Read, P.M. Jenneson, W.B. Gilboy, and E.J. Morton. 2003. Non-invasive imaging of roots with high resolution X-ray microtomographhy. *Plant Sci.* **255**:351–359.

Horikoshi, K. 1998. Barophiles – deep sea micro-organisms adapted to an extreme environment. *Curr. Opin. Microbiol.* **1**:291–295.

Isken, S. and J.A.M. de Bont. 1996. Active efflux of toluene in a solvent resistant bacterium. *J. Bacteriol.* **178**:6056–6058.

Johnson, B.D. 1998. Biodiversity and ecology of acidophilic microorganisms. *FEMS Microbiol. Ecol.* **27**: 307–317.

Jones, B.E., W.D. Grant, A.W. Duckworth, and G.G. Owenson. 1998. Microbial diversity of soda lakes. *Extremophiles* **2**:191–200.

Jones, L.E. 2002. *Myth and Middle-Earth.* Cold Spring Press, Cold Spring Harbor, NY.

Kashefi, K. and D.R. Lovley. 2003. Extending the upper temperature limit for life. *Science* **301**: 934.

Kelley, D.S., J.A. Karson, G.L. Fruh-Green, et al. 2005. A serpentinite-hosted ecosystem: the Lost City hydrothermal field. *Science* **302**:1428–1434.

Ladd, J.N., R.C. Foster, P. Nannipieri, and J.M. Oades. 1996. Soil structure and biological activity. *In*: G. Stotzky and J.-M. Bollag (eds) *Soil Biochemistry*, Vol. 9, pp. 23–78. Marcel Dekker, New York.

Lengeler, J.W., G. Drews, and H.G. Schlegel (eds) 1999. *Biology of Prokaryotes*. Blackwell Science, Stuttgart.

Lindsay, W.L. 1979. *Chemical Equilibria in Soils*. Wiley and Sons, New York.

Lisitzin, A.P. 1996. *Oceanic Sedimentation*. American Geophysical Union, Washington, DC.

Madigan, M.T. and J.M. Martinko. 2006. *Brock Biology of Microorganisms*, 11th edn. Prentice Hall, Englewood Cliffs, NJ.

Madsen, E.L. 1995. Impacts of agricultural practices on subsurface microbial ecology. *Adv. Agron.* **54**: 1–67.

Madsen, E.L. 1996. A critical review of methods for determining the composition and biogeochemical activities of soil microbial communities in situ. *In*: G. Stotzky and J.-M. Bollag (eds) *Soil Biochemistry*, Vol. 9, pp. 287–370. Marcel Dekker, New York.

Madsen, E.L. and W.C. Ghiorse. 1993. Ground water microbiology: subsurface ecosystem processes. *In*: T. Ford (ed.) *Aquatic Microbiology: An ecological approach*, pp. 167–213. Blackwell Scientific Publications, Cambridge, MA.

McBride, M.B. 1994. *Environmental Chemistry of Soils*. Oxford University Press, New York.

Miller, G.T., Jr. 2004. *Living in the Environment*, 13th edn. Thomson Brooks/Cole, Pacific Grove, CA.

Morel, F. and J. Hering. 1993. *Principles and Applications of Aqueous Geochemistry*. Wiley and Sons, New York.

Nunan, N., K. Wu, I.M. Young, J.W. Crawford, and K. Ritz. 2002. *In situ* spatial patterns of soil bacterial populations, mapped at multiple scales, in an arable soil. *Microbial Ecol.* **44**:296–305.

Nunan, N., K. Wu, I.M. Young, J.W. Crawford, and K. Ritz. 2003. Spatial distribution of bacterial communities and their relationships with the microscale. *FEMS Microbiol. Ecol.* **44**:203–215.

Price, P.B. and T. Sowers. 2004. Temperature dependence of metabolic rates for microbial growth, maintenance, and survival. *Proc. Natl. Acad. Sci. USA* **101**:4631–4636.

Ringrose-Voase, A.J. and G.S. Humphreys (eds) 1994. *Soil Micromorphology: Studies in management and genesis*. Elsevier Science, New York.

Schaechter, M., J.L. Ingraham, and F.C. Niedhardt. 2006. *Microbe*. American Society for Microbiology Press, Washington, DC.

Schleper, C., G. Puehler, I. Holz, et al. 1995. *Picrophilus* gen nov, fam nov – a novel aerobic, heterotrophic, thermoacidophilic genus and family comprising Archaea capable of growth around pH 0. *J. Bacteriol.* **177**:7050–7059.

Schlesinger, W.H. 1997. *Biogeochemistry: An analysis of global change*, 2nd edn. Academic Press, New York.

Schut, F., R.A. Prins, and J.C. Gottschal. 1997. Oligotrophy and pelagic marine bacteria: facts and fiction. *Aquat. Microbiol. Ecol.* **12**:177–202.

Schwarzenbach, R.P., P. Gschwend, and D.M. Imboden. 2002. *Environmental Organic Chemistry*, 2nd edn. Wiley and Sons, New York.

Seyfried, W.E. and M.J. Mottl. 1995. Geologic setting and chemistry of deep sea hydrothermal vents. *In*: D.M. Karl (ed.) *The Microbiology of Deep-sea Hydrothermal Vents*, pp. 1–34. CRC Press, Boca Raton, FL.

Smith, R.L. 1990. *Ecology and Field Biology*, 4th edn. Harper and Row, New York.

Sposito, G. 1989. *The Chemistry of Soils*. Oxford University Press, New York.

Stewart, J.A. 1960. *The Myths of Plato*, 2nd edn. Southern Illinois University Press, Carbondale, IL.

Stumm, J.J. and W. Morgan. 1996. *Aquatic Chemistry*, 3rd edn. Wiley and Sons, New York.

Thieme, J., G. Schneider, and C. Knochel. 2003. X-ray tomography of a microhabitat of bacteria and other soil colloids with sub-100 nm resolution. *Micron* **34**:339–344.

Tisdall, J.M. and J.M. Oades. 1982. Organic matter and water-stable aggregates in soil. *J. Soil Sci.* **33**:141–163.

van de Vossenberg, J.L.C.M., A.J.M. Driessen, W. Zillig, and W.N. Konings. 1998. Bioenergetics and cytoplasmic membrane stability of the extremely acidophilic, thermophilic archaeon *Picrophilus oshimae*. *Extremophiles* **2**:67–73.

van der Wielen, P.W.J.J., H. Bolhuis, S. Borin, et al. and the BioDeep Scientific Party. 2005. The enigma of prokaryotic life in deep hypersaline anoxic basins. *Science* **307**:121–123.

van Loosdrecht, M.C.M., J. Lyklema, W. Norde, and A.J.B. Zehnder. 1990. Influence of interfaces on microbial acivity. *Microbiol. Rev.* **54**:75–87.

Wetzel, R.G. 2001. *Limnology: Lake and river ecosystems,* 3rd edn. Academic Press, San Diego, CA.

Whitman, W.B., D.C. Coleman, and W.J. Wiebe. 1998. Prokaryotes: the unseen majority. *Proc. Natl. Acad. Sci. USA* **95**:6578–6583.

Young, I.M. and J.W. Crawford. 2004. Interactions and self-organization in the soil–microbe complex. *Science* **304**:1634–1637.

5

Microbial Diversity: Who is Here and How do we Know?

This book is devoted to microorganisms and ways to understand their origins, evolution, ecology, and roles in governing the biogeochemical status of Earth. Prior chapters have focused on planetary history and conditions (Chapters 1–4), and on bold themes in prokaryotic and eukaryotic biology (e.g., Section 2.12). This chapter is designed to deliver to the reader fundamental facts and principles about the diversity of microorganisms that dwell in the biosphere. We begin with operational definitions of cultured versus uncultured microorganisms. To take a census, we need to be able to sample, recognize, and classify different microorganisms. Recognition can only occur when we know what microorganisms are and how to distinguish one from another. We must ask and answer two key questions: "What is a microbial species?" and "How does the species concept apply if the only known microbial trait is a 16S rRNA gene sequence from the environment?" Molecular phylogeny is discussed and the small subunit rRNA-based tree of life is used to portray biotic diversity. An overview is presented of the major traits and diversity of microbial life: eukaryotic (algae, fungi, protozoa, and slime molds), prokaryotic (Bacteria and Archaea), and viral.

Chapter 5 Outline

5.1 Defining cultured and uncultured microorganisms
5.2 Approaching a census: an introduction to the environmental microbiological "toolbox"
5.3 Criteria for census taking: recognition of distinctive microorganisms (species)
5.4 Proceeding toward census taking and measures of microbial diversity
5.5 The tree of life: our view of evolution's blueprint for biological diversity
5.6 A sampling of key traits of cultured microorganisms from the domains *Eukarya*, *Bacteria*, and *Archaea*
5.7 Placing the "uncultured majority" on the tree of life: what have nonculture-based investigations revealed?
5.8 Viruses: an overview of biology, ecology, and diversity
5.9 Microbial diversity illustrated by genomics, horizontal gene transfer, and cell size

5.1 DEFINING CULTURED AND UNCULTURED MICROORGANISMS

Microscopy provides a foundation for establishing, testing, and honing our understanding of microorganisms in waters, sediments, soils, and other habitats. The first microscopes used by R. Hooke and A. van Leeuwenhoek (see Section 1.3) relied on reflected lamp light for illumination and on both the structure and motility of the microbes to verify their vitality. During more than three centuries that have elapsed since the pioneering observations of "wee animalcules", the technology has advanced in resolution (e.g., electron microscopy, atomic force microscopy) and in ways to observe and probe the mechanisms of cellular processes (Bonnell, 2001; Murphy, 2001; Braga and Ricci, 2004; Hibbs, 2004; Darby and Hewitson, 2006; Taatjes and Mossman, 2006).

One insightful and widely used approach for visualizing microorganisms in environmental samples is epifluorescent microscopy (Figure 5.1). Figure 5.1a shows an image of soil solids after dilution and spreading onto a glass surface. This type of imaging allows individual particles to be distinguished from one another. But soils are composed of inorganic components and organic materials (both alive and dead) of widely ranging sizes (see Chapter 4). The vast majority of prokaryotes in soil are not morphologically distinctive – they resemble 1 μm-sized "specks" – as do many inorganic particles and bits of detritus. Fortunately, the unique traits of life (especially the biomarker, DNA) provide an often facile way to distinguish prokaryotes from inert soil particles. Double-stranded DNA has a high affinity for a variety of fluorescent compounds such as acridine orange (AO) and DAPI (4′,6-diaminidino-2-phenylindole). When environmental samples are properly stained with AO or DAPI, individual microorganisms can be imaged and enumerated (Figure 5.1b). Adding 1 g of soil to 99 ml of sterile (cell-free) phosphate buffer disperses the soil particles and the majority of cells become suspended in this 100-fold dilution (Figure 5.2). A 1000-fold dilution of the soil microorganisms can be prepared by transferring 10 ml of the first dilution to 90 ml of sterile buffer. Known (microliter) volumes of such dilutions can be examined using epifluorescent microscopy. Epifluorescent microscopy cell counts in such preparations, adjusted according to the degree of dilution, serve as the basis for microbial abundance data such as those reported in Chapter 4 (Section 4.5, Tables 4.6–4.10).

The soil dilutions described above for an epifluorescent microscopy assay can also be used in cultivation-based isolation and enumeration procedures (Figure 5.2). For these, 0.1 ml of a given soil dilution is spread onto the surface of a solid agar medium in a Petri dish. If 10 colonies grow from 0.1 ml of a 1000-fold dilution of 1 g of soil, then we infer a viable plate count of 10^5 microorganisms per gram of soil (each colony is presumed to be derived from a single cell; each of the 10 colonies

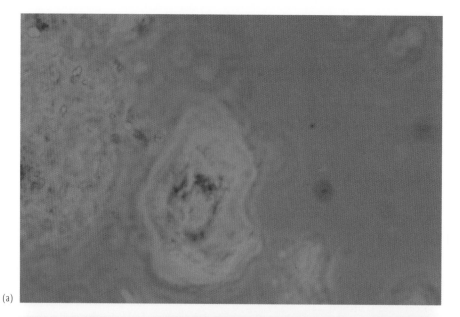

(a)

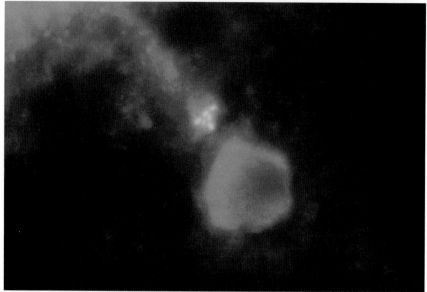

(b)

Figure 5.1 Microscopic images of a soil sample after dilution and smearing onto a glass slide. (a) Phase-contrast image showing soil solids. (b) Epifluorescence image of the same field after staining with a dye (acridine orange) that specifically binds to nucleic acids. Microorganisms are revealed as bright fluorescent cells. (From W.C. Ghiorse, Cornell University, with permission.)

represented are one ten-thousandth of the microorganisms in the original 1 g of soil). Thus, viable plate counts of microorganisms can be directly compared to epifluorescent microscopic counts of microorganisms from identical environmental samples. Inevitably when this comparison is made, the number of cultured cells on agar plates is 100–1000 times fewer than the total number of microscopic cells (Figure 5.2).

Figure 5.3 presents a scheme that operationally defines six categories of microorganisms in nature based on culturability. The small subset of the total microorganisms that form colonies on agar plates are, by definition, ones that grow. As we learned in Chapter 3, microbial growth requires proper resources, especially a carbon source, nutrients, electron donors, and electron acceptors. As we learned in Chapters 3 and 4, biosphere resources providing selective pressures for microbial growth over evolutionary time are diverse, complex, and poorly characterized. Furthermore, the resources may be metabolized via cooperation among microbial populations. Thus, the first dichotomous branch in the scheme of Figure 5.3 reflects how well microbiologists can devise growth media that match the needs of members of the microbial world.

1 *Cultured microorganisms* are those that have been successfully isolated and purified in the laboratory. These are represented in culture collections such as those in the United States and Germany, Belgium, the Netherlands, Japan, China, the United Kingdom, France, and Poland [e.g., American Type Culture Collection (ATCC) and Deutsche

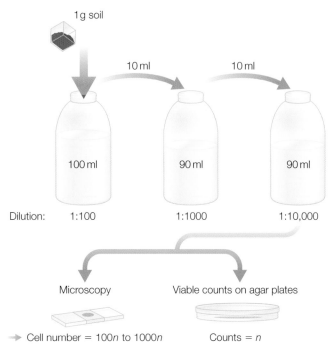

Figure 5.2 Preparation of soil dilutions in sterile buffer. The contrasts between the number of microscopic microorganisms and the number that grow on solid agar media has been termed the "great plate count anomaly".

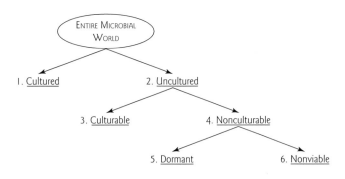

Figure 5.3 Six categories of culturability for microorganisms in nature. The categories are operationally defined – based on techniques of detection that include microscopy and both traditional and novel procedures for growth and isolation. See text for details.

Table 5.1

Tabulation of cultured prokaryotic taxa (from Garrity, G.M., T.G. Libum, and J.A. Ball. 2005. The revised roadmap to the manual. *In*: D.J. Brenner, N.R. Krieg, and J.T. Staley (eds) *Bergey's Manual of Systemic Bacteriology*, Vol. 2, Part A, 2nd edn, table 1, p. 163. Springer-Verlag, New York. With kind permission of Springer Science and Business Media.)

Taxonomic category	Domain		
	Bacteria	*Archaea*	**Total**
Phyla	24	3*	27
Class	32	9	41
Subclass	5	0	5
Order	75	13	88
Suborder	17	0	17
Family	217	23	240
Genera	1115	79	1194
Species	6185	281	6466

* Number reflects the addition of *Nanoarchaeota*.

Sammlung von Mikroorganismen und Zellkulturen (DSMZ)]. Information produced from these cultured, model organisms is the basis for ~99% of what is known about microbiology. The textbooks are written about these organisms and their structures, growth characteristics, physiology, pathogenicity, genetics, etc. The tally and diversity of cultured microorganisms is constantly growing because microbiologists are constantly devising new ways to meet the nutritional and physiological needs of microorganisms moved from their native habitats (e.g., soil, sediment, waters, gastrointestinal tracts of animals) to artificial laboratory media. As of 2006, the number of distinctive cultured prokaryotic microorganisms (e.g., species of *Bacteria* and *Archaea*) stood at 6466 – falling into 1194 genera and 240 taxonomic families (Table 5.1; Garrity et al., 2005).

2 *Uncultured microorganisms* are the remainder. No one knows the full extent of microbial diversity in the biosphere. Estimates (see Section 5.4) are that the number of uncultured species (the "uncultured microbial majority", Rappé and Giovannoni, 2003) may be 10^4 times that of the cultured minority. Uncultured microorganisms are the ones for which no appropriate growth medium has been devised. But conceptually, the uncultured category can be further dissected (see second dichotomous branch in Figure 5.3).

3 *Culturable microorganisms* are the ones that will become cultured when a clever microbiologist devises a growth medium that matches the organ-

ism's nutritional needs. Key physical and chemical growth conditions must also be provided.

4 *Nonculturable microorganisms* are ones present in soils, sediment, waters, or other habitats whose physiological state prevents them from being cultured. The third dichotomous branch of Figure 5.3 splits "nonculturables" into two categories: dormant and nonviable.

5 *Dormant cells* (previously defined in Sections 3.4 and 3.5) may have been quiescent for so long, with such a small proportion of the cellular components intact, that growth is not possible. A poorly understood resuscitation step may be required before dormant cells can be cultured (del MarLleo et al., 2005; Oliver, 2005; Vora et al., 2005).

6 *Nonviable cells* are ones that cannot be resuscitated. They may be visible in a microscope but are irreparably damaged or moribund. These cells are, effectively, dead.

5.2 APPROACHING A CENSUS: AN INTRODUCTION TO THE ENVIRONMENTAL MICROBIOLOGICAL "TOOLBOX"

- **How would you take a census of cookies in a bakery?**
- **How would you document the diversity of the plant species in a 100 m2 of forest?**

To achieve these two seemingly simple goals, you need at least three basic abilities: (i) a way to sample the cookies or plants; (ii) a way to recognize the cookies or plants – to differentiate them from other things; and (iii) a way to distinguish different classes of cookies or plants from their peers. Once the three abilities are established, you simply count the different classes of distinctive cookies or plants as they are being sampled. Here it is helpful to recognize and define several aspects of census taking:

- The sampling step is unlikely to be 100% effective; thus, we estimate the total based upon the sampled subset.
- One measure of diversity is the *richness*: the total number of different types of taxa (cookies, plants, or microbial species).
- Another measure of diversity is *evenness*: this encompasses both the variety of taxa and their relative abundances.

In Section 4.5 we learned that soil habitats possess $\sim 10^9$ microorganisms in a single gram. Furthermore (see Section 5.1), the vast majority of microorganisms have not yet been cultured.

- **Given this situation, is it reasonable to even attempt an assessment of microbial diversity?**
- **Do we microbiologists have anything resembling the three essential abilities (sample, recognize, classify) required to conduct a census of cookies and plants?**

Box 5.1 introduces the four types of fundamental tools (methodological approaches) available to environmental microbiologists: microscopy, cultivation, physiological incubations, and biomarkers. All four of the approaches can be brought to bear on the goal of taking a census of microorganisms present in a liter of lake water, in a gram of soil, and/or in the biosphere at large.

Box 5.1

Introducing the four fundamental methodological approaches in environmental microbiology

There are four fundamental methodological approaches that generate information addressing environmental microbiological questions and issues. These approaches are:

1 *Microscopy*: direct imaging at the microscale to verify the presence of microbial cells. Microscopy provides direct, irrefutable data on the total abundance of microorganisms in environmental samples. Recent technological advancements in the resolution and types of information gathered during imaging aim to identify individual cells and probe their biogeochemical activities.

2 *Cultivation*: environmental matrixes (e.g., soil, sediment, water) are diluted and suspended microorganisms are transferred to liquid or solid media where nutrients are provided – with hopes that the cells present will grow. Assays for assessing growth include colony formation (on agar plates), increased cell numbers in liquid media, and chemical endpoints indicative of a physiological process. By counting individual colonies growing from known dilutions of environmental samples, different types of microorganisms can be enumerated.

3 *Physiological incubations*: microbial populations occur as complex communities in environmental samples. Environmental samples can be brought into the laboratory, sealed in vessels, and subjected to assays that demonstrate the physiological potential of the microorganisms, such as production of CO_2 or CH_4, nitrogen fixation, sulfate reduction, or metabolism of pollutant compounds.

4 *Biomarker extraction and analysis*: prokaryote-specific molecular structures (e.g., nucleic acids, proteins, lipids) can be directly extracted from environmental samples. After analysis, these provide insightful clues about the presence and activity of microorganisms (see also Section 2.2).

No single approach leads to a thorough understanding or answer to a given question. Information from all four approaches can complement and confirm one another. When this confluence occurs, the discipline of environmental microbiology is advanced.

1 *Microscopy* (see discussion above and Box 5.1) allows images of microorganisms to be obtained directly from environmental samples. Such data are essential for enumeration of total cells in environmental samples. Technological innovations in microscopy, such as flow cytometry and other procedures (see Section 6.7 and Table 6.4), have the potential to gather additional information about the types of cells present (such as size and RNA content).

2 *Cultivation* (see discussion above and Box 5.1) allows microorganisms to be grown and isolated when their nutritional requirements are met. The growth response of quantitatively diluted microorganisms can reveal the abundances of particular microbiological types.

3 *Physiological incubations* (Box 5.1) document the presence and biogeochemical potential of microorganisms in vessels containing either diluted or undiluted environmental samples. Under controlled conditions, physical (e.g., growth displayed as turbidity) or chemical (e.g., consumption of a carbon source or an electron acceptor; production of CO_2, acidity, or metabolites) changes characteristic of physiological processes can be easily recognized. The incubations display endpoints for a given physiological process (e.g., methanogenesis or sulfate reduction). When environmental samples are quantitatively diluted and used in such incubations, the abundance of a given functional type of microorganism can be quantitatively assessed. (If the endpoint is reached in highly dilute preparations, then the number of active organisms initially present in the environmental sample was high.) All such physiological assays have the same caveat as cultivation-based approaches because if nutritional needs are not met, the physiological potential of microorganisms cannot be expressed.

4 *Biomarkers* have revolutionized our understanding of the microbial world during the last two decades. Biomarkers go hand in hand with microscopy, as the mainstay of *noncultivation-based procedures* for taking a microbiological census in natural habitats such as water, sediment, and soil. We saw in Chapter 2 (Section 2.1, Table 2.2, and Box 2.2) that biomarkers (especially stable isotopic ratios and molecular fossils) have proven essential for discovering clues about the ancient history of prokaryotic life. The biomarkers that have proven most insightful for contemporary microorganisms are nucleic acids – especially the sequences of taxonomically and evolutionarily insightful ribosomal RNA genes [i.e., small subunit 16S rRNA (in prokaryotes) and 18S rDNA (in eukaryotes); see Section 5.5]. Nucleic acid biomarkers have had a profound impact on environmental microbiology for three major reasons: (i) fine distinctions between nucleotide sequences can be made when many of these extracted biomarkers are compared; (ii) the sensitivity for detecting (cloning and sequencing) nucleic acid is very high due to the applicability of the polymerase chain reaction (PCR) in recovering sequences derived from environmental samples; and (iii) identical

molecular criteria for recognizing and comparing microorganisms apply to sequences derived from both cultivated and noncultivated sources.

5.3 CRITERIA FOR CENSUS TAKING: RECOGNITION OF DISTINCTIVE MICROORGANISMS (SPECIES)

Cultured microorganisms (defined in Section 5.1) are analogous to domesticated animals – these are the ones we can breed, manage, and study. It follows that uncultured microorganisms are analogous to wild animals – we get fleeting glimpses of them during our forays into their habitats. We know that the fundamentals of domesticated animal reproduction, speciation, and taxonomy apply to wild animals. So, too (to the degree possible), microbial ecologists apply the taxonomic rules of domesticated microorganisms to those that are not yet cultured.

The concept of "microbial species" becomes crucial in our efforts to take a census of microorganisms in nature. The criteria for species allows us to recognize and classify, hence document the presence and abundance of, different types of microorganisms in soils, sediments, waters, and other habitats. For cultured microorganisms, Rosselló-Mora and Amann (2001) have stated that species categories should be based on both phylogenetic and phenotypic traits. Each species is:

> a genomically coherent cluster of individual organisms that show a high degree of overall similarity with respect to many independent characteristics, and is diagnosable by a discriminative phenotypic property.

A perceptive reader will notice that the above definition is imbued with legalistic connotations. The wording of the definition must be defensive and precise because microbial (and other) taxonomists are imposing an artificial system of categories and nomenclature on the continuum of biological variations delivered by evolution. Debates on how best to define microbial species are ongoing (e.g., Ward, 1998; Rosselló-Mora and Amann, 2001; Cohan, 2002; Stackebrandt et al., 2002; Kassen and Rainey, 2004; Gevers et al., 2005). Nonetheless, there is widespread agreement on the genetic criteria that define a species of cultivated bacteria (Box 5.2). Genomic hybridization determines the degree to which DNA strands from two different bacteria hybridize to one another relative to themselves (Stackebrandt et al., 2002). If the percent hybridization (a direct measure of sequence similarity) is 70–100%, the bacteria are deemed to be the same species. If the degree of hybridization is 25–70%, they are designated the same genus. If the hybridization is less than 25%, the test bacteria are deemed unrelated genera.

Such DNA–DNA hybridization assays using pure-culture microorganisms are clearly impossible to carry out with uncultured microorganisms

Box 5.2

Criteria to establish if microorganisms belong to the same species

There are three main taxonomic and molecular criteria used for deciding that two cultivated microorganisms belong to the same species (Rosselló-Mora and Amann, 2001; Stackebrandt et al., 2002; Gevers et al., 2005):

1 *DNA–DNA hybridization*: (i) 70% or greater DNA–DNA relatedness in whole genome reannealing tests; and (ii) <5°C difference in the temperature of DNA helix dissociation (Δ*T*m) for the two strains.

2 *16S rRNA gene*: >97% identity in sequence. This criterion is not absolute, as many phenotypically distinctive species have been discovered to have >97% 16S rRNA gene sequence identity.

3 *Whole genome sequence comparison*: here the criteria are not yet clear. Many strains of the same species can differ substantially in genome size and content. Identifying the critical phenotypic and genotypic traits that define "species" is an ongoing challenge, largely because subjective judgments must be made by taxonomists. A multigene strategy, known as multilocus sequence analysis (MLSA) has immense promise for developing new definitive criteria.

– often characterized only by the sequence of 16S rRNA genes retrieved from a particular sample of soil, sediment, or water. When the 16S rRNA biomarker (sequence) is the primary measure of identity and taxonomy, the term OTU (for operational taxonomic unit) applies. The 16S rRNA gene-based OTUs can be grouped at a "species-like" level if their identities are at least 97% identical (Box 5.2). Comparisons of 16S rRNA sequences among uncultured microorganisms should be viewed as an insightful, yet temporary and approximate, way to estimate taxonomy and phylogeny. Whenever a microorganism becomes cultured (domesticated), the additional phenotypic and genotypic determinations are sure to refine prior "species" and "OTU" designations based solely on the 16S rRNA gene.

Of course, the ultimate means of genetic classification and comparison between microorganisms is the whole genome sequence (see Section 3.2 and Box 5.2). With whole genomes, everything is revealed. But if thousands of different genes (genetic loci) can be compared, distinctions must be made between those that are taxonomically and evolutionarily insightful and those that are not. In this regard, choosing the right genes to efficiently assist in assembling taxa ("coherent groups of individuals that comprise species", see above) is a challenge. Very likely, the future of microbial taxonomy and species designations will rely on sequence comparisons of many prudently selected genetic loci – a technique known as multilocus sequence analysis (MLSA; Coenye et al., 2005; Gevers et al., 2005).

Science and the citizen

When taxonomic and genetic diversity really matters: *Escherichia coli* **as a normal intestinal inhabitant and as a dangerous food-borne pathogen**

Headline news from the United States Centers for Disease Control and your local newspaper

In the fall of 2006, packages of fresh spinach, grown in California and distributed nationally, were contaminated with *E. coli* O157:H7. Approximately 200 people in 26 states (Figure 1) were infected. Symptoms included severe sudden abdominal cramps and bloody diarrhea lasting from 1 to 8 days.

In 1993, a similar outbreak of painful, bloody diarrhea led to kidney failure in a large group of children in Seattle, Washington. Some of the children died. The illnesses were also caused by *E. coli* O157:H7, contracted from contaminated hamburgers sold to the public by a single fast-food restaurant chain.

Outbreaks of *E. coli* O157:H7 infection occur regularly and have been both large and small in localized areas and across several states. Transmission of food-borne *E. coli* was first associated with contaminated ground beef but has also been spread through: unpasteurized milk and fruit juices; spinach, lettuce, sprouts, salami, and contaminated drinking water; swimming in or drinking sewage-contaminated water; contact with infected animals (such as in

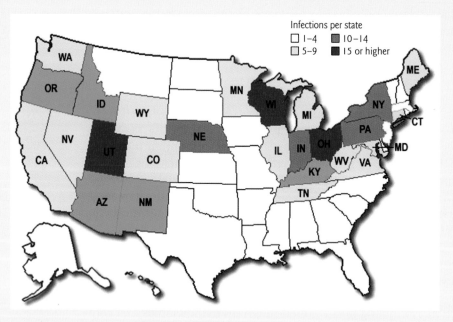

Figure 1 Occurrence of *E. coli* O157:H7 infections caused by spinach consumption in the United States, fall 2006. (From Centers for Disease Control, with permission, http://www.cdc.gov/foodborne/ecolispinach/100606.htm.)

petting zoos); and person to person, especially among children in day care centers. The way *E. coli* O157:H7 is transmitted changes over time, which is why the United States Centers for Disease Control work closely with state health departments to monitor and investigate cases and outbreaks of *E. coli* O157:H7.

SCIENCE: A harmless inhabitant of the human intestine evolves into a pathogen

Background
E. coli O157:H7 is one of hundreds of strains of the bacterium *E. coli*. Although most strains are harmless, strain O157:H7 produces powerful toxins that can cause severe illness. *E. coli* O157:H7 has been found in the intestines of healthy cattle, deer, goats, and sheep. *E. coli* O157:H7 was first recognized as a cause of human illness in 1982 during an outbreak of severe bloody diarrhea; the outbreak was traced to contaminated hamburgers. Since then, more infections in the United States have been caused by eating undercooked ground beef than by any other food.

How did E. coli *O157:H7 develop?*
Strain O157:H7's entire genome has been sequenced – as has the genome of the benign inhabitant of the human intestine *E. coli* strain K-12 (see Section 3.2). The pathogen's genome is encoded in 5.62 Mb of DNA, while strain K-12's genome is 4.74 Mb; 4.1 Mb is shared in common (Ohnishi et al., 2002). Much of the 1.5 Mb of DNA characteristic of strain O157:H7 carries virulence genes that encode factors for attachment to host cells, production of molecules that interfere with host signaling pathways, and production of two different types of toxins (Shiga toxins) that disrupt protein synthesis in the host. The bacterium also contains a large virulence-associated plasmid. Approximately two-thirds of strain O157:H7's virulence genes are associated with virus-like genetic elements (Ohnishi et al., 2002). This strongly suggests that horizontal gene transfer is the mechanism that gradually converted the original benign *E. coli* into a powerful pathogen (see Sections 8.5, 8.7, and 9.2 for additional discussion of mechanisms of evolution).

Evolutionary analysis has led to a model of strain O157:H7's development. Figure 2 shows a scheme of the likely final stages in a series of genetic acquisitions and deletions that led to the contemporary pathogenic *E. coli* strain O157:H7 Sakai.

A major lesson here is that microbial diversity is not necessarily reflected in the naming (the taxonomy) of a bacterium. Both strains K-12 and O157:H7 are E. coli. *But hidden behind their common species designation are big differences, sometimes deadly ones.*

Research essay assignment
The genetic basis of some diseases in some pathogens (of plants, animals, humans) has been traced to "pathogenicity islands". These are large, unstable regions of a microorganism's chromosome that have been acquired through horizontal gene transfer (see Section 5.9). Encoded in the genes are biosynthetic pathways that lead to virulence factors that include cell surface structures, toxins, or other traits. Based on a survey of the scientific literature, write an essay defining pathogenicity islands and the role they play in at least one type of disease.

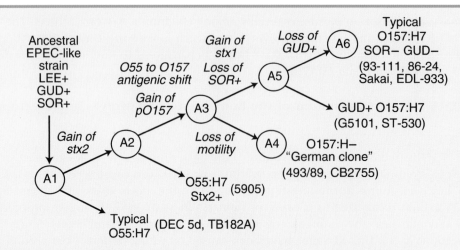

Figure 2 Stepwise genomic changes leading to the emergence of *E. coli* O157:H7. The model proposes six stages (A1–A6) for the evolution of *E. coli* O157:H7 from an enteropathogenic *E. coli*-like ancestor. "Gain" and "loss" refer to the acquisition and ejection (respectively) of DNA encoding clusters of genes encoding particular traits. EPEC, enteropathogenic *E. coli*; GUD, β-glucuronidase utilization; LEE, locus of enterocyte effacement; pO157, large virulence plasmid; SOR, sorbitol utilization; stx2, Shiga toxin. (From Wick, L.M., W. Qi, D.W. Lacher, and T.S. Whittam. 2005. Evolution of genomic content in the stepwise emergence of *Escherichia coli* O157:H7. *J. Bacteriol.* **187**:1783–1791 and Feng, P., K.A. Lampel, H. Karch, and T.S. Whittam. 1998. Genotypic and phenotypic characterization of *Escherichia coli* O157:H7. *J. Infectious Dis.* **177**:1750–1753. American Society for Microbiology and University of Chicago Press, with permission.)

5.4 PROCEEDING TOWARD CENSUS TAKING AND MEASURES OF MICROBIAL DIVERSITY

Now we return to the task begun in Section 5.2.

- **How do we take a census that documents the number and diversity of microorganisms in a natural habitat?"**

While all four broad methodologies described in Box 5.1 can contribute information to a census, it is nucleic acid biomarkers, especially DNA sequences encoding the small subunit rRNA gene, that are widely recognized as the procedure of choice for conducting a census of microorganisms that occur in nature. 16S rRNA genes can be *sampled*, and they afford both *recognition* and *classification* (see Sections 5.2 and 5.3) of the prokaryotic hosts of the genes. Recovering 16S RNA gene sequences from soil or sediments is like recovering license plates from an automobile wrecking yard – even if the automobile is lost in the heap, the automobile may be accurately accounted for by its license plate. Several insightful reviews on microbial diversity and census have recently appeared (Hughes et al., 2001; Ward, 2002; Bohannan and Hughes, 2003; Schloss and Handelsman, 2004; Curtis and Sloan, 2005). These are the basis for the following discussion and the summary on diversity measures shown in Box 5.3.

Rarefaction curves

A rarefaction curve approach to microbial diversity (a common one in plant and animal ecology) measures the accumulation of distinctive taxa as a function of sampling effort. This approach presumes that, as sampling proceeds, members of the rarest species will gradually be added to the total (Box 5.3). Thus, the asymptote in the rarefaction accumulation curve estimates total diversity in the habitat sampled. When Tringe et al. (2005) sampled 16S rRNA genes from soil, the pool of 1700 clones revealed 847 sequence types from more than a dozen far-ranging taxonomic phyla. Moreover, the rarefaction curve built from the data failed to indicate an asymptote. Thus, Tringe et al.'s (2005) data set (like the rarefaction data in Box 5.3) did not approach a complete count of sequence types. This means that microbial diversity was very high and that it was incompletely sampled.

Schloss and Handelsman (2004) applied rarefaction analysis to 78,166 16S rRNA gene sequences in the Ribosomal RNA Project II's database. The goal was to use asymptotes of rarefaction curves as quantitative estimates of microbial diversity. Very steep slopes of rarefaction curves reflect both incomplete sampling and high diversity in microbial populations. Based on the mild upward slopes of the rarefaction curves, the authors concluded that "either current sampling methods (that produced the database) are not adequate to identify 10^7 to 10^9 different species . . . or these estimates (of diversity) are high".

DNA hybridization

DNA–DNA reassociation rates are another way to estimate community diversity. DNA can be extracted from environmental samples and after the strands of the double helix are denatured at high temperature, their rate of reassociation (or re-annealing) is determined by the size and complexity of the DNA. Torsvick et al. (1990) reasoned that pooled genomic DNA from soil might abide by the same rehybridization rules as large genomic DNA from purified cultures (Box 5.3). The data from Torsvick et al. (1990) suggested that soil DNA reassembled so slowly that it was 7000 times as complex as the genome of a single bacterium. Subsequent estimates of species diversity in soil by Torsvick et al. (2002) suggest there are 40,000 species per gram.

Diversity estimation theory was recently taken even further by Gans et al. (2005) who reinterpreted DNA reassociation curves from soil microbial communities. The authors recast equations describing DNA

Box 5.3

Estimating microbial diversity: let me count the ways

Diversity estimation approaches	Example	Comments	References
Rarefaction curves*	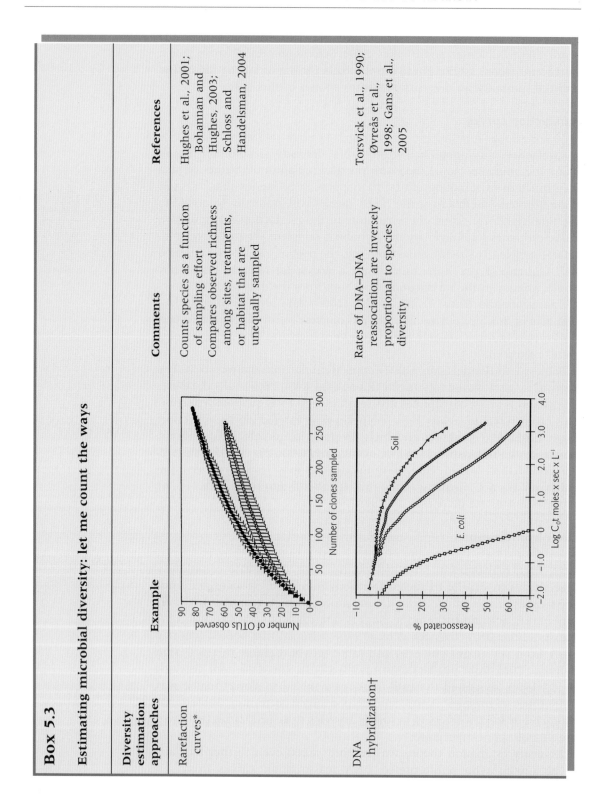	Counts species as a function of sampling effort. Compares observed richness among sites, treatments, or habitat that are unequally sampled	Hughes et al., 2001; Bohannan and Hughes, 2003; Schloss and Handelsman, 2004
DNA hybridization†		Rates of DNA–DNA reassociation are inversely proportional to species diversity	Torsvick et al., 1990; Øvreås et al., 1998; Gans et al., 2005

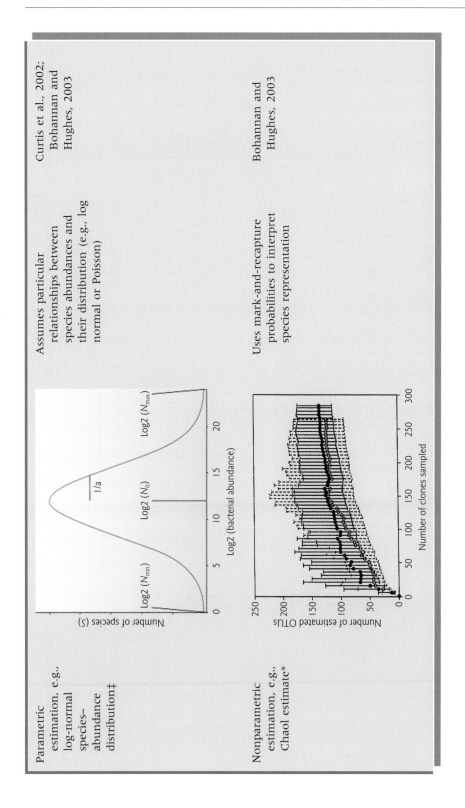

Parametric estimation, e.g., log-normal species–abundance distribution‡

Assumes particular relationships between species abundances and their distribution (e.g., log normal or Poisson)

Curtis et al., 2002; Bohannan and Hughes, 2003

Nonparametric estimation, e.g., Chaol estimate*

Uses mark-and-recapture probabilities to interpret species representation

Bohannan and Hughes, 2003

Box 5.3 Continued

Diversity estimation approaches	Example	Comments	References
Phylogenetic trees§		The diversity of sequences graphically depicted in phylogenetic trees is reflected in the number of taxa (leaves of a tree), branch lengths, and branching patterns	Bohannan and Hughes, 2003; Robertson et al., 2005
Ecological diversity indices, e.g., Shannon–Weiner Index	$H = -\sum\limits_{i=1}^{S} (Pi)(\log_2)(Pi)$ Assumes random sampling from large community in which total species number is known	H = index of species diversity S = number of species Pi = proportion of total sample belonging to the ith species	Atlas, 1984; Krebs, 2001

* Figure from Hughes, J.B., J.J. Hellmann, T.H. Ricketts, and B.J.M. Bohannan. 2001. Counting the uncountable: statistical approaches to estimating microbial diversity. *Appl. Environ. Microbiol.* **67**:4399–4406. With permission from the American Society for Microbiology.

† Figure from Øvreås, L., S. Jensen. F.L. Daae, and V. Torsvik. 1998. Microbial changes in a perturbed agricultural soil investigated by molecular and physiological approaches. *Appl. Environ. Microbiol.* **64**:2739–2742. With permission from the American Society for Microbiology.

‡ Figure reprinted from Curr. Opin. Microbiol. Vol. 6. Bohannan, B.J.M. and J. Hughes. 2003. New approaches to analyzing microbial biodiversity data. *Curr. Opin. Microbiol.* **6**:282–287. Copyright 2003, with permission from Elsevier.

§ Figure from Morrell, V. 1997. Microbial biology: microbiology's scarred revolutionary. *Science* **276**:699–702. Reprinted with permission from AAAS.

annealing rates to allow quantitative comparison of different ecological species–abundance models. The analysis showed that a power law best described the abundance distribution of prokaryotes. The authors concluded that more than one million genomes occurred in pristine soil – exceeding previous estimates by two orders of magnitude.

Parametric estimation

Parametric diversity estimation procedures were used by Curtis et al. (2002), who drew upon ecological theory to devise a way to calculate total diversity in a microbial community. Standard measures of biotic diversity in plant and animal ecology rely upon two fundamental pieces of information: the number of species and the number of individuals in each species. Such data are often presented as "species–abundance curves" in which the number of species is plotted versus the number of individuals per species. Curtis et al. (2002) postulated that, by assuming a particular type of statistical distribution (log normal), species–abundance curves apply to microorganisms in nature (Box 5.3). The investigators were able to relate the total diversity of prokaryotic communities to the ratio of two potentially measurable variables: the total number of individuals and the abundance of the most abundant species. Using this approach, Curtis et al. (2002) reported the number of species in ocean water was 160 per milliliter; in soil the species abundance was 6400–38,000 per gram.

Nonparametric estimation

A nonparametric estimation of the microbial diversity, the Chao1 statistic, requires no assumptions about species distribution models (Box 5.3). Instead, mark–release–recapture statistics are used. These were developed for estimating the size of animal populations. The approach documents the proportion of species (or OTUs) that have been observed before (recaptured), relative to those that have been observed only once. Samples from diverse communities are predicted to contain few recaptures.

Chao1 estimates total species richness as:

$$S_{\text{Chao1}} = S_{\text{obs}} + n_1^2/2n_2$$

where S_{obs} is the number of observed species, n_1 is the number of singletons (species captured once), and n_2 is the number of doubletons (species captured twice).

The Chao1 statistic estimates both the total diversity of a given habitat and the precision of the estimates (Bohannan and Hughes, 2003), thus facilitating comparison of diversities between habitats. The index is well suited for data sets skewed toward low-abundance species, as is likely for microbial communities (Hughes et al., 2001).

Phylogenetic trees

Phylogenetic trees are routinely prepared as a graphic way to display contrasts between sequences of nucleic acid biomarkers (especially the genes encoding small subunit rRNA; see Box 5.3). After alignment of the sequences, a variety of computer-based software algorithms (see Sections 5.5 and 5.7) can precisely assess the degree of relatedness between sequences. The results are dendrograms (or phylogenetic trees) in which closely related sequences appear as neighboring "leaves" or "clusters of twigs", while unrelated sequence are placed on far-removed branches. The linear distances between any two positions on the tree are proportional to their degree of sequence dissimilarity. The branching patterns of a given phylogenetic tree display distinctive identities of the taxa. Branching patterns can be qualitatively and quantitatively compared between trees (Martin, 2002; Bohannan and Hughes, 2003; Robertson *et al.*, 2005).

Ecological diversity indices

Ecological diversity indices have been devised by ecologists studying plants and animals, constituting a wide variety of species–diversity measures (Atlas, 1984; Atlas and Bartha, 1998; Krebs, 2001). These include richness, dominance (Simpson Index), and equitability (Shannon–Wiener Index) indices (see Box 5.3). Though it is tempting to apply these standard ecological indices to microbial communities, many of the calculated parameters make presumptions (e.g., random sampling, complete knowledge of total species, and particular statistical species–abundance distributions) that are unlikely to apply to naturally occurring microbial communities.

Comparative approach

Hong et al. (2006) recently adopted an integrated approach to compare the effectiveness of four parametric and two nonparametric procedures for estimating species richness using a single 16S rRNA gene library from marine sediment. Their data consisted of >500 clones and >400 OTUs. Hong et al. (2006) argued that the validity of all prior estimation approaches was questionable. The investigators applied consistent, rigorous criteria (such as statistical "goodness-of-fit" and minimized standard error) to all six statistical models. The study selected the "Pareto" parametric model as the most powerful means to evaluate their data set; the model estimated the total diversity the marine sediment to be 2434 ± 542 bacterial species. Census taking will be discussed further in Section 6.7.

5.5 THE TREE OF LIFE: OUR VIEW OF EVOLUTION'S BLUEPRINT FOR BIOLOGICAL DIVERSITY

Human knowledge of the diversity of life forms has undergone at least four major revolutions. In 1866, Haeckel presented a three kingdom view of plants, protists (single-celled eukaryotes), and animals. In 1937, Chatton recognized that fundamental cellular architecture justified only two main kingdoms: prokaryotes and eukaryotes. In the late 1950s, Whittaker again used phenotypic traits to develop the five kingdom model: bacteria, fungi, protists, plants, and animals. In the 1970s, C. R. Woese (Figure 5.4) and colleagues devised a far more direct and quantitative way to discover relationships between forms of life. These researchers realized that genotypic information (e.g., the sequencing of proteins and nucleic acids) is far superior to phenotypic information for discovering evolutionary relationships. Phenotypic comparisons are made based on subjective judgments; while sequences of nucleotides or amino acids can be translated into objective, precise, mathematically defined measures of true phylogenetic relatedness (Woese, 1987).

Woese and colleagues sought a *molecular chronometer* that would reveal evolutionary relationships between life forms. Molecular chronometers exhibit sequence changes, caused by random mutations, that tick away

Figure 5.4 C. R. Woese, molecular microbiologist at the University of Illinois, whose phylogenetic analyses led to the three domain tree of life. (From University of Illinois at Urbana-Champaign, with permission.)

(ideally at a constant rate) like the second hand of a clock. The amount of sequence change in the molecule carried by two ancestrally related lineages (species) is termed *sequence divergence*. The degree of divergence between the two lineages reflects both the rate that mutations are fixed in the molecule and the time over which the mutational changes have occurred. To be useful in molecular phylogeny, the chronometer of choice must: (i) have clock-like random mutations; (ii) change at rates commensurate with the evolutionary distance of interest; and (iii) be rich enough in information to allow fine distinctions to be made, while maintaining stability in overall structure. Woese and colleagues realized that rRNAs (especially small subunit rRNAs) meet the criteria for an ideal molecular chronometer: (i) they occur in all cellular organisms that carry out protein biosynthesis; (ii) they feature a high degree of functional constancy; (iii) despite functional constraints, different positions in the three-dimensional structure of rRNAs change at different rates – allowing both distant and close phylogenetic relationships to be charted; (iv) their size offers at least 50 helical stalks that constitute a rich source for comparative analysis; and (v) the genes that encode rRNAs can be sequenced relatively easily (Woese, 1987).

Figure 5.5 shows secondary structures of small subunit rRNA molecules representative of life's three domains. Each of the domains (*Bacteria* and *Archaea* carry 16S rRNA; *Eukarya* carry 18S rRNA) exhibits common resemblances and characteristic differences. To illustrate sequence- and structural-level differences that afford relatively fine distinctions to be

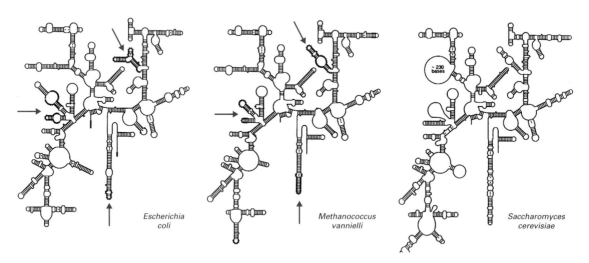

Escherichia coli

Methanococcus vannielli

Saccharomyces cerevisiae

Figure 5.5 Representative small subunit rRNA secondary structures for the three primary domains: from left to right, *Bacteria*, *Archaea*, and *Eukarya*. Arrows pointing within the *Escherichia coli* and *Methanococcus vannielli* structures identify three key locales where *Bacteria* and *Archaea* characteristically differ. (From Woese, C.R. 1987. Bacterial evolution. *Microbiol. Rev.* **51**:221–271. With permission from the American Society for Microbiology.)

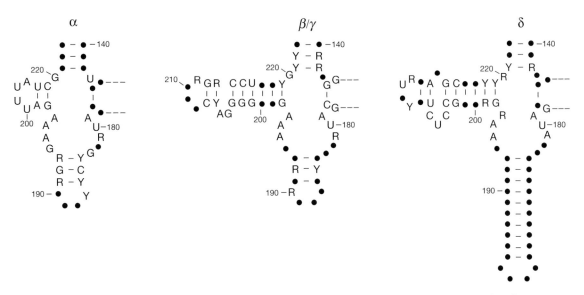

Figure 5.6 Comparison of the secondary structures of two helical domains of rRNA for three subdivisions of the Proteobacteria – dots are shown in locations of nucleotides that are not highly conserved. (From Woese, C.R. 1987. Bacterial evolution. *Microbiol. Rev.* **51**:221–271. With permission from the American Society for Microbiology.)

drawn, consider Figure 5.6, which focuses upon two helical domains in the bacterial 16S rRNA molecule that help define α, δ, and β/γ subdivisions of the Proteobacteria (this taxonomic class will be discussed in Section 5.6). A major distinction between the δ-Proteobacteria (Figure 5.6, right) and the other three subdivisions is a 10 base pairs (bp) extension in the stalk of the downward-pointing helix. Another clear distinction in the structures is in the left-pointing helix. The stalk of this helix consists of 2 bp in the α-Proteobacteria and 8 bp in the β/γ- and δ-Proteobacteria.

Using structural molecular contrasts described above and computer-based algorithms that compare the sequence of genes encoding small subunit rRNA molecules, Woese and colleagues created a Tree of Life that displays true, evolution-based phylogenetic relationships among organisms (Figure 5.7). The power of molecular phylogenetic techniques cannot be overstated. Life consists of three major domains (*Bacteria, Archaea,* and *Eukarya*) and many of the lineages at the base of the tree are adapted to high temperatures (hyperthermophilic), suggesting that the last universal common ancestor (see Section 2.7) emerged from a hot primeval habitat.

Implicitly, the three-domain concept supersedes and negates prior historical paradigms in biology (three kingdoms, prokaryote/eukaryote, and five kingdoms; mentioned above). *Archaea* are distinctive from *Bacteria* (they are like night and day). Grouping these two domains within a single term, "prokaryote" is considered by many microbiologists to be unwise and erroneous. Despite this fact, "prokaryote" has a utilitarian aspect

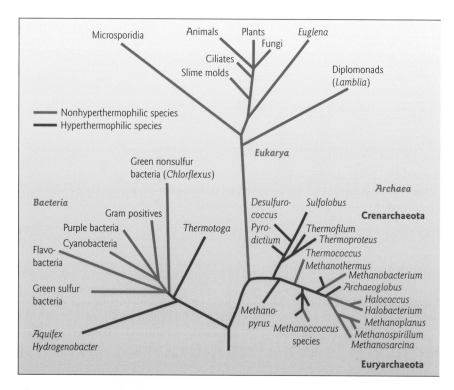

Figure 5.7 Tree of life based on small subunit rRNA sequence analysis. Phyla residing at the base of the tree (in purple) are thermophiles. Phyla evolving later (in green) are nonthermophilic. More detailed taxonomic subgroups (class, order, family, genera, and species) are not shown. (From Morrell, V. 1997. Microbial biology: microbiology's scarred revolutionary. *Science* **276**:699–702. Reprinted with permission from AAAS.)

that is likely to allow it to linger in the microbiological lexicon many years into the future.

5.6 A SAMPLING OF KEY TRAITS OF CULTURED MICROORGANISMS FROM THE DOMAINS *EUKARYA, BACTERIA,* AND *ARCHAEA*

Early evolutionary events on Earth were presented in Chapter 2, which summarized current ideas and facts regarding stages of development in early life: prebiotic Earth, the iron/sulfur world, the RNA world, the "last common universal ancestor", early division of *Bacteria* from *Archaea*, the rise of oxygen, endosymbiotic theory, and the emergence of *Eukarya*. The long series of evolutionary and biogeochemical successions has delivered to us the contemporary tree of life (Figure 5.7). Woesean analysis of small subunit rRNA molecules produced both the broad three-domain tree of life and the many major divisions (phyla) with finer subdivisions within each domain. The branches, twigs, stems, and leaves on the tree (not shown in Figure 5.7) roughly correspond to the progressively refined taxonomic categories of class, order, family, genera, and species (see Table 5.1 and Section 5.3). The version of the tree of life shown in Figure 5.7 portrays phylogenetic and taxonomic relationships between representative *cultured members* of all three domains. Because the cultured *Bacteria*, *Archaea*, and *Eukarya* shown in Figure 5.7 have been subjected to phenotypic, genotypic, and ecological analyses, a great deal is known about their biology. We now survey below the key phenotypic traits of selected members of the three domains.

The domain *Eukarya*

Several fundamental structural, cellular, genetic, and phenotypic differences between eukaryotes and prokaryotes were presented in Chapter 2 (see Section 2.12, Table 2.3, and Box 2.3). Eukaryotic microorganisms often feature larger cell sizes (10–25 μm versus 1 μm of prokaryotes), organelles [e.g., nucleus, mitochondria, chloroplasts, plastids, Golgi bodies, multicellularity, tissue development, flexible cell walls, complex sexual reproduction involving gametes produced by meiosis, and a high degree of metabolic specialization (an absence of metabolic versatility)]. In surveying branches on the tree of life (see Figure 5.7), it is clear that maximum temperatures for *Eukarya* (generally ~50–60°C) fall far below the extreme temperature adaptations (>80°C) exhibited by prokaryotes. Furthermore, unicellular *Eukarya* (diplomonads, microsporidia, slime molds, ciliates, and *Euglena*) emerged early in evolution – they reside at the base of the *Eukarya* trunk. Fungi, plants, and animals emerged later in evolution. The reader may recall that the early, phenotype-based (and anthropocentric) versions of the tree of life incorrectly placed plants, animals, and fungi in their own taxonomic kingdoms. In terms of evolutionary distance gauged by branch lengths in the small subunit rRNA tree, the distinctions between animals, plants, and fungi are minor when compared to the branch lengths of other phyla within and between domains. Table 5.2 provides an overview of many of the key traits of eukaryotic microorganisms.

Algae

Algae are water-dwelling, oxygen-evolving phototrophs that can be unicellular or multicellular. Figure 5.8 shows a photomicrograph of typical algal cells. All algae have single-cell reproductive structures. Evolutionarily, algae appear to be the ancestor of land plants – the latter have developed sophisticated vascular systems and reproductive structures that lessen their dependence upon free water for growth, buoyancy, and reproduction. With the exception of oceanic brown algae (known as kelp), algae are structurally far less complex than land plants. Chlorophyll *a* is the

Table 5.2

Broad overview of eukaryotic microorganisms and their traits (compiled from Madigan and Martinko, 2006)

Group	Nutrition	Motility	Structure	Reproduction
Fungi	Osmotrophic (uptake of soluble nutrients through cell wall)	Nonmotile (except some spores)	Filamentous, coenocytic (unicellular)	Complex patterns of haploid, diploid, dikaryotic, and/or heterokaryotic states
Algae	Photosynthetic	Flagella (amoeboid)	Unicellular, filamentous, multicellular	Sexual, asexual; alternation of generations – haploid and diploid stages
Slime molds	Phagotrophic	Amoeboid	Coenocytic, unicellular	Haploid, diploid
Protozoa: ciliates, *Euglena*, diplomonads, microsporidia	Phagotrophic, osmotrophic, photosynthetic	Flagella, cilia, amoeboid	Unicellular	Complex patterns of haploid and diploid states

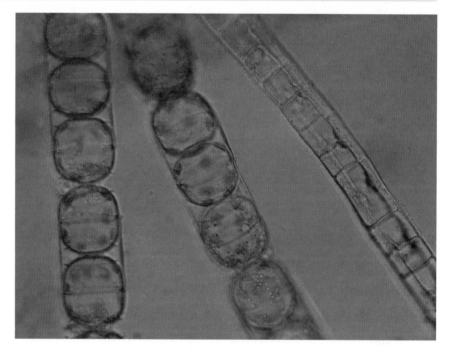

Figure 5.8 Image of the green alga, *Melosira*, found in a salt marsh, Heron's Head Park, San Francisco, ×800 magnification. (From W. Lanier, with permission.)

pivotal photosynthetic pigment used in the chloroplasts of all algae. Taxonomic divisions within the algae rely upon: accessory photosynthetic pigments (e.g., carotenoids and phycobilins), morphology, life cycle, food reserve materials (such as polysaccharides), and ecology. Table 5.3 provides a summary of the properties of the seven major groups of algae. Algae play key roles as primary producers (via photosynthesis) in the aquatic habitats where they reside. In addition to the free-living photosynthetic lifestyle, some algae form symbiotic associations with other forms of life (see Section 8.1). These associations include lichens, coral reefs (where the density of unicellular dinoflagellates may be 30,000/ml), marine sponges, protozoa, flatworms, and mollusks (nudibranchs) that become photosynthetic after consuming algae.

A comprehensive treatment of algae is beyond the scope of this chapter. For additional information, see Gualtieri and Barsanti (2006), Wehr and Sheath (2003), Larkum et al. (2003), Graham and Wilcox (2000), Sze (1998), Stevenson et al. (1996), and Chaudhary and Agrawal (1996).

Fungi

Fungi are filamentous (tube-dwelling), nonphotosynthetic osmotrophs that contribute significantly to heterotrophic metabolism, especially to the cycling of plant biomass carbon, in many ecosystems. The three generally recognized morphological types for fungi are yeasts (unicellular), molds (surface coatings), and mushrooms (macroscopic fruiting bodies). Figure 5.9 shows a photograph representative of tree-dwelling fungi. Fungi possess cell walls, often composed of chitin (a polymer of *N*-acetyl-glucosamine also found in the exoskeletons of insects and crustaceans). At the cellular level, the tubular walls that house the cytoplasm, nucleus, and other organelles form filaments that are known as hyphae. "Coencoytic" fungi are those whose hyphae lack regular partitions (septa) that divide the cytoplasm into uninucleate cells. The occurrence of septa and their morphology are key distinguishing traits in fungal taxonomy. A network of hyphae constitutes a mycelium. Furthermore, a large mass of mycelia is termed a thallus. Terrestrial fungi exist largely as extensive microscopic hyphal networks that penetrate large volumes of soil. Mushrooms (thalli) that appear on the soil surface are fruiting bodies – specialized structures for the dissemination of fungal propagules known as spores. The filamentous nature of many fungi, combined with biochemical equipment (enzymes) for digesting plant cell walls, allow them to be successful as plant pathogens and as heterotrophs that digest the biomass of cellulose- and lignin-rich fallen trees (see Section 7.3). Fungi also infect humans – causing medical mycoses that include opportunistic diseases of the skin, lungs, and other tissues. Life cycles and sexual reproduction in fungi may be highly complex; including processes such as: (i) dikaryotization, where mycelia of different strains may fuse without nuclear fusion; (ii)

Table 5.3

Properties of the major groups of algae (modified from MADIGAN, M. and J. MARTINKO. 2006. *Brock Biology of Microorganisms*, 11th edn, p. 475. Prentice Hall, Upper Saddle River, NJ. Copyright 2006, reprinted by permission of Pearson Education, Inc., Upper Saddle River, NJ)

Algal group	Common name	Morphology	Pigments	Typical representative	Carbon reserve materials	Cell wall	Major habitats
Chlorophyta	Green algae	Unicellular to leafy	Chlorophylls *a* and *b*	*Chlamydomonas*	Starch (α-1,4-glucan), sucrose	Cellulose	Freshwater, soils, a few marine
Euglenophyta	Euglenoids	Unicellular, flagellated	Chlorophylls *a* and *b*	*Euglena*	Paramylon (β-1,2-glucan)	No wall present	Freshwater, a few marine
Dinoflagellata	Dinoflagellates	Unicellular, flagellated	Chlorophylls *a* and *c*, xanthophylls	*Gonyaulax, Pfiesteria*	Starch (α-1,4-glucan)	Cellulose	Mainly marine
Chrysophyta	Golden-brown algae, diatoms	Unicellular	Chlorophylls *a* and *c*	*Nitzschia*	Lipids	Two overlapping components made of silica	Freshwater, marine, soil
Phaeophyta	Brown algae	Filamentous to leafy, occasionally massive and plant-like	Chlorophylls *a* and *c*, xanthophylls	*Laminaria*	Lammarin (β-1,3-glucan) mannitol	Cellulose	Marine
Rhodophyta	Red algae	Unicellular, filamentous to leafy (coralline)	Chlorophylls *a* and *d*, phycocyanin, phycoerythrin	*Polysiphonia*	Floridean starch (α-1,4-glucan and α-1,6-glucan	Cellulose ($CaCo_3$)	Marine
Xanthophyta	Yellow-green algae	Unicellular, amoebae, coenocytic filaments	Chlorophyl *c*	*Vaucheria*	Leucosin	Cellulose, pectin (silica)	Freshwater

Figure 5.9 Photograph of fungal fruiting bodies on a forest tree in Ithaca, NY. (Courtesy of E. L. Madsen.)

karyogamy, nuclear fusion; (iii) meiosis, creating specialized haploid spores; and (iv) growth of haploid mycelia. Table 5.4 provides a summary of the properties of the five major groups of fungi. Their common habitats range from soil to water to decaying plant material. One crucial ecological role of fungi, mycorrhizae (literally "fungus root"), is thought to stem from coevolution with land plants. Mycorrhizal symbiotic relationships between soil fungi and root tissues lead to enhanced nutrient (especially phosphorus) uptake by the fungal symbiotic partner in exchange for plant photosynthate (provided by the plant partner; see Section 8.1).

For additional information on fungi, see Dugan (2006), Xu (2005), Spooner and Roberts (2005), Lundquist and Hamelin (2005), Dighton et al. (2005), Petersen et al. (2004), Hirt and Horner (2004), Dighton (2003), Burnett (2003), and Kirk et al. (2001).

Protozoa

Protozoa are unicellular forms of life whose structures and behavior set them apart from the two prokaryotic domains and from other branches in the *Eukarya* (especially slime molds, fungi, and algae; see Figure 5.7). The biology and taxonomy of traditional protozoology (based on phenotypic

Table 5.4

Properties of the major groups of fungi (from MADIGAN, M. and J. MARTINKO. 2006. *Brock Biology of Microorganisms*, 11th edn, p. 469. Prentice Hall, Upper Saddle River, NJ. Copyright 2006, reprinted by permission of Pearson Education, Inc., Upper Saddle River, NJ)

Group	Common name	Hyphae	Typical representative	Type of sexual spore	Habitats	Common diseases caused
Ascomycetes	Sac fungi	Septate	*Neurospora, Saccharomyces, Morchella* (morels)	Ascospore	Soil, decaying plant material	Dutch elm, chestnut blight, ergot, rots
Basidiomycetes	Club fungi, mushrooms	Septate	*Amanita* (poisonous mushroom), *Agaricus* (edible mushroom)	Basidiospore	Soil, decaying plant material	Black stem, wheat rust, corn smut
Zygomycetes	Bread molds	Coenocytic	*Mucor, Rhizopus* (common bread mold)	Zygospore	Soil, decaying plant material	Food spoilage; rarely involved in parasitic disease
Oomycetes	Water molds	Coenocytic	*Allomyces*	Oospore	Aquatic	Potato blight, certain fish diseases
Deuteromycetes	Fungi imperfecti	Septate	*Penicillium, Aspergillus, Candida*	None known	Soil, decaying plant material, surfaces of animals bodies	Plant wilt, infections of animals such as ringworm, athlete's foot, and other dermatomycoses, surface or systemic infections (*Candida*)

traits such as size, structure, and behavior) is revealed by the tree of life to span an enormous evolutionary distance – from "ancient" representatives residing close to the evolutionary emergence from prokaryotes (diplomonads and microsporidia) to sophisticated, highly evolved ciliates. Biological information about the protozoa is still organized around traditional phenotypic groupings, not molecular phylogeny. The range of structural, ecological, and nutritional protozoan traits is vast (Table 5.5). Figure 5.10 shows a diagram of a ciliated protozoan. The one feature common to all protozoa is a membrane-bound nucleus – but other common assumptions about widespread eukaryotic features stop here. The less evolved protozoa (diplomonads and microsporidia) lack mitochondria and other common organelles. Instead of mitochondria, these often anaerobic cells contain "hydrogenosomes" [mitochondria-like organelles that convert pyruvate to adenosine triphosphate (ATP), H_2, acetate, and CO_2] and/or a small, poorly understood, sack-like structure termed a mitosome (Madigan and Martinko, 2006). Phylogenetic analysis of hydrogenosome genes indicates that, like mitochondria and chloroplasts, hydrogenosomes were bacterial endosymbionts (related to the present-day bacterial genus *Clostridium*) engulfed by ancient anaerobic eukaryotic cells. Another feature emphasizing the evolutionarily transitional place of microsporidia and diplomonads is that they contain 16S, not 18S, rRNA! Recent molecular phylogenetic analyses using genes other than small subunit rRNA have argued for taxonomic relocation of microsporidia in the tree of life to a position very close to the fungi. Clearly, many issues about the biology and evolution of protozoa are still in transition.

Motility via flagella (whip-like structures) is thought to have been an early step in protozoan evolution. Flagella are a readily recognizable distinguishing phenotype – giving rise to the terms "flagellates" and "mastigophora" in protozoology (Table 5.5). However, phylogenetically, flagella are of little significance because the trait is widespread among the many protozoan taxa. From a developmental standpoint, it is reasonable to presume (Sleigh, 1989) that ancestral flagellated protozoa made the transition to both amoeboid motility (via loss of flagella) and to ciliated motility (via multiplication and diminution of flagella). Another key force in protozoan evolution was incorporation of the chloroplast organelle. In this regard, the *Euglena* line of descent was reasonably successful (see Figure 5.7); these organisms are phototrophic and highly motile via flagella. This mixture of animal-like motility and plant-like nutrition (photosynthesis) blurs the taxonomic and disciplinary lines in protozoan biology. Another example of blurry taxonomy is the golden-brown alga, *Ochromonas*: this organism is flagellated and photosynthetic. Is *Ochromonas* a protozoan or an alga?

Even the most highly evolved ciliates (Table 5.5) are a challenge to understand because of their unique cellular features. Many ciliates carry both a micronucleus (diploid, critical for inheritance and sexual reproduction,

Table 5.5

Properties of the major groups of protozoa (from MADIGAN, M. and J. MARTINKO. 2006. *Brock Biology of Microorganisms*, 11th edn, p. 464. Prentice Hall, Upper Saddle River, NJ. Copyright 2006, reprinted by permission of Pearson Education, Inc., Upper Saddle River, NJ)

Traditional grouping based on phenotype	Groupings based on small subunit RNA	Typical representative	Nutrition	Habitats	Common diseases caused
Mastigophore (flagellates)	None (flagellated phenotype is widespread)	*Trypanosoma, Leishmania, Trichomonas*	Osmotroph, phagotroph	Freshwater; soil; animal parasites	African sleeping sickness, leishmaniasis
Euglenoids (phototrophic flagellates)	Euglena	*Euglena*	Phototroph	Freshwater, soil, some marine	None known
Sarcodina (amoebae)	Closely related to slime molds	*Amoeba, Entoamoeba*	Phagocystosis	Freshwater, soil, marine; animal parasites	Amoebic dysentery (amoebiasis).
Ciliophora (ciliates)	Ciliates	*Balantidium, Paramecium, Tetrahymena*	Ingestion through mouth-like opening	Freshwater, soil, marine; animal parasites; rumen	Dysentery
Apicomplexa (sporozoans)	Variable	*Plasmodium, Toxoplasma, Cryptosporidium*	Osmotroph	Primarily animal parasites, insects (vectors for parasitic diseases)	Malaria, toxoplasmosis, cryptosporidiosis
Diplomonad (mitochondria absent)	Diplomonad	*Giardia*	Osmotroph	Freshwater; obligate animal parasites	Giardiasis
Microsporidia (mitochondria absent)	Microsporidia	*Encephalitozoon*	Osmotroph	Freshwater; obligate animal parasites	Tissue infections (muscles, lungs, gastrointestinal tract)

but transcriptionally inactive) and a macronucleus (derived from the micronucleus, polyploid, and the sole site of gene expression, via RNA transcription, for the cell). Another structural oddity of some protozoa is the "apical complex", a unique cluster of rings and micro-tubules involved in biosynthetic processes and attachment. The presence of the apical complex is the traditional defining trait for the Apicomplexa (sporozoan) group of protozoa (Table 5.5). Members of the Apicomplexa are obligate parasites of insects and animals that produce resistant sporozoite stages that facilitate the organism's survival during transmission from one animal host to the next.

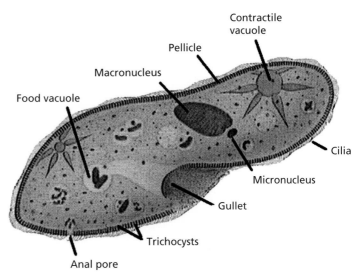

Figure 5.10 Diagram of the cellular architecture of the ciliate protozoan, *Paramecium*. (From Purves, W.K., G.H. Orians, and H.C. Heller. 1992. *Life: The science of biology*, 3rd edn. Sinauer Associates, Inc./W.H. Freeman and Co., Sunderland, MA. Reprinted with permission.)

The diverse examples of endo-symbionts, nutritional constraints, and structural modification in protozoan biology are impressive. So also are the ecological and medical impacts of protozoa (Table 5.5). Colonies of protozoa (e.g., *Volvox* and marine protozoa known as radiolarians) can be relatively large in size (mea-sureable in centimeters) and exhibit spatial specialization resembling tis-sues in higher *Eukarya*. As free-living bacterivores in soil and aquatic habitats, predatory protozoa often play important roles in microbial food chains. Grazing upon prokaryotic cells can enhance their growth rate, and hence stimulate the biogeochemical processes that prokaryotes catalyze (Fenchel, 1987). Furthermore, the protozoan biomass is, itself, an impor-tant component of the trophic pyramid (see Section 7.4), as protozoa are fed upon by zooplankton (in aquatic habitats) and nematodes, mites, or insects (in soil). From a medical perspective, a plethora of important human and animal diseases (from dysentery to malaria) are caused by protozoa.

For additional information on protozoa, see Sterling and Adam (2004), Clarke (2003), Lee et al. (2000), Smith and Parsons (1996), Margulis et al. (1993), Sleigh (1989), and Fenchel (1987).

Slime molds

Slime molds are eukaryotes closely related to both amoeboid protozoa and fungi. 18S rRNA analysis shows that slime molds are more ancient than fungi but more evolved than *Euglena* (see Figure 5.7). Figure 5.11

Figure 5.11 Photograph of slime mold fruiting bodies on forest foliage in Olympic National Park, Washington State, USA. (From Olympic National Park, with permission.)

shows a photograph representative of slime molds on forest foliage. The key trait of slime molds is that they exhibit colonial behavior in which large motile masses of amoebae move to engulf and feed on decaying plant material and bacteria in soil or on fallen trees. The colonial stage alternates with a morphologically specialized stage that produces spore-forming structures. After release, the spores germinate into haploid swarm cells. Compatible swarm cells fuse into diploid vegetative amoebae that grow, reproduce, and cycle back to the colonial stage.

Based on morphology, especially the presence or absence of cell walls, there are two types of slime molds: cellular and acellular. In the acellular slime molds (e.g., *Physarum*) the vegetative masses of indefinite size are termed "plasmodia"; whereas, in the cellular slime molds (e.g., *Dictyostelium*) the aggregated cells are termed "pseudoplasmodia". The cellular slime molds have been extensively studied at the biochemical and genetic level, as a model for understanding cellular differentiation. The ecological impact of slime molds is not well explored (e.g., Landolt et al., 2004). For additional information on slime molds see Margulis et al. (1993) and Stephenson and Stempen (1994).

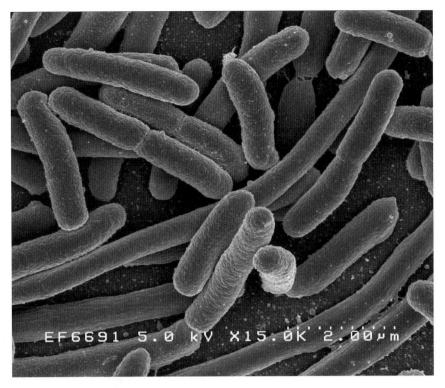

Figure 5.12 Scanning electron micrograph of the γ-Proteobacteria, *E. coli*, grown in culture and adhered to a glass surface. (From Rocky Mountain Laboratories, NIAID, NIH, with permission.)

The domain *Bacteria*

Bacteria are prokaryotes that are distinctive from *Archaea* (Figure 5.7). The critical distinction is in the 16S rRNA gene sequences. Figure 5.12 shows a photograph representative of the common γ-Proteobacteria, *Escherichia coli*. Additional phenotypic distinctions include: ester linkages in membrane lipids, muramic acid in cell walls, protein synthesis initiated by formylmethionine tRNA, a single type of four-subunit RNA polymerase, Pribnow box-type promoter structure for transcription of genes, sensitivity to certain protein-synthesis inhibitors (chloroamphenicol, streptomycin, kanamycin), and absence of growth above 100°C; some members carry out chlorophyll-based photosynthesis (see Table 2.3). As mentioned in Section 5.1 and Table 5.1, the most recent version of *Bergey's Manual of Systematic Bacteriology* (Garrity et al., 2005) lists 24 phyla of *Bacteria*. Descriptions of all 24 phyla appear in the manual and are expanded upon at lower taxonomic levels in other related works (e.g., Brenner et al., 2005;

Dworkin et al., 2006; Madigan and Martinko, 2006). A complete survey of all 24 phyla is beyond the scope of this chapter; however, descriptions based on Garrity et al. (2005) of the eight bacterial phyla shown in Figure 5.7 appear below. Distinctions between Gram-negative and Gram-positive microorganisms were presented in Box 2.3.

Aquifex

Aquifex/Hydrogenobacter is the deepest and earliest branching phylum of *Bacteria*. All members are Gram-negative, nonsporulating rods or filaments with optimum growth in the range of 65–85°C. These are chemolithoautotrophs or chemolithoorganotrophs using H_2, S^0, or $S_2O_3^{2-}$ as electron donors and O_2 or NO_3^- as electron acceptors.

Thermotoga

Thermotoga is also an early, deeply branching phylum. All members are Gram-negative, nonsporulating, rod-shaped bacteria that possess a characteristic sheath-like outer layer or "toga". *Thermotoga* are strictly anaerobic heterotrophs, utilizing a broad range of organic compounds as carbon sources and electron donors. Thiosulfate and S^0 are electron acceptors.

Green nonsulfur bacteria

Green nonsulfur bacteria, another deep-branching bacterial lineage, are Gram-negative, filamentous organisms exhibiting gliding motility. Members of this phylum are physiologically diverse. Some (*Chloroflexus*) contain bacteriochlorophyll and are obligate or facultative anoxygenic phototrophs, exhibiting what may be the earliest form of photosynthetic reaction centers and CO_2 fixation mechanisms (see Section 2.10). The electron donors for CO_2 reduction are H_2 and H_2S; sulfur granules do not accumulate. Other members of this phylum do not contain bacteriochlorophyll and are chemoheterotrophs.

Green sulfur bacteria

Green sulfur bacteria (also known as *Chlorobi*) are Gram-negative, spherical, ovoid, straight, curved, or rod-shaped cells that are strictly anaerobic and obligately phototrophic. Cells grow preferentially on fixed organic compounds in the light (photoheterotrophy). Some species may utilize sulfide or thiosulfate as electron donors for CO_2 fixation. Sulfur granules accumulate on the outside of the cells when grown on sulfide in the light, and sulfur is rarely oxidized further to sulfate. Ammonia and N_2 gas are used as nitrogen sources.

Flavobacteria

Flavobacteria (also known as Bacteroides) share a common phylogenetic root with green sulfur bacteria, yet the branching depth within each group warrants the phylum category. The phylum is split into three major sublineages, Bacteroides, Flavobacterium, and Sphingobacteria, that range from obligate anaerobes to obligate aerobes. The broad phenotypic characteristics of this group includes: Gram-negative aerobic or microaerophilic rods; Gram-negative anaerobic rods; nonphotosynthetic, nonfruiting, gliding bacteria; bacterial symbionts of invertebrate species; sheathed bacteria; and nonmotile or rarely motile, curved Gram-negative bacteria.

Cyanobacteria

Cyanobacteria are Gram-negative, unicellular, colonial, or filamentous oxygenic and photosynthetic bacteria exhibiting complex morphologies and life cycles. The principal characters that define all members of this phylum are the presence of two photosystems (PSI and PSII; see Section 2.10) and the use of H_2O as the reductant in photosynthesis. All members contain chlorophyll *a* with or without other light-harvesting pigments. Although facultative photoheterotrophy or chemoheterotrophy may occur in some species or strains, all known members are capable of photoautotrophy (using CO_2 as the primary source of cell carbon).

Purple bacteria

Purple bacteria (also known as Proteobacteria) constitute the largest phylogenetic group within the *Bacteria*, divided into five major subdivisions (α, β, γ, δ, and ε), composed of more than 520 genera. Phenotypic groups within the Proteobacteria include: Gram-negative aerobic or microaerophilic rods and cocci; anaerobic straight, curved, and helical Gram-negative rods; anoxygenic phototrophic bacteria; nonphotosynthetic, nonfruiting, gliding bacteria; aerobic chemolithotrophic bacteria and associated genera; facultatively anaerobic Gram-negative rods; budding and/or appendaged nonphototrophic bacteria; symbiotic and parasitic bacteria of vertebrate and invertebrate species; and fruiting, gliding bacteria. As a group, these are all Gram-negative, show extreme metabolic diversity, and represent the majority of Gram-negative bacteria of medical, industrial, and agricultural significance.

Gram-positive bacteria

Gram positives are found in two phyla, the Firmicutes and Actinobacteria.

The Firmicutes constitute an extensive phylum of >1500 species within >220 genera spanning three main subdivisions: the Clostridia, the Mollicutes, and the Bacilli. This phylum, though dominated by Gram-positive microorganisms, is phenotypically diverse and includes some members with Gram-negative cell walls. Phenotypic groups of the Firmicutes include thermophilic bacteria; anaerobic, straight, curved, and helical Gram-negative rods; anoxygenic phototrophic bacteria; nonphotosynthetic, nonfruiting, gliding bacteria; aerobic nonphototrophic chemolithotrophic bacteria; dissimilatory sulfate- or sulfite-reducing bacteria; symbiotic and parasitic bacteria of vertebrate and invertebrate species; anaerobic Gram-negative cocci; Gram-positive cocci; endospore-forming Gram-positive rods and cocci; regular and irregular nonsporulating Gram-positive rods; mycoplasmas; and thermoactinomycetes.

Actinobacteria are the second phylum of Gram-positive *Bacteria* – phylogenetically related to, but distinct from, the Firmicutes. Actinobacteria feature a high level of morphological, physiological, and genomic diversity that falls within this phylum's six major taxonomic orders and 14 suborders. The phylum can be broadly divided into two major phenotypic groups: unicellular, nonsporulating Actinobacteria and the filamentous, sporulating Sporoactinomycetes. The unicellular Actinobacteria include some Gram-negative aerobic rods and cocci; aerobic, sulfur oxidizing, budding and/or appendaged bacteria; Gram-positive cocci; regular nonsporulating, Gram-positive rods; irregular non-sporulating, Gram-positive rods; and mycobacteria. The Sporoactinomycetes include noncardioform actinomycetes, actinomycetes with multicellular sporangia, actinoplanetes, *Streptomyces* and related genera, maduromycetes, *Thermomonospora* and related genera, and other sporoactinomycete genera.

The domain *Archaea*

Archaea are prokaryotes that are distinctive from *Bacteria* (Boone and Castenholz, 2001). The critical distinction is in the 16S rRNA gene sequences. Figure 5.13 shows a photograph of *Methanosarcina*, a member of the *Archaea*. Additional phenotypic distinctions include: histone proteins associated with chromosomal DNA; absence of muramic acid in cell walls; ether-linked membrane lipids (see Section 2.7); protein synthesis initiated by methionine tRNA; several types of RNA polymerases (8–12 subunits each); TATA box-type promoter structure for transcription of genes; insensitivity to bacterial-type protein synthesis inhibitors; some members are methanogenic; no members carry out chlorophyll-based photosynthesis; and some members are able to grow at temperatures above 100°C (see Section 2.12 and Table 2.3).

As mentioned in Section 5.1 and Table 5.1, the most recent version of *Bergey's Manual of Systematic Bacteriology* (Garrity et al., 2005) lists three cultured phyla of the *Archaea*. Two of these, the Crenarchaeota and

Figure 5.13 Epifluorescent photomicrograph of a representative of the *Archaea, Methanosarcina,* a methanogen. Note the clusters of four green fluorescent cells characteristic of this genus. (From S. H. Zinder, Cornell University, with permission.)

Euryarchaeota appear in the tree of life (see Figure 5.7). A third phylum (Nanoarchaeota) was recently established to accommodate the discovery and recent isolation of the tiny parasitic, hyperthermophilic microorganism deemed *Nanoarchaeum equitans* (Huber et al., 2003). Brief descriptions of the three phyla and several of the pertinent taxonomic subdivisions shown in Figure 5.7 appear below (Garrity et al., 2005; Madigan and Martinko, 2006).

Crenarchaeota

The Crenarchaeota consist of a single phylum, the Thermoproteus which is composed of four orders: *Thermoproteales, Caldisphaerales, Desulfuro-coccides,* and *Sulfolobales.* Members of the phylum are morphologically diverse, including rods, cocci, filamentous forms, and disk-shaped cells that stain Gram negative. Motility is observed in some genera. The organisms are obligately thermophilic, with growth occurring at temperatures ranging from 70 to 113°C. The organisms are acidophilic and are aerobic, facultatively anaerobic, or strictly anaerobic chemolithoautotrophs or chemoheterotrophs. Most metabolize S^0. Chemoheterotrophs may use organic compounds as electron donors and S° as an electron acceptor (reducing it to H$_2$S). Members of this phylum have been isolated from many of the most extreme environments on Earth – hot springs, sulfataras, hydrothermal vents, etc. For this reason the cultured Crenarcheota have provided a wealth of information about physiological and metabolic diversity.

Euryarchaeota

The Euryarchaeota is a phylum consisting of eight classes: *Methanobacteria, Methanococci, Methanomicrobia, Halobacteria, Thermoplasmata, Thermococci, Archaeoglobi,* and *Methanopyri*. In phylogenetic analyses (e.g., Figure 5.7), many phyla share a common root; therefore, several phylogenetic relationships are ambiguous (Garrity et al., 2005). The Euryarchaeota are morphologically diverse and occur as rods, cocci, irregular cocci, lancet shaped, spiral shaped, disk shaped, triangular, or square cells. Cells stain Gram positive or Gram negative based on the absence or presence of pseudopeptidoglycan (a peptidoglycan analog) in cell walls. In some cases, cell walls consist entirely of protein or may be entirely absent. Five major physiological groups have been described: methanogenic *Archaea*, extremely halophilic *Archaea*, *Archaea* lacking a cell wall, sulfate-reducing *Archaea*, and extremely thermophilic S^0 metabolizers.

Nanoarchaeota

The Nanoarchaeota represent a deep and novel lineage within the *Archaea*. Although the secondary structure of 16S rRNA of the Nanoarchaeota conforms to that expected for *Archaea*, the primary structure (sequence) is so novel that precise placement on the phylogenetic tree is difficult. Figure 5.14 shows details of the archaeal phylogenetic tree prepared by Huber et al. (2003), the discoverers of the only isolated representative of Nanoarchaeota, named *Nanoarchaeum equitans*. Shown on the tree in Figure 5.14 are bacterial and eukaryal lines (for reference) and four archaeal phyla. The fourth phylum (Korarchaeota), consisting of only noncultivated sequences, will be discussed in Section 5.7.

Nanoarchaeum equitans occurs as very small cocci (0.35–0.5 µm) singly or in pairs. The cell volume is approximately 1% that of typical *E. coli* cells. The recently sequenced genome is also very small (0.49 Mbp; see Sections 3.2 and 5.9). *N. equitans* cannot yet be grown in pure culture. The bacterium can be co-cultured with a member of the Crenarchaeota in the genus *Ignicoccus*. *N. equitans* was found attached to the outer surface of *Ignicoccus* cells that originated from hydrothermal sediment samples in Iceland. The host, *Ignicoccus*, is a strict anaerobe, growing optimally at 90°C: *N. equitans* may be a parasite or a symbiont. Clues from the genome sequence promise to help explain its physiology.

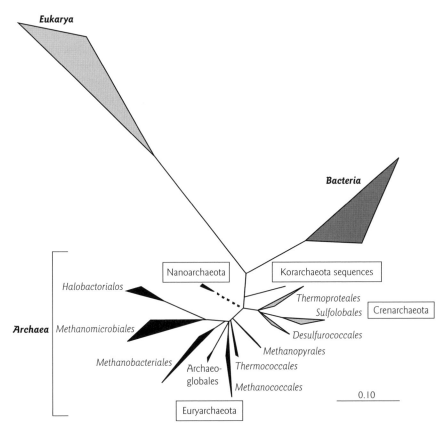

Figure 5.14 Phylogenetic tree based on 16S rRNA comparisons showing the Nanoarchaeota in relation to the other three archaeal phyla and the two other domains. At the time of publication of this book, the Korarchaeota had no cultured representatives. The scale bar represents 0.1 nucleotide change per position. (Reprinted from Huber, H., M.J. Hohn, K.O. Stetter, and R. Rachel. 2003. The phylum *Nanoarchaeota*: present knowledge and future perspectives of a unique form of life. *Res. Microbiol.* **154**:165–171. Copyright 2003, with permission from Elsevier.)

5.7 PLACING THE "UNCULTURED MAJORITY" ON THE TREE OF LIFE: WHAT HAVE NONCULTURE-BASED INVESTIGATIONS REVEALED?

Information presented in Chapter 4 (Section 4.5) provided estimates of the Earth's total prokaryotic biomass, as distributed across various habitats. Two related matters, how to operationally define and estimate microbial diversity, have been discussed in Sections 5.2–5.4. The current state of both eukaryotic and prokaryotic microbial taxonomy, based on cultured microorganisms, was presented in Sections 5.5 and 5.6. To strengthen the emerging portraits of the diversity of microorganisms, we now turn to information on uncultured prokaryotic microorganisms (defined in Section 5.1) using nonculture-based procedures.

An analysis of *Bacteria*

In 2003, Rappé and Giovannoni collected and analyzed accumulating 16S rRNA gene sequences directly derived from samples of freshwater, seawater, sediment, and soil. The authors' goal was to assess the impact of environmental libraries of 16S rRNA genes on our understanding of the diversity, evolution, and phylogeny of *Bacteria*. Figure 5.15 presents a picture of total bacterial phylogenetic diversity. The phylogenetic tree (Rappé and Giovannoni, 2003) integrates 16S rRNA sequence data from both cultured microorganisms (e.g., Figure 5.7) and noncultured microorganisms (Sections 5.1 and 5.2). The figure displays 52 phyla (major lineages) within the domain *Bacteria*. Each phylum is displayed as a "wedge" or "leaf" at the end of a branch. The size of the wedge representing each phylum reflects the number of accumulated sequences and the shading shows: (i) the 12 cultured phyla originally described by C. Woese in 1987 (black); (ii) the 14 additional phyla that have been discovered through cultivation procedures since 1987 [bringing the total to 26 (white)]; and (iii) the additional 26 phyla discovered through nonculture-based sequencing of environmental 16S rRNA gene libraries (gray). The message from the information shown in Figure 5.15 is astounding. At a glance a reader can see that in two decades, *phylum-level* assessment of bacterial diversity roughly doubled using cultivation-based techniques and *doubled again using 16S rRNA environmental gene-cloning procedures* devised by N. Pace and colleagues (Pace et al., 1986).

Figure 5.15 (*opposite*) Phylogenetic tree illustrating the major lineages (phyla) of the domain *Bacteria* (see text for details). Wedges shown in black represent the 12 original phyla, as described by Woese (1987); in white are the 14 phyla with cultivated representatives recognized since 1987; and in gray are the 26 candidate phyla that contain no known cultivated representatives. Horizontal wedge distances indicate the degree of divergence within a given phylum. The scale bar corresponds to 0.05 changes per nucleotide position. Phylum names are designated by selecting the first applicable option out of the following: (i) their convention in *Bergey's Manual of Systematic Bacteriology*, if it exists (Garrity et al., 2005); (ii) the first described representative genus within the phylum if it has cultivated representatives; (iii) the first label given to a candidate phylum if previously published; or (iv) the first clones or environment where the first clones were retrieved, for previously unnamed candidate phyla. This evolutionary distance dendogram was constructed by the comparative analysis of over 600 nearly full-length 16S ribosomal RNA gene sequences using the ARB sequence analysis software package, selected from a larger database of over 12,000 sequences. A modified version of the "Lane mask" was employed in this analysis, along with the Olsen evolutionary distance correction and neighbor-joining tree-building algorithm. (From Rappé, M.S. and S.J. Giovannoni. 2003. The uncultured microbial majority. *Annu. Rev. Microbiol.* **57**:369–394. Reprinted with permission from *Annual Reviews of Microbiology*, Vol. 57. Copyright 2003 by Annual Reviews, www.annualreviews.org.)

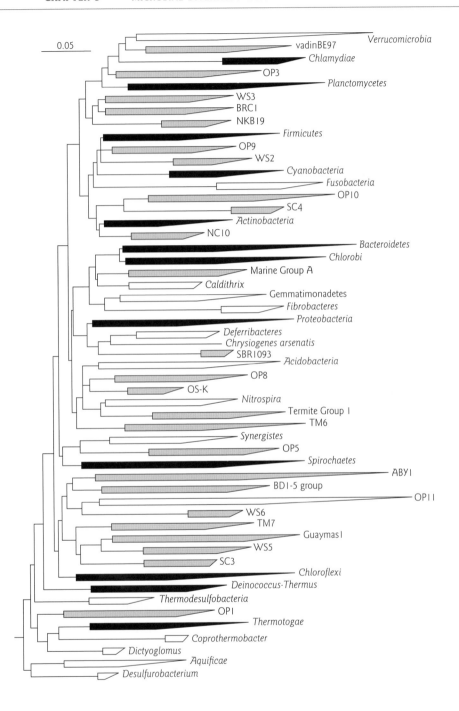

An analysis of *Archaea*

Environmental clone libraries of 16S rRNA genes have also significantly advanced knowledge of archaeal phylogenetic diversity. One way to appreciate such contributions is to refer to Figure 5.14, depicting four archaeal phyla. One of these phyla, the Korarchaeota was discovered and established by creating a clone library of 16S rRNA genes from DNA extracted from obsidian pool hot-spring sediments (74–93°C) in Yellowstone National Park (Barns et al., 1996). A decade later, the Korarchaeota still have no cultured representatives.

Environmental clone libraries have also documented unsuspected diversity within the Crenarchaeota and Euryarchaeota phyla. Despite the fact that cultivated members of these two phyla often have thermophilic phenotypes, archaeal 16S rRNA gene sequences have been discovered under moderate environmental conditions that prevail in habitats such as soil, lake sediments, anaerobic digesters, and tissues of marine animals (DeLong, 1998). Moreover, both Crenarchaeota and Euryarchaeota occur extensively in cold, low-nutrient, marine waters and sediments (DeLong et al., 1992; Fuhrman et al., 1992; DeLong, 1998). *Detection of the 16S rRNA sequences in unexpected habitats raises new questions about the potential impact and ecological roles of uncultivated Archaea.* A potential answer to one such prominent question was recently provided by Könneke et al. (2005), who isolated an autotrophic ammonia-oxidizing member of the Crenarchaeota that appears to represent an important component of these mysterious ocean-dwelling cells whose 16S rRNA sequences were discovered by Fuhrman and DeLong. Clues that *Archaea* may carry out nitrification were initially obtained from 16S rRNA clone libraries derived from nitrifying estuary sediments and marine aquaria. After repeated transfers on enrichment media containing ammonia as an electron donor, oxygen as an electron acceptor, and bicarbonate as a carbon source, Könneke et al. (2005) isolated *Nitrosopumilus maritimus*, the first member of *Archaea* that can carry out nitrification. This new microorganism, which is phylogenetically closely related to ubiquitous marine prokaryotes, provides a window into the possible biogeochemical role of marine Crenarchaeota in cycling both carbon and nitrogen (see Sections 7.3–7.6).

Finally, a recent comprehensive re-examination of archaeal 16S rRNA sequences (Robertson et al., 2005) reveals both vast uncultured diversity and a potential need for phylogenetic restructuring of the *Archaea*. Improved computational abilities in both computer hardware and sequence-processing software allowed Robertson et al. (2005) to refine the details of branching patterns and contrast cultured versus uncultured sources of 16S rRNA sequences. Figure 5.16 displays what may become the next (revised) phylogeny of the *Archaea*. In the new analysis, the archaeal phylogeny retains two main branches (phyla), Crenarchaeota and

(a)

(b)

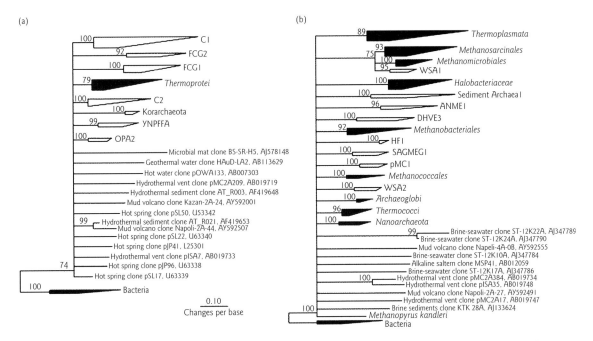

Figure 5.16 Phylogenetic tree illustrating the two major lineages (phyla) of the domain *Archaea*: (a) Crenarchaeota, and (b) Euryarchaeota. Solid colored groups have at least one cultured representative; others are known only from environmental samples. The sequences were downloaded from GenBank in February 2005 and were manually aligned using the ARB software. Three bacterial sequences were used as outgroups. PHYLIP software was used to generate 100 bootstrap data sets and to build the consensus tree that resulted from running the 100 data sets through RAxML. Any nodes in the tree that had less than 70% bootstrap support were deleted. ARB and PHYLIP are commonly used phylogenetic software tools. Phylogenetic relationships were generated with the RAxML software from 712 archaeal rRNA sequences that were at least 1250 nucleotides long. (Reprinted from Robertson, C.E., J.K. Harris, J.R. Spear, and N.R. Pace. 2005. Phylogenetic diversity and ecology of environmental *Archaea. Curr. Opin. Microbiol.* **8**:638–642. Copyright 2005, with permission from Elsevier.)

Euryarchaeota. These, respectively, absorb the Korarchaeota and Nano-archaeota phyla (shown previously in Figure 5.14). Within each of the two main phyla shown in Figure 5.16, many evenly spaced taxonomic divisions emanate from a single branch. This radiating branching pattern known as "polytomy" (Robertson et al., 2005) may be adopted in future versions of archaeal taxonomy and phylogeny.

Regardless of the forms of the trees in Figure 5.16, the contrasts between cultivated and noncultured sources of sequences delivers the same astounding message for *Archaea* as was delivered for *Bacteria* in Figure 5.15. Taxa on the trees represented by solid symbols have cultivated representatives; while open symbols are from uncultured environmental clone libraries. Thus, within the Crenarchaeota 20 of the 21 lines of descent are not represented in culture. Within the Euryarchaeota, the proportion of uncultured lines of descent is 8 out of 25.

Clearly, the application of small subunit rRNA gene-cloning techniques to naturally occurring microbial communities has made us aware of a vast, unexplored diversity of prokaryotic life.

5.8 VIRUSES: AN OVERVIEW OF BIOLOGY, ECOLOGY, AND DIVERSITY

Viruses are one of the five classic forms of microbiological life (the others are prokaryotes, fungi, algae, and protists). Viruses, like prokaryotes, have domesticated representatives, upon which virtually all of the science of virology is based. Also, as for prokaryotes, molecular environmental surveys have been performed to reveal a broad diversity of previously unknown viruses. But because viruses are obligate intracellular parasites (that lack protein synthesis capabilities), small subunit rRNA genes, essential for ribosomes, do not occur in viruses. Thus, there is no branch for viruses on the tree of life shown in Figure 5.7. However, molecular phylogenetic approaches have been recently applied to viruses (Hurst, 2000; Rohwer and Edwards, 2002; Mayo et al., 2005; Koonin et al., 2006). Below is a brief summary of virus biology, ecology, and diversity.

Viruses are acellular agents of heredity and disease. They have two complementary definitions:

1 Genetic elements containing DNA or RNA that replicate in cells but have an extracellular state.
2 Obligate intracellular parasites whose DNA or RNA is encapsidated in proteins encoded by the viral genomes; the parasites have evolutionary histories independent of their hosts.

Viruses are particles in the size range of ~100 nm. A complete virus particle is termed a virion, and is composed of the genetic nucleic acid core (genome) surrounded by a capsid shell formed from virus-encoded proteins. Viruses that infect bacteria are termed bacteriophages. Figure 5.17 shows an electron micrograph of a virus particle (Figure 5.17b) and the zones of clearing caused by a virus infecting a uniform layer (lawn) of its host bacterium (Figure 5.17a). The genome (DNA or RNA), housed in the capsid, is injected into the host bacterium after specific molecular recognition between components of the viral capsid and lipopolysaccharide in the outer membrane of the bacterium.

There are no traces or fossils of viruses in the geologic record. Preserved historical samples of virus-bearing tissue date back only to the early 1900s. Therefore, clues to the evolutionary history of viruses are limited to molecular phylogeny and educated speculation. There are at least three provocative (and untested) theories about the origin of viruses.

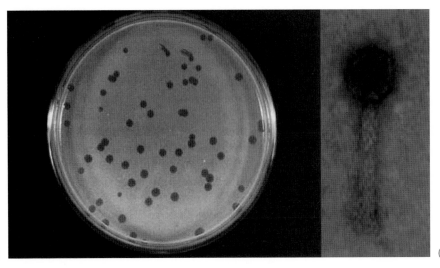

(a)   (b)

Figure 5.17 (a) Photograph of zones of clearing caused by viral infections of bacteria growing on an agar plate, and (b) an electron micrograph of the virus revealing the head and tail structures. (From T. Nakai, Hiroshima University, with permission.)

1 *Regressive theory.* A parasite may have colonized a host, become intracellular, completely dependent upon host, and then gradually lost genetic traits.
2 *"Rebel" DNA or RNA.* Normal cellular components (DNA, RNA) may have gained the ability to replicate autonomously, and hence evolve; replicating nucleic acids in this family include viroids, virusoids, plasmids, and transposons.
3 *Ancient virus world.* Favored by Koonin et al. (2006), contemporary viruses may be remnants of ancient, self-replicating RNA precursors to modern life; these may have preceded cellular life and become intracellular parasites after *Bacteria* and *Archaea* evolved.

Viruses are sophisticated, highly evolved entities; and their evolution continues today. As intracellular parasites, they generally have limitations on their size and shape. For RNA viruses (see below), there may also be limitations on genome size and structure imposed by the frequency of error during cellular RNA replication.

Key structural aspects of virus biology include: their hereditary material (DNA or RNA; or DNA and RNA, as genetic material, at different life-cycle stages); whether or not the nucleic acids are single stranded or double stranded, linear, or circular; capsid architecture, size, and shape; and the presence and type of lipid envelope that may surround the capsid. Medical aspects of viruses include characteristics such as disease type, host type (animals, plants, prokaryotes, protists), means of transmission, life cycle, and infection specificity. Molecular mechanisms underlying the details

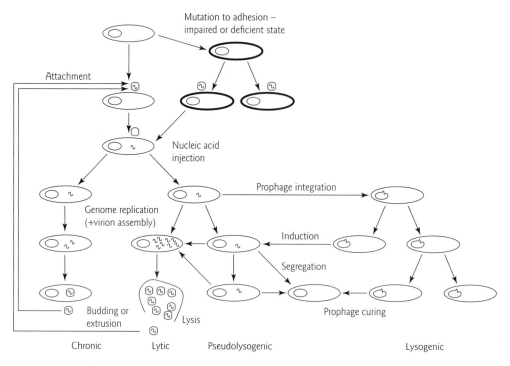

Figure 5.18 Types of viral life cycles. The model shows a typical bacterial cell with a circular chromosome (top), attachment by a virus or impaired attachment, and various stages of infection and intracellular viral behavior. See text for details. (From Weinbauer, M.G. 2004. Ecology of prokaryotic viruses. *FEMS Microbiol. Rev.* **28**:127–181. With permission from Blackwell Publishing, Oxford, UK.)

of all aspects of viral biology, biochemistry, and replication deepen our understanding of life processes and may lead to treatment strategies that prevent infection (e.g., medicines) and cure diseases (e.g., gene therapy).

Figure 5.18 shows a typical life cycle for bacterial viruses and introduces the concept of lysogeny. Viruses can exhibit several types of life cycles: chronic, lytic, lysogenic, and pseudolysogenic. The chronic and lytic cycles are diagrammed on the left-hand side of Figure 5.18. They both proceed in five key steps in which the virus enters the cell and redirects host metabolism to create new virus particles released after host cell lysis. The five steps are: (i) attachment of the virion to a susceptible host cell; (ii) penetration (injection of the virion or its nucleic acid into the cell); (iii) alteration of the cell biosynthetic apparatus so that virally encoded enzymes and nucleic acids are generated; (iv) assembly of capsid shells and packaging of nucleic acids within them; and (v) release of mature virions from the cell. Note that, if attachment does not occur (Figure 5.18, upper center), infection is thwarted. As shown in Figure 5.18, the chronic and lytic cycles differ in the extent of intracellular virion assembly and the mode of release (nondestructive budding versus lysis and death of the host cell, respectively). In the lysogenic cycle (Figure 5.18, right), the viral genome integrates into the host chromosome and replicates along with it, as if it were a normal component of the host chromosome. The genetically integrated phage DNA is known as a "prophage" because it resides in the host as a dormant precursor to an excised, potentially lytic, phage. When a suitable set of cellular triggers (a molecular dialog between environment, host cell, and prophage) is activated, the prophage excises from the host chromosome and has the potential to begin the lytic cycle. Pseudolysogeny is the term applied to an intermediary state (between lysogenic and lytic stages) in which an extrachromosomal virus replicates in synchrony with the host chromosome (much like a plasmid).

The ecology of viruses and their impact on biogeochemical processes are active areas of research in environmental microbiology (e.g., Fuhrman, 1999; Wommack and Colwell, 2000; Weinbauer, 2004; Suttle, 2005). Microscopic and isolation-based surveys from environmental samples have found that viral abundances exceed prokaryotic abundances and that a significant fraction of naturally occurring prokaryotic communities are infected by viruses. Figure 5.19 shows a seawater sample and its resident viruses. The abundance of bacterial viruses in soils, sediments, freshwaters, and ocean waters varies greatly and is related to host abundance and activity (Weinbauer, 2004). Recent estimates of abundances of viruses in seawater (from 10^5/ml in deep waters to 10^8/ml in productive coastal waters) have been extrapolated to yield several rather astonishing observations (Suttle, 2005):

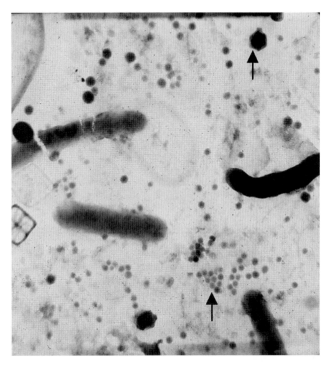

Figure 5.19 Transmission electron micrographs showing viruses and bacteria in a water sample from the northern Adriatic Sea. Viruses and bacteria were stained with uranyl acetate. Arrows point to viruses of different size. The large virus in the upper right corner has a head diameter of 150 nm. (From Weinbauer, M.G. 2004. Ecology of prokaryotic viruses. *FEMS Microbiol. Rev.* **28**:127–181. With permission from Blackwell Publishing, Oxford, UK.)

1 Ocean waters may contain a total of ~4×10^{30} viruses (compare this to prokaryote numbers in the oceans, see Section 4.5).

2 Assuming that each virus contains 0.22 fg of carbon, viral biomass in the oceans amounts to 200×10^9 kg of carbon, second only to the biomass of prokaryotes (see Section 4.5).

3 Assuming an average length of 100 nm for each virus, if aligned end to end they would span a distance of 10 million light years.

Recent nonculture-based efforts that prepare and sequence clone libraries (see Section 6.7) from sampled human feces, ocean water, and sampled marine sediments (Edwards and Rohwer, 2005) have provided evidence for ~1000 viral genotypes in human feces, ~5000/200 L in seawater, and up to 10^6 in marine sediment. The vast majority of these environmental virus sequences differ immensely from those of domesticated viruses. Lytic viruses threaten to obliterate prokaryotic populations, whereas lysogenic and chronic infections represent parasitic interactions. Virus-induced mortality of prokaryotes is highly variable in time and space – dependent especially on habitat characteristics, host abundances, and the host's physiological state. Because viral infections are host-specific and density-dependent, it is widely accepted that infections can control the composition of prokaryotic communities. Competitively dominant prokaryotes can be kept in check because viruses "kill the winner" (Weinhauer, 2004). Regarding biogeochemistry, viruses are influential via at least two mechanisms: (i) lytic cycles can reduce the sizes, hence physiological activity of prokaryotes that catalyze individual steps in the cycling of carbon, nitrogen, sulfur, and other elements; and (ii) via the "viral shunt", nutrients can be moved from the cell-associated nutrient pool to the particulate and dissolved nutrient pool. Figure 5.20 illustrates the viral shunt for an oceanic habitat. In the classic view of trophic dynamics, microbial phytoplankton are grazed by zooplankton, which serve as food for carnivores. If lytic viruses divert phytoplankton biomass into the pool of dissolved and particulate organic matter, there are two potential consequences: the organic matter is utilized as a carbon source by heterotrophic bacteria (producing CO_2) and the flux of particular carbon (detritus) reaching and stored in marine sediments is diminished (Suttle, 2005). Both of these consequences may contribute to global warming (see Section 7.2). For additional information on viruses, see Granoff and Webster (2005), Cann (2005), Villarreal (2005), Knipe et al. (2001), and Hurst (2000).

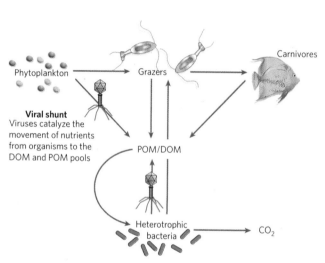

Figure 5.20 Viral influences on biogeochemistry. Viruses cause the lysis of cells, converting them into particulate organic matter (POM) and dissolved organic matter (DOM). This reduces the rate at which carbon sinks from the surface layer into the deep ocean where the carbon is trapped for millennia (biological carbon pump). Instead the carbon is retained in the surface waters where it is photooxidized and respired, in chemical equilibrium with the atmosphere. The net effect is a faster rate of CO_2 buildup in the atmosphere than would occur if the POM were "exported" to the deep ocean. (Reprinted by permission from Macmillan Publishers Ltd: *Nature*, from Suttle, C.A. 2005. Viruses in the sea. *Nature* **437**:356–361. Copyright 2005.)

5.9 MICROBIAL DIVERSITY ILLUSTRATED BY GENOMICS, HORIZONTAL GENE TRANSFER, AND CELL SIZE

As discussed in Chapter 3 (Section 3.2), the genetic blueprint of an organism, its genome sequence, is a comprehensive window into the evolutionary history and contemporary physiological ecology of the organism. We now reside in the genomic and postgenomic age. As of 2007, ~450 bacterial, ~35 archaeal, ~12 fungal, ~5 protist, and ~1700 viral genome sequences had been completed with many more in progress (US Department of Energy, Joint Genome Institute, http://img.jgi.doe.gov). Careful inspection of patterns within and between genomes (comparative genomics) is a new and novel tool for understanding life on Earth and the evolutionary processes that have operated. Comparative genomics, though still a rather young science, offers several key lessons about the diversity of microorganisms and traits shared between different taxonomic groups.

Genomic sizes of *Archaea, Bacteria, Eukarya*, and viruses do not sort into distinctive, predictable groups (Ward and Fraser, 2005). Information portrayed in Figure 5.21 shows that some specialized members of the *Eukarya* (e.g., the protozoan parasite, *Cryptosporidium*, and the filamentous fungus, *Ashbya gossypii*) have streamlined genomes that roughly match the size of the larger bacterial genomes. Moreover, the photosynthetic protozoan ciliate, *Guillardia theta*, has a genome that is *smaller* than that of most bacteria. Perhaps, as expected, the sizes of bacterial and archaeal genomes have approximately the same range. However, some bacterial genomes (left-hand cluster in Figure 5.21) are very small (<1 Mb), reflecting highly specialized lifestyles. Remarkably, two viral genomes, those of the extremely large (400 nm) *Mimivirus* and *Cotesia congragata bracovirus* (CcBv), are as large or larger than several bacterial genomes (Ward and Fraser, 2005).

Comparative genomics has also been a key contributor to the knowledge of horizontal gene transfer (HGT), the acquisition of heredity (especially DNA) from phylogenetically distant organismal lineages. In classic genetic theory, heredity elements (genes) are passed on "vertically" from generation to generation, from parental cell directly to progeny. HGT mechanisms [e.g., in prokaryotes the cellular uptake of DNA from the environment ("transformation"), intercellular plasmid transfer (conjugation), and virus-mediated genetic exchange (transformation)] are thought to contribute to genetic innovation in the organisms whose reproductive strategies lack eukaryotic sexual recombination. Through retrospective examination and comparison of genes among phylogenetically distant organisms, comparative genomics has made some startling discoveries. These include: (i) the presence of 81 archaeal-like genes clustered in 15 regions in the genome of *Thermotoga maritima* (domain *Bacteria*; Nelson et al., 1999); (ii) the presence of a bacterial genome fragment from the insect

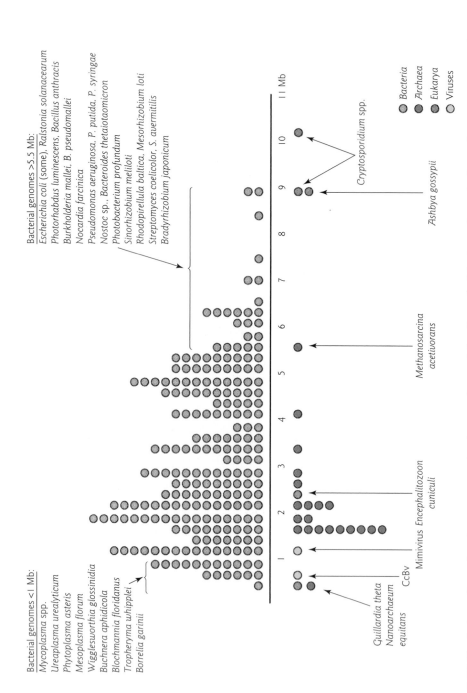

Figure 5.21 Depiction of overlapping genome size in members of the *Bacteria* (blue), *Archaea* (red), *Eukarya* (green), and viruses (yellow), in the size range (approximately 0.5–10.5 Mb) in which this overlap has been found to occur. The number of circles at a given point on the scale indicate the number of completed genomes that possess a specific size. Circles that represent unusually small (<1 Mb) or large (>5.5 Mb) bacterial genomes are labeled with the species name. (Reprinted from Ward, N. and C.N. Fraser. 2005. How genomics has affected the concept of microbiology. *Curr. Opin. Microbiol.* **8**:564–571. Copyright 2005, with permission from Elsevier.)

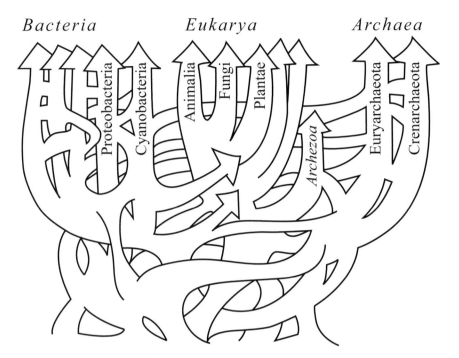

Figure 5.22 A version of the tree of life (based on small subunit rRNA sequences) that incorporates horizontal gene transfer processes in shaping the genetic composition of the three domains, *Eukarya*, *Archaea*, and *Bacteria*. (From Doolittle, R.F. 1999. Phylogenetic classification and the universal tree. *Science* **284**:2124–2129. Reprinted with permission from AAAS.)

endosymbiont, *Wolbachia*, in the X chromosome of the insect host (Kondo et al., 2002); and (iii) the presence of bacteriophytochrome genes in the genome of the fungus, *Neurospora crassa* (Galagan et al., 2003). The means by which the above genes were exchanged remains largely unknown. Nonetheless, such HGT mechanisms are vital in contributing to biological diversity. The impact of HGT on both the form and validity of the tree of life has been hotly debated for many years (e.g., Doolittle, 1999; Gogarten et al., 2002; Lawrence and Hendrickson, 2005). A version of the tree of life showing HGT is shown in Figure 5.22.

Another major trait for assessing biotic diversity is cell size (e.g., physical dimensions, shape, surface-to-volume ratio). This trait has major implications for the rate of material fluxes both within cells and between cells and their habitats. Ultimately cell size and shape reflect selective pressures over evolutionary history. Figure 5.23 presents a comparison of the size of known microorganisms (Ward and Fraser, 2005).

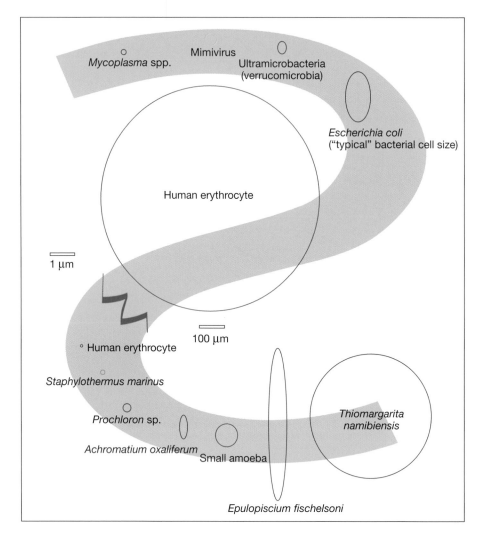

Figure 5.23 Depiction of overlapping cell size in members of the *Bacteria* (blue), *Archaea* (orange), *Eukarya* (green), and viruses (yellow). The diagram is divided into two different scales, with the upper portion showing the relative cell sizes of small and "normal" bacteria, as well as a mimivirus, in relation to a human erythrocyte (~9 μm in diameter). The same erythrocyte is used in the magnified lower portion of the diagram, to demonstrate the large cell size present in certain bacteria, culminating in the extremely large cells (average 500 μm diameter) of the giant sulfur bacterium *Thiomargarita namibiensis*. (Reprinted from Ward, N. and C.N. Fraser. 2005. How genomics has affected the concept of microbiology. *Curr. Opin. Microbiol.* **8**:564–571. Copyright 2005, with permission from Elsevier.)

Depicted on two different size scales are some of the smallest microorganisms (Figure 5.23, top) and some of the largest (Figure 5.23, bottom). The pattern in the range of physical sizes of organisms across domains is much like the pattern for genome sizes depicted in Figure 5.21.

Representatives of *Eukarya*, prokaryotes, and viruses do not necessarily sort predictably by size: they have overlapping physical dimensions. The mimivirus particle is larger than the smallest *Mycoplasma* (a bacterial parasite that lacks a cell wall). Moreover, at the other end of the size spectrum, the oxygenic phototroph, *Prochloron* (a symbiont of marine invertebrates) and the sulfur-oxidizing chemolithotroph, *Achromatium* (from freshwater sediments) are much larger than a human blood cell (erythrocyte). Two prokaryotes are nearly unbelievably large: *Epulopiscium* (a chemosynthetic heterotroph from fish intestines) and *Thiomargarita* (a sulfur-oxidizing chemolithotroph) dwarf the two eukaryotic cells shown in Figure 5.23, the human blood cell (erythrocyte), and a small free-living amoeba.

STUDY QUESTIONS

1 The illustration in Figure 5.1 shows one way (DNA staining) to microscopically distinguish microbial cells from other particulate materials in environmental samples. Use your knowledge of biology to suggest additional ways. Please state the approach, generally (based on broad principles) and provide three specific examples or strategies.

2 Regarding uncultured organisms (Figures 5.2, 5.3), the big challenge is trying to bridge the gap between "artificial laboratory media" and the metabolic resources that actually occur in nature. Can you suggest a way to create growth media that might successfully meet the challenge?

3 Envision this. You are the proprietor of a new bakery. The store offers many types of cookies. There are many variations on the themes of chocolate chip cookie recipes. Some have raisins, others oatmeal, others peanut butter, others are wheat-free, others are low calorie, others nondairy, while others offer mixed fruits, nuts, and butterscotch chips. You offer 250 permutations of the available combinations. In anticipation of the arriving crowds of customers, you begin to categorize the cookies so you can arrange and sell particular cookies from particular locations around the shop.

   (A) What signs would you make for displaying the cookies at each location? What criteria would you choose to distinguish one type of cookie from a close relative? Consider not only the ingredients in each recipe but the perceptions and expectations of the consuming public.

   (B) How different is this exercise from microbial taxonomy?

4 Section 5.4 and Box 5.3 describe six different approaches to assess microbial diversity. Devise a microbial diversity research question about a habitat that *you personally* are interested in. Then inspect the six diversity measurement approaches and choose one that best meets your objectives. Explain why you made your choice – justifying it on the basis of your goals, the assumptions behind the procedure, and the effort. Then state the limitations that you would need to understand when interpreting diversity data processed with the chosen approach.

5 In Section 2.7 the idea of a last universal common ancestor (LUCA) was introduced. Where is LUCA represented in Figure 5.7? What phylum shown on that figure is the closest relative to LUCA?

6 If, during evolution, humans had acquired chloroplasts and photosynthetic capabilities, where would we now fall on the tree of life (Figure 5.7)?

7 Information in Section 5.7 and Figures 5.15 and 5.16 suggest that the rate of nonculture-based discovery of new organisms greatly exceeds the pace of culture-based discovery. What factors contribute to this situation? What developments may reverse it?

8 In Section 5.8, a "kill the winner" rule was mentioned as one of the ecological roles of viruses in natural habitats. Does this make sense to you? If so, why? If not, why not?

9 Write a paragraph on the potential impact of lytic bacterial viruses on biogeochemical processes. Include in your discussion the evolutionary logic of parasites killing their hosts.

10 Figure 5.23 shows bacteria that are thousands of times larger than a common *E. coli* cell. Given the discussion of cell size in Section 3.5, can you make any inferences or develop testable hypotheses about the selective pressures, habitat, and bacterial physiologies that lead to large bacterial cells?

REFERENCES

Atlas, R.M. 1984. Diversity of microbial communities. *Adv. Microbial Ecol.* **7**:1–47.

Atlas, R.M. and R. Bartha. 1998. *Microbial Ecology: Fundaments and applications.* Benjamin/Cummings, Menlo Park, CA.

Barns, S.M., E.F. Delwiche, J.D. Palmer, and N.R. Pace. 1996. Perspectives on archael diversity, thermophily and monophyly from rRNA sequences. *Proc. Natl. Acad. Sci. USA* **93**:9188–9193.

Bohannan, B.J.M. and J. Hughes. 2003. New approaches to analyzing microbial biodiversity data. *Curr. Opin. Microbiol.* **6**:282–287.

Bonnell, D.A. (ed.) 2001. *Scanning Probe Microscopy and Spectroscopy: Theory, techniques, and applications*, 2nd edn. Wiley–VCH, New York.

Boone, D.R. and R.W. Castenholz (eds) 2001. *Bergey's Manual of Systematic Bacteriology*, Vol. 1, *The Archaea and the Deeply Branching and Phototrophic Bacteria*, 2nd edn. Springer-Verlag, New York.

Braga, P.C. and D. Ricci (eds) 2004. *Atomic Force Microscopy: Biomedical methods and applications.* Humana Press, Totowa, NJ.

Brenner, D.J., N.R. Krieg, and J.T. Staley. (eds) 2005. *Bergey's Manual of Systematic Bacteriology*, Vol. 2, *The Proteobacteria*, Part A, *Introductory Essays*, 2nd edn. Springer-Verlag, New York.

Burnett, J.H. 2003. *Fungal Populations and Species.* Oxford University Press, New York.

Cann, A.J. 2005. *Principles of Molecular Virology*, 4th edn. Elsevier Academic Press, Boston, MA.

Chaudhary, B.R. and S.B. Agrawal. 1996. *Cytology, Genetics and Molecular Biology of Algae.* SPB Academic Publications, New York.

Clarke, K.J. 2003. *Guide to the Identification of Soil Protozoa–Testate Amoebae.* CEH-Windermere, Ambleside, UK.

Coenye, T., D. Gevers, Y. Van de Peer, P. VanDamme, and J. Swings. 2005. Towards a prokaryotic genome taxonomy. *FEMS Microbiol. Rev.* **29**:147–167.

Cohan, F.M. 2002. What are bacterial species? *Annu. Rev. Microbiol.* **56**:457–487.

Curtis, T.P. and W.T. Sloan. 2005. Exploring microbial diversity – a vast below. *Science* **309**:1331–1333.

Curtis, T.P., W.T. Sloan, and J.W. Scannell. 2002. Estimating prokaryotic diversity and its limits. *Proc. Natl. Acad. Sci. USA* **99**:10494–10499.

Darby, I.A. and T.D. Hewitson. 2006. *In situ Hybridization Protocols*, 3rd edn. Humana Press, Totowa, NJ.

del MarLleo, M., B. Bonato, D. Benedetti, and P. Canepari. 2005. Survival of enterococcal species in aquatic environments. *FEMS Microbiol. Ecol.* **54**:189–196.

DeLong, E.F. 1998. Everything in moderation: Archaea as "non-extremophiles". *Curr. Opin. Genet. Dev.* 8:649–654.

DeLong, E.F., K.Y. Wu, B.B. Prezelin and R.V.M. Jovine. 1992. High abundance of *Archaea* in Atlantic marine picoplankton. *Nature* **371**:695–697.

Dighton, J. 2003. *Fungi in Ecosystem Processes*. Marcel Dekker, New York.

Dighton, J., J.F. White, and P. Oudemans (eds) 2005. *The Fungal Community: Its organization and role in the ecosystem*. CRC Press, Boca Raton, FL.

Doolittle, R.F. 1999. Phylogenetic classification and the universal tree. *Science* **284**:2124–2129.

Dugan, F.M. 2006. *The Identification of Fungi: An illustrated introduction with keys, glossary, and guide to literature*. American Phytopathological Society, St. Paul, MN.

Dworkin, M., M.S. Falkow, E. Rosenberg, K.-H. Schleifer, and E. Stackebrandt. 2006. *The Prokaryotes. A handbook on the biology of bacteria*. Springer-Verlag, New York.

Edwards, R.A. and F. Rohwer. 2005. Viral metagenomics. *Nature Rev. Microbiol.* **3**:504–510.

Fenchel, T. 1987. *Ecology of Protozoa: The biology of free living phagotrophic protists*. Science Tech Publishers, Madison, WI.

Feng, P., K.A. Lampel, H. Karch, and T.S. Whittam. 1998. Genotypic and phenotypic characterization of *Escherichia coli* O157:H7. *J. Infectious Dis.* **177**:1750–1753.

Fuhrman, J.A. 1999. Marine viruses and their biogeochemical and ecological effects. *Nature* **399**:541–548.

Fuhrman, J.A., K. McCallum, and A.A. Davis. 1992. Novel major archaebacterial group from marine plankton. *Nature* **356**:148–149.

Galagan, J.E., S.F. Calvo, K.A. Burkovich, et al. 2003. The genome sequence of the filamentous fungus *Neurospora crassa*. *Nature* **422**:859–868.

Gans, J., M. Wolinsky, and J. Dunbar. 2005. Computational improvements reveal great bacterial diversity and high metal toxicity in soil. *Science* **309**:1387–1390.

Garrity, G.M., T.G. Libum, and J.A. Ball. 2005. The revised roadmap to the manual. *In*: D.J. Brenner, N.R. Krieg, and J.T. Staley (eds) *Bergey's Manual of Systematic Bacteriology*, Vol. 2, Part A, 2nd edn, pp. 159–220. Springer-Verlag, New York.

Gevers, D., F.M. Cohan, J.G. Lawrence, et al. 2005. Re-evaluating prokaryotic species. *Nature Rev. Microbiol.* **3**:733–739.

Gogarten, J.P., W.F. Doolittle, and J.G. Lawrence. 2002. Prokaryotic evolution in light of gene transfer. *Molec. Biol. Evol.* **19**:2226–2238.

Graham, L.E. and L.W. Wilcox. 2000. *Algae*. Prentice Hall, Upper Saddle River, NJ.

Granoff, A. and R.G. Webster (eds) 2005. *Encyclopedia of Virology*, 2nd edn. Elsevier, New York.

Gualtieri, P. and L. Barsanti. 2006. *Algae: Anatomy, biochemistry, and biotechnology*. Taylor and Francis, Philadelphia, PA.

Hibbs, A.R. 2004. *Confocal Microscopy for Biologists*. Kluwer Academic/Plenum Press, New York.

Hirt, R.P. and D.S. Horner (eds) 2004. *Organelles, Genomes, and Eukaryote Phylogeny: An evolutionary synthesis in the age of genomics*. CRC Press, Boca Raton, FL.

Hong, S.-H., J. Bunge, S.-O. Jeon, and S.S. Epstein. 2006. Predicting microbial species richness. *Proc. Natl. Acad. Sci. USA* **103**:117–122.

Huber, H., M.J. Hohn, K.O. Stetter, and R. Rachel. 2003. The phylum *Nanoarchaeota*: present knowledge and future perspectives of a unique form of life. *Res. Microbiol.* **154**:165–171.

Hughes, J.B., J.J. Hellmann, T.H. Ricketts, and B.J.M. Bohannan. 2001. Counting the uncountable: statistical approaches to estimating microbial diversity. *Appl. Environ. Microbiol.* **67**:4399–4406.

Hurst, C. 2000. An introduction to viral taxonomy and the proposal of Akamara, a potential domain for the acellular agents. *In*: C. Hurst (ed.) *Viral Ecology*, pp. 41–62. Academic Press, San Diego, CA.

Kassen, R. and P.B. Rainey. 2004. The ecology and genetics of microbial diversity. *Annu. Rev. Microbiol.* **58**:207–231.

Kirk, P.M., P.F. Cannon, J.C. David, and J.A. Stalpiers (eds) 2001. *Ainsworth and Bisby's Dictionary of the Fungi*, 9th edn. CABI Bioscience, New York.

Knipe, D.M., P.M. Howley, and D.E. Griffin (eds) 2001. *Fundamental Virology*, 4th edn. Lippincott, Williams and Wilkins, Philadelphia, PA.

Kondo, N., N. Nikoh, N. Ijichi, M. Shimada, and T. Kukatsu. 2002. Genome fragment of *Wolbachia* endosymbiont transferred to X chromosome of host insect. *Proc. Natl. Acad. Sci. USA* **99**:14280–14286.

Könneke, M., A.E. Bernhard, J.R. de la Torre, C.B. Walker, J.M. Waterbury, and D.A. Stahl. 2005. Isolation of an autotrophic ammonia-oxidizing marine archaeon. *Nature* **437**:543–546.

Koonin, E.V., T.G. Senkevich, and V.V. Dolja. 2006. The ancient virus world and evolution of cells. *Biol. Direct* **1**:29.

Krebs, C.J. 2001. *Ecology: The experimental analysis of distribution and abundance*, 5th edn. Benjamin Cummings, San Francisco, CA.

Landolt, J.C., J.C. Cavender, and S.L. Stephenson. 2004. Cellular slime molds of the Great Smokey Mountains, National Park, USA. *Syst. Geogr. Plants* **74**:293–295.

Larkum, A.W.D., S.E. Douglas, and J.A. Raven. 2003. *Photosynthesis in Algae*. Kluwer Academic, Boston, MA.

Lawrence, J.G. and H. Hendrickson. 2005. Genome evolution in bacteria: order beneath chaos. *Curr. Opin. Microbiol.* **8**:572–578.

Lee, J.J., G.F. Leedale, and P. Bradbury (eds) 2000. *An Illustrated Guide to the Protozoa: Organisms traditionally referred to as protozoa, or newly discovered groups*, 2nd edn. Society of Protozoologists, Lawrence, KS.

Lundquist, J.E. and R.C. Hamelin (eds) 2005. *Forest Pathology: From genes to landscapes*. American Phytopathological Society, St. Paul, MN.

Madigan, M.T. and J.M. Martinko. 2006. *Brock Biology of Microorganisms*, 11th edn. Prentice Hall, Upper Saddle River, NJ.

Margulis, L., H.I. McKhann, L. Olendzenski, and S. Hiebert (eds) 1993. *Illustrated Glossary of Protoctista: Vocabulary of the algae, apicomplexa, ciliates, foraminifera, microspora, water molds, slime molds and the other protoctists*. Jones and Bartlett Publications, Boston, MA.

Martin, A.P. 2002. Phylogenetic approaches for describing and comparing the diversity of microbial communities. *Appl. Environ. Microbiol.* **68**:3673–3682.

Mayo, M.A., J. Maniloff, U. Desselberger, L.A. Ball, and C.M. Fauquet (eds) 2005. *Virus Taxonomy: Classification and nomenclature of viruses*. 8th Report of the International Committee on Taxonomy. Elsevier/Academic Press, Oxford, UK.

Morrell, V. 1997. Microbial biology: microbiology's scarred revolutionary. *Science* **276**:699–702.

Murphy, D.B. 2001. *Fundamentals of Light Microscopy and Electron Imaging*. Wiley-Liss, New York.

Nelson, K.E., R.A. Clayton, S.R. Gill, et al. 1999. Evidence of lateral gene transfer between *Archaea* and bacteria from genome sequence of *Thermotoga maritima*. *Nature* **399**:323–329.

Ohnishi, M., J. Terajima, K. Kurakawa, et al. 2002. Genomic diversity of enterohemorrhagic *Escherichia coli* O157 revealed by whole genome sequencing. *Proc. Natl. Acad. Sci. USA* **99**:17043–17048.

Oliver, J.D. 2005. The viable but nonculturable state in bacteria. *J. Microbiol.* **43**:93–100.

Øvreås, L., S. Jensen, F.L. Daae, and V. Torsvik. 1998. Microbial changes in a perturbed agricultural soil investigated by molecular and physiological approaches. *Appl. Environ. Microbiol.* **64**:2739–2742.

Pace, N.R., D.A. Stahl, D.J. Lane, and G.J. Olsen. 1986. The analysis of natural microbial populations by ribosomal RNA sequences. *Adv. Microbial Ecol.* **9**:1–55.

Peterson, R.L., H.B. Massicote, and L.H. Melville. 2004. *Mycorrhizas: Anatomy and cell biology*. NRC Research Press, Ottawa, Canada.

Purves, W.K., G.H. Orians, and H.C. Heller. 1992. *Life: The science of biology*, 3rd edn. Sinauer Associates/W.H. Freeman and Co, Sunderland, MA.

Rappé, M.S. and S.J. Giovannoni. 2003. The uncultured microbial majority. *Annu. Rev. Microbiol.* **57**:369–394.

Robertson, C.E., J.K. Harris, J.R. Spear, and N.R. Pace. 2005. Phylogenetic diversity and ecology of environmental *Archaea*. *Curr. Opin. Microbiol.* **8**:638–642.

Rohwer, F. and R. Edwards. 2002. The phage proteomic tree: a genome-based taxonomy for phage. *J. Bacteriol.* **184**:4529–4535.

Rosselló-Mora, R. and R. Amann. 2001. The species concept for prokaryotes. *FEMS Microbiol. Rev.* **25**:39–67.

Schloss, P.D. and J. Handelsman, 2004. Status of the microbial census. *Microbiol. Molec. Biol. Rev.* **68**:686–691.

Sleigh, M.A. 1989. *Protozoa and Other Protists*, 2nd edn. Arnot Publishers, New York.

Smith, D.F. and M. Parsons (eds) 1996. *Molecular Biology of Parasitic Protozoa*. Oxford University Press, Oxford, UK.

Spooner, B. and P. Roberts. 2005. *Fungi*. Collins, London.

Stackebrandt, E., W. Frederiksen, G.M. Garrity, et al. 2002. Report of the ad hoc committee for the re-evaluation of the species definition in bacteriology. *Internatl. J. Syst. Evol. Microbiol.* **52**:1043–1047.

Stephenson, S.L. and H. Stempen. 1994. *Myxomycetes: A handbook of slime molds*. Timber Press, Portland, OR.

Sterling, D.R. and R.D. Adam (eds) 2004. *The Pathogenic Enteric Protozoa: Giardia, Entamoeba,*

Cryptosporidium and Cyclospora. Kluwer Academic, Boston, MA.

Stevenson, R.J., M.L. Bothwell, and R.L. Lowe. 1996. *Algal Ecology: Freshwater benthic ecosystems.* Academic Press, San Diego, CA.

Suttle, C.A. 2005. Viruses in the sea. *Nature* **437**:356–361.

Sze, P. 1998. *A Biology of the Algae*, 3rd edn. McGraw-Hill, Boston, MA.

Taatjes, D.J. and B.T. Mossman (eds) 2006. *Cell Imaging Techniques: Methods and protocols.* Humana Press, Totowa, NJ.

Torsvick, V., J. Goksoyr, and F.L. Daae. 1990. High diversity in DNA of soil bacteria. *Appl. Environ. Microbiol.* **56**:782–787.

Torsvick, V., L. Øvreås, and T.F. Thingstad. 2002. Prokaryotic diversity: magnitude, dynamics, and controlling factors. *Science* **296**:1064–1066.

Tringe, S.G., C. von Mering, A. Kobayashi, et al. 2005. Comparative metagenomics of microbial communities. *Science* **308**:554–557.

Villarreal, L.P. 2005. *Viruses and the Evolution of the Life.* American Society for Microbiology Press, Washington, DC.

Vora, G.J., C.E. Meador, M.M. Bird, C.A. Bapp, J.D. Andreadis, and D.A. Stenger. 2005. Microarray-based detection of genetic heterogeneity, anti-microbial resistance, and the viable but nonculturable state in human pathogenic, *Vibrio* spp. *Proc. Natl. Acad. Sci. USA* **102**:19109–19114.

Ward, B.B. 2002. How many species of prokaryotes are there? *Proc. Natl. Acad. Sci. USA* **99**:10234–10236.

Ward, D.M. 1998. A natural species concept for prokaryotes. *Curr. Opin. Microbiol.* **1**:271–277.

Ward, N. and C.N. Fraser. 2005. How genomics has affected the concept of microbiology. *Curr. Opin. Microbiol.* **8**:564–571.

Wehr, J.D. and R.G. Sheath (eds) 2003. *Freshwater Algae of North America: Ecology and classification.* Academic Press, San Diego, CA.

Weinbauer, M.G. 2004. Ecology of prokaryotic viruses. *FEMS Microbiol. Rev.* **28**:127–181.

Wick, L.M., W. Qi, D.W. Lacher, and T.S. Whittam. 2005. Evolution of genomic content in the step-wise emergence of *Esherichia coli* O157:H7. *J. Bacteriol.* **187**:1783–1791.

Woese, C.R. 1987. Bacterial evolution. *Microbiol. Rev.* **51**:221–271.

Wommack, K.E. and R.R. Colwell. 2000. Virio-plankton: viruses in aquatic ecosystems. *Microbiol. Molec. Biol. Rev.* **64**:69–114.

Xu, J. 2005. *Evolutionary Genetics of Fungi.* Horizon Bioscience, Wymondham, UK.

FURTHER READING

Cohan, F.M. and E.B. Perry. 2007. A systematics for discovering the fundamental units of bacterial diversity. *Curr. Biol.* **17**:R373–R386.

Karl, D.M. 1986. Determination of *in situ* microbial biomass, viability, metabolism, and growth, *In*: J.S. Poindexter and E.R. Leadbetter (eds) *Bacteria in Nature*, Vol. 2, pp. 85–176. Plenum, New York.

Pace, N.R. 1997. A molecular view of microbial diversity and the biosphere. *Science* **276**:734–740.

Retchless, A.C. and J.G. Lawrence. 2007. Temporal fragmentation of speciation in bacteria. *Science* **317**:1093–1096.

Sogin, M.L., H.G. Morrison, J.A. Huber et al. 2006. Microbial diversity in the deep sea and the under-explored "rare biosphere". *Proc. Natl. Acad. Sci. USA* **103**:12115–12120.

6

Generating and Interpreting Information in Environmental Microbiology: Methods and their Limitations*

Using routine human curiosity as a guide, the standard set of questions to pose about naturally occurring microorganisms that dwell in biosphere habitats (e.g., water, sediments, soils, gastrointestinal tracts of animals, and all of the remainder) are: "Who? What? When? Where? How? Why?" Information in Chapter 5 (see Section 5.2) introduced the reader to the broad essentials of the methodological approaches in environmental microbiology (see Box 5.1). This chapter expands and elaborates upon the details of measurement procedures, experimental designs, data generation, and data interpretation. These details are essential for ongoing and future efforts aimed at advancing understanding of environmental microbiology, itself, and of its related disciplines (e.g., environmental engineering, biogeochemistry, ecology, medical microbiology, biotechnology). There are a variety of useful methods-related references that are available (e.g., Kemp et al., 1993; Weaver et al., 1994; Burlage et al., 1998; Rochelle, 2001; Hurst et al., 2002; Kowalchuk et al., 2004; Osborne and Smith, 2005).

Environmental microbiology is a methods-limited discipline. In this book, we have already seen at least two concrete examples of methodological innovations that drastically changed the intellectual landscape of environmental microbiology: (i) the development of nonculture-based procedures (see Sections 5.1, 5.2, 5.4, and 5.7); and (ii) Woesean molecular phylogeny that redefined evolutionary and taxonomic relationships among microbes (see Section 5.5). Some investigators would argue that metagenomics (Section 6.9) represents another major methodological innovation. There is every reason to anticipate additional methods-related paradigm shifts in environmental microbiology's future. Konopka (2006) has argued that a unifying set of theoretical ecological principles would be immensely beneficial in guiding future advancements in microbial ecology and environmental microbiology – such theories have the potential to match or exceed the influence of methodological innovation.

* Sections 6.2 and 6.10 of this chapter feature text reproduced with permission from *Nature Reviews Microbiology* (Madsen, E.L. 2005. Identifying microorganisms responsible for ecologically significant biogeochemical processes. *Nature Rev. Microbiol.* **3**:439–446. Copyright 2005, Macmillan Magazines Ltd, www.nature.com/reviews).

6.1 HOW DO WE KNOW?

Look out the window. Stroll through a woodland or a garden. Scrutinize photographs of the Earth's various biomes. What is happening in those habitats from a biogeochemical standpoint? How do we know what is happening? Box 6.1 illustrates what we can and cannot directly infer from snapshot images of landscapes. Environmental microbiology is a multifaceted discipline that relies upon natural history, inference, deduction, and experimentation to draw conclusions about causality in our world.

Epistemology is a way to approach information and knowledge. *The Oxford Companion to Philosophy* (Honderich, 1995) defines epistemology as "that branch of philosophy concerned with the nature of knowledge, its possibility, scope, and general basis". However, epistemology does not exist only in philosophical realms. Six decades earlier, Cunningham (1930) defined epistemology as "the science which sets forth and establishes the existence of true and certain human knowledge, the means of acquiring such knowledge, and the norm by which we can distinguish such knowledge from falsity". In forging even stronger bonds between scientific inquiry and epistemology, Bateson stated that epistemology is "a branch of biology . . . [it is] the process of the acquisition of information and its storage" (Donaldson, 1991). It is "the science of how we can know anything". Despite the clear impetus for pursuing knowledge of environmental microbiology, throughout its history, methodological limitations have impeded obtaining answers to fundamental questions about microorganisms in the biosphere. The objectives of this chapter are to examine constraints on knowledge of environmental microbiology and to describe how an integration and accrual of new methodologies into a continuum of field and molecular observations progressively advances the epistemological basis of environmental microbiology.

6.2 PERSPECTIVES FROM A CENTURY OF SCHOLARS AND ENRICHMENT-CULTIVATION PROCEDURES

Environmental microbiology may be described as a science for optimists. Driven by a sense of significance and discovery, investigators forge ahead despite adversity. One way to appreciate the role of optimism in environmental microbiology relies upon the metaphor of half a cup of green tea. Pessimists are often defined as the type of people who complain about the cup being

Box 6.1

A typical ecosystem: how do we know what biogeochemical processes are occurring?

Courtesy of E. L. Madsen, Cornell University, with permission.

Left is a photograph of a temperate forest and freshwater stream. What organisms are there? What is happening biogeochemically? Is carbon being fixed? Is carbon being respired? Is nitrogen being fixed? Is the nitrogen in biomass proteins being converted to amino acids, then ammonia, then to nitrate? Is sulfur being cycled? What organisms are there? Plants? Protozoa? *Archaea? Bacteria*? What are they doing? The only definite biogeochemical process evident in the photograph is CO_2 fixation by plants. Without photosynthesis, the plants would be absent.

Illustrating the methodological challenges: higher plants and their biogeochemistry versus microorganisms and theirs

The methodological challenges of discerning microbial activities in soil, sediment, or water may perhaps best be appreciated by considering how we know that higher plants carry out photosynthesis. In surveying a given landscape, humans can gather evidence for photosynthesis simply by noting the location of vegetation. Humans and the vegetation are roughly the same scale (approximately meters); therefore, detecting plants and their spatial relationships to one another and their habitats is facile. Photosynthesis is the major biogeochemical function of higher plants; without it there would be no plants, or food chains based thereupon. Thus, the presence of higher plants provides evidence for conversion of atmospheric CO_2 to biomass, and (because rooted plants are immobile) we simultaneously discern where the photosynthesis has occurred. At a mere glace then, humans gain plant-related biogeochemical knowledge addressing four key questions: "who?" (the plant), "what?" (photosynthesis), "when?" (recent history), and "where?" (the plant's location). To gain knowledge of the remaining two commonly asked key questions, "how?" and "why?", we rely on reductionistic biological disciplines that include physiology, biochemistry, genetics, and molecular biology – some of which can be applied to field-gathered plant samples or be manifest as chambers deployed to field sites. There are many striking contrasts between how the six key questions pertinent to plant photosynthesis are answered and how the same questions are answered for metabolic activities of microorganisms in field habitats (see main text).

half empty. Optimists, on the other hand, emphasize that the cup is half full. In environmental microbiology, research efforts seldom lead to a *complete* data set. There is always some missing information and the missing portion may be immense. The quotes in Table 6.1 provide a historical perspective on environmental microbiology's long-standing methodological limitations. The first entry in Table 6.1 is from S. Winogradsky who worked extensively from the late 1800s into the 1900s (see Section 1.3). The views shown in Table 6.1 span a century and show at least two major themes: (i) traits of microorganisms expressed in the laboratory are not likely to be those found in nature; and (ii) methodological obstacles need to be overcome so that long-standing answers to ecological questions can be obtained.

Enrichment culturing from nature tells us what might happen. Some of the earliest and most influential investigations in the history of environmental microbiology relied on enrichment culturing strategies (Beijerinck, 1888; Winogradksy, 1949; Overmann, 2006) to identify and isolate individual microbial cultures capable of carrying out novel metabolic processes, such as growth on ammonia as an energy source, fixation of atmospheric nitrogen into cell protein, and the use of unusual (perhaps pollutant) organic compounds as carbon and energy sources or final electron acceptors.

Enrichment culturing uses a sample of a naturally occurring microbial community as an inoculum for laboratory-prepared growth medium that is designed to select a small subset of the initial community (Figure 6.1). The logic behind enrichment culturing involves devising growth conditions that allow particular members of the community to multiply and

Table 6.1

Historical excerpts documenting methodological limitations in environmental microbiology (copyright 1996, from Madsen, E.L. 1996. A critical analysis of methods for determining the composition and biogeochemical activities of soil microbial communities *in situ*. *In*: G. Stotzky and J.-M. Bollag (eds) *Soil Biochemistry*, Vol. 9, pp. 287–370. Reproduced by permission of Taylor and Francis Group, a division of Informa plc.)

Excerpt	References
"We have as yet no science of soil microbiology proper, although we possess a great deal of information on various groups of soil microorganisms and what they can do when grown in artificial culture media", said S. Winogradsky, whose exact contributions to our understanding of certain soil microbiological processes can hardly be equaled.	Waksman, 1927
We possess, at the present time, considerable information on the organisms inhabiting the soil and on the chemical processes of many of these organisms, under controlled laboratory conditions; but little is known of the processes carried out in the soil itself, by the numberless representatives of the soil flora and fauna.	Waksman, 1927

Table 6.1 *Continued*

Excerpt	References
When I stand on the litter of a forest floor, or on a thatch of grassland, I like to look down, to ask myself just what do I really know about the microorganisms in that ecosystem. Immediately a whole host of categorical relationships, or definitions and generalizations from textbooks of ecology, come to mind. I feel a certain satisfaction in knowing that moisture and temperature influence microbial activity, but nevertheless, I still feel unable to comprehend the total functional ecology in that ecosystem. I know with reasonable assurance what measurements I can make or what microbes I can isolate or enumerate or identify within my own limited area of competence, and I realize what an exceedingly small area that competence embraces.	Clark, 1973
But, however the organism is obtained, whether from nature or from the laboratory, it is clear that it is not sufficient that the organism or consortium of organisms act . . . in the laboratory. Action must occur in the world at large. If we have learned nothing else from research in microbial ecology, we have learned that microbes do not do the same things in the laboratory that they do in nature.	Brock, 1987
The composition of the microbial community is still uncertain. The old explanations for the low recoveries of bacteria by plate counts and the platitudes concerning the role of particular morphological or physiological groups in soil processes remain.	Alexander, 1991
The stunning impact that new methods are having on aquatic microbial ecology points out that the field is still methods limited. Despite tremendous progress over the last 25 years, we lag well behind other ecological research areas. We have only a primitive ability to describe the organisms present in nature, what these organisms are actually doing, and what controls their activity and growth.	Hobbie, 1993
Both the activity and biomass measurements can be artifacts of the experimental methods employed. To some degree, the whole field is still methods limited. There are many successful methods but we still lack some crucial techniques that would allow a complete picture. Much of the effort in the field has been spent on investigations of the physiology and potentialities of many types of bacteria in the laboratory, yet in the field we do not even know the identity of the species carrying out many of the important processes, or the correct rate of growth of bacteria in water and sediments.	Hobbie and Ford, 1993
We need information on the species carrying out the microbial processes, the controls of these processes, the importance of grazing by protozoans or infection by viruses as controls of the microbial biomass, and microbial growth rates in nature. Although there has been tremendous progress, key techniques for measurement are still needed before we can have faith in the results and begin to apply the techniques over a range of habitats. These techniques will allow a quantitative view of the ecology of aquatic bacteria rather than the largely qualitative and descriptive view we have today.	Hobbie and Ford, 1993
The development of new methodologies to understand microorganisms in their environments has become a science unto itself. The need for these methods comes from the fact that classic methodologies of medical and industrial microbiology developed over the last hundred years produce artifacts when applied to natural populations in the environment.	Paul, 1993

eventually dominate within the mixed populations that were initially present. For instance, if one is interested in finding aerobic microorganisms that can grow on benzene (oxidizing it to CO_2 and incorporating the substrate carbon into new cells), then the enrichment medium would contain benzene as the sole carbon and energy source, and oxygen as the electron acceptor. A 1 g soil inoculum can contain thousands of species (see Section 5.4), although only a small percentage of these would be expected to grow on benzene. After a 1–2-week incubation, benzene de-graders would become dominant. Then, by plating small volumes of the enriched populations onto benzene growth med-ium solidified with agar (Figure 6.1), individual colonies of benzene degraders can be picked, further purified, isolated, and characterized using appropriate physiological, biochemical, and/or gen-etic procedures. *It is important to note that the microorganisms found from enrich-ment culture procedures and the metabolic information they generate may be ecologically irrelevant.* Each microorganism (in the laboratory or in nature) has the genetic potential to carry out a multitude of metabolic processes – each of which is conditionally regulated by their envir-onment. Therefore, the presence in an environmental sample of a particular organism or gene that is capable of catalyzing a particular process cannot be taken as evidence that the process is occurring in situ (Brock, 1987).

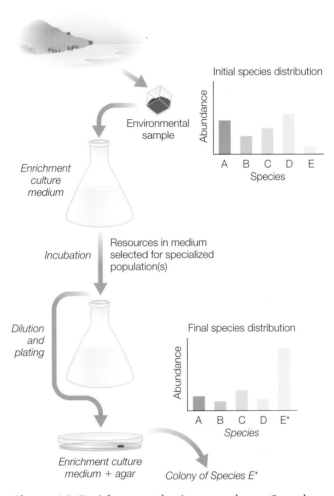

Figure 6.1 Enrichment culturing procedures. Growth conditions in the enrichment-culture medium provides electron donor, electron acceptor, and nutrients that allow a subset of the initial microbial community to flourish. To illustrate population changes during enrichment, five hypothetical species (A through E) are shown before and after enrichment.

6.3 CONSTRAINTS ON KNOWLEDGE IMPOSED BY ECOSYSTEM COMPLEXITY

To know that microorganisms are the agents of geochemical change in soil, sediments, and waters, environmental microbiologists face the

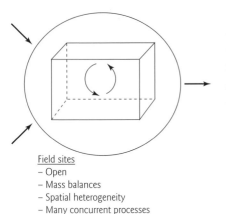

Field sites
– Open
– Mass balances
– Spatial heterogeneity
– Many concurrent processes

Figure 6.2 Diagram illustrating the open, dynamic nature of real-world microbial habitats such as soils, sediments, and waters. Not only do materials flow through the systems (making mass balances difficult to assemble), but the materials can be consumed, or generated by biological or chemical processes with this system.

challenge of documenting both the change (e.g., conversion of plant biomass to CO_2 in sediments, nitrogen fixation in soil, or methane production in wetlands) and the role of microorganisms as causative agents. *Microorganisms live and act in aquatic and terrestrial habitats whose complexities pose obstacles that impede directly measuring in situ activities of residence microorganisms.* As described in Chapters 1 and 4, these are continuous, open systems subject to fluxes of energy (e.g., sunlight, wind, tides) and materials (e.g., aqueous precipitation, erosion, deposition, infiltration, runoff) (Figure 6.2; see also Figures 1.3, 1.4). Thus, accurate accounting of the masses of materials is difficult, if not impossible, in most habitats. Even if accounts of material fluxes through open systems were accurate, another task remains – distinguishing microbial activities from the many other processes (chemical, physical, and physiological or other transformations carried out by higher organisms) that also influence field geochemical parameters. Except for photosynthesis by higher plants, most physiological reactions carried out by higher organisms are of little global significance. Yet in many localized habitats the contribution by plants and animals to the production, consumption, and transformation of geochemical materials cannot be ignored.

Although interactions within food webs can be modeled (see Figure 1.4), comprehensive documentation of the many simultaneous nutrient cycling, trophic, and biochemical interactions in field sites has yet to be achieved (Parton et al., 1988; Kroeze et al., 2003). Physical and chemical (abiotic) processes that must also be considered when measuring geochemical change in field settings include dilution, advection, dispersion, volatilization, sorption, photolysis, alteration by clay surfaces or other inorganic materials, and inorganic and organic equilibria (Voudrais and Reinhard, 1986; Wolfe, 1992; Thibodeaux, 1995; Stumm and Morgan, 1996; Schwarzenbach et al., 2002). Furthermore, within each microbial habitat, complex synergistic geochemical changes are effected by consortia of microbial species (Gottschal et al., 1992; Zinder, 1993; Schink and Stams, 2006; see Section 3.10). Additional complexity of microbial activities in field sites stems from their dynamic changes in time and space (see Section 1.4). The physical, chemical, nutritional, and ecological conditions of microorganisms in field settings are heterogeneous and vary from the micrometer to beyond the kilometer scales (Groffman, 1993; Hobbie, 1993; Hobbie and Ford, 1993; Parkin, 1993). Moreover, the biota and their respective physiological (e.g., growth rate, excretion, differentiation, death) and behavioral (e.g., migration, predation, competition, parasitism, symbiosis; see Sections 8.1 and 8.2) activities respond to climate-induced and/or other environmental changes. As mentioned in Chapter 1 (Section 1.4), if concentrations of any key biogeochemical parameters (e.g., ammonia, nitrate, methane, dissolved organic carbon) are found to fluctuate in lake water or sediments, interpreting such field measurements is very difficult. The changes in nutrient pools at any given moment are controlled by processes of production, consumption, and transport. Clearly, the many compounded intricacies of field habitats and microorganisms make their geochemical activities difficult to decipher.

6.4 ENVIRONMENTAL MICROBIOLOGY'S "HEISENBERG UNCERTAINTY PRINCIPLE": MODEL SYSTEMS AND THEIR RISKS

A Heisenberg-type uncertainty principle is inescapable in environmental microbiology (Madsen, 1991, 1998a, 2005) and must be confronted both in the examination of field-site samples and in exploiting the spectrum of disciplines that contribute to our mechanistic understanding of microbiological processes. When one begins in a field site or with site-derived samples, *the closer microorganisms are examined, the more likely the resultant information is to suffer from artifacts imposed by the measurement procedures.* This dilemma, which acknowledges the linkage between performing measurements on microorganisms and imposing artifacts on resultant data, is analogous to the Heisenberg uncertainty principle in quantum chemistry (Castellan, 1983; Zumdahl, 1986; Madsen, 1991). According to quantum theory, accurate measurements of the position and momentum of an electron are mutually exclusive. In environmental microbiology, the precision of reductionistic procedures and the relevance of information derived therefrom to field biogeochemical processes are often mutually exclusive because artifacts may develop. The basis for such artifacts is habitat disturbance and the responsiveness of both individual microorganisms and entire microbial communities to environmental change implicit in habitat disturbance (Hobbie, 1993; Hobbie and Ford, 1993; Madsen, 1996). Microorganisms are small (on the order of micrometers). The million-fold discrepancy in size between humans and microorganisms ensures that gathering field samples for microscopic and other analyses will physically disturb both the microorganisms and their habitats. Environmental microbiologists generally agree that, given sufficient time, the microbial community present in every environmental sample will change according to selective pressures (resources and environmental conditions) imposed by the removal of samples from their original location in field sites and by all intentional and unintentional laboratory incubation conditions (temperature, oxygen tension, physical disturbance, addition of nutrients or growth substrates, etc.). Removal of samples from a study site is equivalent to embarking on enrichment culturing procedures.

As described in Section 6.2, implicit in enrichment culturing procedures is the ability of microorganisms to respond and change when subjected to environmental perturbations. The nature of the microbial responsiveness during enrichment culturing is clear: resuscitation from dormancy and growth of (often) minor populations during laboratory incubation periods lasting days to years. But even if relatively brief incubations preclude shifts in population dynamics owing to growth and death, microorganisms still respond to environmental change. For instance, intricate biochemical signaling pathways allow cells to sense and respond to key nutrients (e.g., light, O_2, other electron acceptors, carbon sources; Antelmann et al., 2000), stress (e.g., acid, oxidative damage, inhibitory substances; Imlay, 2003), and cell-to-cell signaling molecules (quorum sensing pheromones; Miller and Bassler, 2001; see also Section 8.4 and Box 8.9). Timeframes for these responses range from nanoseconds (light), to milliseconds (O_2, toxicity), to minutes (enzyme synthesis) or hours (sporulation) (see Section 3.5, Table 3.5, and Figure 3.6).

This remarkable propensity of populations within naturally occurring microbial communities to change is a blessing for microbiologists practicing enrichment culture. However, it is a major impediment for those seeking to interpret physiological and ecological measurements performed on laboratory-incubated environmental samples such as water, soil, or sediments. The validity of measurements conducted on microbial communities removed from their original field setting is uncertain, because we cannot be sure that conditions imposed upon the native microorganisms (postsampling and incubation) have not quantitatively or qualitatively altered their populations and physiological reactions. Potentially misleading "bottle effects" are implicit in all measurements performed on sampled microbial communities (e.g., Venrick et al., 1977; Vaulot et al., 1995).

Therefore, for the practicing environmental microbiologist, there is considerable controversy surrounding the amount of time required for microorganisms in environmental samples to respond to sampling and experimentally induced environmental changes. Implicit in many published investigations is the hypothesis that accurate qualitative and quantitative microbial activity determinations of in situ processes can be performed in the laboratory within a "safe period" before artifacts develop (Figure 6.3). This hypothesis has not been adequately tested. Yet, its validity is essential for the extrapolation of results from laboratory incubations to field sites (Staley and Konopka, 1985; Karl, 1986, 1995; Tiedje et al., 1989; Pinckney et al., 1995; Madsen, 1996, 1998a). The alternative (conservative) methodological approach views laboratory incubations of environmental samples, at best, as a means toward estimating field processes. From the conservative viewpoint, quantitative extrapolation from laboratory results to actual field processes is taboo (Karl, 1995; Madsen, 1998a) because the instant an environmental sample is removed from a

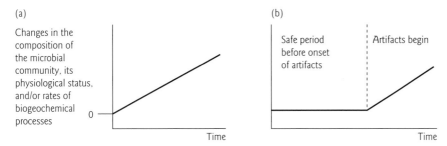

(a)

Changes in the
composition of
the microbial
community, its
physiological status,
and/or rates of
biogeochemical
processes

0

Time

(b)

Safe period
before onset
of artifacts

Artifacts begin

Time

Figure 6.3 Uncertainties in seeking data on in situ biogeochemical
processes from samples removed from the field and incubated in the
laboratory. The two graphs describe the quantitative and/or qualitative
influence of sampling and incubation on biogeochemical processes of
interest. (a) Changes in environmental samples may begin the instant
they are disturbed in a field site, or (b) after some uncertain "safe period"
during which valid measurements may theoretically be completed. (From
Madsen, E.L. 1996. A critical analysis of methods for determining the
composition and biogeochemical activities of soil microbial communities
in situ. *In*: G. Stotzky and J.-M. Bollag (eds) *Soil Biochemistry*, Vol. 9,
pp. 287–370. Copyright 1996. Reproduced by permission of Taylor and
Francis Group, a division of Informa plc.)

field study site, intricate and tightly regulated genetic-, biochemical-, cel-
lular-, and population-level changes may be triggered (see above). It is
the investigator's inability to obtain disturbance-free samples and to fully
characterize, understand, and duplicate field conditions in the laboratory
that undermine the acceptance of laboratory measurements performed
on field samples as valid surrogates for true in situ field processes. Figure 6.1 illustrates how the abundances of microbial populations (hence,
their physiological reactions) can shift during laboratory incubation of environmental samples, even in the absence of selective media.

6.5 FIELDWORK: BEING SURE SAMPLING PROCEDURES ARE COMPATIBLE WITH ANALYSES AND GOALS

Sampling is a methodology whose importance cannot be overestimated.
The most sophisticated methodologies asking "Who? What? When?
Where? How? Why?" deliver erroneous information if environmental sam-
ples containing natural microbial communities are mishandled.

Proper sampling of the source of microorganisms of interest is critical
for achieving valid data in environmental microbiology. Results from microscopic, biomarker, physiological, and cultivation protocols described in this
chapter are only as sound as the investigators' hands in gathering

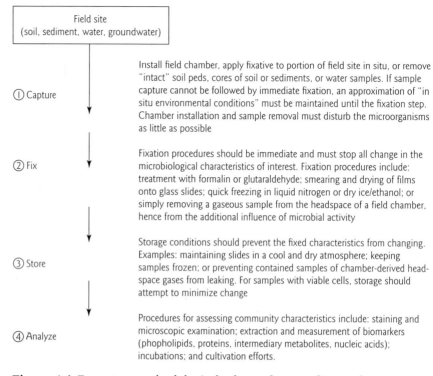

Figure 6.4 Four-step methodological scheme for sampling and processing and generating information from microbial communities in nature. (From Madsen, E.L. 1996. A critical analysis of methods for determining the composition and biogeochemical activities of soil microbial communities *in situ*. *In*: G. Stotzky and J.-M. Bollag (eds) *Soil Biochemistry*, Vol. 9, pp. 287–370. Copyright 1996. Reproduced by permission of Taylor and Francis Group, a division of Informa plc.)

microorganisms that truly represent the sampling site. Aseptic techniques (such as the use of flame-sterilized implements and the enclosure of samples within previously sterilized vessels) are often essential. Because microbiological characteristics of environmental samples are prone to postsampling changes (see Section 6.4), fixation procedures should be carefully scrutinized.

The overarching theme in sampling microbial habitats and in subsequent handling of the samples is to be sure that post-sampling procedures minimize changes in the microbial community while affording acquisition of the sought information. Figure 6.4 provides an overview of the four procedural steps that lead from a field site of interest to information about the native microorganisms. The capture–fix–store–analyze scheme in Figure 6.4 goes hand-in-hand with the paradigm shown in Figure 6.5,

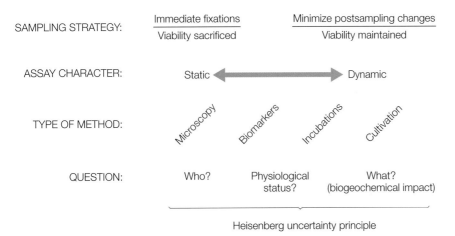

Figure 6.5 Integrated paradigm for methodological approaches in environmental microbiology. The four basic types of methods (microscopy, biomarkers, incubations, cultivation) are shown in relation to the questions being asked, to the character of the assays (static versus dynamic), and to the appropriate sampling strategies. Environmental microbiology's Heisenberg uncertainty principle (see Section 6.4) appears at the base of the diagram because avoidance of methodological artifacts is crucial in the quest to obtain valid, environmentally relevant data.

which integrates the questions posed with both the methods used and the corresponding sampling procedure.

Sampling for microbiological analyses

As shown in Figure 6.5, the environmental microbiological questions being posed, the methods being used, and the information being generated are under constant threat by the Heisenberg uncertainty principle (see Section 6.4). For this reason, the first sampling step in Figure 6.4 is labeled "capture". The idea is to *capture information about the field site's microbial community before the validity of the information drifts away*. Pragmatically, the capture step is the initial physical contact between investigator and the water, sediment, or soil of interest.

- **How do you remove a sample from the field? Is it physically disturbed?**
- **If so, how much time passes between disturbance and stabilizing the information delivered by the final analysis?**

As shown in Figure 6.4, the capture step may involve initiating a field incubation by placing a chamber over the field site so that dynamic changes in headspace gases (e.g., methane, CO_2, N_2O; see Sections 6.8, 7.3, and 7.4) can be monitored. If a field chamber is installed, the investigator must complete the assays rapidly so that the chamber, itself, does not alter microbial activity due to physical disturbances, restriction of gaseous exchange, and/or buildup of greenhouse heat. For static, snapshot-type assays aimed at describing community composition via its biomarkers, viability of cells is not of concern; thus, fixation of cells and their biomarkers in a field-like state should occur immediately after capture (Figure 6.5). Assays that involve incubation of environmental samples so that dynamic physiological measurements can be completed (e.g., sulfate reduction, methanogenesis, biodegradation of environmental pollutants) require viable cells; therefore, fixed samples cannot be used. Under these circumstances, an investigator's best strategy is to complete the dynamic physiological assays as rapidly as possible, while taking steps to mimic in situ conditions.

The "store" step in Figure 6.4 recognizes that time passes between capture of the sample and completing the analytical procedures that produce information about the microbial community. If samples are fixed in the field (bound for geochemical, biomarker, or microscopic assays) they must be stored so they do not change. If samples are bound for physiological assays (Figure 6.5), the most widely recommended sample-holding procedure is cooling of the samples on ice until laboratory processing. The microbial populations present in soil and water samples may begin to shift and change the moment an environmental sample is removed from the field site (see Sections 6.2 and 6.4). These changes continue through cold storage, distribution of the samples, assay vessels, and continued laboratory incubation during the assays. It is this inevitable, intractable set of microbiological changes (as well as our inability to match laboratory to in situ field conditions) that often make it unwise to extrapolate the results of laboratory physiology assays directly to field sites.

Sampling for habitat characterization

The capture–fix–store–analyze scheme of Figure 6.4 applies equally well to samples from field sites aimed at habitat characterization. To develop hypotheses about field selective pressures, resources exploited by microorganisms, electron donor–electron acceptor relationships, and biogeochemical processes catalyzed by microorganisms (see Sections 3.7, 3.8, and 7.4), we must have accurate data describing in situ field geochemistry. To obtain field geochemical databases, individuals conducting site surveys must be sure that results of analyses accurately reflect true field conditions, not postsampling changes in the samples. *Thus, there is a critical need to relate results of geochemical measurements, performed on field samples, directly to processes and conditions in the field.* Therefore, the utmost care must be taken to avoid artifacts that may be imposed on analytical results by imprudent delays in analysis completion, improper sample fixation, or laboratory incubations. Whenever possible, portable field instruments should be used. In subsurface habitats, cone penetrometry (Chiang et al., 1992) and a variety of in situ water analyses and sample gathering and fixation procedures are relevant to understanding on-site biogeochemical processes. As indicated in Table 6.2, however, many measurements cannot be completed in the field. When sample removal from field sites cannot be avoided, a variety of crucial decisions must be made. Selecting the locations within field sites for obtaining representative samples is no simple matter (Parkin and Robinson, 1994; Wollum, 1994).

Table 6.2

Organic and inorganic chemical assays pertinent to understanding the biogeochemistry of field sites (modified from Madsen, E.L. 1998b. Theoretical and applied aspects of bioremediation: the influence of microbiological processes on organic compounds in field sites. *In*: R. Burlage, R. Atlas, D. Stahl, G. Geesey, and G. Sayler (eds) *Techniques in Microbial Ecology*, pp. 354–407. Oxford University Press, New York. By permission of Oxford University Press)

Analytical approach	Sample preparation	Information	References
Portable field meters	None is required if probes can be immersed in soil, sediment, or waters in situ	Temperature, O_2, conductivity, and other measures pending availability of specific-ion electrodes and other analytical probes and standards	APHA (2002) and a variety of commercial manufacturers (e.g., Yellow Springs Instruments, Inc., Yellow Springs, OH)
Instruments in mobile field laboratories (e.g., spectrophotometer, GC, and others) and in base analytical laboratories (e.g., generally higher precision spectrophotometer, GC, GC/MS, HPLC, HPLC/MS, ion chromatography, and others)	Sampling, gathering, and fixation protocols vary with each specific assay. Fixation is designed to avoid chemical artifacts that may develop between the time that samples are removed from the field site and assays are completed	Inorganic nutrients and electron acceptors (e.g., O_2, NO_3^-, Fe^{3+}, Fe^{2+}, Mn^{4+}, Mn^{2+}, NH_4^+, S^{2-}, SO_4^{2-}, H_2, CO_2); organic constituents (e.g., the contaminants, co-contaminants, metabolites, dissolved and total organic carbon)	US Environmental Protection Agency methods as outlined by Christian (2003), Fifield and Haines (2000), Weaver et al. (1994), Wagner and Yogis (1992), Keith (1996), and a variety of commercial manufacturers (e.g., Hach Inc., Loveland, CO)

GC, gas chromatography; HPLC, high performance liquid chromatograph; MS, mass spectrometry.

Site heterogeneity and field analysis of electon-accepting processes

• **Should each sample be considered unique and treated independently?**
• **Or should samples be pooled (averaged) prior to performing measurements?**
Answer: Answers to such questions reside in the experimental goals and the methodologies employed (see below).

Deciding how to gather samples (with a shovel, spoon, pump, or drilling rig with or without aseptic techniques) can be a critical issue. In addition, the vessels for containing environmental samples must be clean, leakproof, and compatible with all intended uses. Once a field sample has been transferred to a container, the sample fixation protocol must be selected carefully to avoid artifacts. Obviously, fixation must prevent the parameter(s) of interest from changing and allows completion of the intended measurement procedures. Fixation may be accomplished by freezing (in liquid nitrogen or by placing samples on dry ice) or by adding biological inhibitors (e.g., formalin) and chemical fixatives (e.g., acid) (APHA, 2002).

Final electron acceptors that dominate the physiological reactions of field sites, or discrete zones therein, provide useful criteria for categorizing biogeochemical regimes (see Sections 3.7 and 3.8). Understanding these ambient conditions that control and respond to in situ physiological processes is essential. Site-specific efforts are required for defining which of the many possible biogeochemical regimes – for example, aerobic, denitrifying, iron-reducing, manganese-reducing, fermentative, dehalogenating, sulfate-reducing, or methanogenic (Zehnder and Stumm, 1988; Morel and Hering, 1993; Hemond and Fechner, 1994; Lovley et al., 1994; Stumm and Morgan, 1996) – actually occur in the field. And, of course, spatial and temporal heterogeneity are the key impediments for successful site characterization. For example, in uniform, well-mixed aquatic habitats, consistent readings from an oxygen probe at a variety of locations and times can be interpreted accurately and extrapolated to the site as a whole. However, in soils and saturated sediments that are spatially heterogeneous, it is very difficult to know precisely where and when particular physiological regimes are established. All site characterization data must be interpreted in terms of the physiological processes that produce and consume geochemical constituents (Smith, 2002). Key insights into in situ microbial physiology can be provided by field measurements of co-reactants and endproducts of microbial metabolism (e.g., CO_2, Fe^{2+}, Mn^{2+}, S^{2-}, NO_2^-, N_2O, NH_4^+, organic acids, methane, and other compounds indicative of electron-accepting processes; see Section 7.3), as well as by concentration gradients of final electron acceptors themselves (e.g., O_2, NO_3^-, Fe^{3+}, Mn^{4+}, SO_4^{2-}; see Sections 7.4 and 7.5) along site transects. In this regard, Chapelle et al. (1997) have devised a gas sampling bulb protocol for anaer-

Table 6.3

Relationships between the in situ partial pressure of hydrogen gas in anaerobic field sites and the dominant terminal electron-accepting process at that site (compiled from Chapelle et al., 1997)

Hydrogen gas partial pressure (nM)	Dominant terminal electron-accepting process
<0.1	Nitrate reduction
0.2–0.8	Iron reduction
1–4	Sulfate reduction
5–30	CO_2 reduction – methanogenesis

obic groundwaters in the field that, in combination with hydrogen gas determinations and Winkler titrations for oxygen (APHA, 2002), provides definitive information on dominant anaerobic redox couples (Table 6.3).

In interpreting field measurements, one must be mindful of the presence of microenvironments, which may allow localized pockets of anaerobiosis to occur in seemingly aerobic habitats. Furthermore, many of the reduced endproducts may diffuse away or be transported from the location where they were produced. For instance, detection of methane in field samples (from a natural gas-free locality) indicates that the highly reducing biogeochemical conditions associated with methanogenesis are operative in the vicinity of the sampling point. However, because methane is a volatile and mobile gas, its detection does not necessarily define the physiological activities in progress at the time and the location of sample removal.

6.6 BLENDING AND BALANCING DISCIPLINES FROM FIELD GEOCHEMISTRY TO PURE CULTURES

Assembling mass balances for geochemical components in field sites, distinguishing microbiological from other processes, and tracing circuitous routes of geochemical materials through food chains and oxidation/reduction reactions are formidable tasks (see Box 6.1 and Figure 6.2; see also Figures 1.3, 1.4). Many environmental microbiologists have confronted this situation and concluded that such adversities are nearly insurmountable in efforts aimed at discerning what microorganisms are doing in field sites (Bull, 1980; Hobbie, 1993; Hobbie and Ford, 1993; Madsen, 1996). The common way to contend with uncertainties of microbial activities is to initiate flask assays in the laboratory that monitor the chemical transformation(s) of interest in samples gathered from field sites. These laboratory assays provide definitive qualitative evidence for *potential* microbial metabolic reactions because sterilized or poisoned treatments can be examined as abiotic controls and mass balances are made possible by performing the assays in sealed vessels. For example, in the 1870s, while examining microbial transformations of nitrogen, Schloesing and Muntz (cited in Waksman, 1927) described a key link in the nitrogen cycle, nitrification (see Sections 7.3 and 7.4), by reporting that nitrate was formed from ammonia in nonsterile, but not in poisoned, columns of sand infiltrated by sewage effluent. However, as described in Sections 6.2 and 6.4, it is critical to acknowledge that measurements performed on laboratory-incubated environmental samples *reveal what may be, but not necessarily what is*, actually occurring in field sites.

Controlled model laboratory experiments allow a logical, reductionistic progression to proceed from field sample, to laboratory incubation, to enrichment cultures, to the isolation of pure cultures, and to elucidation of cellular and subcellular processes (Figure 6.6). This progression is the source of information presently available on ecological, physiological, biochemical, genetic, and molecular aspects of microbially mediated geochemical reactions. These model system approaches are powerful because of the control attained in the laboratory and the use of experimental designs that can address specific hypotheses. Ironically, this reductionism is another basis for environmental microbiology's Heisenberg uncertainty principle. As each layer of reductionism unfolds, the complexity of the experimental system under scrutiny diminishes (Figure 6.6). But with each simplification step, the likelihood of the resultant information being ecologically relevant also diminishes. Perhaps the riskiest step in attempting to gain a mechanistic understanding of biogeochemical processes is the selection of pure cultures for study (see Section 6.2). With only a few exceptions, such as disease-causing agents (see Section 6.10) or endosymbionts (Ruby, 1996, 1999; see Section 8.1), whose ecological niche often allows them to act almost as pure cultures in nature, imperfect methodologies and the complexity of field sites (see Section 6.3) have hampered environmental microbiologists' attempts to know which members of microbial communities are responsible for biogeochemical field processes (see Section 6.10).

The environmental relevance of data from pure-culture studies conducted in the laboratory is suspect for at least two reasons:

1 The in situ biogeochemical process of interest is likely effected by intact naturally occurring microbial communities composed of complex mixtures of cells that often constitute intricate biochemical food webs (Gottschal et al., 1992; Zinder, 1993; Schink and Stams, 2006). Thus, the single organisms examined in pure-culture investigations are unlikely to be active or numerically dominant in nature and, therefore, may not be the correct objects of study.

2 Even if the pure-cultured organism being studied was responsible for the metabolic process in situ, the laboratory conditions used to grow

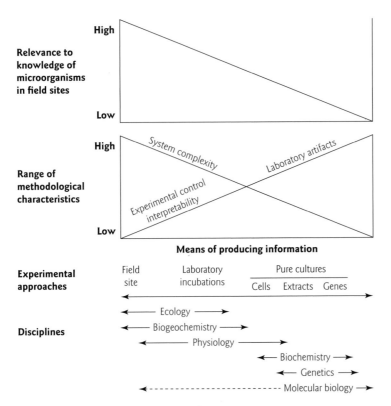

Figure 6.6 Relationships between means of producing information
in environmental microbiology, their methodological characteristics,
and their relevance to knowledge of microorganisms in field sites.
As experimental approaches and corresponding disciplines become
increasingly reductionistic (move from left to right), the relevance of
the result information to microorganisms in field sites has traditionally
diminished. (Reprinted and modified with permission from Madsen, E.L.
1998. Epistemology of environmental microbiology. *Environ. Sci. Technol.*
32:429–439. Copyright 1998, American Chemical Society.)

and characterize the behavior of the organism may depart radically from
the variety of influential in situ environmental factors (surfaces, colloids, gradients in substrate concentration, pH, final electron acceptors,
etc.). The luxurious growth conditions sometimes provided in the laboratory may cause the metabolic process being studied to differ, quantitatively and perhaps qualitatively, from the process in situ where
physical, chemical, and ecological constraints are likely to modify the
organism's expression and regulation of genes (Lindow, 1995).
The void between field conditions in nature and physiological conditions in the laboratory is analogous to the void that separates cultured
and noncultured microorganisms (discussed in Section 5.1). Researchers

on many fronts are actively working to better replicate natural selective and nutritional conditions that allow microorganisms to grow, to be isolated, and to carry out their normal biogeochemical processes in the laboratory (e.g., Button et al., 1993; Ward et al., 1995; Kaeberlein et al., 2002; Rappé et al., 2002; Könneke et al., 2005). These continued efforts will surely lead to increasing numbers of cultured microorganisms that are ecologically relevant (see Section 6.10).

In moving from left to right in the experimental approaches and scientific disciplines depicted in Figure 6.6, environmental microbiologists traverse from highly relevant but uncontrolled and sometimes uninterpretable field site measurements (see Figure 6.2 and Box 6.1) to sophisticated, yet simplified, experimental systems increasingly likely to induce artifacts and hence be of uncertain relevance to microbiological processes in nature. The nucleic acid-based surveys of microorganisms described in Chapter 5 (see Sections 5.2 and 5.7) and Section 6.7 often fail to detect microorganisms obtained via culture-based procedures (Ward et al., 1993; Pace, 1997; Janssen, 2006); thus, the free-living noninfectious, nonendosymbiotic microbial model systems studied in pure culture that supply virtually all of our knowledge of biochemistry, genetics, and molecular biology may not be ecologically significant. This does not mean that the fundamental biochemical and genetic processes revealed by laboratory-grown pure cultures have no bearing on ecological matters; indeed, many cellular processes, such as nucleic acid replication, ribosome structure, and adenosine triphosphate (ATP) generation are universal among virtually all life forms (Kluyver and van Niel, 1954; Neidhardt et al., 1990). Furthermore, laboratory experiments conducted on environmental samples, mixed cultures, and pure cultures have been invaluable in elucidating basic physiological principles of methanogenesis, nitrification, denitrification, and photosynthesis (among others) that control nutrient cycling in field sites (see below and Sections 7.4–7.6). However, biochemical divergence between field and laboratory metabolic processes should not be unexpected for many ecologically significant biogeochemical processes.

6.7 OVERVIEW OF METHODS FOR DETERMINING THE POSITION AND COMPOSITION OF MICROBIAL COMMUNITIES

Microorganisms removed from their native environments can be characterized microscopically (see Section 5.1). However, very little is known about the three-dimensional structure of microenvironments that surround microorganisms in field sites. New approaches such as environmental scanning electron microscopy and X-ray tomography are developing for examining complex environments such as soil. Yet complete microscopic characterization of soil is a distant possibility because the soil biomass occupies only 10^{-3} to $10^{-6}\%$ of the soil volume (see Section 4.6). This means that a multitude of microscopic fields, each surveying a very small volume of soil, would need to be processed to obtain information accurately representing in situ spatial relationships of soil microorganisms. Even in relatively homogenous habitats such as water columns in lakes and oceans, discerning dynamic, three-dimensional relationships within microbial communities has been elusive. Thus, unlike plants in landscapes (see Section 6.1), detailed knowledge of where microorganisms dwell is very difficult to obtain because of scale-related and sampling-related physical characteristics of microhabitats and microorganisms therein.

To answer the question "Who is there?", environmental microbiologists have developed four general types of assays already introduced in Box 5.1 and refined in Section 6.4 and Figure 6.5: (i) viable plate counts and isolation of organisms able to grow on laboratory-incubated selective agar media; (ii) extraction and analysis of nucleic acids, or other cellular biomarkers; (iii) microscopic examination of fixed, stained samples; and (iv) laboratory incubations that assess physiological potential of sampled microorganisms. Each of these methodologies has its own strengths and limitations. Common to the first three is the high probability of overlooking members of microbial communities that may be functionally significant but that may occur in low abundances and therefore be undetected. As we learned in Chapter 5 (Section 5.1), results of viable plate count assays provide information about the small (<1%) proportion of the initially diverse mixture of microorganisms that are able to grow under physiological conditions imposed by limited resources presented to the microorganisms in laboratory-incubated media. Unmet challenges in designing the proper laboratory conditions for growing microorganisms are a major reason for such low cultivation efficiencies, but some microorganisms may also attain an "unculturable" physiological state (see Section 5.1).

The microscopic approach for characterizing naturally occurring microorganisms typically disperses an environmental sample (e.g., soil), preserves it with a chemical fixative, and smears a portion onto a glass slide where the key microbial components (especially DNA, antigenic cell surfaces, or targeted nucleic acid sequences) can be stained (with general nucleic acid-binding dyes, cell-specific antibodies, or with gene-specific oligonucleotide hybridization probes, respectively) to distinguish microbial cells from the inorganic and noncellular organic materials (see Section 5.1). General nucleic acid staining provides information on total microorganisms but usually falls short of providing information about the identity of individual cells because few types of microorganisms are morphologically distinctive. Microscopy can be combined with cell-specific (antibody and nucleic acid) procedures designed to allow particular microorganisms or metabolic processes (as expressed genes) to be recognized. The resulting assays can yield powerful insights into the composition and activity of naturally occurring microbial communities. Microscopic and other assays for process-specific biomarkers are discussed further in Section 6.10. When such microscopy-based probes are used, it remains a challenge to verify the specificity and accuracy of results from cells that probe positively in complex, naturally occurring communities.

Extraction of cell-specific biomarkers has proven to be effective for some cellular components (such as phospholipids fatty acids; Tunlid and White, 1992; Findlay and Dobbs, 1993; Pinkart et al., 2002) but susceptible to inefficiencies and biases for others (such as nucleic acids; Moré et al., 1994; Farrelly et al., 1995; Suzuki and Giovannoni, 1996; Miller et al., 1999). Box 6.2 describes the use of phospholipid fatty acids in environmental microbiology. Nucleic acid extraction followed by cloning and sequencing of phylogenetically revealing 16S rRNA genes (discussed below and in Sections 5.4 and 5.7) has provided evidence for novel residents of many habitats. When applied to a given field site, the results of this phenotype-free means of identifying microorganisms usually contrast strikingly with those of growth-based assays. However, physiological inferences from phenotype-free methodologies can be misleading because prokaryotes that are closely related by small subunit rRNA sequence criteria can display widely different physiological and biogeochemical capabilities (Pace, 1997; see also Section 5.6).

Analysis of extracted nucleic acids, with emphasis on 16S rRNA genes

Although revolutionary insights into naturally occurring microbial communities have been provided by nucleic acid approaches (see Section 5.7), they have their own methodological biases that shape the outcomes of inquiries into microbial ecology (White, 1994; van Wintzingerode, 1997). The basic approach to nucleic acid analysis of naturally occurring microbial communities is depicted schematically in Figure 6.7. "Environment" in Figure 6.7 is any habitat of interest supporting a naturally occurring microbial community. Every link in the chain of events from the top to

---

Box 6.2

Using lipid biomarkers in environmental microbiology

A variety of non-nucleic acid biomarker molecules (e.g., phospholipid fatty acids, muramic acid, chitin, chlorophyll *a*) can be extracted from soil or other environmental samples to provide information about the abundance of specific groups of microorganisms (i.e., *Bacteria*, *Archaea*, fungi, algae). In utilizing these types of procedures, certain assumptions about the content of each biomarker per cell and their extraction efficiency must be carefully evaluated. Featured in this box are lipids, especially phospholipid fatty acids (PLFAs) because of the insights they provide into community composition. These insights are derived from the fact that various taxonomic microbial groups synthesize PLFAs of distinctive architecture. Structurally discernible characteristics of PLFAs [based on gas chromatography/mass spectrometry (GC/MS) analyses] include such traits as total number of carbon atoms per molecule, number of double bonds, position of the double bond relative to the omega end of the molecule, and whether or not the stereochemistry across the double bond is *cis* or *trans* (see below). The profiles of PLFA biomarkers from aquatic, sedimentary, and soil environments have been used to discern relative abundances of a variety of distinctive eukaryotic (e.g., plant, animal, microeukaryote) and prokaryotic (e.g., *Desulfobacter*, anaerobic desaturase pathways, *Bacillus*-type Gram positive) groups (Tunlid and White, 1992; Findlay and Dobbs, 1993; Pinkart et al., 2002). The procedures used to obtain and interpret PLFA profiles are elaborate and perhaps best learned by direct instruction in laboratories where the techniques have been established.

Signature lipid biomarkers can be extracted from a variety of environmental samples and operationally divided into five pools (see below). The pool of greatest utility in environmental microbiology is ester-linked PLFA (of *Bacteria*). The ester-linked PLFAs are essential components of intact cell membranes that are unstable after cell death and lysis. Thus, when intact ester-linked PFLAs are extracted and analyzed, the resulting data describe the composition of viable biomass.

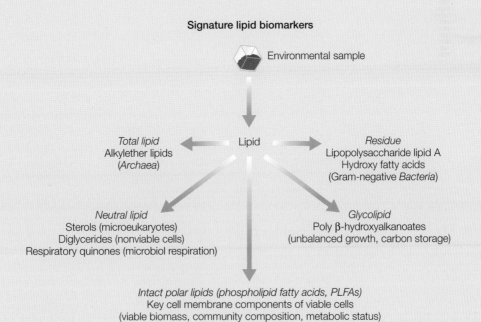

Signature lipid biomarkers

Environmental sample

Lipid

Total lipid
Alkylether lipids
(*Archaea*)

Residue
Lipopolysaccharide lipid A
Hydroxy fatty acids
(Gram-negative *Bacteria*)

Neutral lipid
Sterols (microeukaryotes)
Diglycerides (nonviable cells)
Respiratory quinones (microbiol respiration)

Glycolipid
Poly β-hydroxyalkanoates
(unbalanced growth, carbon storage)

Intact polar lipids (phospholipid fatty acids, PLFAs)
Key cell membrane components of viable cells
(viable biomass, community composition, metabolic status)

Box 6.2 *Continued*

PLFA analysis steps

1. Extract
2. Purify
3. Remove FA side chain
4. Derivatize FA side chain on carboxyl end
5. Analyze by GC/MS

Key characteristics of FA side chains

1. Type of fatty acids
2. Length
3. Site of double bonds
4. Stereochemistry at double bonds

Example

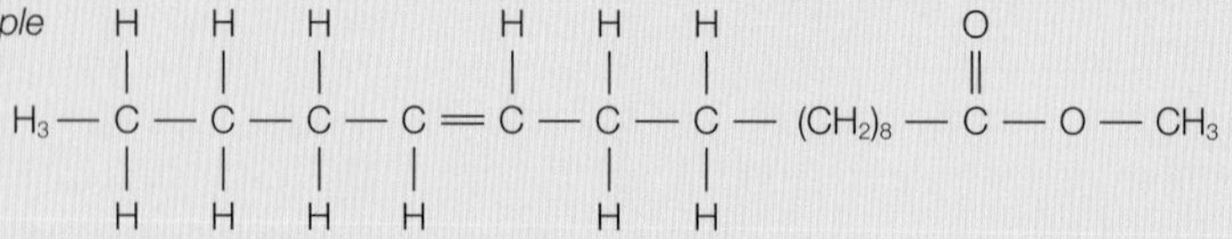

$$H_3 - \overset{\overset{\displaystyle H}{|}}{\underset{\underset{\displaystyle H}{|}}{C}} - \overset{\overset{\displaystyle H}{|}}{\underset{\underset{\displaystyle H}{|}}{C}} - \overset{\overset{\displaystyle H}{|}}{\underset{\underset{\displaystyle H}{|}}{C}} - \overset{\overset{\displaystyle H}{|}}{\underset{\underset{\displaystyle H}{|}}{C}} = \overset{\overset{\displaystyle H}{|}}{C} - \overset{\overset{\displaystyle H}{|}}{\underset{\underset{\displaystyle H}{|}}{C}} - \overset{\overset{\displaystyle H}{|}}{\underset{\underset{\displaystyle H}{|}}{C}} - (CH_2)_8 - \overset{\overset{\displaystyle O}{\|}}{C} - O - CH_3$$

Name : 16 : 1 ω 4t [16-carbon molecule prior to derivatization; 1 double bond in *trans* (t) configuration; located 4 carbons in from the methyl (ω) end of the molecule]

Chemical structure of PLFAs

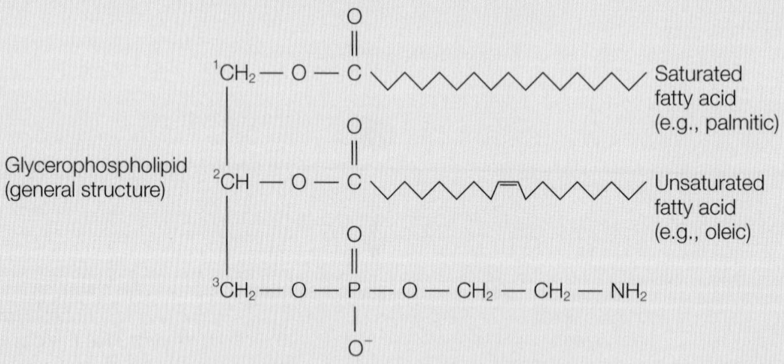

Glycerophospholipid
(general structure)

Head group substituent: Ethanolamine

bottom of Figure 6.7 must be carefully
scrutinized and flawlessly implemented
in order to obtain results that are truly
indicative of the microorganisms native to
the habitat. The environmental sample
may need to be aseptically handled to
avoid microbial contamination from irrelevant sources. Furthermore, whenever
possible, the sample should be frozen
immediately to avoid changes in the
microbial community imposed by physiological perturbations during sample
handling (see Sections 6.4 and 6.5).
Nucleic acid extraction (Figure 6.7) is
accomplished by a variety of physical
and chemical procedures (Ogram, 1998;
Miller et al., 1999) that lyse the cells and
isolate the nucleic acids from accompanying cellular and environmental debris and
solids (e.g., nonmicrobial biomass, detritus,
sand, silt, clay, humic acids). The extraction step may be far less than 100%
efficient and biased against lysis-resistant
microorganisms (Moré et al., 1994).

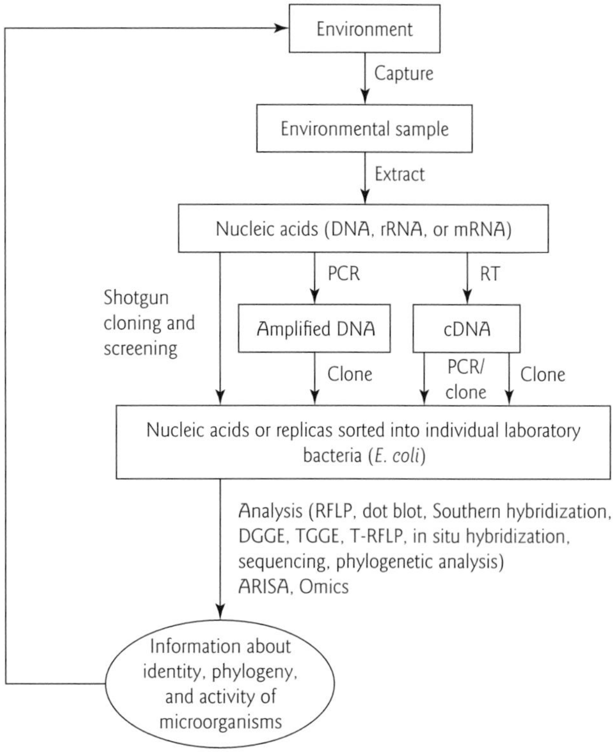

After the nucleic acids have been
extracted and purified, they exist as complex mixtures with individual molecular
fragments (genes, partial genes, or sets of
genes) often in low concentrations. The
cloning process allows these individual
DNA fragments to be incorporated into

Figure 6.7 Stepwise scheme for carrying out nucleic
acid analysis of naturally occurring microorganisms.
For abbreviations, see Table 6.4. (From Madsen E.L.
2000. Nucleic acid procedures for characterizing the
identity and activity of subsurface microorganisms.
Hydrogeol. J. **8**:112–125, fig. 1. With kind permission
of Springer Science and Business Media.)

individual vectors (such as those derived from plasmids or viruses; many
of these cloning vectors have been designed by molecular biologists for
genetic engineering purposes – see below and Section 6.9), which in turn,
are taken up by individual bacteria (e.g., *Escherichia coli*) that replicate in
Petri dishes. The three basic ways to clone (obtain replicas of nucleic acids
sorted into individual bacteria, Figure 6.7) are:

1 *Shotgun cloning.* This approach involves merging each individual extracted nucleic acid fragment with an individual plasmid or other vector,
   and allowing individual bacteria (e.g., *E. coli*) to take up the vector. This
   may be followed by screening of a large number of the recombinant
   bacteria (carrying a wide variety of nucleic acid fragments) for the rare
   gene(s) of interest.
2 *Polymerase chain reaction (PCR) amplification* from extracted DNA using
   primers for the gene of interest. These primers must be designed based

on prior DNA sequence information that indicates highly conserved, specific regions of the gene that may flank variable regions of the gene; such variable regions contain the nucleic acid sequence information of interest that reflects the identity or potential metabolic activity of microorganisms in the environmental sample.

3 *Reverse transcriptase (RT) step* that converts rRNA or mRNA to a complementary DNA (cDNA) template that can then be PCR amplified or directly cloned as described above (Amann et al., 1995; Liu and Stahl, 2002).

Thus, each bacterial colony that grows from the cloning steps shown in Figure 6.7 contains a single nucleic acid fragment originally present in the microbial community captured in the environmental sample.

None of the above three cloning strategies is artifact-free. All involve sequence-specific hybridization and binding for the amplification and merger (ligation) between vector and the sought nucleic acid fragment. Such molecular interactions are likely to favor some but not other sequences (Ward et al., 1995; van Wintzingerode et al. 1997). Sequence bias is likely to occur in all enzymatic processing (RT and PCR) of nucleic acids prior to cloning. Misrepresentation of the original community composition by PCR amplification and/or cloning has also been well documented (Farrelly et al., 1995; Suzuki and Giovannoni, 1996; Polz and Cavanaugh, 1998; Frey et al., 2006) and is analogous to the shifts in before-and-after populations shown by the bar graphs in Figure 6.1. A report by Tanner et al. (1998) demonstrated an additional threat to the validity of PCR-based cloning and sequencing procedures. These researchers discovered that rRNA sequences can be retrieved from reagent-only preparations to which no environmental sample was added. Thus, great care must be taken to avoid mistaking reagent-borne microorganisms for ones dwelling in the habitat of interest.

Subsequent to successful cloning of the nucleic acid of interest, various analytical procedures can yield the sought information describing the identity, phylogeny, and/or potential activity of the sampled microorganisms. Molecular biology offers many sophisticated tools for characterizing nucleic acids. Those listed near the bottom of Figure 6.7 (RFLP, dot blot, Southern hybridization, DGGE, TGGE, T-RFLP, ARISA, in situ hybridization, sequencing, and phylogenetic analysis) have been applied routinely in gathering information about the naturally occurring microorganisms that were the source of the nucleic acids (see Section 5.4). All of the above technical terms are explained in Table 6.4. Because of their significance, the community fingerprinting methods (T-RFLP, DGGE, TGGE, and ARISA) are described in Box 6.3. A key structural feature of the diagram in Figure 6.7 is the long vertical arrow that connects the information generated back to the environment. As has been elucidated by Amann et al. (1995) and Madsen (1998a, 2005), routine application of methods from laboratory experiments to habitats of interest and back again provides a

Table 6.4

Glossary and explanation of molecular biological and other procedures and terms used to gain information about microorganisms by analyzing and manipulating nucleic acids (from Ausubel et al., 1999; Alberts et al., 2002; Primrose and Twyman, 2006; and the Lyons website http://seqcore.brcf.med.umich.edu/doc/educ/dnapr/mbglossary/mbgloss.html)

Term	Meaning and use
16S rRNA	Small subunit RNA in ribosomes, whose sequence is the basis for molecular phylogeny and taxonomy of prokaryotes
454 pyrosequencing	The technology uses a massively parallel sequencing-by-synthesis (SBS) system capable of sequencing roughly 20 megabases of raw DNA sequence per 4.5 h run of the sequencing instrument. The system relies on fixing nebulized and modified DNA fragments to small DNA-capture beads in a water-in-oil emulsion. The DNA fixed to these beads is then amplified by PCR. Finally, each DNA-bound bead is placed into a small well on a fiber optic chip. A mix of enzymes is also packed into the well and the four nucleotides (TAGC) are washed in series over the chip. During the nucleotide flow, each of the hundreds of thousands of beads with millions of copies of DNA is sequenced in parallel. If a nucleotide complementary to the template strand is flowed into a well, the polymerase extends the existing DNA strand by adding one (or more) nucleotide(s). The addition of one or more nucleotides results in a reaction that generates a light signal that is recorded by the instrument. The patterns of emitted light are processed by a computer algorithm into the sequences of the original template. Prior to assembly, the length of each continuous strand of sequenced DNA is ~200 bp
ARISA	Automated ribosomal intergenic spacer analysis is a molecular fingerprinting procedure, related to T-RFLP, used to characterize microbial communities. ARISA relies on PCR to amplify the region of DNA that resides between 16S and 23S rRNA genes (for prokaryotes). Like T-RFLP, one of the PCR primers bears a fluorescence tag. However, because the intergenic region is highly variable in length (150–1200 bp), no restriction digest is required
BAC	Bacterial artificial chromosome: a cloning vector capable of receiving between 100 and 300 kb of target sequence. BACs are propagated as a minichromosome in a bacterial host. The size of the typical BAC is ideal for use as an intermediate in large-scale genome-sequencing projects. Entire genomes can be cloned into BAC libraries, and entire BAC clones can be shotgun-sequenced fairly rapidly
Biomarker probe	Fluorescently tagged molecule that binds specifically to biomarkers
Biomarkers	Cell components (e.g., membranes, cell walls, enzymes, nucleic acids) whose detection is evidence for the identity and/or activity of microorganisms
BLAST	Basic local alignment search tool: a computer program that identifies sequence similarities among genetic sequences. The BLAST family of programs compares and provides a similarity score for DNA or protein sequences after matching them to sequences in huge compilations of DNA or protein-sequence databases, such as Genbank and Swiss-Prot

Table 6.4 *Continued*

Term	Meaning and use
Blotting	A technique for detecting one RNA within a mixture of RNAs (a northern blot) or one type of DNA within a mixture of DNAs (a Southern blot) or one type of protein within a mixture of proteins (a western blot). A blot can prove that one species of RNA or DNA or protein is present, how much there is, and its approximate size. Basically, blotting involves gel electrophoresis, transfer to a blotting membrane (typically nitrocellulose or activated nylon), and incubating with a radioactive probe. Exposing the membrane to X-ray film produces darkening at a spot correlating with the position of the DNA or RNA or protein of interest. The darker the spot, the more targeted molecule there was
cDNA	Complementary DNA: a piece of DNA copied from an RNA molecule (usually mRNA, but also rRNA) using the reverse transcriptase (RT) enzyme
Cosmid	A type of cloning vector used to clone large pieces (35–45 kb) of DNA. These are plasmids that have been modified to resemble virus particles (they have a "cos" site). Cosmids can be packaged into bacteriophage heads (a reaction which can be performed in vitro) and then efficiently introduced into bacteria
DGGE	Denaturing gradient gel electrophoresis: specialized 16S RNA gene PCR primers amplify a portion of the gene (~400 bp) and include a high "G + C clamp" that anchors one end of the double-stranded DNA to itself. The gene fragments are all the same size, but they move to different locations in a denaturing gradient gel because different sequences are immobilized (due to denaturation) at different locations in the gel. This is a DNA fingerprinting procedure that is applied to PCR-amplified 16S rRNA genes from microbial communities. The result is a series of horizontal bands in a single lane of an electrophoresis gel. After separation, the separated individual bands (DNA fragments) can be isolated and sequenced
Dot blot	A procedure that distributes and fixes a variety of DNA standards in a matrix of distinct locations (as "dots") onto a nylon membrane. The dots are hybridized to unknown mixtures of labeled DNA. Locations where hybridization is strong reveals sequence similarity between known sequences and the unknown ones. Used to identify extracted or cloned DNA fragments, relative to known standards (see Blotting)
FISH	Fluorescent in situ hybridization: microscopic detection of cells whose biomarkers (e.g., DNA, rRNA) hybridize to fluorescently tagged probe molecules of known binding specificity
Flow cytometery	Analysis of biological material by detection of the light-absorbing or fluorescing properties of cells or subcellular fractions (i.e., chromosomes) passing in a narrow stream through a laser beam. An absorbance or fluorescence profile of the sample is produced. Automated sorting devices, used to fractionate samples, sort successive droplets of the analyzed stream into different fractions depending on the fluorescence emitted by each droplet

Table 6.4 *Continued*

Term	Meaning and use
Fluorescent antibody	Antibody raised in the immune system of a rabbit (for example) to recognize a particular antigen, such as a protein specific to a cell or an enzyme involved in a cellular process. After purification, the antibody can be linked to a fluorescent marker so that the targeted cell or protein can be recognized using in a sample of interest using fluorescence microcopy
Fosmid	A cloning vector. A fosmid is a type of cosmid (f-factor), which is like a plasmid, but is capable of containing much larger pieces of DNA, up to 50 kb compared to about 10 kb in a plasmid. Like plasmids, fosmids are circular. However, unlike plasmids, *E. coli* cannot carry multiple copies of a fosmid
Hybridization	The reaction by which the pairing of complementary strands of nucleic acid occurs. DNA is usually double stranded, and when the strands are separated they will rehybridize under the appropriate conditions. Hybrids can form between DNA–DNA, DNA–RNA, or RNA–RNA. The hybrids can form between a short strand and a long strand containing a region complementary to the short one. Imperfect hybrids can also form, but the more imperfect they are, the less stable they will be (and the less likely to form). To "anneal" two strands is the same as to "hybridize" them
Immunofluorescent probe	Microscopically visualized fluorescently tagged antibodies that recognize cell-specific antigens
In situ hybridization	See FISH
Microarray	Microarrays, otherwise known as "genechips", are tools developed originally by eukaryotic biologists to survey gene expression. The mRNA pool from a given tissue or cell line is used to generate a labeled sample which is hybridized in parallel to as many as 8000 DNA sequences that are immobilized onto a solid surface in an ordered two-dimensional array. Microarray technology can reveal patterns of genes that are activated under experimentally manipulated conditions
Microautoradiography	Use of microscopy to visualize silver grains created in photoemulsions by radioactive substances incorporated into microbial cells
mRNA	Messenger RNA: an RNA molecule, transcribed from DNA, which contains sequences translated by ribosomes into proteins
Northern hybridization	Transfer of RNA to a solid membrane where, after fixation, it can be hybridized against a variety of labeled nucleic acid templates so that the sought RNA fragment(s) can be identified (see Blotting)
Omics	The family of disciplines (e.g., genomics, proteomics, transciptomics, metabolomics, etc.) that systematically compile, analyze, and interpret bioinformatics data
PCR	Polymerase chain reaction: a technique for replicating a specific piece of DNA in vitro, even in the presence of excess nonspecific DNA. Primers are added (which initiate the copying of each strand) along with nucleotides and Taq polymerase. By cycling the temperature, the target

Table 6.4 *Continued*

Term	Meaning and use
	DNA is repetitively denatured and copied. A single copy of the target DNA, even if mixed in with other undesirable DNA, can be amplified to obtain billions of replicates. PCR can be used to amplify RNA sequences if they are first converted to DNA via reverse transcriptase. This two-phase procedure is known as RT-PCR (see RT)
Phylogenetic analysis	A set of quantitative approaches (manifest as computer algorithms) that infer evolutionary relationships among DNA and protein sequences. Often multiple alignments are analyzed – allowing groups of related sequences ("clades") to be displayed in relation to their relatives or neighbors on evolutionary trees. Commonly used computer programs include PHYLIP and PAUP
Plasmid	Cloning vector and naturally occurring extrachromosomal circular DNA of bacteria. *E. coli*, the usual bacterium used in molecular genetics experiments, will replicate plasmid DNA, as long as the plasmids have an "origin of replication". Plasmids carry inserted (cloned) DNA and produce millions of copies of the cloned insert
Quantitative PCR (qPCR)	A technology that allows a user to estimate the initial number of DNA templates that were present at the beginning of any given PCR reaction. Procedures rely upon spectrophotometric monitoring of PCR reactions as they occur in real time within individual wells of microtiter plates. Specialized PCR primers generate fluorescent signals with each PCR cycle. Calibration curves allow the user to relate the initial number of target DNA sequences to the number of PCR cycles required to reach a selected intensity of fluorescence
Real-time PCR	See Quantitative PCR
RFLP	Restriction fragment length polymorphism: a pattern-generating procedure that uses gel electrophoresis to separate DNA fragments that result from the recognition and cutting of an initially continuous piece of DNA by one or more DNA-cutting (restriction) enzymes
RT	Reverse transcriptase: an enzyme that will make a DNA copy of an RNA template. A DNA-dependent RNA polymerase. RT is used to make cDNA
Sequence	As a noun, "sequence" is the order of monomeric subunits in DNA or RNA or protein molecules. For DNA, the sequence is the arrangement of A, T, G, and C bases it contains. As a verb, "to sequence" is to determine the structure of a particular DNA, RNA, or protein molecule: for instance, the specific sequence of nucleotides in a piece of DNA
Shotgun sequencing	A way of determining the sequence of a large DNA fragment. The large fragment is broken into many small pieces (~3 kb), and then each is taken up by a plasmid in an *E. coli* host and sequenced. By finding out where the 3 kb pieces overlap, the sequence of the larger DNA fragment becomes apparent. Many regions of the original fragment will be sequenced several times; this overlap or "clone coverage" is necessary and allows assembly of the pieces

Table 6.4 *Continued*

Term	Meaning and use
SIP	Stable isotope probing: following compounds bearing stable isotopic atoms (signatures) into and through microbial communities to tag, separate, and, later, detect populations involved in metabolism of the compound
Stable isotopic signature	Distinctive ratio of heavy versus light atoms of a given element in a given compound; discernible by mass spectrometry
TGGE	Thermal gradient gel electrophoresis: analogous to DGGE, except the denaturing gradient is determined by temperature instead of salts that alter the binding strength of nucleic acid strands to one another. This is a DNA fingerprinting procedure that is applied to PCR-amplified 16S rRNA genes from microbial communities
Transposon	Transposons are one of three types of transposable elements that facilitate the mobilization of genes from one location to another within a given chromosome. Transposase enzymes facilitate insertion of transposon DNA in new locations. Transposon mutagenesis is a molecular technique that intentionally introduces mutations, as a means of genetic analysis
T-RFLP	Terminal restriction fragment polymorphism: PCR amplification of 16S rRNA genes inserts a fluorescent tag on one end of each of the amplified genes. After digestion with restriction enzyme(s), a single fluorescent molecule is formed from each amplified gene and the size of each fragment is governed by the gene sequence and the particular restriction enzyme(s) used in cutting the DNA. This is a DNA fingerprinting procedure that is applied to PCR-amplified 16S rRNA genes from microbial communities. The DNA analysis instrument detects the size and intensity of fragments – resulting in a chromatography-like fingerprint of the community
Western blot	A technique for analyzing mixtures of proteins to show the presence, size, and abundance of one particular type of protein. Similar to Southern or northern blotting, except that a protein mixture is electrophoresed in an acrylamide gel, and the "probe" is an antibody which recognizes the protein of interest, followed by a radioactive secondary probe (see Blotting)

means to refine information, and to develop and test new hypotheses about microorganisms in nature (see Section 6.10).

Developments in genomic techniques, including large-scale nucleic acid sequencing and the use of microarrays for surveying microbial communities for gene content and expression, are discussed in Section 6.9.

Summary

Despite substantial sophistication in many of the above procedures assessing "Who is there?" in naturally occurring microbial communities,

Box 6.3

16S rRNA community fingerprinting: DGGE, TGGE, T-RFLP, and ARISA

DGGE (denaturing gradient gel electrophoresis)

Specialized 16S RNA gene PCR primers amplify a portion of the gene [~400 base pairs (bp)] and include a high "G + C clamp" that anchors one end of the double-stranded DNA to itself. The gene fragments are all the same size, but they move to different locations in a denaturing gradient gel because different sequences are immobilized (via denaturation) at different locations in the gel.

This is a DNA fingerprinting procedure that is applied to PCR-amplified 16S rRNA genes from microbial communities. The result is a series of horizontal bands in a single lane of an electrophoresis gel. After separation, the separated individual bands (DNA fragments) can be isolated and sequenced.

Figure 1 shows a photograph displaying a DGGE analysis of six different environmental samples, each containing distinctive microbial communities (Muyzer et al., 1993).

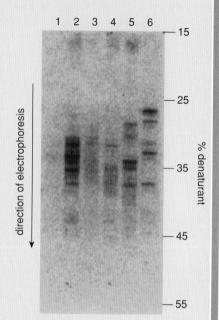

Figure 1 Denaturing gradient gel electrophoresis. (From Muyzer, G., E.C. DeWall, and A.G. Uitterlinden. 1993. Profiling of complex microbial populations by denaturing gradient gel electrophoresis analysis of polymerase chain reaction-amplified genes coding for 16S rRNA. *Appl. Environ. Microbiol.* **59**:695–700. With permission from the American Society for Microbiology.)

TGGE (thermal gradient gel electrophoresis)

Analogous to DGGE, except that the denaturing gradient is determined by temperature instead of salts that alter the binding strength of nucleic acid strands to one another.

This is a DNA fingerprinting procedure that is applied to PCR-amplified 16S rRNA genes from microbial communities.

T-RFLP (terminal restriction fragment length polymorphism)

PCR amplification of 16S rRNA genes inserts a fluorescent tag on one end of each of the amplified genes. After digestion with restriction enzyme(s), a single fluorescent molecule is formed from each amplified gene and the size of each fragment is governed by the particular gene sequence and the restriction enzyme(s) used in cutting the DNA.

This is a DNA fingerprinting procedure that is applied to PCR-amplified 16S rRNA genes from microbial communities (Abdo et al., 2006). The DNA analysis instrument detects the size and intensity of fragments – resulting in a chromatography-like fingerprint of the community.

Fingerprints from four different environmental samples, each containing distinctive microbial communities, are shown in Figure 2 (Liu et al., 1997).

ARISA (automated rRNA intergenic spacer analysis)

This is a molecular fingerprinting procedure, related to T-RFLP, used to characterize microbial communities (Fisher and Triplett, 1999). ARISA relies on PCR to amplify the region of DNA that resides between 16S and 23S rRNA genes (for prokaryotes). ARISA has also been developed and applied in characterizing members of eukaryotic microbial communities such as fungi (bearing 18S rRNA). Like T-RFLP, one of the PCR primers is labeled with a fluorescent tag. But because the intergenic region is highly variable in length (150–1200 bp), no digestion with restriction enzymes is required to develop and compare fingerprints of community composition. A typical freshwater microbial community exhibits ~40 peaks in an electropherogram.

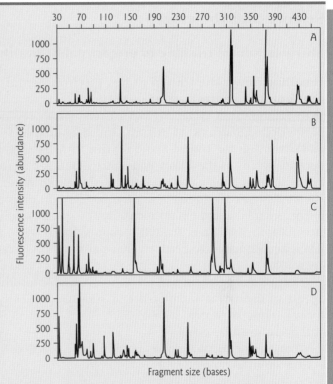

Figure 2 Terminal restriction fragment length polymorphism. (From Liu, W.-T., T.L. Marsh, H. Cheng, and I.J. Forney. 1997. Characterization of microbial diversity by determining terminal restriction fragment length polymorphisms of genes encoding 16S rRNA. *Appl. Environ. Microbiol.* **63**:4516–4522. With permission from the American Society for Microbiology.)

a complete census has yet to be successfully accomplished in any environment (see Section 5.4). Furthermore, of the millions of species of bacteria believed to exist globally, less than 7000 have been characterized in traditional culture collections, and more than 10 times that number of rRNA genes from field-extracted nucleic acids have been sequenced. Thus, nucleic acid-based procedures provide a profound reminder that there is much knowledge yet to be gained about the microbial world. Additional information on biomarker-based assays of microorganisms and their potential metabolic activity is presented in Section 6.10.

Science and the citizen

Microbial source tracking to protect public health

Headline news: the United States Environmental Protection Agency (EPA) must monitor and protect water quality

In 1972, the Clean Water Act was passed by Congress and signed into law by the president of the United States (R. M. Nixon). The legislation set the goal of assuring that lakes, rivers, and streams in the United States remain "fishable and swimmable" (Simpson et al., 2002). Section 303(d) of the Clean Water Act requires that total maximum daily loads be established for undesirable materials that detract from water quality. Such materials include: (i) chemicals (e.g., fertilizers, pesticides, metals); (ii) physical parameters (e.g., turbidity, heat, discharged solids); and (iii) microorganisms (especially pathogenic bacteria, protozoa, or viruses). According to Simpson et al. (2002), approximately 35%, 45%, and 44% of the assessed rivers, lakes, and estuaries, respectively, have at times been classified as "impaired", based on concentrations of one or more of the three types of pollutants.

Effective management and protection of public health requires that waterborne microbial agents of disease be monitored and controlled. Clearly, in order to know the microbiological status of water, reliable and effective practices for detecting potential microbial pathogens must be established:

- What are the methodologies used in these practices?
- How do they contend with uncultured microorganisms?
- Do the practices help identify the sources of microbiological contamination so that the sources can be identified and curtailed?

SCIENCE: Microbial source tracking is a continuously advancing discipline that uses both conventional (growth-based) and molecular tools to detect microbiological hazards and their sources (Santo Domingo and Sadowsky, 2006)

Background
For many decades, public health authorities have relied upon fecal coliform bacteria (such as *E. coli*) as an index for the potential presence of potent pathogenic microorganisms commonly transmitted via feces. Coliforms (operationally defined as "facultatively aerobic, Gram-negative, nonspore-forming, rod-shaped *Bacteria* that ferment lactose with gas formation with 48 h at 35°C) are common inhabitants of the intestinal tracts of humans and animals. Because coliforms are not normal inhabitants of soils, sediments, or natural waters, coliforms are used as *indicators* of fecal contamination of water. Coliforms are particularly useful indicators because they can be recovered from environmental samples as culturable colonies that grow on specialized media where they can be readily classified and characterized. If detected, coliforms signal that additional detective work is appropriate to identify the source of the contamination and that pathogen-specific tests may be warranted.

Challenge
Detection of microbial pollution in a given water body tells us that pollution has occurred – but what we really need is a way to distinguish between different types of culturable coliforms and match them to source reservoirs.

Meeting the challenge with culture-based procedures
The culturable fecal microorganisms found in waters can be subcultured and their resistance to different antibiotics can be determined (see Section 8.7). The phenotypic trait of antibiotic resistance is largely habitat-based. Human-derived fecal bacteria have greater resistance to antibiotics used in human medicine; while animal-derived fecal bacteria show their own spectrum of resistances. In a given watershed, there are a finite number of potential sources of microbial contaminants (e.g., animal production facilities, septic systems, sewage outfalls). By surveying patterns of antibiotic resistance in the source reservoirs and matching them to those found in contaminated waters, the offending culprits can be identified and forced to improve their waste-management practices.

Meeting the challenge with molecular biology procedures
Microbial source tracking (MST) practitioners are fully aware that indicator coliforms, though useful, do not tell the whole story. The main drawbacks are: (i) the detected indicator organisms are, themselves, not pathogens; and (ii) "uncultured" microorganisms (see Sections 5.1, 5.7, and 6.7) are overlooked.

As a first step toward refining current microbial monitoring efforts, molecular techniques have been successfully applied in characterizing cultured fecal coliforms and other microbial contaminants. The procedures that have been used include RFLP, PCR using primers that produce high-resolution fingerprints known as "rep-PCR", and T-RFLP (for explanations of these terms, see Section 6.7 and Table 6.4). These methods offer a superior degree of discrimination power, compared to patterns of antibiotic resistance.

The ultimate advancement in MST would assemble a network of highly sensitive *nonculture*-based monitoring procedures that target the entire spectrum of potential water-borne microbial pathogens (from viruses to prokaryotes to protozoa). Table 1 below shows currently available candidate methodological procedures that may strengthen MST approaches (Simpson

Table 1

Comparison of molecular typing methods that may be considered for MST (reprinted with permission from Simpson, J.M., J.W. Santo Domingo, and D.J. Reasoner. 2002. Microbial source tracking: state of the science. *Environ. Sci. Technol.* **36**:5279–5288. Copyright 2002, American Chemical Society)

Method	Description	Advantages	Disadvantages
RFLP	Electrophoretic analysis where DNA is detected with probes after Southern blotting	Reproducible	Technically demanding
		Most strains typeable	Many probes needed to achieve adequate discrimination
Ribotyping	Southern hybridization of genomic DNA cut with restriction enzymes, probed with ribosomal sequences	Works with most strains	Slow
		Automated	Complex procedure Inconclusive results

Table 1 *Continued*

Method	Description	Advantages	Disadvantages
Phage typing	Testing for susceptibility to different types of phages	Does not require electrophoresis	Need access to phage libraries
		High level of host specificity	Not all strains typeable
			Technically demanding
			Inconclusive results
rep-PCR	PCR is used to amplify palindromic DNA sequences	Discriminatory	Cell culture required
	Couple with electrophoretic analysis	Does not require knowledge of genomic structure	Requires large database of isolates
		Reproducible	Variability increases as database increases
DGGE	Electrophoretic analysis of PCR products based on melting properties of the amplified DNA sequence	Works on isolates and total DNA community	Technically demanding
		Reproducible	Time-consuming
			Limited simultaneous processing
LH-PCR	Separates PCR products for host-specific genetic markers based upon length differences	Does not require culturing	Expensive equipment required
		Does not require database	Technically demanding
T-RFLP	Uses restriction enzymes coupled with PCR in which only fragments containing a fluorescent tag are detected	Does not require culturing	Expensive equipment required
		Does not require database	Technically demanding
PFGE	DNA fingerprinting using rare cutting restriction enzymes coupled with electrophoretic analysis	High discrimination	Long assay time
		Works with most strains	Limited simultaneous processing
		Reproducible	
		Conclusive results	
AFLP	DNA fingerprinting using both rare and frequent cutting restriction enzymes coupled with PCR amplification	High discrimination	Technically demanding
		Works with most strains	Expensive equipment required
		Reproducible	
		Conclusive results	
		Automatable	

RFLP, restriction fragment length polymorphism; rep-PCR repetitive palindromic polymerase C reaction; DGGE, denaturing gradient gel electrophoresis; LH-PCR, length heterogeneity PCR; T-RFLP, terminal RFLP; PFGE, pulsed field gel electrophoresis; AFLP, amplified fragment length polymorphism.

et al., 2002). Initial applications of the procedures listed in the table will be restricted to scientific research. Under the Clean Water Act, thousands upon thousands of water samples are gathered and tested annually; thus, logistical and economic constraints will determine which, if any, of the procedures shown in the table will be routinely implemented on a broad scale.

Research essay assignment

If you were head of the World Health Organization (WHO), or the Centers for Disease Control (CDC), or the US Environmental Protection Agency (EPA), you would be facing the enormous task of protecting the public against potential disease-causing microorganisms. Yet, you would also be faced with limited financial budgets and limited personnel needed to implement programs. How would you set priorities for deciding which potential environmentally transmitted pathogenic agents should receive attention? List the criteria (up to 10) that you would use in prioritizing your agency's focus on environmental monitoring for disease prevention.

   Now use the world wide web and the published scientific literature to find the criteria actually used by the WHO, the CDC, or the EPA. Prepare an essay comparing your criteria with theirs.

6.8 METHODS FOR DETERMINING IN SITU BIOGEOCHEMICAL ACTIVITIES AND WHEN THEY OCCUR

Of the millions of microorganisms found in each cubic centimeter of soil, sediment, and water, there are thousands of species, each with complex genomes conferring the potential to carry out a variety of biogeochemical processes (see Chapters 3, 5, and 7). Furthermore, many naturally occurring microorganisms exist as spores or other resting, dormant, or nonviable forms (see Section 3.5). Thus, unlike the clear link between the presence of higher plant and photosynthetic activity (see Section 6.1), the presence of microorganisms in environmental samples provides few clues about their specific physiological functions in situ.

   The question "What are microorganisms doing?" can be subdivided into "What is the general physiological status of the cells?" and "What specific geochemical activities are the cells engaged in?" To assess the general physiological status of microorganisms in field sites, environmental microbiologists again rely on samples that are usually physically disturbed by removal from the field. And, similar to procedures inquiring about the composition of microbial communities (Section 6.7), information about physiological status can be obtained from measurements conducted on laboratory-incubated samples, from biomarkers extracted from the samples, and/or via microscopic techniques. Several key indicators of the

Table 6.5

Biomarkers used to assess the physiological status of microorganisms in microbial communities (from Staley and Konopka, 1985; Karl, 1986, 1993; Tunlid and White, 1992; Findlay and Dobbs, 1993; Madsen, 1996; Pinkart et al., 2002)

Biomarker	Information conveyed
Ribosome content	Indicative of protein synthesis activity
Intracellular energy reserves (poly-β-hydroxyalkanoates, ATP)	Overall energy charge and nutritional status
Proportions of *trans/cis* or cyclopropyl phospholipid fatty acids or electron transport carriers	These membrane components reflect nutritional status and starvation
Time-course measurements of cell elongation, uptake of physiological substrates, or the reduction of dyes indicative of respiratory activity	These assays, performed on laboratory-incubated environmental samples, provide an indication of metabolic activity

physiological status of microbial cells are shown in Table 6.5. These kinds of assays are insightful, but each is limited in the information provided and each carries artifactual risks associated with the Heisenberg uncertainty principle (Karl, 1986, 1995; Madsen, 1996; see also Figure 6.5 and below).

Methods for inquiring into the specific in situ geochemical activities catalyzed by microorganisms seek to document the impact of microbial activities on the chemical composition of soils, sediments, waters, and atmosphere. For some microbial activities, the geochemical materials of interest or related microbial metabolites are volatile gases; hence, the underlying net microbial processes are measurable using field chambers placed over the surface of habitats being studied (Figure 6.8) (e.g., Yavitt et al., 1990; Conrad, 1996). Examples include fluxes of nitrous oxide (see Sections 7.2–7.4) released from wetlands or fluxes of atmospheric methane consumed by microorganisms at the soil surface. However, when neither the geochemical materials nor their metabolic products are volatile, documenting field metabolic processes requires more elaborate strategies that include physiologically guided chemical analysis of field samples such as those shown in Table 6.6. The credibility of such biogeochemical activity measures varies on a case-by-case basis with the habitat studied, the means of procedural implementation, and the microbiological process of interest. Most are influenced by Heisenberg-type uncertainties discussed above (see Sections 6.4 and 6.5).

Accurate knowledge of temporal aspects of microbial activity in field sites, addressing the question, "When are the microorganisms active?", is difficult to obtain. If field samples are fixed the moment they are gathered, then information subsequently gleaned after analysis completion, can be considered indicative of the status of the microbial community at the time of sampling (see Section 6.4 and Figure 6.4; Madsen, 1996). This real-time type characterization of microorganisms in field sites is implicit in most field chamber (Figure 6.8; Conrad, 1996) and microelectrode (Glud et al., 1994) investigations and has been applied to biodegradation of environmental contaminants (Wilson and Madsen, 1996; Wilson et al., 1999). Another approach has been direct extraction and analysis of active enzymes (e.g., Ogunseitan, 1997). But knowledge of when microorganisms carry out key biogeochemical reactions is often uncertain – inferred after the fact. Just as net photosynthesis in higher plants is inferred by the presence of plant biomass, microbial decay processes in steady-state ecosystems (e.g., salt marshes, forests, grasslands) can be inferred from the steady state itself (Heal and Harrison, 1990; Likens and Bormann, 1995), but the scale of spatial resolution may not address the mechanistic biochemical intricacies sought by microbiologists.

Figure 6.8 Field-chamber approach for determining in situ biogeochemical activity. Shown is a cluster of open-ended stainless steel cylinders inserted into the soil surface. After being capped with red rubber septa, the cylinders allow changes in the concentration of head-space gases to be monitored over time. The gas-sampling syringe (shown) delivers the sampled gas to a gas chromatograph for analysis. The chamber pictured here (and larger ones, sometimes fashioned from plexiglass) is typically installed temporarily over soil or water habitats. Large chambers may be fitted with an internal battery-operated fan to insure uniform mixing of gases (e.g., CO_2, N_2O, and CH_4). (From J. Yagi, Cornell University, with permission.)

6.9 METAGENOMICS AND RELATED METHODS: PROCEDURES AND INSIGHTS

A major theme of this chapter (and, indeed, this entire book), is that environmental microbiology is a methods-limited discipline. When new types of measurements arise, the resulting novel information shapes the discipline's intellectual landscape and allows investigators to forge new frontiers. "Metagenomics" is a recent example.

Metagenomics

Metagenomics (also known as "environmental genomics" or "ecological genomics"; Handelsman, 2004; Allen and Banfield, 2005) is a methodology,

Table 6.6

Analytical chemistry procedures to assess the physiological processes carried out by microorganisms in field sites (from Karl, 1986, 1995; Revsbech and Jørgensen, 1986; Levin et al., 1992; Kemp et al., 1993; Pichard and Paul, 1993; Glud et al., 1994; Weaver et al., 1994; Hodson et al., 1995; Ogram et al., 1995; Madsen, 1996; Wilson and Madsen, 1996; Burlage et al., 1998; Wilson et al., 1999; Grossman, 2002; Hurst et al., 2002)

Chemical assay	Information conveyed
Stable isotope ratios and fractionation patterns in naturally occurring compounds	Signature ratios of stable isotopes can link pools of microbial substrates (e.g., carbon compounds) to their metabolic products (e.g., carbon dioxide or methane)
Release of stable, isotopically labeled materials	Imposed stable isotopic labels can be used to track the flow of pools of microbial substrates (e.g., $^{15}N\text{-}NH_4^+$) to their metabolic products (e.g., $^{15}N\text{-}NO_3^-$, when nitrification is being examined)
Isolating a portion of the habitat for hypothesis-driven manipulations	Chemical analyses performed on upstream and downstream samples can reveal processes that occur between sampling locations (e.g., metabolism of a groundwater pollutant such as toluene)
In situ microelectrode measurements of chemical gradients	Gradients of substances, especially electron donors and electron acceptors (e.g., oxygen or nitrate), are created by microbial respiratory processes
mRNAs and/or enzymes indicative of gene expression (e.g., fixation of carbon dioxide, nitrogen fixation, biodegradation of organic pollutants)	Measuring cellular precursors to biogeochemical processes support hypotheses that the processes are in progress in a field site. Results are convincing when experimental designs include data from adjacent, inactive control sites
Conducting physiological assays indicative of the metabolic activity of interest on laboratory-incubated field samples (e.g., methanogenesis, denitrification, or biodegradation of environmental pollutants)	Short-term incubations of field samples exposed to hypothetical substrates can confirm metabolic processes – especially when no lag time is found between addition of the substrate and subsequent metabolism

sometimes creating vast data sets, that advances our understanding of biotic diversity and evolution. New hypotheses about the function of micro-organism and microbial communities can arise from metagenomic data. Metagenomics is a logical extension of biomarker analysis, already described within environmental microbiology's "tool box" (see Figure 6.5 and Box 5.1). Metagenomics applies the large-scale cloning and sequencing procedures (that have led to complete DNA blueprints of single organisms; see Section 3.2) to DNA from environmental samples that contain entire microbial communities. Figures 6.9 and 6.10 provide a broad overview of the methods used in metagenomic studies. If you compare the nucleic acid-based community analysis already described in Figure 6.7 with Figures 6.9 and 6.10, the major contrasts are clear:

1 In metagenomics there is an expansion of interest from small subunit rRNA to all genes.

2 In metagenomics there is no PCR step that can distort the relative abundances of DNA fragments.

3 The size of the DNA fragment sorted into cloning vectors can be small [~3 kilobases (kb)] in the shotgun cloning approach or very large (~40–100 kb) in the bacterial artificial chromosome (BAC) and cosmid/fosmid cloning approaches.

4 The large (BAC or cosmid/fosmid) DNA fragments are subcloned, sequenced, and assembled into a single genomic fragment representing a portion of the genome from a single microorganism. Thus, associations between genes (especially 16S rRNA and functional genes) can be determined; these gene associations reflect those in a single uncultured host residing in the sampled community.

5 The short (shotgun) DNA fragments are derived from genetically heterogeneous populations. Even with hundreds or thousands of 3 kb pieces of cloned DNA (that contribute to thorough "clone coverage"), the sequences from shotgun cloning may not be able to be assembled because the 3 kb sequencing lengths from complex communities may not share sufficient sequence similarity to be aligned to create coherent contiguous genomic fragments (contigs).

As shown in Figure 6.9, after the environmental DNA has been incorporated into a cloning vector and a bacterial host, there are three distinctive strategies for continuing the metagenomic analysis (Riesenfeld et al., 2004):

1 The random (shotgun) sequencing approach uses plasmids as vectors, and thousands of ~3 kb inserts represented in the clone library are sequenced. Such high throughout sequencing projects (e.g., first three entries in Table 6.7) utilize instrumentation and facilities developed to sequence the human genome. The monetary expense of sequencing does not influence project implementation – the investigators "do it all". It follows then, that the shotgun approach to metagenomics generates immense DNA-sequencing data sets. Consistent with this, note that the clones obtained in entries 1–3 of Table 6.7 were not screened for

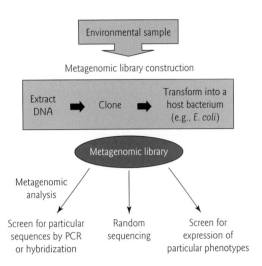

Figure 6.9 Overview of metagenomics: construction of a DNA library from mixed microbial populations in an environmental sample and three approaches for analyzing the sequences to reach conclusions about gene content. Both taxonomic (i.e., small subunit rRNA) and functional gene sequences are analyzed. The function of some genes can be confirmed by cloning and expressing them (as proteins) in bacterial hosts such as *E. coli*. PCR, polymerase chain reaction. (From Riesenfeld, C.S., D.D. Schloss, and J. Handelsman. 2004. Metagenomics: genomic analysis of microbial communities. *Annu. Rev. Genet.* **38**:525–552. Reprinted with permission from *Annual Review of Genetics*, Vol. 38. Copyright 2004 by Annual Reviews, www.annualreviews.org.)

Table 6.7

Examples of metagenomic investigations: their strategies, goals, and outcomes

Habitat sampled	Cloning strategy	Library screen	Vector	Size of inserted DNA (kb)	Total DNA sequenced (kb)	Outcome	References
1. Sargasso Sea (SS) and transit from SS through Panama Canal to Eastern Pacific Ocean	Shotgun	None	Plasmid	2.6	1.63×10^6 and 6.3×10^6	Mega-scale environmental genomics projects recovered $\sim 1.2 \times 10^6$ genes (many novel) in the 2004 effort and 6.12×10^6 protein in 2007. Extremely useful database. Assembly of contigs from heterogenous populations is a challenge	Venter et al., 2004; Rusch et al., 2007
2. Acid mine drainage	Shotgun	None	Plasmid	3.2	7.6×10^4	Extremely simple microbial community was amenable to high clone coverage and assembly of sequences as two near-complete genomes and partial recovery of three others. Community metabolism was modeled	Tyson et al., 2004
3. Eastern Pacific Ocean	Shotgun	None	Plasmid		$\sim 7.4 \times 10^2$	Thousands of previously unknown, undiscovered virus sequences were found	Breitbart et al., 2002

4. Pacific Ocean	Large insert	16S rRNA	BAC	80	60	Novel *Archaea* and associated genes found in ocean waters	Beja et al., 2000
5. Beetle endosymbionts	Large insert	Antibiotic production	Cosmid	–	54	Novel pathway for synthesizing a polyketide antibiotics was discovered	Piel, 2002
6. Soil	Large insert	16S rRNA	Fosmid	33–44	34	Novel link between uncultured soil *Archaea* (identified by the 16S rRNA genes) and associated genes	Quaiser et al., 2002
7. Soil	Large inset	Expression of antibiosis activity	Cosmid	–	~4	Discovery of genes that encode synthesis of antimicrobial compounds	Brady and Clardy, 2000
8. Soil	Small insert	Expression of lipase activity	Plasmid	2–7	613	Discovery of genes that encode lipolytic enzymes	Henne et al., 2000
9. Feces and soil	Large insert	Expression of biotin production	Cosmid	30–40	~11	Discovery of genes that confer biotin synthesis	Entcheva et al., 2001

BAC, bacterial artificial chromosome.

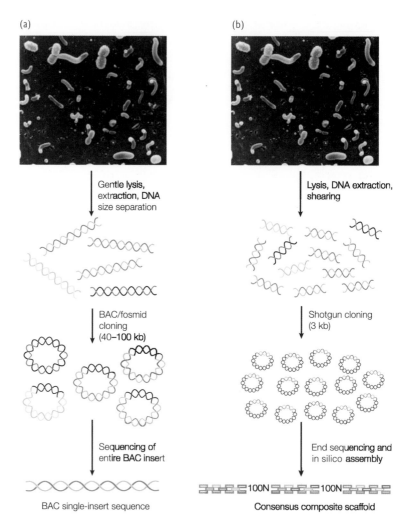

Figure 6.10 A glimpse of details of metagenomic methodology: microbial community DNA sequencing. Schematic diagram of common approaches for retrieving genomic sequence information from natural microbial populations. (a) One approach uses large DNA inserts recovered in bacterial artificial chromosomes (BACs) that are each derived from an individual cell. Subsequent sequencing and assembly results in a contiguous DNA sequence that is derived from a single cell in the original populations. (b) Another approach is based on recovery of small inserts, and attempts subsequent assembly from cloned DNA derived from a genetically heterogeneous population. The end result is an assembly of DNA sequence that is derived from many different cells. (Reprinted by permission from *Nature Reviews Microbiology*, from DeLong, E.F. 2005. Microbial community genomics in the ocean. *Nature Rev. Microbiol.* **3**:459–469. Copyright 2005, Macmillan Publishers Ltd, www.nature.com/reviews.)

particular genes of interest prior to sequencing; also, note that the numbers of base pairs sequenced is astonishingly high.

2 When a metagenomic investigation is focused upon retrieving large, intact DNA fragments (e.g., BAC cloning) derived from a single uncultured host bacterium, then the library of cloned DNA is normally screened prior to sequencing (Figure 6.9). If a known DNA sequence is being sought, then the screen for the clone library can be based on sequence-specific PCR or hybridization reactions performed on individual bacterial hosts carrying the cloned DNA. When the desired fragment is found, subcloning and sequencing is initiated (Figures 6.9, 6.10). Entries 4–6 in Table 6.7 provide examples of metagenomic libraries prepared from various habitats (ocean water, insect endosymbionts, soil) that led to the recovery of novel genes from uncultured hosts that shed light on microbial diversity (associations between hosts and their genes, or pathways of antibiotic production).

3 The third type of metagenomic strategy shown in Figure 6.9 (entries 7–9 in Table 6.7) seeks new genes or gene clusters from clone libraries based on the encoded phenotype. Thus, the cloned genes are transferred to a host bacterium (e.g., *E. coli*, *Pseudomonas*, or *Streptomyces*). If the genes are expressed in the new host and the sought phenotype is detected, the cloned DNA is then sequenced and analyzed. This phenotype-based approach is generally aimed at discovering the genetic basis for commercially promising products such as antibiotics, nutritional supplements, or enzymes. The screening step in such investigations is crucial: if novel, potentially valuable, cloned genes are not expressed in the screening host, the genes will go undetected. A prominent series of metagenomic investigations is summarized in Box 6.4.

Note: An approach, not shown in Figure 6.9, bypasses the cloning step completely – utilizing 454 pyrosequencing technology (see Table 6.4) to generate thousands of short (~200 bp) sequencing reads. The first application of this strategy in environmental microbiology was by Edwards et al. (2006). As 454 pyrosequencing and related new technologies become prevalent in future years, metagenomics is likely to gain prominence.

Other genome-enabled procedures

Now that we have entered the "genomic and postgenomic" age in biology, there is an almost predictable progression in techniques used to advance our understanding of individual organisms:

1 The genome sequence provides a blueprint and rigorous basis for hypotheses about function, regulatory networks, and, ultimately, the behavior and ecology of the sequenced organism.

2 Hypotheses from point 1 can be tested using techniques that conduct comprehensive surveys of transcribed genes [i.e., the transcriptome assessed with microarrays (see below) prepared from the genome

Box 6.4

Metagenomics and the role of proteorhodopsin in the oceans: more than "chasing sequences"

One of the most remarkable scientific success stories arising from metagenomic investigations is the discovery of light-dependent proton pumps, known as proteorhodopsins, in microorganisms that occur throughout the oceans. The ongoing series of investigations have developed as follows.

Beja et al. (2000) discovered on a 140 kb BAC clone from the eastern Pacific Ocean (Monterey Bay) a new class of photosynthetic genes in the rhodopsin family (named proteorhodopsin). Also on the BAC clone was the 16S rRNA gene of an organism known as "SAR86". Taxonomically, the SAR86 bacterium was an α-proteobacterium known only as a 16S rRNA gene cloned from the Sargasso Sea (hence "SAR"). Nothing was known about SAR86 biology or associated genes until the BAC clone linked the proteorhodopsin gene to its host. When the proteorhodopsin gene was transferred into an *E. coli* host and expressed, the protein product of the gene actually functioned in *E. coli* as a light-driven proton pump. Prior to this report, rhodopsins had only been known to occur in extremely halophilic *Archaea*. *Thus, the function, the host, and the habitat for light-driven pumps (potentially able to pump protons, hence contribute to cell metabolism such as ATP production) were all very exciting news. Metagenomics had possibly discovered an unknown and ecologically significant metabolic function.*

Beja et al. (2001) next reported from surveys of Monterey Bay waters, that both the genes encoding proteorhodopsin and the proteorhodopsin proteins, themselves, were widespread. To test whether proteorhodopsin-like molecules were functional in the planktonic microbial populations, the investigators analyzed membrane preparations from bacteria collected from surface water using a laser flash-photolysis technique (normally applied by biochemists to membrane preparations from pure cultures of photoactive organisms grown in the laboratory). The proteorhodopsin was there in cells native to the ocean. Moreover, new metagenomically produced BAC libraries (from Monterey Bay, the central Pacific, and the Southern Oceans) were found to contain proteorhodopsin-like genes. Sequence analysis of these proteorhodopsin genes showed structural variations in the encoded proteins. Furthermore, when expressed in *E. coli* the distinctive proteins exhibited distinctive differences in their light absorption spectra – one for shallow waters and one for deep waters.

Thus, the report by Beja et al. (2001) established the ubiquity of the novel proteorhodopsin gene in oceanic bacterioplankton, and suggested that the depth-dependent light intensities in the oceans had selected for specialized variants in the genes and proteins. *Overall, the global biogeochemical significance of proteorhodopsin-mediated phototrophy was becoming clearer: the hosts and proteins are widespread, and the genes are stratified in depth-selected populations. Therefore, proteorhodopsin-based phototrophy is likely to have a significant impact on carbon and energy flux in the ocean.*

The next major installment in the proteorhodopsin story came from a series of initially independent investigations. 16S rRNA gene surveys had been used to discover another 16S rRNA gene sequence from the Sargasso Sea (deemed "SAR 11"; Giovannoni et al., 1990) that was later shown to constitute ~24% of the rRNA genes in major ocean waters. Rappé et al. (2002) were able to culture and isolate a bacterium representative of the SAR11 group of ubiquitous oceanic bacteria. The genome sequence of the recently cultured representative of Sargasso Sea-type ("SAR") bacteria, deemed *Pelagibacter ubique*, revealed proteorhodopsin genes (Giovannoni et al., 2005a). In laboratory culture of *P. ubique*, the proteorhodopsin genes were expressed and the proteins exhibited absorption spectra similar to other previously characterized proteorhodopsins. However, when grown in filtered ocean water, *P. ubique* cells showed *no differences in growth rates or cell yields when in light or in darkness* (Giovannoni et al., 2005b). *Thus, the phenotype conferred by proteorhodopsin genes remains a mystery. There may not be one at all. Alternatively, the biochemical role of proteorhodopsin may be as a light sensor and its physiological impact may be subtle – important under special environmental conditions not yet simulated in the laboratory.*

Summary

Beja et al. (2000, 2001) reported elegant experiments describing the biogeography of proteorhodopsin genes and their encoded proteins. While proteorhodopsin phenotypes had been explored by expressing the proteins in *E. coli*, the physiological role of proteorhodopsins in their ocean-derived host cells had not been confirmed.

The genome-enabled physiological studies by Giovannoni et al. (2005b) of the SAR11 bacterium, which fortuitously was found to carry proteorhodopsin genes, showed that, despite being expressed, proteorhodopsin has no known physiological impact. New developments are expected after publication of this text.

Lesson

Nonculture-based and cultured-based procedures go hand-in-hand. The studies by Beja et al. (2000, 2001) and Giovannoni (2005a, 2005b) revealed the remarkable power of nonculture-based metagenomic inquiry. Hypotheses about previously unknown organisms and processes only arose from nonculture-based inquiry. But to test the hypotheses, culture-based physiological (and other) assays are required.

sequence], translated genes [i.e., the proteome, assessed with chromatography and mass spectrometry procedures that identify proteins extracted from cells using open reading frames (ORFs) in the genome as a guide for protein identification], and metabolites (i.e., the metabolome, the constellation of metabolites present in a cell at a given time, as assessed by chromatography and mass spectrometry).

3 Results from the "omics"-based procedures from point 2 are extended and confirmed using more traditional physiological and genetic (gene knockout and complementation) assays.

Part or all of the above three-step paradigm for advancing knowledge of individual organisms in pure culture has begun to be applied to intact, naturally occurring microbial communities. Three brief examples follow.

Example 1: metagenomic-enabled proteomics study

A metagenomic-enabled proteomics study has recently been completed on a microbial community dwelling in low pH (~0.8) mine drainage waters (Ram et al., 2005). This extends the work listed in entry 2 of Table 6.7 from a DNA blueprint of an extremely simple microbial community (approximately five microorganisms) to a snapshot of the physiological functioning of the community manifest as 2033 proteins distributed among 60 metabolic gene categories.

Example 2: microarrays

Microarrays, otherwise known as "genechips", are tools developed originally by eukaryotic biologists to survey gene expression (Sharkey et al., 2004). The mRNA pool from a given tissue or cell line is used to generate a labeled sample which is hybridized in parallel to a large number of DNA sequences that are immobilized onto a solid surface in an ordered two-dimensional array. Commercially available microarrays are treated glass slides with 8000 or more genes or probes per square centimeter. Gene expression levels are normally presented as ratios of hybridization signals found at a given location (representing a given gene) in the array, for treated versus control cells (Sharkey et al., 2004). When applied to pools of mRNA, microarray technology can reveal patterns of genes that are activated under particular conditions (ranging from invasion of a human by a pathogen to denitrification by a soil bacterium); thus linking the activity of particular genes to processes catalyzed by their hosts.

In environmental microbiology, application of microarray technology is appealing because it has the potential to monitor changes in community composition [via 16S rRNA genes (Loy et al., 2002; DeSantis et al., 2003) or whole genomes (Wu et al., 2004; Bae et al., 2005)] and changes in functional genes associated with processes that cycle nutrients such as carbon, nitrogen, or sulfur (Dennis et al., 2003; Tarancher-Oldenburg et al., 2003; Zhou, 2003; see Chapter 7). The degree of signal intensity from the probed microbial community reflects experimentally chosen hybridization conditions and gene abundances relative to negative-control genes on the array. Like all promising techniques in environmental microbiology, microarray approaches must be applied prudently because they have strengths and weaknesses. Recent efforts suggest that microarray hybridization technology may be successfully used to analyze functional genes in entire microbial communities after a relatively unbiased "multiple displacement amplification" procedure (Dean et al., 2001; Wu et al., 2006).

Example 3: phosphonates

Phosphonates are a form of phosphorus in marine systems long known to chemists, but the physiological impact of phosphonates was unknown until a clue was provided in the genome of the marine cyanobacterium,

Trichodesmium. Under conditions of phosphate starvation, *Trichodesmium* was found to express genes and corresponding proteins involved in phosphonate uptake and metabolism (Dyhrman et al., 2006). Thus, genomics-based assays were essential in revealing a metabolic response by *Trichodesmium* to a previously unrecognized nutrient in ocean water. This discovery may explain *Trichodesmium*'s prevalence in low-phosphate marine waters and why microorganisms lacking the genes for phosphonate utilization may be restricted to high-phosphorous habitats.

6.10 DISCOVERING THE ORGANISMS RESPONSIBLE FOR PARTICULAR ECOLOGICAL PROCESSES: LINKING IDENTITY WITH ACTIVITY

It is appropriate, at this point, to recall the remarks from various scholars shown in Table 6.1 about methodological limitations in environment microbiology. Given the independent challenges of answering "Who is there?" and "What are they doing?" it may seem glaringly optimistic to ask "Can the microorganisms responsible for particular biogeochemical processes in the field be identified?" After all, procedures for documenting in situ biogeochemical activity (without discerning the particular responsible party) are rather demanding (see Sections 6.4 and 6.8). Thus, studies pursuing the goal of identifying the microorganisms responsible for particular biogeochemical field processes are ambitious. Yet, they are crucial for advancing environmental microbiology and improving our ability to manage the biogeochemical processes that maintain the biosphere.

This section presents a perspective on past and current attempts to discover the identity of microorganisms that are responsible for catalyzing the key biogeochemical reactions of in situ soils, sediments, and waters. Insights are sought by contrasting ways of documenting causality in medical microbiology – Koch's postulates – with those of environmental microbiology. We will rely upon Figure 6.11, which presents a model for the generation and interpretation of environmental microbiological information that integrates five key considerations: (i) complexity of the experimental system under study; (ii) the path of inquiry taken by investigators to generate information; (iii) methodologies; (iv) data/information generated by the methods; and (v) procedures to assure ecological validity of the data.

Koch's postulates in medical microbiology and environmental microbiology

In 1884, Robert Koch developed fundamental criteria for proving that a particular microorganism (*Bacillus anthracis*) was responsible for a particular process (anthrax disease) in a particular habitat (sheep). This generalized four-step guideline, known as Koch's postulates, is as follows:

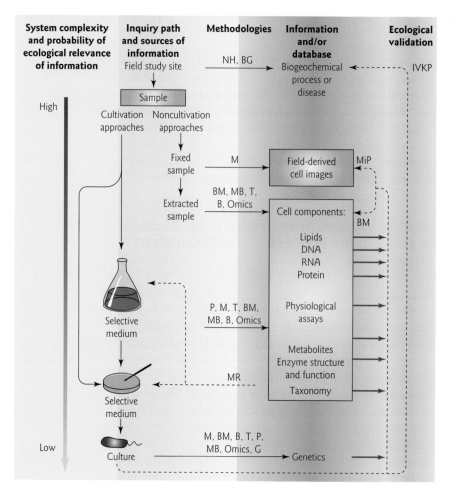

Figure 6.11 Model for the generation and interpretation of environmental microbiological information, with emphasis on field relevance and ecological validation of data. Column 1 provides a scale for evaluating the likely ecological relevance of information in the other four columns. With each successive methodological step away (down) from direct field measurements, the risk of artifacts (ecologically misleading data) increases. Column 2 provides an outline of microbiological procedures (cultivation-based or nonculture-based) that are used as sources of information about microorganisms in nature. Column 4 shows the types of information created by various methodological procedures (column 3). The dashed arrows in column 5 show the main feedback pathways that can be used to validate the ecological relevance of microbiological data. Dashed arrows connecting column 4 to 2 show a means for improving growth media, as guided by field-derived "omics" information. B, biochemical characterization, BG, biogeochemistry; BM, biomarkers (e.g., 16S rRNA genes, lipids); G, genetic characterization (e.g., operons, regulation); IVKP, inoculation to verify Koch's postulates; M, microscopy; MB, molecular biological characterization (e.g., cloning and sequencing); MIP, microscopic probing (immuno- and omics-based visualization in field-fixed cells); MR, medium refinement based on expressed genes and other biomarkers discovered in field samples; NH, natural history; Omics, genomics, proteomics, metabolomics, transcriptomics, and so on; P, physiological characterization; T, taxonomic characterization. (Reproduced and modified with permission from *Nature Reviews Microbiology*, from Madsen, E.L. 2005. Identifying microorganisms responsible for ecologically significant biogeochemical processes. *Nature Rev. Microbiol.* **3**:439–446. Macmillan Magazines Ltd, www.nature.com/reviews.)

1 The microorganism should be found in all cases of the disease in question, and the microorganism's distribution in the body should be in accordance with the lesions observed.

2 The microorganism should be grown in pure culture in vitro (or outside the body of the host) for several generations.

3 When such a pure culture is inoculated into susceptible animal species, the typical disease must result.

4 The microorganism must again be isolated from the lesions of such experimentally produced disease.

Koch's postulates have been the gold standard in medical microbiology for establishing causality and have survived intact to the present, with minor modifications that accommodate recent molecular biological techniques (Brooks et al., 2004; Falkow, 2004). The fifth column in Figure 6.11's paradigm provides a schematic mechanism for ecological validation of data, including inoculation to verify Koch's postulates (IVKP). For microbiologists concerned with ecological processes, linking a microorganism's identity to its activity in its habitat has, with several exceptions, proven to be an evasive goal. Below are suggestions why medical microbiologists have been far more successful than environmental microbiologists in identifying causative agents.

Table 6.8 compares and contrasts for medical and environmental microbiology four key factors that influence the determination of causality: complexity of the habitat plus its inhabitants, the process of interest, identifying a potential agent, and linking the agent to actual field processes. As stated in Table 6.8, human disease is readily recognized in the field (afflicted humans in society convey contagious agents) and has an enormous detrimental impact. Thus, the impetus for understanding and intervening is also enormous. In contrast, the impetus for the discovery and management of ecologically important biogeochemical reactions has been less pressing – perhaps because biogeochemical processes in field habitats are not facile to discern and because such processes generally proceed regardless of intervention.

"Culturability" is the other major factor that has likely allowed medical microbiology to flourish while environmental microbiologists have perhaps fallen out of step. Culturability is a direct reflection of two interacting issues: (i) the relative ratio of target to nontarget organisms in the initial inoculum; and (ii) an ability to accurately simulate the native habitat in media. When Robert Koch embarked down the cultivation-based path (Figure 6.11, column 2), his initial field sample (blood from a diseased sheep) was essentially a monoculture containing a "large number of regular, rod-shaped, colorless, immotile structures" (Koch, 1884) that were microscopically discernible. Compare this to the vast, confusing zoo of candidates (e.g., thousands of species and 10^9 cells per gram of soil) that confront a soil microbiologist. Furthermore, Robert Koch found that the blood-borne bacilli readily reproduced on solid media containing

Table 6.8

Contrasts between information on causality in medical and environmental microbiology (reproduced and modified with permission from *Nature Reviews Microbiology*, from Madsen, E.L. 2005. Identifying microorganisms responsible for ecologically significant biogeochemical processes. *Nature Rev. Microbiol.* **3**:439–446. Macmillan Magazines Ltd, www.nature.com/reviews)

Subdiscipline within microbiology	Traits of habitats studied	Characteristics of microbial processes	Steps to identifying potential causative agents	Ways of linking identity to field processes
Medical microbiology of human disease (e.g., anthrax, tuberculosis, SARS)	Habitat is human body Globally distributed Evolutionarily stable Consistent uniform resources for microbial colonization Reliably simulated in laboratory media or animal model Low-diversity microbial community offers few background organisms that confound isolation of causative agent	Diseases are reliably recognized in the field Immense negative impact on host; intervention essential Huge impetus for scientific study (disease prevention)	Pathogens are often culturable because habitats (hosts) are well simulated in laboratory media or animal model Disease specificity and habitat uniformity assure that a single agent is cause of global problem Relatively high chance of isolating correct organism because it comes from low-diversity community	Koch's postulates are well established for medical microbiology

Microbiology of ecologically important biogeochemical reactions (e.g., nitrification, sulfate reduction, methane oxidation, metabolism of environmental pollutants)	Soils, sediments, and waters are globally distributed, but show high physical and geochemical variability in time and space Highly variable resources; severe but unpredictable nutritional limitation is the rule Unreliably simulated in the laboratory because geochemical complexity defies characterization Extremely high community diversity; thousands of background organisms that can be mistaken for causative agents	Biogeochemical reactions are often difficult to document in the field; geochemical footprints of processes may not be apparent in open field sites Robust, reliable processes often have positive impacts, regardless of human understanding or intervention Historically, little impetus for scientific study, relative to human disease	Biogeochemical agents often have not yet been cultured because habitats are so poorly understood, so difficult to simulate Large- and small-scale habitat diversity may select for many different agents within flexible ecological guilds that carry out processes Relatively low chance of isolating ecologically significant agents because community diversity is immense Process may stem from many cooperating populations	Owing to habitat complexity, community complexity, culturing challenges, and perhaps functional redundancy in communities, Koch's postulates rarely apply Topic of ongoing multidisciplinary research involving microscopy, biomarker probes, stable isotopic signatures, autoradiography, and stable isotopic probing

"nutrient gelatin or boiled potato" (Koch, 1884). Facile culturability is not a given in medical microbiology [e.g., *Treponema pallidum* (syphilis) and *Mycobacterium leprae* (leprosy) cannot yet be grown in vitro; Brooks et al., 2004], but uniform, stable, globally distributed nutrient conditions of the human body are undeniably easy to mimic in growth media relative to the uncharacterized, site-specific, heterogeneous complexity of soils, sediments, and waters (see, for example, Sections 3.6, 4.6, and 6.2 and Box 6.1). Many biogeochemical processes are not catalyzed by individual microorganisms; but instead by cooperating populations (consortia). Moreover, it seems likely that guilds of physiologically equivalent microorganisms in different habitats may be compositionally distinctive (Table 6.8).

Thus, identifying ecologically significant microorganisms using Koch's postulates has been evasive because of a combination of impetus, community complexity, and limitations of cultivation techniques. Fortunately, other paths toward ecological validation exist that do not require cultivation of microorganisms. These paths (horizontal arrows in column 5 of Figure 6.11) often rely upon microscopic probing of field-fixed cell images for DNA, RNA, or other biomarkers indicative of cell identity and/or activity (see below).

Linking field biogeochemical processes to responsible agents

Recent progress has been made on many fronts that contribute to successful identification of ecologically significant microorganisms. These fronts include impetus for inquiry, deciphering community complexity, improving cultivation procedures, as well as development of new strategies and techniques that largely substitute for Koch's postulates (during the interim while microorganisms in biosphere habitats remain uncultured). Substitutes for Koch's postulates are outlined in the fifth column, bottom row of Table 6.8 and detailed examples are discussed below, especially as presented in Table 6.9.

Increasing impetus for understanding microbially mediated environmental processes probably reflects the growing public and governmental awareness of the frailty of our planet (e.g., Raven 2002; Sugden et al., 2003) under stresses of population growth, climate change, pollution, and disease transmission. Understanding the complexity of habitats and naturally occurring microbial communities is implicit in current research areas and exemplified by geochemical characterization of deep ocean hydrothermal vents (Reysenbach and Shock, 2002) and Lake Vostok buried deep beneath polar ice (Jouzel et al., 1999), and by recent metagenomic whole community genome-sequencing efforts (Hallam et al., 2004; Tyson et al., 2004; Venter et al., 2004; Rusch et al., 2007). Cultivation strategies have already taken a significant leap forward via efforts in which minimally altered environmental samples are used to meet the complex

Table 6.9

Selected examples of efforts in environmental microbiology to identify microorganisms responsible for field biogeochemical processes (reproduced and modified with permission from *Nature Reviews Microbiology*, from Madsen, E.L. 2005. Identifying microorganisms responsible for ecologically significant biogeochemical processes. *Nature Rev. Microbiol.* **3**:439–446. Macmillan Magazines Ltd, www.nature.com/reviews)

Process	Microorganism	Setting	Strategy	Commentary	References
1. Nitrogen fixation	*Rhizobium*	Nodule on roof of legume plant in field	Inoculate soil lacking native *Rhizobia*	Success for Koch's postulates. Root infection process selects for the bacterial symbiont. Nitrogen fixation results from inoculation and nodulation	Fred et al., 1932
2. Biodegradation of trichloroethene (TCE)	*Dehalococcoides*	Groundwater beneath Air Force base	Inoculate subsurface habitat where TCE persists	A version of Koch's postulates succeeds. Metabolism only occurred after inoculation	Major et al., 2002
3. Glutamate uptake and DNA synthesis	Unknown	Water samples from Long Island Sound and Narragansett Bay	Microbiology combined with microautoradiography; incubated in the laboratory	First attempt to capture microscopic image of cells incorporating radiolabeled compounds. A highly contrived laboratory setting was required for use of radiolabeled substrates (^{14}C glutamate and tritiated thymidine)	Brock and Brock, 1966
4. Nitrification (and $^{14}CO_2$ fixation)	*Nitrobacter* (autotroph)	Sediment samples from Mammoth Cave, KY, enriched on nitrate incubated in the laboratory	Microscopy combined with microautoradiography and immunofluorescent detection of cells	First attempt to apply both fluorescent antibodies (identity) and autoradiography (activity) to soil microorganisms; highly contrived laboratory setting required for use of radiolabeled substrate ($^{14}CO_2$); $^{14}CO_2$ incubation did not directly assay nitrification	Fliermans and Schmidt, 1975

Table 6.9 *Continued*

Process	Microorganism	Setting	Strategy	Commentary	References
5. Nitrification (and $^{14}CO_2$ fixation)	*Nitrosococcus, Nitrosomonas* (autotrophs)	Seawater in bottles incubated aboard ship	Microscopy combined with microautoradiography and immunofluorescent detection of cells	Incubation in bottles with radioactive $NaHCO_3$ was brief under "field-like conditions"; two known nitrifiers were probed with fluorescent antibodies; $^{14}CO_2$ incorporation did not directly assay nitrification	Ward, 1984
6. Amino acid assimilation, DNA synthesis	Alpha-Proteobacteria, *Cytophaga-Flavobacterium* group	Coastal California seawater samples	Microscopy combined with microautoradiography and 16S rRNA FISH detection of cells	Samples (40 ml) of seawater were incubated in the laboratory for 3 h. Uptake of added tritiated glucose and amino acids was measured and imaged via autoradiography. 16S rRNA-based FISH identified active cells	Ouverney and Fuhrman, 1999
7. Organic and inorganic nutrient assimilation	Beta-Proteobacteria	Activated sewage sludge samples	Microscopy combined with microautoradiography and 16S rRNA FISH detection of cells	Samples (2 ml) of sewage sludge were incubated in the laboratory for 2–3 h. Uptake of added ^{14}C acetate, butyrate, bicarbonate, and ^{32}P-phosphate were measured and imaged via autoradiography. 16S rRNA-based FISH identified active cells	Lee et al., 1999

8. Glucose and acetate assimilation	Candidatus *Meganema perideroedes*	Microscopy combined with quantitative microautoradiography and 16S rRNA FISH detection of cells	Samples (2 ml) of sewage sludge were harvested the day before and kept at 4°C, and then experiments were incubated in the laboratory for 1 h at 21°C. Uptake of added ^{14}C acetate and glucose were measured and imaged via autoradiography. 16S rRNA-based FISH identified active cells	Nielsen et al., 2003
9. Methane oxidation	*Methylomonas*	Simultaneous FISH probing of cellular rRNA (identity) and expression of methane monooxygenase expression (mRNA)	Demonstrated principle of using fixed samples for determining both identity and activity. The probed community had been enriched on methane and incubated in the laboratory for 4 weeks. Methane oxidation was not geochemically confirmed	Pernthaler and Amann, 2004
10. Anaerobic methane oxidation	*Archaea* and sulfate reducers	Follow stable isotopic signature of the ^{13}C methane into community biomarkers, cells, and site carbonate deposits	All biomarker, microscopy, and geochemical assays were performed on field-fixed samples. Resultant data support a single explanation; methane is oxidized anaerobically by a consortium of bacteria related to methanogens and sulfate reducers	Hinrichs et al., 1999; Boetius et al., 2000; Orphan et al., 2001; Michaelis et al., 2002

Table 6.9 *Continued*

Process	Microorganism	Setting	Strategy	Commentary	References
11. Assimilation of acetate and methane	*Desulfotomaculum acetoxidans*, type I methanotrophs	Samples of sediments from Tamar mud flat and Lake Loosdrecht	Stable isotopic probing, following ^{13}C-labeled substrates into lipid biomarkers	Small sediment cores incubated in the laboratory for 8 h (acetate) and 14 days (methane). Polar lipid-derived fatty acids were extracted and analyzed by gas chromatography/isotope ratio mass spectrometry	Boschker et al., 1998
12. Assimilation of methanol	Alpha-Proteobacteria, *Acidobacterium*	Sample of oak forest soil	Stable isotopic probing, following ^{13}C-labeled substrate into DNA	Sieved, air-dried soil (10 g samples) fed ^{13}C methanol at 0.5% concentration for 44 days. DNA was then extracted, separated by ultracentrifugation, and the ^{13}C DNA fraction was sequenced and analyzed for 16S rDNA sequences	Radajewski et al., 2000
13. Phenol biodegradation	*Thauera*	Sample from laboratory bioreactor	RNA stable isotope probing, ^{13}C-labeled RNA was extracted, reverse transcribed, and sequenced	First demonstration that the RNA pool can be rapidly labeled. ^{13}C atoms were traced into the ribosome fraction of the community within 24–72 h. Sequencing of reverse transcribed RNA revealed identity of active microbes	Manefield et al., 2002

14. Methanol oxidation by denitrifying microorganisms	*Methylobacillus, Methylophilus*	Sample from laboratory bioreactor	DNA stable isotope probing, confirmed by both FISH and microautoradiography	^{13}C atoms were traced into a DNA clone library during a 24 h incubation. 16S rDNA sequences in library were confirmed by FISH, these, in turn, were confirmed by microautoradiography using radioactive methanol	Ginige et al., 2004
15. Naphthalene biodegradation	*Polaromonas naphthalenivorans*	Contaminated field sediment in South Glens Falls, NY	Field-based DNA stable isotope probing	Addition of ^{13}C naphthalene to field site sediment; respiration assay confirmed in situ biodegradation; extraction and sequencing of the 16S rRNA genes in ^{13}C DNA identified the responsible population. A representative of the population was cultured	Jeon et al., 2003
16. Uranium reduction and immobilization	*Geobacter*	Contaminated subsurface sediment in Rifle, CO	Addition of electron donor (^{13}C and ^{12}C acetate) to field for stimulation of U(VI) reduction; molecular biomarker analyses and field stable isotope probing	Weight of evidence shows very strong association between stimulated U(VI) reduction and an increase in *Geobacter* biomarkers	Anderson et al., 2003; Chang et al., 2005

Table 6.9 Continued

Process	Microorganism	Setting	Strategy	Commentary	References
17. Anaerobic ammonia oxidation (anammox)	Candidatus *Kuenenia stuttgartiensis*	Anaerobic biorector and marine sediments (Costa Rica, Black Sea, Africa)	Document biomarkers in field samples (unique ladderane lipids, 16S rRNA sequences via cloning and microscopy) and incubations of field samples with $^{15}NH_4$ followed by mass spectrometric analysis of the resulting N_2	The case is very convincing: combinations of anammox-specific biomarkers, FISH microscopy, and a physiological assay that distinguishes N_2 production via nitrate reduction versus ammonia oxidation	Dalsgaard et al., 2003; Kuypers et al., 2003, 2005; Schmid et al., 2005
18. Nitrogen fixation	*Trichodesmium*	Ocean water	Recognizable filamentous colonies were collected by filtration and assayed, physiologically, in bottles aboard ship	Weight of evidence leaves no doubt that *Trichodesmium* fixes nitrogen in ocean waters; nitrogen fixation assays relied upon closed-bottle incubations; field activity has been confirmed by immunodetection of nitrogenase enzyme in field samples	Paerl et al., 1989; Capone et al., 1997; Montoya et al., 2004
19. Photosynthesis	*Prochlorococcus*	Ocean water	Rates of cell division, and photosynthesis, were inferred from circadian cell cycles at many depths	Field-fixed cells were analyzed by flow cytometry; specific growth rate was estimated by analyzing circadian cell cycle patterns; CO_2 fixation was implicit in growth of this photosynthetic microorganisms	Vaulot et al., 1995

FISH, fluorescent in situ hybridization.

and subtle nutritional needs of naturally occurring microorganisms (Button et al., 1993; Kaeberlein et al., 2000; Rappé et al., 2002; Zinder, 2002; Lead-better, 2003; see also Sections 6.2 and 6.6).

Using Figure 6.11 as a map for visualizing steps toward progress in environmental microbiology, there are three obvious avenues for increasing the ecological validity of information. First, if the media used in the flask and Petri dish assays (cultivation-based inquiry path, column 2, Figure 6.11) improve in ecological relevance, then the microorganisms eventually isolated are far more likely to be active in nature (the culture will fall closer to the top of column 1). Second, as analyses of field-fixed extracted samples (column 4) deliver increasingly sophisticated information about expressed genes and proteins used by microorganisms in their native habitats, inferences can be made about ambient physiological conditions, carbon substrates, and nutritional needs. Such information can guide the design of media so that new organisms can be cultured. Finally, the several paths of information flow (column 5) for validating data need to be thoroughly utilized. These validation paths are: (i) following Koch's postulates by the addition of cultures to field sites; (ii) the use of pure culture-derived "omics"-based biomarkers to guide analyses of extracted samples; and (iii) using microscopy and biomarker probes to confirm field relevance of information from both pure cultures and extracted samples.

Selected examples of past and current investigations aimed at linking identity of microorganisms to their field activity are shown in Table 6.9 (a glossary for technical terminology is presented in Table 6.4). Entries in Table 6.9 were chosen to be representative of the types of strategies, techniques, challenges, and breakthroughs that have occurred in environmental microbiology over the last several decades. Emphasis is upon identifying microorganisms and being sure their biogeochemical reactions were catalyzed in situ – in real-world field sites containing soil, sediment, or water. The first two entries (symbiotic nitrogen fixation and biodegradation of trichloroethene in contaminated groundwater) reveal that medical microbiology's paradigm (Koch's postulates) can be powerful and insightful. Koch's postulates are only applicable in limited contexts because the active microorganisms must be cultured and initially be in low numbers or absent from the inoculated habitat. The next six entries (3–8) in Table 6.9 illustrate the foundations and later developments in microscopy-based attempts to link identity to activity without using Koch's postulates. Microscopy and microautoradiography were initially used to see which cells in mixed microbial communities incorporated radio-labeled substrates. Later, microautoradiography was combined with cell-specific probing: fluorescent antibodies targeting cell surface antigens of cultured bacteria or fluorescent oligonucleotides targeting sequences of taxonomically revealing ribosomal RNA, often derived from uncultured microorganisms. Recent efforts (entry 9, Table 6.9) portend another strategy that has the potential of avoiding all laboratory incubations by using microscopic fluorescent in situ hybridization (FISH) procedures to probe naturally occurring microorganisms for both identity (rRNA sequence) and for activity (indirectly, via hybridization with the mRNA of expressed functional genes).

Another promising methodological development is stable isotope probing (SIP; entries 10–16, Table 6.9). The strategy follows the stable isotopic signature of an assimilated substrate (e.g., carbon source) into the populations responsible for substrate metabolism in complex microbial communities. Because the assimilated substrate has a distinctive signature mass (e.g., density or $^{13}C : {}^{12}C$ ratio), cells or biomarkers derived therefrom can be separated and/or analyzed in ways that reveal the identity of active cells (Madsen, 2006). Without question, the most elegant example of SIP to date is from a series of investigations documenting anaerobic methane oxidation in deep waters adjacent to methane sources in the Black Sea and in coastal California and Oregon (entry 10, Table 6.9; see also Sections 7.4 and 7.5). These investigations were successful because the field study sites contained a substrate (methane) fortuitously labeled with a unique stable isotopic signature. Such situations are rare. To implement SIP in other contexts, a stable isotopically labeled (e.g., ^{13}C) substrate is dosed to a community and later retrieved in biomarkers. Such biomarkers have included phospholipid fatty acids (whose molecular structure is taxonomically informative; see Box 6.2), DNA, and rRNA (the latter two are sources of 16S rRNA gene sequences; see Section 6.7). Early SIP studies established "proof of principle" for the dosing approach – however these investigations were carried out on enrichment cultures (laboratory-based model soils exposed to high concentrations of ^{13}C-labeled substrates for many weeks). More recently, refinements in the SIP approach have included analyzing the labeled RNA fraction (RNA is rapidly turned over in cells and labeling does not require that the populations undergo growth) and verification of SIP-discovered rRNA sequences with FISH. SIP was applied in a field situation (naphthalene-contaminated sediment) – leading to the discovery and later cultivation of an ecologically significant bacterium, *Polaromonas naphthalenivorans* (entry 15, Table 6.9). Field-based SIP has also been used [in conjunction with DGGE (denaturing gradient gel electrophoresis), 16S rRNA gene phylogeny, and phospholipid fatty acid analysis] to show that *Geobacter* species are likely the causative agents of uranium reduction in a uranium bioremediation field project (entry 16, Table 6.9).

Entries 17 and 18 of Table 6.9 elucidate nitrogen transformations in marine habitats. The fundamentals of anaerobic ammonia oxidation (anammox) have been documented only in the last decade (Strous and Jetten, 2004; see Sections 7.4 and 7.5). Members of an unusual uncultured group of microorganisms in the *Planctomyces* phylum carry out the anammox reaction. Armed with biomarkers and physiological tools derived from microorganisms in bioreactors that use ammonium as an electron donor and nitrite as an electron acceptor, the anammox process was found to occur in marine sediments in the Black Sea and off the coasts of Costa Rica and Africa. *Trichodesmium* (entry 18, Table 6.9) is a long-studied photosynthetic and nitrogen-fixing cyanobacterium found in ocean waters. *Trichodesmium* forms relatively large, filamentous, morphologically recognizable colonies whose global presence and potential for N_2 fixation are undeniable. By the strict criteria developed here, N_2 fixation by *Trichodesmium* has not yet been directly demonstrated because the N_2 fixation assay relied upon ship-incubated water samples. Nonetheless, biomarker studies performed on field-fixed samples have shown that nitrogen fixation genes was transcribed and translated in situ.

The final entry of Table 6.9 focuses on *Prochlorococcus*, another widely distributed ocean inhabitant that is recognizable by flow cytometry. Representatives of *Procholorococcus* have been cultured and their genomes have been sequenced (Rocap et al., 2003). In situ photosynthesis by *Prochlorococcus* was demonstrated by field-based monitoring of cell replication.

Outlook

The ultimate goals of environmental microbiology are to understand mechanistic relationships between habitat characteristics, evolutionary pressures, microbial diversity, biochemical processes, and their genetic controls. Processes carried out by microorganisms in soils, sediments, oceans, lakes, and groundwaters have a major impact on environmental quality, agriculture, and global climate change. Thus, environmental microbiological insights have ecological and technological applications able to harness microbial processes that maintain ecosystems, locally and globally.

Identifying ecologically significant microorganisms is like finding a needle in an unusual haystack – a haystack whose individual pieces can, during the search, change themselves into misleading needles. For more than a century, environmental microbiologists have been confronted by vast unknown microbial diversity (the "haystack"), by population responsiveness (the misleading "needles"), by an enormous size differential between humans (~1 m) and microorganisms (~1 μm), and by the evasive task of documenting the geochemical impact of microorganisms in open, heterogeneous field sites. *The complexity of natural systems has, almost without exception, made it impossible to directly observe the identity of microorganisms and their activities in waters, sediments, and soils. Instead, indirect approaches have emerged.*

As the frontiers of environmental science and microbial ecology advance, we are assured of an astonishing supply of new hypotheses relating microbial diversity to mechanisms of ecologically significant physiological adaptation. Current examples include the challenge of discovering the role of uncultured microbes in oceans (Church et al., 2003) and soils (Felske et al., 2000) and other microorganisms that are widely dispersed but whose metabolic functions are a mystery (see Section 5.7). The new bioinformatic tools and feedback-based investigative strategies available to environmental microbiologists (see Figure 6.11) guarantee complementation and convergence of information generated by cultivation- and noncultivation-based procedures. Future inquiries will surely accelerate progress linking ecologically important microorganisms to their activity in real-world habitats.

STUDY QUESTIONS

1 Prior to reading this chapter, were you familiar with the term, "epistemology"? Have you noticed that direct observations (e.g., "the sun is up" or "the temperature is 28°C") are easy to "know" about? In contrast, many types of scientific knowledge cannot be attained through direct observation. Instead, indirect observations, often involving theory and inference, are used to understand many scientific phenomena. Use library publications to research a topic and then write a two page essay describing some area of science other than environmental microbiology (e.g., biology, physics, chemistry, engineering, medicine) where indirect observations and inference have been successfully and/or routinely used in inquiry. In considering topics for this essay, "think big". Consider issues such as:
   (A) Global warming in environmental science.
   (B) AIDS, mad cow disease, or prions in medical microbiology.
2 Scrutinize Figure 6.1. Enrichment culturing suffers from the potential weakness that, in the end, you have isolated a microorganism that is physiologically interesting but possibly ecologically irrelevant.
   (A) Redraw the two bar graphs in Figure 6.1 in a way that shows an outcome of ecological relevance.
   (B) If you suspect that you have isolated a bacterium that is ecologically important, how would you prove it?
   Formulate your answer as a hypothesis. State the alternative hypothesis or hypotheses. Next state how you would test the hypotheses (see especially Section 6.10).
3 Which one of the quotes in Table 6.1 is most striking to you? Why?
4 Section 6.3 argues that an inability to assemble mass balances hinders our ability to know that microorganisms are responsible for biogeochemical change in a given habitat.
   (A) Can you think of a real-world situation where, contrary to the general rule, mass balances *can* be obtained? (Hint: consider the sewage treatment system for a typical small city. Materials enter and leave the treatment plant and this engineered system should allow measurement of concentrations of materials, i.e. total organic carbon or individual pollutant compounds, and their flows – thus allowing mass balances to be assembled.)
   (B) Describe a scenario at a sewage treatment plant or another suitable chosen system where monitoring of inputs and outputs can successfully document a microbial process in the plant. What other measurements would strengthen your case for the process?
   (C) Next, reconsider the system as a general way to demonstrate microbial metabolic processes. Why might some microbial processes be difficult or impossible to document in your system? (Hint: consider a wide variety of materials that will typically be entering

and leaving the system. Physical and chemical properties of the compounds acted upon and produced as metabolites by microbial processes may not be amenable to mass balance strategies.)

5 In Section 6.4, the term "bottle effects" is mentioned.

   (A) What do you think "bottle effects" are? In answering, consider the detailed physical and chemical conditions that microorganisms experience at the scale of micrometers.

   (B) If you were a free-living, planktonic bacterium in a lake and you were moved into a glass bottle, what physical, chemical, nutritional, and ecological changes might be imposed upon you?

6 (A) In devising a sampling scheme for a microbiological study (see Section 6.5), when would it be absolutely essential to use sterile sampling tools to place the sample into sterile containers?

   (B) Can you envision one or more situations in which aseptic sampling techniques would be unnecessary?

7 Does the "capture–fix–store–analyze" scheme in Figure 6.4 make sense to you? If so, why? If not, why not?

8 At the end of the text in Section 6.5, a methane-rich site is mentioned. How would you distinguish between microbiologically created methane and methane of geologic origin (from natural gas)? To answer this question, consider information provided in all prior chapters – especially Chapter 2 (Section 2.1 and Box 2.2) and Section 6.8. Include arguments using both analytical chemistry procedures and physiological considerations of the habitat.

9 Section 6.7 advances the idea that a microorganism in low abundance might be physiologically and ecologically important.

   (A) Can you explain this idea?

   (B) Do active microorganisms necessarily have high growth yields? In preparing an answer, consider information provided in Chapter 3 – especially the thermodynamic "compass" used to identify processes that may have low free-energy yields.

   (C) Find one or more electron donor–electron acceptor reaction pairs (e.g., Figure 3.9 and Tables 3.7, 3.8) that you consider to be good candidates for processes carried out by microorganisms in ecosystems that meet both of the following criteria: (i) small microbial populations; and (ii) large fluxes of substrate turnover. What are the processes you have chosen?

10 In Figure 6.7, one experimental scheme converts rRNA (initially present as intracellular ribosomes) to cDNA prior to cloning and sequencing. Explain why investigators would choose this reverse transcriptase-based step rather than an experimental approach that uses rRNA genes (DNA) that encode the rRNA for PCR amplification, cloning, and sequencing.

11 Section 6.7 mentions the possibility of finding rRNA genes in reagent-only treatments. In this regard, "control" treatments in experimental designs are crucial. It would be tragic to mistake microorganisms in the PCR and cloning reagents for those in the habitat you are sampling. If you were analyzing rRNA genes from microorganisms collected from a low-biomass habitat, such as air in a hospital, what experimental controls would help you have confidence in the data you produce? To answer this question, assume that 100 L volumes of hospital air are sampled using 0.1 µm pore-size membrane filters housed in a canister.

12 Boxes 6.2 and 6.3 describe lipid biomarkers and 16S rRNA fingerprinting procedures, respectively. Information in Figure 6.5 places these biomarker procedures within the larger context

of the environmental microbiology "toolbox". If you wanted to answer the question "What microorganisms live on my hand?", what methods would you use? Justify your choices. Presume that you have access to all instruments and assays described in this chapter and the rest of this book.

13 Having completed the exercise in question 12, assume that you now have a description of the microbial community that dwells on your hand. Now you are confronted with the task of discovering "What are they doing?" and "When are they doing it?" Devise a series of experimental procedures for answering these two questions. Begin with clear hypotheses, then decide how the hypotheses should be tested, and then describe the experimental steps that would need to be implemented. Take these steps through to completion – beginning with sampling and ending with final analysis of the data. Be careful to consider alternative hypotheses for interpreting each set of measurements and to take steps that avoid potential artifacts in your procedures. In devising this experimental plan, use Figure 6.5 as an overall guide and feel free to utilize analytical, molecular, cultivation, and metagenomic procedures.

14 Information in Box 6.4 describes a series of metagenomic and other investigations aimed at uncovering a previously undocumented, light-driven ecological process: proteorhodopsin-based ATP production in the oceans. In the final installment of the investigations described in Box 6.4 (other installments may have been written and published since the release of this book), light did not seem to influence the physiology of *Pelagibacter ubique*.

(A) What if light *had* accelerated cell growth of the bacterium? Would the ecological significance of proteorhodopsin-based ATP generation be certain?

(B) What other measures (see especially Section 6.10 and Figure 6.11) would still be needed to be completed to fully document and explore the impact of proteorhodopsin-based photosynthesis in the oceans?

REFERENCES

Abdo, A., U.M.E. Schette, S.J. Bent, C.J. Williams, L.J. Forney, and P. Joyce. 2006. Statistical methods for characterizing diversity of microbial communities by analysis of terminal restriction fragment length polymorphisms of 16S rRNA genes. *Environ. Microbiol.* **5**:929–938.

Alberts, B., A. Johnson, J. Lewis, M. Raff, K. Roberts, and P. Walter. 2002. *Molecular Biology of the Cell*, 4th edn. Garland Science, Taylor and Francis Group. New York.

Alexander, M. 1991. Soil microbiology in the next 75 years: Fixed, flexible, or mutable? *Soil Sci.* **151**:35–40.

Allen, E.E. and J.F. Bonfield. 2005. Community genomics in microbial ecology and evolution. *Nature Rev. Microbiol.* **3**:489–498.

Amann, R., W. Ludwig, and K.-H. Schleifer. 1995. Phylogenetic identification and *in situ* detection of individual microbial cells without cultivation. *Microbiol. Rev.* **59**:143–169.

Anderson, R.T., H.A. Vrionis, I. Ortiz-Bernad, et al. 2003. Stimulating the in situ activity of *Geobacter* species to remove uranium from groundwater or a uranium-contaminated aquifer. *Appl. Environ. Microbiol.* **69**:5889–5891.

Antelmann, H., C. Scharf, and M. Hecker. 2000. Phosphate starvation-inducible proteins of *Bacillus subtilis*: proteomics and transcriptional analysis. *J. Bacteriol.* **182**:4478–4490.

APHA (American Public Health Association) 2002. *Standard Methods for the Examination of Water and Wastewater*, 20th edn. APHA, Washington, DC.

Ausubel, F.M., R. Brent, R.E. Kingston, D.B. et al. 1999. *Short Protocols in Molecular Biology*, 4th edn. Wiley and Sons, New York.

Bae, J.W., S.-K. Rhee, J.R. Park, et al. 2005.

Development and evaluation of genome-probing microarrays for monitoring lactic acid bacteria. *Appl. Environ. Microbiol.* **71**:8825–8835.

Beijerinck, M.W. 1888. Anhaufungsversuche mit Urembakterien. *Centralblatt f. Bakteriologie* Part II, **7**:33–61. (English translation in T.D. Brock (ed.) 1961. *Milestones in Microbiology*, pp. 234–237. Prentice Hall, Englewood Cliffs, NJ.)

Beja, O., L. Aravind, E.V. Koonin, et al. 2000. Bacterial rhodopsin: evidence for a new type of phototrophy in the sea. *Science* **289**:1902–1906.

Beja, O., E.N. Spudich, J.L. Spudich, M. Leclerc, and E.F. DeLong. 2001. Proteorhodopsin: phototrophy in the ocean. *Nature* **411**:786–789.

Boetius, A., K. Ravenschlag, C.J. Schubert, et al. 2000. A marine microbial consortium apparently mediating anaerobic oxidation of methane. *Nature* **407**: 623–626.

Boschker, H.T.S., S.C. Nold, P. Wellsbury, et al. 1998. Direct linking of microbial populations to specific biogeochemical processes by ^{13}C-labelling of biomarkers. *Nature* **392**:801–805.

Brady, S.F. and J. Clardy. 2000. Long-chain N-acyl amino acid antibiotics isolated from heterologously expressed environmental DNA. *J. Am. Chem. Soc.* **122**:12903–12904.

Breitbart, M., P. Salamon, B. Andersen, et al. 2002. Genomic analysis of uncultured marine viral communities. *Proc. Natl. Acad. Sci. USA* **99**:14250–14255.

Brock, T.D. 1987. The study of microorganisms in situ: progress and problems. *In*: M. Fletcher, T.R.G. Gray, and J.G. Jones (eds) *Ecology of Microbial Communities*, pp. 1–17. Forty-first Symposium of the Society for General Microbiology, University of St. Andrews. Cambridge University Press, New York.

Brock, T.D. and M.L. Brock. 1966. Autoradiography as a tool in microbial ecology. *Nature* **209**:723–736.

Brooks, G.F., J.S. Butel, and S.A. Morse. 2004. *Medical Microbiology*, 23rd edn. McGraw-Hill, New York.

Bull, A.T. 1980. Biodegradation: some attitudes and strategies of microorganisms and microbiologists. *In*: D.C. Ellwood, J.N. Hedger, M.J. Latham, J.M. Lynch, and J.H. Slater (eds) *Contemporary Microbiology Ecology*, pp. 107–136. Academic Press, New York.

Burlage, R.S., R. Atlas, D. Stahl, G. Geesey, and G. Sayler. 1998. *Techniques in Microbial Ecology.* Oxford University Press, New York.

Button, D.K., F. Schut, P. Quang, R. Martin, and B. Robertson. 1993. Viability and isolation of marine bacteria by dilution culture: theory, procedure, and initial results. *Appl. Environ. Microbiol.* **59**: 881–891.

Capone, D.A., J.P. Zehr, H.W. Paerl, B. Bergman, and E.J. Carpenter. 1997. *Trichodesmium*, a globally significant marine cyanobacterium. *Science* **276**: 1221–1229.

Castellan, G.W. 1983. *Physical Chemistry,* 3rd edn. Addison-Wesley, Reading, MA.

Chapelle. F.H., D.A. Vroblesky, J.C. Woodward, and D.R. Lovley. 1997. Practical considerations for measuring hydrogen concentrations in groundwater. *Environ. Sci. Technol.* **31**:2873–2877.

Chang, Y.-J., P.E. Long, R. Geyer, et al. 2005. Microbial incorporation of ^{13}C-labeled acetate at the field scale: detection of microbes responsible for reduction of U(VI). *Environ. Sci. Technol.* **39**: 9039–9048.

Chiang, C.Y., K.R. Loos, and R.A. Klopp. 1992. Field determination of geological–chemical properties of an aquifer by cone penetrometry and headspace analysis. *Ground Water* **30**:428–436.

Christian, G.D. 2003. *Analytical Chemistry*, 6th edn. Wiley and Sons, New York.

Church, M., E.F. DeLong, H.W. Ducklow, M.B. Karner, C.M. Preston, and D.M. Karl. 2003. Abundance and distribution of planktonic Archaea and Bacteria in the waters west of the Antarctic Peninsula. *Limnol. Oceanograph.* **48**:1893–1902.

Clark, F.E. 1973. Problems and perspectives in microbial ecology. *In*: T. Rosswall (ed.) *Modern Methods in the Study of Microbial Ecology: Proceedings of a symposium*, Vol. 17, pp. 13–16. Bull. Ecol. Res. Committee. Pulished by the Ecological Research Committee of the Swedish Natural Science Research Council, Stockholm, Sweden.

Conrad, R. 1996. Soil microorganisms as controllers of atmospheric trace gases (H$_2$, CO, CH$_4$, N$_2$O, and NO). *Microbiol. Rev.* **60**:609–640.

Cunningham, W.F. 1930. *Notes on Epistemology.* Declan X. McMullen Co., Inc., New York.

Dalsgaard, T., D.E. Canfield, J. Petersen, B. Thamdrup, and J. Acuna-Gonzalez. 2003. N$_2$

production by anammox reaction in the anoxic water column of Golfo Dulce, Costa Rica. *Nature* **422**: 606–608.

Dean, F.B., J.R. Nelson, T.L. Giesler, and R.S. Lasken. 2001. Rapid amplification of plasmid and phage DNA using Phi29 DNA polymerase and multiply-primed rolling circle amplification. *Genome Res.* **11**:1095–1099.

DeLong, E.F. 2005. Microbial community genomics in the ocean. *Nature Rev. Microbiol.* **3**:459–469.

Dennis, P., E.A. Edwards, S.N. Liss, and R. Fulthorpe. 2003. Monitoring gene expression in mixed microbial communities by using DNA microarrays. *Appl. Environ. Microbiol.* **69**:769–778.

DeSantis, T.Z., I. Dubosarskiy, S.R. Murray, and G.L. Andersen. 2003. Comprehensive aligned sequence construction for automated design of effective probes (CASCADE-P) using 16S rDNA. *Bioinformatics* **19**:1461–1468.

Donaldson, R.E. (ed.) 1991. *A Sacred Unity: Further steps to an ecology of mind.* Harper Collins, New York.

Dyhrman, S.T., P.D. Chappell, S.T. Haley, et al. 2006. Phosphonate utilization by the globally important marine diazotroph, *Trichodesmium. Nature* **439**:68–71.

Edwards, R.A., B. Rodrigues-Brio, L. Wegley et al. 2006. Using pyrosequencing to shed light on deep mine microbial ecology. *BMC Genomics* **7**:57.

Entcheva, P., W. Liebl, A. Johann, T. Hartsch, and W.R. Streit. 2001. Direct cloning from enrichment cultures, a reliable strategy for isolation of complete operons and genes from microbial consortia. *Appl. Environ. Microbiol.* **67**:89–99.

Falkow, S. 2004. Molecular Koch's postulates applied to bacterial pathogenicity – a personal recollection 15 years later. *Nature Rev. Microbiol.* **2**:67–72.

Farrelly, V., F.A. Rainey, and E. Stackebrandt. 1995. Effect of genome size and rrn gene copy number on PCR amplification of 16S rRNA genes from a mixture of bacterial species. *Appl. Environ. Microbiol.* **61**:2798–2801.

Felske, A., A. Wolterink, R. Van Lis, W.M. De Vos, and A.D.L. Akkermans. 2000. Response of a soil bacterial community to grassland succession as monitored by 16S rRNA levels of the predominant ribotypes. *Appl. Environ. Microbiol.* **66**:3998–4003.

Fifield, F.W. and P.J. Haines (eds) 2000. *Environmental Analytical Chemistry.* Blackwell Scientific Publications, Malden, MA.

Findlay, R. and F.C. Dobbs. 1993. Quantitiative description of microbial communities using lipid analysis. *In*: P.F. Kemp, B.F. Sherr, E.B. Sherr, and J.J. Cole (eds) *Handbook of Methods in Aquatic Microbial Ecology*, pp. 271–284. Lewis Publishers, Chelsea, MI.

Fisher, M.M. and E.W. Triplett. 1999. Automated approach for ribosomal intergenic spacer analysis of microbial diversity and its application to freshwater bacterial communities. *Appl. Environ. Microbiol.* **65**:4630–4636.

Fliermans, C.B. and E.L. Schmidt. 1975. Autoradiography and immunofluorescence combined for autoecological study of single cell activity with *Nitrobacter* as a model system. *Appl. Microbiol.* **30**: 676–684.

Fred, E.B., I.L. Baldwin, and E. McCoy. 1932. *Root Nodule Bacteria and Leguminous Plants.* University of Wisconsin Studies in Science, No. 5. University of Wisconsin, Madison, WI.

Frey, J.C., E.R. Angert, and A.N. Pell. 2006 Assessment of biases associated with profiling simple, model communities using terminal-restriction fragment length polymorphism-based analyses *J. Microbiol. Methods* **67**:9–19.

Ginige, M.P., P. Hugenholtz, H. Daims, M. Wagner, J. Keller, and L.L. Blackall. 2004. Use of stable-isotope probing, full-cycle rRNA analysis, and fluorescence in situ hybridization–microautoradiography to study a methanol-fed denitrifying microbial community. *Appl. Environ. Microbiol.* **70**: 588–596.

Giovannoni, S.J., T.B. Britschgi, C.L. Moyer, and K.G. Field. 1990. Genetic diversity on Sargasso Sea bacteriophytoplankton. *Nature* **345**:60–63.

Giovannoni, S.J., L. Bubbs, J.-C. Cho, et al. 2005a. Proteorhodopsin in the ubiquitous marine bacterium SARII. *Nature* **438**:82–85.

Giovannoni, S. J., L. Bubbs, J.-C. Cho, et al. 2005b. Genomic streamlining in a cosmopolitan oceanic bacterium. *Science* **309**:1242–1245.

Glud, R.N., J.K. Gundersen, N.P. Revsbech, and B.B. Jorgensen. 1994. Effects on the benthic diffusive boundary layer imposed by microelectrodes. *Limnol. Oceanogr.* **39**:462–467.

Gottschal, J.C., W. Harder, and R.A. Prins. 1992. Principles of enrichment, isolation, cultivation, and preservation of bacteria. *In*: A. Balows, H.G. Trüper, M. Dworkin, W. Harder, K.-H. Schleifer (eds) *The Prokaryotes*, 2nd edn, pp. 149–196. Springer-Verlag, New York.

Groffman, P.M. 1993. Soil microbiology: contributions from the gene to global scale. *In*: J.T. Sims (ed.) *Agriculture Research in the Northeastern United States: Critical review and future perspectives*, pp. 19–26. American Society of Agronomy, Madison, WI.

Grossman, E.L. 2002. Stable carbon isotopes as indicators of microbial activities in aquifers. *In*: C.J. Hurst, R.L. Crawford. G.R. Knudsen, M.I. McInerney, and L.D. Stetzenbach (eds) *Manual of Environmental Microbiology*, 2nd edn, pp. 728–742. American Society for Microbiology, Washington, DC.

Hallam, S.J., N. Putnam, C.M. Preston, et al. 2004. Reverse methanogenesis: testing the hypothesis with environmental genomics. *Science* **305**:1457–1462.

Handelsman, J. 2004. Metagenomics: applications of genomics to uncultured microorganisms. *Microbiol. Molec. Biol. Rev.* **68**:669–685.

Heal, O.W. and A.F. Harrison. 1990. Keynote paper. Turnover of nutrients: a technological challenge. *In*: A.F. Harrson, P. Ineson, and O.W. Heal (eds) *Nutrient Cycling in Terrestrial Ecosystems: Field methods, application and interpretation*, pp. 170–178. Elsevier Applied Science, London.

Hemond, H.F. and E.J. Fechner. 1994. *Chemical Fate and Transport in the Environment*. Academic Press, New York.

Henne A., R.A. Schmitz, M. Bomeke, G. Gottchalk, and R. Daniel. 2000. Screening of environmental DNA libraries for the presence of genes conferring lipolytic activity in *Escherichia coli*. *Appl. Environ. Microbiol.* **66**:3113–3116.

Hinrichs, H.-U., J.M. Hayes, S.P. Sylva, P.G. Brewer, and E.F. DeLong. 1999. Methane-consuming archaebacteria in marine sediments. *Nature* **398**: 802–805.

Hobbie, J.E. 1993. Introduction. *In*: P.F. Kemp, B.F. Sherr, E.B. Sherr, and J.J. Cole (eds) *Handbook of Methods in Aquatic Microbial Ecology*, pp. 1–5. Lewis Publishers, Boca Raton, FL.

Hobbie, J.E. and T.E. Ford. 1993. A perspective on the ecology of aquatic microbes. *In*: T.E. Ford (ed.) *Aquatic Microbiology*, pp. 1–14. Blackwell Scientific Publications, Boston, MA.

Hodson, R.E., W.A. Dustman, R.P. Garg, and M.A. Moran. 1995. In situ PCR for visualization of microscale distribution of specific genes and gene

products in prokaryotic communities. *Appl. Environ. Microbiol.* **61**:4074–4082.

Honderich, T. (ed.) 1995. *The Oxford Companion to Philosophy*: Oxford University Press, Oxford, UK.

Hurst, C.J., R.L. Crawford, G.R. Knudsen, M.I. McInerney, and L.D. Stetzenbach (eds) 2002. *Manual of Environmental Microbiology*, 2nd edn. American Society for Microbiology, Washington, DC.

Imlay, J.A. 2003. Pathways of oxidative damage. *Annu. Rev. Microbiol.* **57**:395–418.

Janssen, P.H. 2006. Identifying the dominant soil bacterial taxa in libraries of 16S rRNA and 16S rRNA genes. *Appl. Environ. Microbiol.* **72**:1719–1728.

Jeon, C.-O., W. Park, P. Padmanabhan, C. DeRito, J.R. Snape, and E.L. Madsen. 2003. Discovery of a novel bacterium, with distinctive dioxygenase, that is responsible for in situ biodegradation in a contaminated sediment. *Proc. Natl. Acad. Sci. USA* **100**:13591–13596.

Jouzel, J., J.R. Petit, R. Souchez, et al. 1999. More than 200 meters of lake ice above subglacial Lake Vostok, Antarctica. *Science* **286**:2138–2141.

Kaeberlein, T., K. Lewis, and S.S. Epstein. 2002. Isolating "uncultivable" microorganisms in pure culture in a simulated natural environment. *Science* **296**:1127–1129.

Karl, D.M. 1986. Determination of in situ microbial biomass, viability, metabolism, and growth. *In*: J.S. Poindexter and E.R. Leadbetter (eds) *Bacteria in Nature*, Vol. 2, pp. 85–176. Plenum Press. New York.

Karl, D.M. 1993. Adenosine triphosphate (ATP) and total adenine nucleotide (TAN) pool turnover rates as measures of energy flux and specific growth rates in natural populations of microorganisms. *In*: P.F. Kemp, B.F. Sherr, E.B. Sherr, and J.J. Cole (eds) *Handbook of Methods in Aquatic Microbial Ecology*, pp. 483–494. Lewis Publishers, Chelsea, MI.

Karl, D.M. 1995. Ecology of free-living hydrothermal vent microbial communities. *In*: D.M. Karl (ed.) *The Microbiology of Deep Sea Hydrothermal Vents*, pp. 35–124. CRC Press, New York.

Keith, L.H. (ed.) 1996. *Principles of Environmental Sampling*, 2nd edn. ACS Professional Reference Book. American Chemical Society, Washington, DC.

Kemp, P.F., B.F. Sherr, E.B. Sherr, and J.J. Cole (eds) 1993. *Handbook of Methods in Aquatic Microbial Ecology*. Lewis Publishers, Boca Raton, FL.

Kluyver, A.J. and C.B. van Niel. 1954. *The Microbe's Contribution to Biology.* Harvard University Press, Cambridge, MA.

Koch, R. 1884. *Mittbeilungen aus dem Kaiserlichen Gesundheitsamte,* 2, 1–88. (English translation in T.D. Brock (ed.) 1961. *Milestones in Microbiology,* pp. 116–118. Prentice Hall, Englewood Cliffs, NJ.)

Könneke, M., A.E. Bernhard, J.R. de la Torre, C.B. Walker, J.M. Waterbury, and D.A. Stahl. (2005). Isolation of an autotrophic ammonia-oxidizing marine archaeon. *Nature* **437**:543–546.

Konopka, A. 2006. Microbial ecology: searching for principles. *Microbe* **1**(4):175–179. American Society for Microbiology, Washington, DC.

Kowalchuk, G.A., F.J. de Bruijn, I.M. Head, A.D.L. Akkermans, and J. D. van Elsas (eds) 2004. *Molecular Microbial Ecology Manual,* 2nd edn. Springer-Verlag, New York.

Kroeze, C., R. Aerts, N. van Breemen, et al. 2003. Uncertainties in the fate of nitrogen. I: An overview of sources of uncertainty illustrated with a Dutch case study. *Nutr. Cycling Agroecos.* **66**:43–69.

Kuypers, M.M.M., G. Lavik, D. Woebken, et al. 2005. Massive nitrogen loss from the Benguela upwelling system through anaerobic ammonia oxidation. *Proc. Natl. Acad. Sci. USA* **102**:6478–6483.

Kuypers, M.M.M., A.G. Sliekers, G. Lavik, et al. 2003. Anaerobic ammonium oxidation by anammox bacteria in the Black Sea. *Nature* **422**:608–611.

Leadbetter, J.R. 2003. Cultivation of recalcitrant microbes: cells are alive, well, and revealing their secrets in the 21st century laboratory. *Curr. Opin. Microbiol.* **6**:274–281.

Lee, N., P.H. Nielsen, K.H. Andreasen, et al. 1999. Combination of fluorescent in situ hybridization and microautoradiography – a new tool for structure-function analyses in microbial ecology. *Appl. Environ. Microbiol.* **65**:1289–1297.

Levin, M.A., R.J. Siedler, and M. Rogul (eds) 1992. *Microbial Ecology: Principles, methods and applications.* McGraw-Hill., New York.

Likens, G.E. and F.H. Bormann. 1995. *Biogeochemistry of a Forested Ecosystem.* Springer-Verlag, New York.

Lindow, S.E. 1995. The use of reporter genes in the study of microbial ecology. *Molec. Ecol.* **4**:555–566.

Liu, W.-T., T.L. Marsh, H. Cheng, and L.J. Forney. 1997. Characterization of microbial diversity by determining terminal restriction fragment length polymorphisms of genes encoding 16S rRNA. *Appl. Environ. Microbiol.* **63**:4516–4522.

Liu, W.-T. and D.A. Stahl. 2002. Molecular approaches for the measurement of density, diversity, and phylogeny. *In*: C.J. Hurst, R.L. Crawford, G.R. Knudsen, M.J. McInerney, and L.D. Stetzenbach (eds) *Manual of Environmental Microbiology,* 2nd edn, pp. 114–134. American Society for Microbiology Press, Washington, DC.

Lovley, D.R., F.H. Chappelle, and J.C. Woodward. 1994. Use of dissolved H_2 concentrations to determine distribution of microbially catalyzed redox reactions in anoxic groundwater. *Environ. Sci. Technol.* **28**:1205–1210.

Loy, A., A. Lehner, N. Lee, et al. 2002. Oligonucleotide microarray for 16S rRNA gene-based detection of all recognized lineages of the sulfate-reducing prokaryotes in the environment. *Appl. Environ. Microbiol.* **68**:5064–5081.

Madsen, E.L. 1991. Determining *in situ* biodegradation: facts and challenges. *Environ. Sci. Technol.* 25:1662–1673.

Madsen, E.L. 1996. A critical analysis of methods for determining the composition and biogeochemical activities of soil microbial communities *in situ*. *In*: G. Stotzky and J.-M. Bollag (eds) *Soil Biochemistry,* Vol. 9, pp. 287–370. Marcel Dekker, New York.

Madsen, E.L. 1998a. Epistemology of environmental microbiology. *Environ. Sci. Technol.* **32**:429–439.

Madsen, E.L. 1998b. Theoretical and applied aspects of bioremediation: the influence of microbiological processes on organic compounds in field sites. *In*: R. Burlage, R. Atlas, D. Stahl, G. Geesey, and G. Sayler (eds) *Techniques in Microbial Ecology,* pp. 354–407. Oxford University Press, New York.

Madsen, E.L. 2000. Nucleic acid procedures for characterizing the identity and activity of subsurface microorganisms. Invited contribution to special issue of *Hydrogeology Journal,* B. Bekins, ed. *Hydrogeol. J.* **8**:112–125.

Madsen, E.L. 2005. Identifying microorganisms responsible for ecologically significant biogeochemical processes. *Nature Rev. Microbiol.* **3**:439–446.

Madsen, E.L. 2006. The use of stable isotope probing techniques in bioreactor and field studies on bioremediation. *Curr. Opin. Biotechnol.* **17**:92–97.

Major, D.W., M.L. McMaster, E.E. Cox, et al. 2002. Field demonstration of successful bioaugmentation to achieve dechlorination of tetrachloroethene to ethene. *Environ. Sci. Technol.* **36**:5106–5116.

Manefield, M., A.S. Whiteley, R.I. Griffiths, and M.J. Bailey. 2002. RNA stable isotope probing, a novel means of linking microbial community function to phylogeny. *Appl. Environ. Microbiol.* **68**: 5367–5373.

Michaelis, W., R. Seifert, K. Hauhaus, et al. 2002. Microbial reefs in the Black Sea fueled by anaerobic oxidation of methane. *Science* **297**:1013–1015.

Miller, D.N., J.E. Bryant, E.L. Madsen, and W.C. Ghiorse. 1999. Evaluation and optimization of DNA extraction and purification procedures for soil and sediment samples. *Appl. Environ. Microbiol.* **65**:4715–4724.

Miller, M.B. and B.L. Bassler. 2001. Quorum sensing in bacteria. *Annu. Rev. Microbiol.* **55**:165–199.

Montoya, J.P., C.M. Holl, J.P. Zehr, A. Hansen, T.A. Villareal, and D.G. Capone. 2004. High rates of N_2 fixation by unicellular diazotrophs in the oligotrophic Pacific Ocean. *Nature* **430**:1027–1031.

Moré, M.I., J.B. Herrick, M.C. Silva, W.C. Ghiorse, and E.L. Madsen. 1994. Quantitative cell lysis of indigenous microorganisms and rapid extraction of microbial DNA from sediment. *Appl. Environ. Microbiol.* **60**:1572–1580.

Morel, F.M.M. and J.G. Hering. 1993. *Principles and Applications of Aquatic Chemistry*. Wiley and Sons, New York.

Muyzer, G., E.C. DeWall, and A.G. Uitterlinden. 1993. Profiling of complex microbial populations by denaturing gradient gel electrophoresis analysis of polymerase chain reaction-amplified genes coding for 16S rRNA. *Appl. Environ. Microbiol.* **59**:695–700.

Neidhardt, F.C., J.L. Ingraham, and M. Schaechter. 1990. *Physiology of the Bacterial Cell*. Sinauer Associates, Sunderland, MA.

Nielsen, J.L., D. Christensen, M. Kloppenborg, and P.H. Nielsen. 2003. Quantification of cell-specific substrate uptake by probe-defined bacteria under in situ conditions by microautoradiography and fluorescence in situ hybridization. *Environ. Microbiol.* **5**:202–211.

Ogram, A. 1998. Isolation of nucleic acids from environmental samples. *In*: R.S. Burlage, R. Atlas, D. Stahl, G. Geesey, and G. Sayler (eds) *Techniques in Microbial Ecology*, pp. 273–288. Oxford University Press, New York.

Ogram, A., W. Sun, F.J. Brockman, and J.K. Fredrickson. 1995. Isolation and characterization of RNA from low-biomass deep-subsurface sediments. *Appl. Environ. Microbiol.* **61**:763–768.

Ogunseitan, O.A. 1997. Direct extraction of catalytic proteins from microbial communities. *J. Microbiol. Methods* **28**:55–63.

Orphan, V.J., C.H. House, K.-U. Hinrich, K.D. McKeegan, and E.F. DeLong. 2001. Methane-consuming archaea revealed by directly coupled isotopic and phylogenetic analysis. *Science* **293**: 484–487.

Osborne, M.A. and C.J. Smith (eds) 2005. *Molecular Microbial Ecology*. Taylor and Francis, New York.

Ouverney, C.C. and J.A. Fuhrman. 1999. Combined microautoradiography-16S rRNA probe technique for determination of radioisotope uptake by specific microbial cell types in situ. *Appl. Environ. Microbiol.* **65**:1746–1752.

Overmann, J. 2006. Principles of enrichment, isolation, cultivation, and preservation of prokaryotes. *In*: M. Dworkin, S. Falkow, E. Rosenberg, K.-H. Schleifer, and E. Stackebrandt (eds) *The Prokaryotes: A handbook on the biology of* bacteria, Vol. 1, *Symbiotic Associations, Biotechnology, Applied Microbiology*, 3rd edn, pp. 80–136. Springer-Verlag, New York.

Pace, N.R. 1997. A molecular view of microbial diversity in the biosphere. *Science* **276**:734–740.

Paerl, H.W., J.C. Priscu, and D.L. Brawner. 1989. Immunochemical localization of nitrogenase in marine *Trichodesmium* aggregates: relationship to N_2 fixation potential. *Appl. Environ. Microbiol.* **55**: 2965–2975.

Parkin, T.B. 1993. Spatial variability of microbial processes in soil – a review. *J. Environ. Qual.* **22**:409–417.

Parkin, T.B. and J.A. Robinson. 1994. Statistical treatment of microbial data. *In*: R.W. Weaver, S. Angle, P. Bottomley, et al. (eds) *Methods of Soil Analysis*, Part 2, pp. 15–40. Soil Science Society of American, Madison, WI.

Parton, W.J., J.W.B. Stewart, and C.V. Cole. 1988. Dynamics of C, N, P, and S in grassland soils: a model. *Biogeochemistry* **5**:109–131.

Paul, J.H. 1993. The advances and limitations of methodology. *In*: T.E. Ford (ed.) *Aquatic Microbiology*, pp. 15–46. Blackwell Scientific, Boston, MA.

Pernthaler, A. and R. Amann. 2004. Simultaneous fluorescence in situ hybridization of mRNA and rRNA in environmental bacteria. *Appl. Environ. Microbiol.* **70**:5426–5433.

Pichard, S.L. and J.H. Paul. 1993. Gene expression per gene dose, a specific measure of gene expression in aquatic microorganisms. *Appl. Environ. Microbiol.* **59**:451–457.

Piel, J. 2002. A polyketide synthase–peptide synthase gene cluster from an uncultured bacterial symbiont of *Paederus* beetles. *Proc. Natl. Acad. Sci. USA* **99**: 14002–14007.

Pinckney, J., H.W. Paerl, and M. Fitzpatrick. 1995. Impacts of seasonality and nutrients on microbial mat community structure and function. *Mar. Ecol. Prog. Ser.* **123**:207–216.

Pinkart, H.C., D.B. Ringelberg, Y.M. Piceno, S.J. McNaughton, and D.C. White. 2002. Biochemical approaches to biomass measurements and community structure analysis. *In*: Hurst, C.J., R.L. Crawford, G.R. Knudsen, M.I. McInerney, and L.D. Stetzenbach (eds) *Manual of Environmental Microbiology*, 2nd edn, pp. 91–101. American Society for Microbiology, Washington, DC.

Polz, M.F. and C.M. Cavanaugh. 1998. Bias in template-to-product ratios in multitemplate PCR. *Appl. Environ. Microbiol.* **64**:3724–3730.

Primrose, S.B. and R.M. Twyman. 2006. *Priciples of Gene Manipulation and Genomics*, 6th edn. Blackwell Science, Oxford, UK.

Quaiser, A., T. Ochsenreiter, H.P. Klenk, et al. 2002. First insight into the genome of an uncultivated crenarchaeote from soil. *Environ. Microbiol.* **4**:603–611.

Radajewski, S., P. Ineson, N.R. Parekh, and J.C. Murrell. 2000. Stable-isotope probing as a tool in microbial ecology. *Nature* **403**:646–649.

Ram, R.J., N.C. Ver Berkmoes, M.P. Thelan, et al. 2005. Community proteomics of natural microbial biofilm. *Science* **308**:1915–1920.

Rappé, M.S., S.A. Connon, K.L. Vergin, and S.J. Giovannoni. 2002. Cultivation of the ubiquitous SAR11 marine bacteriorplankton clade. *Nature* **418**:630–633.

Raven, P.H. 2002. Science, sustainability, and the human prospect. *Science* **297**:954–958.

Revsbech, N.P. and B.B. Jørgensen. 1986. Microelectrodes: their use in microbial ecology. *Adv. Microbial Ecol.* **9**:293–352.

Reysenbach, A.-L. and E. Shock. 2002. Merging genomes with geochemistry in hydrothermal ecosystems. *Science* **296**:1077–1082.

Riesenfeld, C.S., D.D. Schloss, and J. Handelsman. 2004. Metagenomics: genomic analysis of microbial communities. *Annu. Rev. Genet.* **38**:525–552.

Rocap, G., F.W. Larimer, J. Lamerdin, et al. 2003. Genome divergence in two *Prochlorococcus* ecotypes reflects oceanic niche differentiation. *Nature* **424**:1042–1047.

Rochelle, P.A. (ed.) 2001. *Environmental Molecular Microbiology: Protocols and applications*. Horizon Scientific Press, Wymondham, UK.

Ruby, E.G. 1996. Lessons from a cooperative bacterial–animal association: the *Vibrio fischeri–Euprymna scolopes* light organ symbiosis. *Annu. Rev. Microbiol.* **50**:591–624.

Ruby, E.G. 1999. The *Euprymna scolopes–Vibrio fischeri* symbiosis: a biomedical model for the study of bacterial colonization of animal tissue. *J. Molec. Microbiol. Biotechnol.* **1**:13–21.

Rusch, D.B., A.L. Halpern, G. Sutton et al. 2007. The Sorcerer II global ocean sampling expedition: Northwest Atlantic through eastern tropical Pacific. *PLOS Biol.* **5**:398–431.

Santo Domingo, J.W. and M. J. Sadowsky (eds) 2006. *Microbial Source Tracking*. American Society for Microbiology Press, Washington, DC.

Schink, B. and A.J.M. Stams. 2006. Syntrophism among prokaryotes. *In*: M. Dworkin, S. Falkow, E. Rosenberg, K.-H. Schleifer, and E. Stackebrandt (eds) *The Prokaryotes: A handbook on the biology of bacteria*, Vol. 2, 3rd edn, pp. 309–335. Springer-Verlag, New York.

Schmid, M.C., B. Maas, A. Dapena, et al. 2005. Biomarkers for in situ detection of anaerobic ammonia-oxidizing (anammox) bacteria. *Appl. Environ. Microbiol.* **71**:1677–1684.

Schwarzenbach, R.P., P.M. Gschwend, and D.M. Imboden. 2002. *Environmental Organic Chemistry*, 2nd edn. Wiley Interscience, New York.

Sharkey, F.H., I.M. Banat, and R. Marchant. 2004. Detection and quantification of gene expression in environmental bacteriology. *Appl. Environ. Microbiol.* **70**:2795–2806.

Simpson, J.M., J.W. Santo Domingo, and D.J. Reasoner. 2002. Microbial source tracking: state of the science. *Environ. Sci. Technol.* **36**:5279–5288.

Smith, R.L. 2002. Determining the terminal electron-accepting reactions in the saturated subsurface. *In*: C.J. Hurst, R.L. Crawford, G.R. Knudsen, M.J. McInerney, and L.D. Stetzenbach (eds) *Manual of Environmental Microbiology*, 2nd edn, pp. 743–752. American Society for Microbiology Press, Washington, DC.

Staley, J.T. and A. Konopka. 1985. Measurement of in situ activities of nonphotosynthetic microorganisms in aquatic and terrestrial habitats. *Annu. Rev. Microbiol.* **39**:321–346.

Strous, M. and M.M.M. Jetten. 2004. Anaerobic oxidation of methane and ammonium *Annu. Rev. Microbiol.* **58**:99–117.

Stumm, W. and J.J. Morgan. 1996. *Aquatic Chemistry*, 3rd edn. Wiley and Sons, New York.

Sugden, A., C. Ash, B. Hanson, and J. Smith. 2003. Where do we go from here? *Science* **302**:1906.

Suzuki, M.T. and S.J. Giovannoni. 1996. Bias caused by template annealing in the amplification of mixtures of 16S rRNA genes by PCR. *Appl. Environ. Microbiol.* **62**:625–630.

Tanner, M.A., B.M. Goebel, M.A. Dojka, and N.R. Pace. 1998. Specific ribosomal DNA sequences from diverse environmental settings correlate with experimental contaminants. *Appl. Environ. Microbiol.* **64**:3110–3113.

Tarancher-Oldenburg, G., E.M. Briner, C.A. Francis, and B.B. Ward. 2003. Oligonucleotide microarray for the study of functional gene diversity in the nitrogen cycle in the environment. *Appl. Environ. Microbiol.* **69**:1159–1171.

Thibodeaux, L.J. 1995. *Environmental Chemodynamics*, 2nd edn. Wiley and Sons, New York.

Tiedje, J.M., S. Simkins, and P.M. Groffman. 1989. Perspectives on measurement of denitrification in the field including recommended protocols for acetylene-based methods. *Plant Soil* **115**:261–284.

Tunlid, A., and D.C. White. 1992. Biochemical analysis of biomass, community structure, nutritional status, and metabolic activity of microbial communities. *In*: G. Stotzky and J.-M. Bollag (eds) *Soil Biochemistry*, Vol. 7, pp. 229–262. Marcel Dekker, New York.

Tyson, G.W., J. Chapman, P. Hugenholtz, et al. 2004. Community structure and metabolism through reconstruction of microbial genomes from the environment. *Nature* **428**:37–43.

van Wintzingerode, F.V., J.B. Goebel, and E. Stackebrandt. 1997. Determination of microbial diversity in environmental samples: pitfalls of PCR-based rRNA analysis. *FEMS Microbiol. Rev.* **21**:213–229.

Vaulot, D., D. Marie, R.J. Olson, and S.W. Chisholm. 1995. Growth of *Prochlorococcus*, a photosynthetic prokaryote, in the equatorial Pacific Ocean. *Science* **268**:1480–1482.

Venrick, E.L., J.R. Beers, and J.F. Heinbokel. 1977. Possible consequence of containing microplankton for physiological rate measurements. *J. Exp. Marine Biol. Ecol.* **26**:55–76.

Venter, J.C., K. Remington, J.F. Heidelberg, et al. 2004. Environmental genome shotgun sequencing of the Sargasso Sea. *Science* **304**:66–74.

Voudrais, E.A. and M. Reinhard. 1986. Abiotic organic reactions at mineral surfaces. *In*: J.A. Davis and K.F. Hayes (eds) *Geochemical Processes at Mineral Surfaces*, pp. 463–486. ACS Symposium Series No. 323. American Chemical Society, San Diego, CA.

Wagner, R.E. and G.A. Yogis. 1992. *Guide to Environmental Analytical Methods*. Benium Publishing, Schenectady, NY.

Waksman, S.A. 1927. *Principles of Soil Microbiology*. Williams and Wilkins, Baltimore, MD.

Ward, B.B. 1984. Combined autoradiography and immunofluorescence for estimation of single cell activity by ammonium-oxidizing bacteria. *Limnol. Oceanogr.* **29**:402–410.

Ward, D.M., M.M. Bateson, R. Weller, and A. Ruff-Roberts. 1993. Ribosomal RNA analysis of microorganisms as they occur in nature. *Adv. Microbial Ecol.* **12**:219–286.

Ward, N., F.A. Rainey, G. Goebel, and E. Stackebrandt. 1995. Identifying and culturing the "unculturables": a challenge for microbiologists. *In*: D. Allsopp, R.R. Colwell, D.L. Hawksworth (eds) *Microbial Diversity and Ecosystem Function*, pp. 89–110. CAB International, Wallingford, UK.

Weaver, R.W., S. Angle, P. Bottomley, et al. (eds) 1994. *Methods of Soil Analysis*, Part 2. Soil Science Society of America, Madison, WI.

White, D.C. 1994. Is there anything else you need to understand about the microbiota that cannot be derived from analysis of nucleic acids? *Microbial Ecol.* **28**:163–166.

Wilson, M.S., C. Bakermans, and E.L. Madsen. 1999. In situ, real-time catabolic gene expression: extraction and characterization of naphthalene dioxygenase mRNA transcripts from groundwater. *Appl. Environ. Microbiol.* **65**:80–87.

Wilson, M.S. and E.L. Madsen. 1996. Field extraction of a unique intermediary metabolite indicative of real time *in situ* pollutant biodegradation. *Environ. Sci. Technol.* **30**:2099–2103.

Winogradsky, S. 1949. *Microbologie du sol, problémes et methods; cinquante ans de recherches. Oeuvres completes.* Masson, Paris.

Wolfe, N.L. 1992. Abiotic transformations of pesticides in natural waters and sediments. *In*: J.L. Schnoor (ed.) *Fate of Pesticides and Chemicals in the Environment*, pp. 93–104. Wiley and Sons, New York.

Wollum, A.G., II. 1994. Soil sampling for microbiological analysis. *In*: R.W. Weaver, S. Angle, P. Bottomley, D. Bezdicek, et al. (eds) *Methods of Soil Analysis*, Part 2, pp. 2–14. Soil Science Society of American, Madison, WI.

Wu, L., X. Liu, C. Schadt, and J. Zhou. 2006. Microarray-based analysis of subnanogram quantities of microbial community DNAs by using whole-community genome amplification. *Appl. Environ. Microbiol.* **72**:4931–4941.

Wu, L., D.K. Thompson, X. Liu, et al. 2004. Development and evaluation of microarray-based whole genome hybridization for detection of microorganisms within the context of environmental applications. *Environ. Sci. Technol.* **38**:6775–6782.

Yavitt, J.B., G.E. Lang, and A.J. Sexstone. 1990. Methane fluxes in wetland and forest soils, beaver ponds, and low order streams of a temperate forest ecosystem. *J. Geophys. Res.* **95**:22463–22474.

Zehnder, A.J.B. and W. Stumm. 1988. Geochemistry and biogeochemistry of anaerobic habitats. *In*: A.J.B. Zehnder (ed.) *Biology of Anaerobic Microorganisms*, pp. 1–37. Wiley and Sons, New York.

Zhou, J. 2003. Microarrays for bacterial detection and microbial community analysis. *Curr. Opin. Microbiol.* **6**:288–294.

Zinder, S.H. 1993. Physiological ecology of methanogens. *In*: J.G. Ferry (ed.) *Methanogenesis: Ecology, physiology, biochemistry and genetics*, pp. 128–206. Chapman and Hall, New York.

Zinder, S.H. 2002. The future for culturing environmental organisms: a golden era ahead? *Environ. Microbiol.* **4**:14–15.

Zumdahl, S.S. 1986. *Chemistry*. D.C. Health, Lexington, MA.

FURTHER READING

Allen, E.E., G.W. Tyson, R.J. Whittaker et al. 2007. Genome dynamics in a natural archaeal population. *Proc. Natl. Acad. Sci. USA* **104**:1883–1888.

DeLong, E.F., C.M. Preston, T. Mincer et al. 2006. Community genomics among stratified microbial assemblages in the ocean's interior. *Science* **311**:496–503.

Lo, I., V.J. Denef, N.C. Ver Berkmoes et al. 2007. Strain-resolved community proteomics reveals recombing genomics of acidophilic bacteria. *Nature* **446**:537–541.

Microbial Biogeochemistry:
a Grand Synthesis

- *What is the "stuff of life"?*
- *What materials are passed among inhabitants of the biosphere as they are born, grow, replicate, and die?*
- *What developments have occurred over evolution that incorporate elements of rocks and minerals into biological systems?*
- *How dynamic are the pools of nutrients? How are they consumed and how are they regenerated?*
- *What drives vital processes of metabolism, heredity, and evolution?*

Biogeochemistry helps to answer these and other questions related to biosphere function and maintenance. In earlier chapters of this text, broad issues of biogeochemistry were introduced. These issues have included the following.

Chapter 1:
- *Earth habitats as heterogeneous gradients of environmental conditions.*
- *An overview of nutrient cycling and microbial physiological processes (see Tables 1.2, 1.4).*
- *The complexity of dynamic real-world habitats and the notion of budgets and models to help understand them (see Figures 1.3, 1.4).*

Chapter 2:
- *Events that occurred during the Earth's geochemical development over 4.6×10^9 years (see Table 2.1).*
- *The role of hydrophobic metal-sulfide mineral surfaces both in catalyzing energy-capturing reactions and in assembling membrane-bound compartments crucial for early life (see Sections 2.4–2.6 and Figures 2.3–2.6).*
- *The rise of oxygen in the biosphere, including the geologic evidence and biological and evolutionary consequences (see Sections 2.8–2.11, Table 2.1, and Figures 2.1, 2.2, 2.7–2.12).*

Chapter 3:
- *The role of carbon and energy sources in determining five types of microbial nutrition (see Tables 3.3, 3.4).*

- Ecosystem nutrient fluxes that regulate the physiological status and composition of microbial communities (see Table 3.4 and Figure 3.5).
- The biosphere as complex mixtures in thermodynamic disequilibrium (see Figures 3.7, 3.8).
- The thermodynamic hierarchy of half reactions that makes sense of the biosphere's many possible reaction pathways (see Tables 3.7, 3.8, Figures 3.8, 3.9, and Boxes 3.3–3.8).

Chapter 4:

- A survey of the biosphere's biomes, soil types, and aquatic and subsurface characteristics (see Tables 4.1–4.3 and Figures 4.1–4.11).
- A description of the geochemistry and microscale topography of the soil habitat (Figures 4.12–4.14).

Against the backdrop of the above issues from Chapters 1–4, we surveyed the diversity of microbial life (Chapter 5) and the methods used to ask key questions in environmental microbiology (Chapter 6).

Here in Chapter 7, we assess six unifying themes in microbial biogeochemistry: (i) the critical role of minerals (especially metals) in the key life processes of enzymatic reactions and toxicity; (ii) the foundations of global climate change; (iii) the major biosphere pools of nutrients and their turnover; (iv) details of several biogeochemical cycles (carbon, sulfur, nitrogen); (v) how and why to explore fundamental biochemical mechanisms of microbial biogeochemical processes; and (vi) global mass balances for carbon, sulfur, and nitrogen.

Chapter 7 Outline

7.1 MINERAL CONNECTIONS: THE ROLES OF INORGANIC ELEMENTS IN LIFE PROCESSES

Chapter 2 (Sections 2.5 and 2.6) highlighted current theory that metal-sulfide surfaces within porous hydrothermal vent deposits were likely the site of early biosynthetic reactions and assembly of organic molecules. In contemporary prokaryotic (and eukaryotic) enzymes, iron-sulfur clusters (Figure 7.1) may be the biochemical remnants of ancient physiology. Manifest as reaction centers, iron-sulfur clusters are extremely important in a large variety of proteins involved in oxidation/reduction reactions [essential to electron transport, respiration, and adenosine triphosphate (ATP) generation], as well as proteins that add water and oxygen (as hydratase and oxygenase enzymes, respectively) to organic molecules (Wackett et al., 1989).

One way to develop an appreciation for the importance of inorganic elements (minerals) in microbial processes is based simply on bulk cellular composition. Elemental analysis of dehydrated microbial cells shows that carbon, oxygen, nitrogen, hydrogen, and phosphorus comprise

95% of cell mass (Wackett et al., 1989). Incorporated into the integrated organic matrix of a cell, the supply of inorganic nutrients can control (limit) the amount of biomass produced during microbial growth. In this way, nutritionally essential elements are defined (Table 7.1). Table 7.2 provides a summary of the elemental components of microbial cells and examples of their physiological roles. While the macronutrients (C, H, O, N, P, S, K, Mg, Ca, Na) play rather obvious roles as components of major subcellular structures or in osmotic balance, it is the micronutrients that reveal, in striking ways, how the unique reactivities of mineral matter (especially metals) contribute to vital life processes.

Metals are required as cofactors in approximately two-thirds of all enzymes (Devlin, 2001). This fact is nothing short of remarkable. Please take a moment to ponder and appreciate this as it is restated: catalysts that drive two-thirds of your own metabolic processes (from nerve transmission to respiratory ATP production to muscle contractions) and two-thirds of the microbially mediated processes that maintain the biosphere *require* the types of metallic micronutrients listed in the lower portion of Table 7.2. *Without inclusion of the metals within the enzymatic structures, the catalysts would be ineffective.* Thus, there are extremely strong links between inorganic components of the geosphere and the evolution of life [from prebiotic Earth to the iron-sulfur world to the RNA and DNA worlds (see Chapter 2)] through to contemporary biosphere function.

> • **What properties of metals make them so useful for enzymatic catalysis?**

The answer is found in their fundamental atomic characteristics – including atomic mass, ionic radius, charge, oxidation/reduction properties, and the configurations of electrons. Each metal atom possesses a unique combination of electrons in its atomic orbitals. These establish the element's oxidation/reduction properties and its three-dimensional configuration for bonding (e.g., octahedral, tetrahedral, square pyramidal) with functional groups in enzymes (amino acid moieties, other ligands). In turn, each particular combination of organic–metal complex in a given enzyme influences the

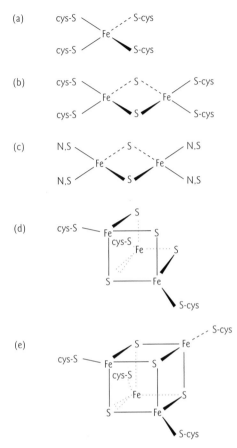

Figure 7.1 Iron-sulfur clusters serve as crucial reactive sites in the enzymes of contemporary organisms. Shown are: (a) a one-iron cluster as found in the rubredoxin protein from *Pseudomonas oleovorans*; (b) a spinach-type 2Fe2S cluster containing four cysteine-sulfur bridging groups (ligands); (c) a Rieske-type 2Fe2S cluster that contains two nitrogen-containing amino acid bridging groups; (d) a 3Fe4S cluster; and (e) a 4Fe4S cluster. (From Wackett, L.P., W.H. Orme-Johnson, and C.T. Walsh. 1989. Transition metal enzymes in bacterial metabolism. *In*: T.J. Beveridge and R.J. Doyle (eds) *Metal Ions and Bacteria*, pp. 165–246. Wiley and Sons, New York. Reprinted with permission from John Wiley and Sons, Inc., New York.)

Table 7.1

Bacterial mineral nutrition and growth: approximate amount of a given element (g) to give 100 g dry biomass (from Hughes, M.N. and R.K. Poole. 1989. *Metals and Micro-organisms*, table 1.1, p. 3. Chapman and Hall, London. With kind permission of Springer Science and Business Media)

Element	g/100 g
K	1.7
Mg	0.1–0.4
Ca	0.1
Mn	0.005
Fe	0.015
Co	0.001
Cu	0.001
Zn	0.005
Mo	0.001

specificity of substrate binding and the reactions catalyzed (Hughes and Poole, 1989). There is a broad selection of inorganic elements (their atomic numbers increase monotonically across the periodic table). There is an even broader, widely varying array of biochemically synthesized organic structures available to form metalloenzymes and other metal–organic molecular complexes within cells. Ultimately, the catalytic and other properties of metal–organic complexes reflect the many available combinations of both inorganic elements and their associated organic molecules.

Wackett et al. (2004) have recently provided a comprehensive overview of the role of all chemical elements in biological, especially microbiological, processes (Figure 7.2). The color-coded patterns in Figure 7.2 expand upon the information provided in Table 7.2 by displaying elements that are "transported, reduced, and/or methylated in some microbes", in addition to those that serve as macro- and micronutrients. When all the biologically active elements are tallied (all but the white boxes in Figure. 7.2), the total is 57. Please note that the periodic table in Figure 7.2 omits elements with atomic numbers above 86 (e.g., francium 87) and both the lanthanide and actinide series.

The presence in microorganisms of element-specific, genetically encoded transport mechanisms further confirms the biological and evolutionary significance of the targeted elements. Well-characterized nutrient-uptake transport systems in prokaryotes include those for K^+, Mg^{2+}, Fe^{3+}, Mn^{2+}, Zn^{2+}, Na^+, PO_4^{3-}, and SO_4^{2+} (Silver, 1998). In contrast, a broad variety of naturally occurring elements can be toxic to microbial (and other) cells. These pose selective pressures of another kind: eliminate toxicity by transforming the element or by mobilizing the element away from the cell.

Table 7.2

Overview of the approximate elemental composition of microbial cells and the physiological function of each element (from MADIGAN, M. and J. MARTINKO. 2006. *Brock Biology of Microorganisms*, 11th edn, table 5.2, p. 105. Prentice Hall, Upper Saddle River, NJ. Reprinted by permission of Pearson Education, Inc., Upper Saddle River, NJ. With contributions from Stanier et al., 1986; Hughes and Pool, 1989; Neidhardt et al., 1990; Wackett et al., 2004; Schaechter et al., 2006)

Element	% dry weight	Examples of cellular function
Macronutrients		
Carbon (C)	50	Building blocks of all macromolecules, carbohydrates, organic acids, proteins, lipids, cell walls, cell membranes, etc.
Hydrogen (H)	8	
Oxygen (O)	20	
Nitrogen (N)	14	Proteins, nucleic acids
Phosphorus (P)	3	Nucleic acids, phospholipids, ATP
Sulfur (S)	1	Amino acids (cysteine, methionine), vitamins, coenzyme A
Potassium (K)	1	Osmotic control, enzyme cofactor, ion balance
Magnesium (Mg)	0.5	Stabilization of macromolecular structure (ribosomes, membranes, nucleic acids), enzyme cofactor
Calcium (Ca)	0.75	Cell wall stability, enzyme cofactor
Sodium (Na)	1	Osmotic control, nutrient transport
Micronutrients*		
Iron (Fe)	0.2	Cytochromes, catalases, peroxidases, iron-sulfur proteins, oxygenases, all nitrogenases
Boron (B)	<0.01	Present in autoinducers for quorum sensing in bacteria; also found in some polyketide antibiotics
Chromium (Cr)	<0.01	Required by mammals for glucose metabolism; no known microbial requirement
Cobalt (Co)	<0.01	Vitamin B_{12} transcarboxylase (propionic acid bacteria)
Copper (Cu)	<0.01	Respiration, cytochrome *c* oxidase; photosynthesis, plastocyanin, some superoxide dismutases
Manganese (Mn)	<0.01	Activator of many enzymes; present in certain superoxide dismutases and in the water-splitting enzyme in oxygenic phototrophs (photosystem II)
Molybdenum (Mo)	<0.01	Certain flavin-containing enzymes, some nitrogenases, nitrate reductases, sulfite oxidases, DMSO-TMAO reductases, some formate dehydrogenases
Nickel (Ni)	<0.01	Most hydrogenases, coenzyme F_{430} of methanogens, carbon monoxide dehydrogenases, urease
Selenium (Se)	<0.01	Formate dehydrogenase, some hydrogenases, amino acid selenocysteine
Tungsten (W)	<0.01	Some formate dehydrogenases, oxotransferases of hyperthermophiles
Vanadium (V)	<0.01	Vanadium nitrogenase, bromoperoxidase
Zinc (Zn)	<0.01	Carbonic anhydrase, alcohol dehydrogenase, RNA and DNA polymerases, many DNA-binding proteins

DMSO, dimethylsulfoxide; TMAO, trimethylamine oxide.
* Not every micronutrient listed is required by all cells; some metals listed are found in enzymes present in only specific microorganisms.

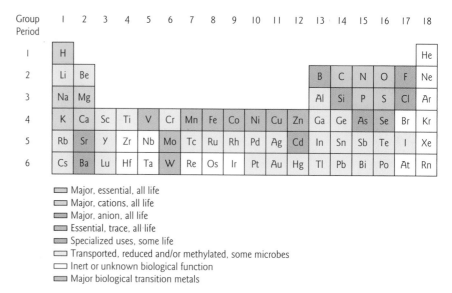

Figure 7.2 Classification of the elements of the periodic table based on their role in biological, especially microbiological, processes. (From Wackett, L.P., A.G. Dodge, and L.B.M. Ellis. 2004. Microbial genomics and periodic table. *Appl. Environ. Microbiol.* **70**:647–655. With permission from the American Society for Microbiology.)

The detoxification mechanisms are manifest as efflux pumps, reduction, and/or methylation (Silver, 1998). A group of elements drawn from the light blue and green areas of Figure 7.2 that are detoxified by well-characterized specific, genetically encoded processes include: Hg, As, Sb, Cd, Zn, Co, Ni, Ag, Cu, Cr, Fe, and Pb (Table 7.3).

7.2 GREENHOUSE GASES AND LESSONS FROM BIOGEOCHEMICAL MODELING

One of the most important twentieth century developments in global biogeochemistry has been alterations in the chemical composition of Earth's atmosphere: a buildup of tropospheric (low-altitude) greenhouse gases. Microorganisms significantly influence this issue. Reviewing the greenhouse effect, global warming, and accompanying climate change provides a means to illustrate critical principles that apply to all biogeochemical concerns.

Problem: global warming

The Earth's lower atmosphere, the troposphere (up to ~15 km thick; Schlesinger, 1997), consists of a blanket-like layer of gases that keeps the

Table 7.3

Nine classes of inorganic compounds that are toxic to microorganisms and their corresponding genetic and physiological detoxification mechanisms (from Silver, S. 1998. Genes for all metals – a bacterial view of the periodic table. *J. Indust. Microbiol. Biotechnol.* **20**:1–12, fig. 2. With kind permission of Springer Science and Business Media)

Class	Element/compound	Mechanism of detoxification or resistance*
1	Hg^{2+}	*mer.* Hg^{2+} and organomercurials are enzymatically detoxified via cleavage redox reactions that mobilize the compounds away from the cell (see Table 7.4)
2	AsO_4^{3-}, AsO_2^-, SbO^+	*ars.* Arsenate is enzymatically reduced to arsenite by ArsC. Arsenate and antimony are "pumped" out by the membrane protein ArsB that functions chemiosmotically alone or with the additional ArsA protein as an ATPase
3	Cd^{2++}	*cadA.* Cd^{2+} (and Zn^{2+}) are pumped from Gram-positive bacteria by a P-type ATPase with a phosphoaspartate intermediate
4	Cd^{2+}, Zn^{2+}, Co^{2+}, and Ni^{2+}	*czc.* Cd^{2+}, Zn^{2+}, Co^{2+}, and Ni^{2+} are pumped from Gram-negative bacteria by a three-polypeptide membrane complex that functions as a divalent cation/$2H^+$ antiporter. The complex consists of an inner membrane protein (CzcA), an outer membrane protein (CzcC), and a protein associated with both membranes (CzcB)
5	Ag^+	*sil.* Ag^+ resistance results from pumping from bacteria by three-polypeptide chemiosmotic exchangers plus a P-type ATPase
6	Cu^{2+}	*cop.* Plasmid Cu^{2+} resistance results from a four-polypeptide complex, consisting of an inner membrane protein, an outer membrane protein, and two periplasmic copper-binding proteins. In *Pseudomonas*, Cop results in periplasmic sequestration of Cu^{2+}. In addition, chromosomally encoded P-type ATPases provide partial resistance by effluxing Cu^{2+} or Cu^+
7	CrO_4^{2-}	*chr.* Chromate resistance results from a single membrane polypeptide that causes reduced net cellular uptake, but efflux has not been demonstrated
8	TeO_3^{2-}	*tel.* Tellurite resistance results from any of several genetically unrelated plasmid systems. Although reduction to metallic Te^0 frequently occurs, this does not seem to be the primary resistance mechanism
9	Pb^{2+}	*pum.* Lead resistance appears to be due to an efflux ATPase in Gram-negative bacteria and the accumulation of intracellular $Pb_3(PO_4)_2$ in Gram-positive bacteria

* The involved genetic system is italicized (e.g., *mer* are mercury-resistance genes, and *ars* are arsenic-resistance genes). Enzymes encoded by particular genes are shown in nonitalicized letters: for example ArsC is the structural enzyme, arsenate reductase, encoded by the *arsC* gene. Resistance systems await understanding for bismuth (Bi), boron (B), thallium (Tl), and tin (Sn).

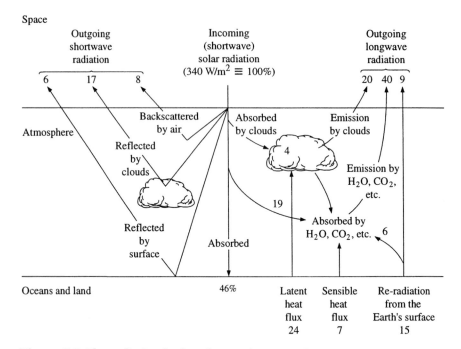

Figure 7.3 The radiation budget for Earth. Latent heat is heat stored in water as it changes from liquid to the vapor phase. Sensible heat is convection, especially from the oceans to the atmosphere. See text for details. (Reprinted from Schlesinger, W.H. 1997. *Biogeochemistry: An analysis of global change*, 2nd edn. Academic Press, New York. Copyright 1997, with permission from Elsevier.)

Earth warm. Figure 7.3 provides an overview of the radiation budget for our planet. Incoming light delivers energy at a rate (or power) of 340 W/m² to the outer atmosphere. A portion of the incident light (~30%) is reflected back into outer space, largely unaltered. The remainder interacts with global materials (gases, clouds, water, land, biota), being entirely absorbed or absorbed and then released back to space as outgoing light of longer wavelength (~70% of the incident power). Without the naturally occurring greenhouse gases to retain heat, Earth's temperature would drop by 33°C (Henshaw et al., 2000), causing substantial conversion of liquid water to ice.

As shown in Figure 7.3, the major greenhouse gases contributing to storage of heat in the troposphere are H_2O (approximately two-thirds of storage) and CO_2 (about one-third). But during the last century, several naturally occurring greenhouse gases (e.g., CO_2, methane, nitrous oxide, ozone) as well as a variety of synthetic compounds known as chlorofluorocarbons (CFCs) have increased significantly above historical levels (Figure 7.4; Houghton et al., 2001).

The overall rise in tropospheric carbon dioxide (~10%) and methane (~9%) during the last quarter century (Figure 7.4) display annual sinusoidal oscillations reflecting a seasonal shift between processes that produce and consume each of the gases. In the case of CO_2, the two opposing processes are clearly photosynthesis and respiration; the former causing an annual decline in CO_2 concentration in the northern hemisphere each summer (Schlesinger, 1997). The long-term rising trend in CO_2 concentration is the result of fossil-fuel burning (largely coal and petroleum products) – which releases CO_2 at a rate that exceeds the rate of return to nonatmospheric global carbon reservoirs. In the case of methane, neither its rise in concentration, nor its annual oscillation have been adequately explained (Schlesinger, 1997). A variety of human-managed systems [e.g., wetland rice production, burning of vegetation, herds of ruminant livestock (whose anaerobic digestive tracts harbor methanogenic bacteria), and mining operations] have the potential to be driving the global increase in tropospheric methane – though the specific cause has not yet been identified. A major potential mechanism of methane consumption is microbial metabolism: both aerobic and anaerobic microorganisms in soils, waters, and sediments use methane as a source of energy (ATP production) and/or carbon (see Sections 3.11, 7.4, and 7.5). Because these microbial processes are subject to annual activation by warm temperatures, they likely contribute to the annual oscillations shown in Figure 7.4. The other major (and nonseasonal) loss mechanism for tropospheric methane involves scavenging in the atmosphere by gas-phase hydroxyl radicals (Schlesinger, 1997).

The perceptive reader will note that the vertical scale units for CO_2 concentration is parts per million (ppm) in Figure 7.4, while the scale for the other four compounds is smaller by a factor of 10^3 (parts per billion) or 10^6 (parts per trillion).

> • **Why would relatively small rises in tropospheric concentrations of nitrous oxide (N_2O) and synthetic gases used by the refrigeration industry (CFC-12 and CFC-11; CCl_2F_2 and CCl_3F, respectively) be of global significance?**

The reason the greenhouse effect responds to trace levels of CH_3, N_2O, and CFCs lies in two key details of their infrared absorption spectra (the main mechanism of atmospheric heat retention; Charlson, 2000):

1 These gases (Box 7.1) absorb strongly within that part of the infrared spectrum where water vapor and CO_2 do not already absorb.
2 This portion of the spectrum is also a major avenue of Earth's loss of radiant energy.

The five gases shown in Figure 7.4 (plus trophospheric ozone, see below) are estimated to account for nearly all of the global warming

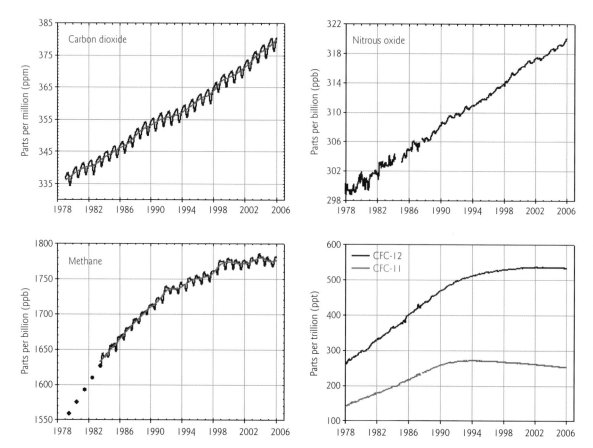

Figure 7.4 Historical records of the concentrations of five atmospheric gases important for global warming that occur in the troposphere. CFC-12 (CCl_2F_2) and CFC-11 (CCl_3F) also contribute to stratospheric ozone destruction. Data are for the northern hemisphere. Oscillations (CO_2, methane) reflect how the balance between destruction and production fluctuate annually. See text for details. (From D. Hofmann, NOAA, with permission.)

that has occurred since industrialization began in ~1750 (Charlson, 2000; Houghton et al., 2001). Information in Figure 7.5 (from the Intergovernmental Panel on Climate Change organized by the World Meteorological Organization and the United Nations Environmental Programme) provides an integrated glimpse of the diversity of forces that can alter the Earth's radiation balance (radiative forcing in Figure 7.5 has units of W/m² of the globe's surface areas). Greenhouse gases and tropospheric ozone constitute only two of the dozen other factors that may cool or warm the atmosphere. Among the others are aerosols, aviation-induced clouds, land use, and variability in incident solar energy. As is shown on the horizontal axis of Figure 7.5, scientific understanding is higher for the influence of greenhouse gases than any of the other factors.

Box 7.1

Greenhouse gases: on a molar basis their effects are not equal (from Rodhe, H. 1990. A comparison of the contribution of various gases to the greenhouse effect. *Science* **248:1217–1219. Reprinted with permission from AAAS)**

Type of gas	Contribution, relative to CO_2, of various greenhouse gases to the greenhouse effect
CO_2	1
CH_4	25
N_2O	200
Ozone (O_3)	2,000
CFC-11	12,000
CFC-12	15,000

The impact of each particular greenhouse gas on global warming results from a combination of:
• The compound's infrared absorption spectrum.
• The amount of energy at a given wavelength already absorbed by water and CO_2.
• The intensity of potential radiation loss from Earth to outer space at a given wavelength.
(See text for details.)

Lessons from modeling biogeochemical cycles

A variety of naturally occurring processes at the interface between the geosphere, atmosphere, and hydrosphere are cyclic in nature:
• Water evaporates, and then condenses as precipitation.
• Continents erode into the oceans, sediments are then uplifted by crustal movements.
• Photosynthesis removes carbon dioxide from the atmosphere, then respiration returns it.
Models are tools that allow cyclic processes to be formally recognized. Then, information about particular phenomena can be systematically assembled and used to: (i) interpret historical events; and (ii) predict future events (Rodhe, 2000). By formally declaring facts about and relationships between components of biogeochemical systems, models (especially quantitative models) are heuristic – they reveal knowledge gaps and inaccuracies, and identify new areas of inquiry. In turn, the newly created information can be funneled into improved models that lead to an increased understanding of the system of interest.

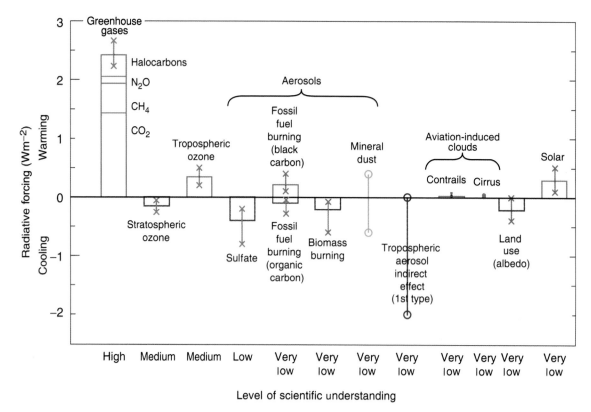

Figure 7.5 Twelve interacting factors (including greenhouse gases and tropospheric ozone) that govern the radiation balance in the Earth's atmosphere. The horizontal axis assesses the level of scientific understanding for each of the 12 factors. (From Houghton, J.T., Y. Ding, D.J. Griggs, et al. 2001. *Climate Change 2001: The science basis.* Cambridge University Press, Cambridge, UK. With permission from the Intergovernmental Panel on Climate Change.)

On paper, conceptual biogeochemical models (e.g., of a lake or a rainforest or the entire globe) resemble a series of boxes connected by arrows. An example is shown in Figure 7.6 – the visual depiction of carbon and nitrogen compounds as they function in terrestrial habitats. A variety of ecosystem processes cause carbonaceous and nitrogenous substances to be passed between vegetative biomass and soil detritus. In Figure 7.6, carbon mineralization (conversion to CO_2) is shown, as is nitrogen mineralization to inorganic forms such as ammonia and nitrate (the available nitrogen pool, N_{AV}). One of the key challenges for biogeochemical modeling is to quantify the types and sizes of nutrient pools and the fluxes (arrows in Figure 7.6) between them. Information in Box 7.2 elaborates on some of the terms and concepts pertinent to developing biogeochemical models.

Xiao et al. (1997) have woven the terrestrial ecosystem model shown in Figure 7.6 (itself, a highly complex computer program) into a tapestry of other computer programs aimed at describing the global response of plant communities to climate change across 18 different biomes. The terrestrial ecosystem model works on a spatial grid of >62,000 cells that map the Earth's surface (each ~55 × 55 km in size). In the computer program, calculations for this set of cells are fed into a series of additional atmospheric chemistry and climate models that are coupled together (Figure 7.7). The mega model is designed to integrate the detailed interactions between vegetation, soil, and greenhouse gases – thereby predicting how an anticipated rise in tropospheric CO_2 concentration will influence the global carbon balance – especially the distribution of carbon in pools of plant biomass and soil detritus (soil organic carbon and humus).

Keeping in mind the above overview of environmental compartments, elemental pools, and fluxes between pools, we now turn to the detailed microbiological processes that drive many of the fluxes.

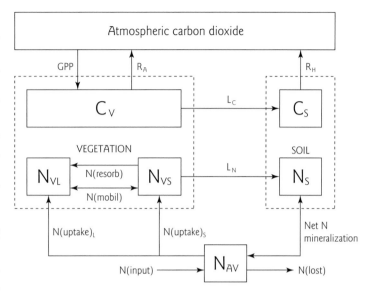

Figure 7.6 Conceptual biogeochemical model of carbon and nitrogen flow in global terrestrial habitats. This terrestrial ecosystem model consists of boxes representing three pools of carbon and four pools of nitrogen dispersed across three environmental compartments: the atmosphere, vegetation, and soil. C_S, C_V, carbon in soil or vegetation; L_C, L_N, litter carbon or nitrogen; GPP, gross primary productivity; N_S, N_V, nitrogen in soil or vegetation; N_{VL}, labile nitrogen in vegetation; N_{VS}, structural nitrogen in vegetation; N_{AV}, available inorganic nitrogen in soil; R_A, R_H, respiration by autotrophs or heterotrophs. (From Xiao, S., D.W. Kicklighter, J.M. Melillo, A. D. McGuire, P. H. Stone, and A. P. Sokolov. 1997. Linking a global terrestrial biogeochemical model and a 2-dimensional climate model: implications for the global carbon budget. *Tellus* **49B**:18–37. With permission from Blackwell Publishing, Oxford, UK.)

7.3 THE "STUFF OF LIFE": IDENTIFYING THE POOLS OF BIOSPHERE MATERIALS WHOSE MICROBIOLOGICAL TRANSFORMATIONS DRIVE THE BIOGEOCHEMICAL CYCLES

Conceptually, the "stuff of life" is two things: (i) it is the vital biomass, the bodies of functioning creatures (e.g., viruses, *Bacteria*, *Archaea*, fungi, protozoa, nematodes, insects, plants, animals, humans; see Sections 4.5 and 5.5–5.9) presently alive on Earth; and (ii) it is the substances (light and thermodynamically unstable mixtures of materials; see Sections 3.6–

Box 7.2

Issues common to models of all biogeochemical processes (compiled from Rodhe, 2000)

- Reservoirs (pools of materials) = M.
- Fluxes = mass transferred between reservoirs per unit time.
- Source (flux into reservoir) = Q.
- Sink (flux out of reservoir) = S.
- Budget = balance sheet of sources, sinks.
- Reservoir turnover time $T = M/S = M/Q$ in steady state.
- Residence time (of a molecule in a given pool of material) = probability-based calculation of how long a molecule of substance will take to exit the system once it has entered.
- Response time (of a reservoir) = timescale that characterizes the adjustment to equilibrium after a sudden change in the system.
- Linearity of system response (to perturbations): responses may be *linear* (directly proportional to input rates or to the degree of change), or responses may be *nonlinear* (by departing from linear responses in unpredictable ways).
- Is the system in steady state?
- What are the system transport and mixing characteristics?
- The system must be well defined, including: boundaries, timescales, and spatial scales.

3.11) that fuel biochemical reactions used to keep the creatures alive and well. From the view point of microbiology, *the selective pressures that have driven evolution operate via specific organic and inorganic compounds.* These are the resources of our planet that have been (and continue to be; day-by-day, minute-by-minute) utilized for catabolic and anabolic purposes. Particular compounds occur in particular geochemical contexts. Remember, evolution acts at the level of mutations in DNA and their manifestation as altered enzyme recognition, specificity, and function.

Information in Table 7.4 provides a listing of many of the substances in the biosphere that drive biogeochemical processes. The listing is not comprehensive, but it is representative of biogeochemically important pools and processes. The substances are sorted by dominant elemental cycle (C, S, N, etc.) and type (organic or inorganic). Included in the table is an overview of the source of the compounds, mechanisms of biochemical utilization, and examples of key biota involved in substance utilization. The organic compounds, shown in columns 2 and 3 of Table 7.4, are, in essence, a compositional analysis of the major biomolecules in current life forms or what they have become (peat, humic substances, kerogen, petroleum, coal). The inorganic compounds (also columns 2 and 3 of

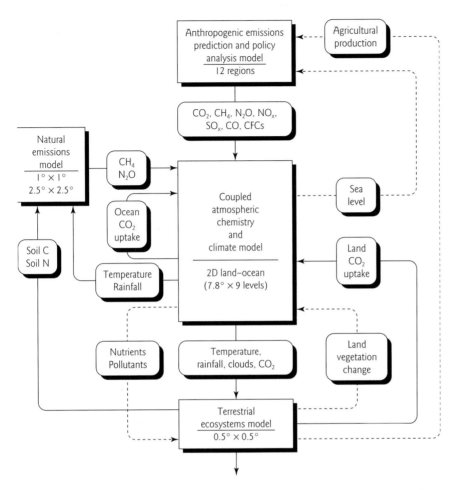

Figure 7.7 Global climate change mega model created by the integrated assessment framework at Massachusetts Institute of Technology. The goal of the mega model is to assess and predict interactions between terrestrial ecosystems and many climate-related processes. Note that the terrestrial ecosystem model (Figure 7.6) is embedded within the mega model. The numbers show the dimension of the surface area grids used to assemble the model (units are degrees latitude by degrees longitude). (From Xiao, S., D.W. Kicklighter, J.M. Melillo, A.D. McGuire, P.H. Stone, and A.P. Sokolov. 1997. Linking a global terrestrial biogeochemical model and a 2-dimensional climate model: implications for the global carbon budget. *Tellus* **49B**:18–37. With permission from Blackwell Publishing, Oxford, UK.)

Table 7.4) are gases or minerals in soil, sediments, waters, and/or the atmosphere that participate in metabolic processes as electron donors, as electron acceptors, as building blocks of biomolecules, or as potential toxicants (in the case of Hg).

Table 7.4

Selected compounds in the biosphere that participate in major microbial biogeochemical processes (compiled from Zabel and Morrell, 1992; Buist et al., 1995; Lengeler et al., 1999; Ehrlich, 2002; Zahrl et al., 2005)

Dominant elemental cycle	Type of compound	Compounds in reservoirs	Production and location of compounds	Key process for compound utilization*	Key organisms active in utilization
Carbon (C)	Organic (live and dead biomass)	Cellulose	Plants; in soils, sediments, water	Hydrolysis via cellulases; glucosidases, cellulosomes; and respiration of subunits	Fungi, prokaryotes
		Lignin	Plants; in soils, sediments, water	Enzymatic combustion via Fe-ligninase, Mn-peroxidase,Cu-laccase	Fungi
		Hemicellulose	Plants; in soils, sediments, water	Hydrolysis via hemicellulase; and respiration of subunits	Fungi, prokaryotes
		Starch	Plants; in soils, sediments, water	Hydrolysis via amylases; and respiration of subunits	All life
		Chitin	Fungi, insects, marine zooplankton; in soils, sediments, water	Hydrolysis via chitinases and N-acetyl glucosaminidase; and respiration of subunits	Prokaryotes, fungi
		Peptidoglycan	Prokaryotes; in soils, sediments, water	Hydrolysis via peptidoglycan hydrolase, lytic transglycosylase, proteases; and respiration of subunits	Prokaryotes, fungi
		Lipids	All life; in soils, sediments, water	Hydrolysis via lipase; and respiration of subunits	All life
		Protein	All life; in soils, sediments, water	Hydrolysis via proteases; and respiration of subunits	All life
		Nucleic acids	All life; in soils, sediments, water	Hydrolysis via nucleases; and respiration of subunits	All life
		Humic substances	Random biotic and abiotic synthesis in soils, sediments	Prokaryotic and fungal enzymes, physical and chemical weathering	Prokaryotic heterotrophs
		Peat, keragen, coal	Diagenetic processes in shallow to deep geologic formations	Resistant	—

Element		Form	Reservoir/processes	Process	Organisms
Carbon (C)		Petroleum	Diagenetic processes in deep geologic formations	Hydroxylation by oxygenase enzymes, β-oxidation of alkanes, dioxygenase cleavage of aromatic rings; anaerobic activation to form conjugated organic acids leading to reduction and hydrolysis; and respiration	Aerobic and anaerobic prokaryotes, and fungi
	Inorganic	CO_2	Acidification of carbonate minerals, biotic respiration; in atmospheric pool	Final electron acceptor; photosynthesis	Methanogenic *Archaea*; plants, algae, chemolithotrophic prokaryotes
		CH_4	Diagenetic processes in geologic formations; methanogenesis in anaerobic microbial communities; in atmospheric pool, gas hydrate "clathrate" deposits in ocean floor	Used as carbon and energy source (methanotrophy)	Aerobic and anaerobic prokaryotic methanotrophs
		CO	Lignin catabolism, fossil fuel combustion; in atmosphere	Used as electron donor for ATP generation	Carboxydotrophic, acetogenic, and methanogenic prokaryotes
Sulfur (S)	Organic (live and dead biomass)	Protein (especially methionine and cysteine)	All life; in soils, sediment, water	Hydrolysis via proteases; and respiration of subunits	All life
		Humic substances	Random biotic and abiotic synthesis in soils, sediments	Prokaryotic and fungal enzymes, physical and chemical weathering	Heterotrophic prokaryotes, fungi
	Inorganic	Sulfate	Salt dissolution, sulfide oxidation, elemental sulfur oxidation; in soils, sediment, water (especially oceans)	Used as final electron acceptor in dissimilatory sulfate reduction, reduced during sulfate reduction by ATP sulfurylase enzyme	Heterotrophic anaerobic prokaryotes

Table 7.4 *Continued*

Dominant elemental cycle	Type of compound	Compounds in reservoirs	Production and location of compounds	Key process for compound utilization*	Key organisms active in utilization
		Thiosulfate	Chemical oxidation of sulfur minerals; in soils, sediment, waters	As electron acceptor, reduced to sulfide As electron donor, oxidized to sulfate	Sulfate-reducing bacteria Sulfur-oxidizing chemolithotrophic bacteria
		Dimethyl sulfoxide (DMSO)	Oxidation of DMS; in marine systems	Reduced as electron acceptor to DMS; used as electron donor, oxidized to sulfate	Prokaryotic anaerobic heterotrophs, aerobic heterotrophs
		Elemental S	Sulfide oxidation; in soils, sediment, water	Electron acceptor (reduced to sulfide) and electron donor (oxidized to sulfate) for ATP generation and for phototrophic CO_2 reduction (oxidized to sulfate)	Anaerobic prokaryotes, chemolithoautotrophs, phototrophic S bacteria
		Dimethyl sulfide, (DMS)	Release from decaying marine algae, reduction of DMSO; in marine systems	Electron donor for ATP generation with oxygen, nitrate and sulfate as electron acceptor; and for CO_2 reduction and methanogenesis	Chemoorganotrophs, chemolithotrophs, phototrophs, methanogens
		Hydrogen sulfide, mineral sulfides	Geothermal gases, sulfate reduction; in geologic deposits, sulfate-rich anaerobic sediments	Electron donor for ATP generation and for phototrophic CO_2 reduction	Chemolithoautotrophs, phototrophic S bacteria
Nitrogen (N)	Organic (live and dead biomass)	Protein	All life; in soils, sediments, water	Hydrolysis via proteases; and respiration of subunits	Prokaryotic and eukaryotic heterotrophs
		Humic substances	Random biotic and abiotic synthesis in soils, sediments	Prokaryotic and fungal enzymes, physical and chemical weathering	Prokaryotic heterotrophs, fungi
		Trimethylamine N-oxide (TMAO)	Nitrogenous excretion by fish	Used as electron acceptor, reduced to trimethylamine	Prokaryotic anaerobes

			Assimilation†: nutrient uptake	Plants, microorganisms
Inorganic	NH₃	Dinitrogen fixation by prokaryotes; ammonification from amino acids; dissimilatory nitrate reduction to ammonia; in soils, sediment, water	Assimilation†: nutrient uptake Dissimilation‡: used as electron donor for ATP generation; nitrification, anammox	Plants, microorganisms Prokaryotic chemolithotrophs (nitrifiers, anammox)
	NO_3^-	Nitrification; in soils, sediment, water	Assimilation: nutrient uptake Dissimilation: used as electron acceptor for ATP generation; denitrification to N_2 or dissimilatory nitrate reduction to NH_4^+	N-assimilation pathways in plants, fungi, prokaryotes Nitrate-reducing prokaryotes
	NO_2^-	Nitrification and denitrification; in soils, sediment, water	Used as electron acceptor reactant in anammox, also reduced to NO in denitrification Oxidized to nitrate in nitrification	Anammox and denitrifying prokaryotes Nitrifying prokaryotes
	NO	Intermediary metabolite in denitrification; in soils, sediment, water	Used as electron acceptor (reduced to N_2O)	Prokaryotic denitrifiers (see Table 7.6 and Box 7.7)
	N_2O	Intermediary metabolite in denitrification; in soils, sediment, water	Used as electron acceptor (reduced to N_2)	Prokaryotic denitrifiers (see Table 7.6 and Box 7.7)
	N_2	Metabolic endproduct of denitrification; in soils, sediment, water; atmospheric pool	Reduced to NH_3, during nitrogen fixation by nitrogenase	Nitrogen-fixing prokaryotes
Phosphorus (P)	Organic (live or dead biomass) Phospholipids	All life; in soils, sediment, water	Hydrolysis, assimilation, catabolism	All life
	Nucleic acids	All life; in soils, sediment, water	Hydrolysis, assimilation, catabolism	All life
	ATP	All life; in soils, sediment, water	Catabolism	All life
	Inorganic Mineral phosphates	Mineral reservoirs; mineralization of biomass	Uptake, assimilation	All life

Table 7.4 Continued

Dominant elemental cycle	Type of compound	Compounds in reservoirs	Production and location of compounds	Key process for compound utilization*	Key organisms active in utilization
Hydrogen (H)	Inorganic	H_2 gas	Geothermal formation; anaerobic food chains in sediments and in animal guts; atmospheric pool	Electron donor for ATP generation for aerobes and anaerobes	Chemolithotrophic prokaryotes (e.g., aerobic hydrogen oxidizers, methanogens)
Iron (Fe)	Organic (live and dead biomass)	Heme, metalloenzymes	Anabolism in all life	Catabolism	All life
	Inorganic	Fe^{2+} (aq) and in minerals	Minerals; reduction of Fe^{3+}; in soils, sediment, water	Biotic and abiotic (O_2) oxidation of Fe^{2+} to Fe^{3+}	Iron-oxidizing bacteria, which include heterotrophs (e.g., *Leptothrix*) and autotrophs (e.g., *Thiobacillus ferrooxidans*)
				Electron donor for chemolithoautotrophs using O_2 and NO_3^- as final electron acceptors	
				Electron donor for anoxygenic photosynthetic bacteria	Purple, nonsulfur bacteria
		Fe^{3+} [e.g. Fe_2O_3, $Fe(OH)_3$]	Minerals, chemical and microbial oxidation of Fe^{2+}; in soils, sediment, water	Anaerobic final electron acceptor in respiratory ATP generation	Anaerobic prokaryotes (e.g., *Geobacter*)
Mercury (Hg)	Organic	CH_3Hg^+, $(CH_3)_2Hg$	Methylation predominantly by sulfate-reducing bacteria in anaerobic sediments	Demethylation and reduction to Hg^{2+} and Hg^0	UV light, prokaryotic heterotrophs, methanogens fungi (see Sections 8.3 and 8.6)
	Inorganic	Hg^0	Minerals, reduction of Hg^{2+}, demethylation of methyl-Hg	Methylation	Sulfate-reducing prokaryotes, fungi,
		Hg^{2+}	Minerals, photo-oxidation of Hg^{2+}	Detoxification via reduction; methylation	Prokaryotes with *mer* genes; sulfate-reducing prokaryotes

* Major biological mechanism(s) of compound utilization include enzymatic steps in biogeochemical cycles.

† Assimilation: nutrient uptake – incorporated into cellular components and biomass.

‡ Dissimilation: used as final electron acceptor – reduced and released as metabolic waste; not incorporated into cell biomass.

Hydrolytic breakdown of polymers

Most of the organic compounds shown in column 3 of Table 7.4 are the products of cellular biosynthetic reactions creating essential structural polymers (e.g., cellulose, lignin, lipids, chitin, protein; as membranes, cell walls, exoskeletons, muscle tissue, etc.; see Section 3.9). A key step in the recycling of these high molecular weight (insoluble, often particulate) materials is conversion via hydrolysis to constituent monomers, that can be transported to the interior of microbial cells to fuel catabolic and/or anabolic reactions. "Hydrolysis" is, by definition, the nucleophilic attack of a water molecule across a C-C bond: H is added to one side and OH to the other as two molecules are released from the site of bond cleavage. Box 7.3 describes the major biopolymers of the biosphere and their metabolism.

Box 7.3

Major polymers of the biosphere and their metabolism

Biopolymer	Monomer	Number of monomers	Glucosidic and other bonds
Starch	Glucose	100–1000	α-1,4 and α-1,6
Hemicellulose	Xylose, glucose, mannose, galactose, uronic acids	~200	β-1,4 and cross links
Cellulose	Glucose	3,000–26,000	β-1,4
Lignin	Phenyl propane units (p-hydroxycinnamyl alcohols)	~10,000	Random carbon–carbon and ether linkages
Chitin	N-acetyl glucosamine	~10,000	β-1,4
Peptidoglycan	N-acetylglucosamine, N-acetyl muramic acid, amino acids	4,000	β-1,4 tetrapeptide cross links

Several biochemical properties of the first six biopolymers in Table 7.4 are shown above. Constituent monomers, chain length, and the type of chemical bonds between monomers have major effects on biopolymer properties.

Starch
The most common storage material of plants, is composed of glucose monomers, linked in a branching pattern that connects the carbon number 1 of a given glucose molecule to either carbon number 4 or number 6 of an adjacent glucose molecule. Thus, the bonds between the two sugars are "1,4-" and "1,6-" glucosidic bonds. The "α" in column 4 (above) refers to the geometry of the C-O-C linkage between molecules. Starch molecules composed of up to 1000 monomers are readily susceptible to hydrolysis by amylosaccharide (or "amylase") enzymes that are widespread among organisms (Lengeler et al., 1999).

Box 7.3 *Continued*

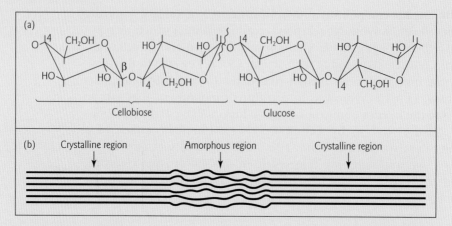

Figure 1 Structure of cellulose: (a) β-glucosidic bonds, and (b) schematic structure of a cellulose fibril. (From Lengeler, J.W., G. Drews, and H.G. Schlegel. 1999. *Biology of Prokaryotes*, fig. 9.3. Blackwell Science, Stuttgart. With permission from Blackwell Science, Stuttgart.)

Cellulose

Though cellulose is also formed from glucose units with bonds between carbons number 1 and 4, the geometry of the C-O-C bond linkage is in the opposite (β) configuration from those in starch. The β configuration confers high tensile strength and causes a 180° rotation, of each glucose molecule relative to its neighbor (Figure 1). In addition, the combination of very long chains (~10,000 monomers) and hydrogen bonding between adjacent chains causes the polymers to form fibrils that are rigid, insoluble, and crystalline. These properties render cellulose susceptible to attack by only a highly specialized set of enzymes encoded by a limited set of bacteria and fungi.

Plant tissue

Plant tissue, especially wood, is a major pool of organic carbon in the biosphere (see Sections 7.4 and 7.6). Biochemically, wood is termed "lignocellulose" (Hammel, 1997; Deobold and Crawford, 2002) because cellulose and lignin are in intimate contact. Functionally, lignin adds strength and rigidity to cellulose fibrils. During tree growth, wood biosynthetic reactions encrust the cellulose fibril within a matrix of lignin monomers (benzene rings with three-carbon side chains, known as "phenyl propane" units). Peroxidase and laccase enzymes contribute to lignin biosynthesis by activating the lignin monomers as free radicals and they polymerize randomly and spontaneously (Hammel, 1997). Thus, lignin (Figure 2) in trees bears a close resemblance to, and is a precursor of, humic substances in soil. Clearly, as a habitat for microbial growth, wood presents many challenging characteristics: it is insoluble, crystalline, dry, of low nutrient (low proportions of N, P, Fe, etc.), and a structurally integrated matrix of cellulose and lignin (with other polysaccharides). Fungi, particularly *Basidiomycetes* and *Ascomycetes* (see Section 5.6) are considered the major players in lignocellulose decomposition (Zabel and Morrell, 1992; Hammel, 1997). Though bacteria

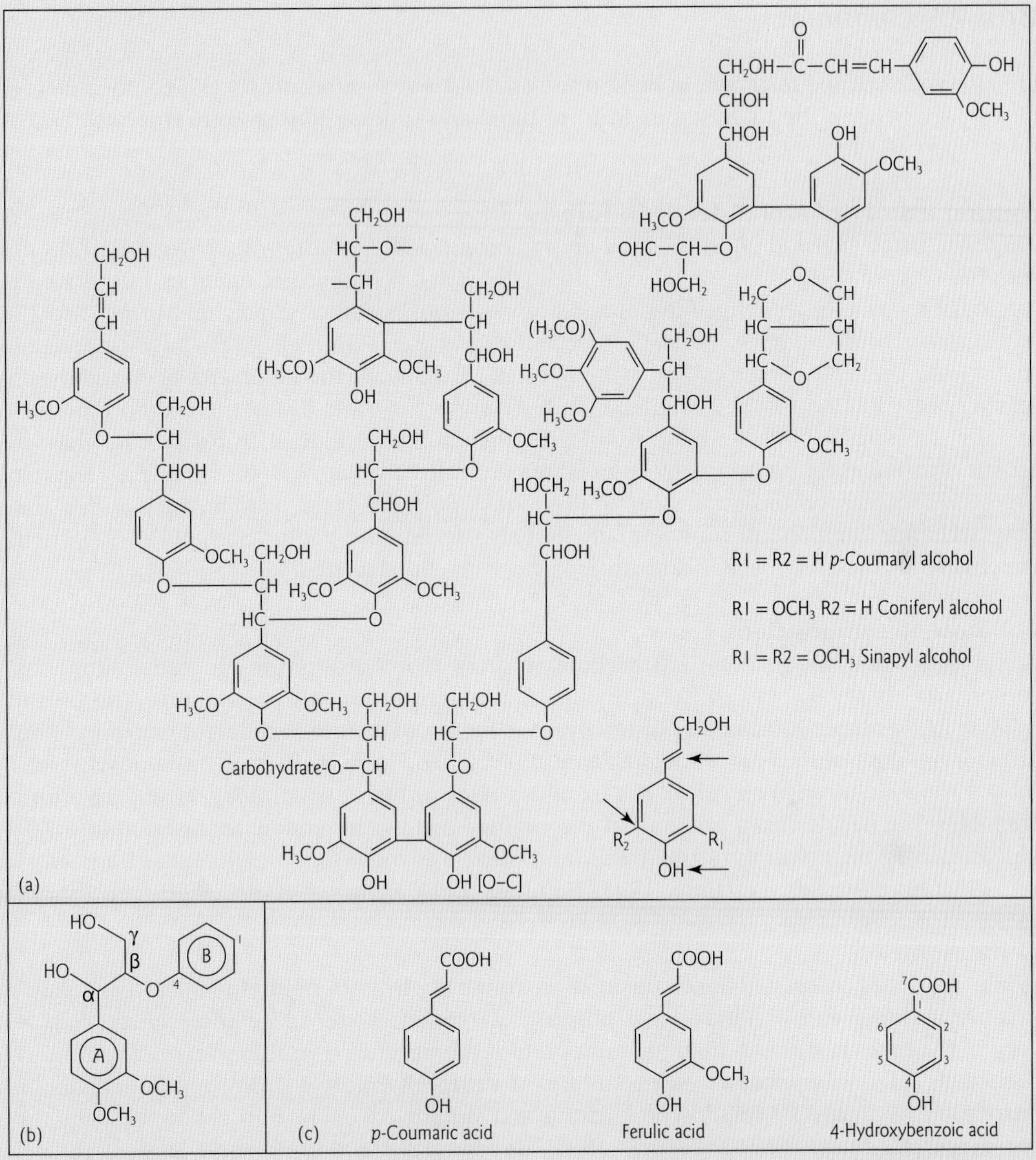

Figure 2 Lignin structure and its constituents. (a) The three precursor alcohols of lignin are shown on the lower right. One electron oxidation of these lignin building blocks and subsequent polymerization reactions produce the three-dimensional lignin structure shown. The arrows indicate the position of intermonomeric bonds in lignin, including arylether, biphenyl, and diphenylether bonds. (b) Lignin model compound with β-O-4 ether linkage. (c) Aromatic acids found in lignin–carbohydrate linkages. (From Lengeler, J.W., G. Drews, and H.G. Schlegel. 1999. *Biology of Prokaryotes*, fig. 9.9. Blackwell Science, Stuttgart. With permission from Blackwell Science, Stuttgart.)

Box 7.3 *Continued*

are also well adapted to carry out cellulose metabolism in some contexts (especially anaerobically, see below), the filamentous form of fungi and unique ligninase enzyme system set them apart (Hammel, 1997).

Natural history of wood decay in forests

There are three outward types of wood decay decomposition: white rot, brown rot, and soft rot. These three designations are based upon the visual and textural qualities of the wood [an initial mixture of white (cellulosic) and brown (argininic) tissues; Zabel and Morrell, 1992]. After *white rot* fungi (largely *Basidiomycetes* and *Ascomycetes*; Hammel, 1997) have completed their work, the remaining logs are white because virtually all the lignin has been removed, leaving only remnants of the original cellulose. *Brown rot* fungi decay the cellulose preferentially – leaving behind the majority of the original brown lignin-rich tissue. The characteristic of *soft rot* is that it occurs often in water-saturated (though aerobic) habitats – leaving the woody tissue porous and structurally weak. Soft rot fungi (*Ascomycetes*) attack the polysaccharide components preferentially, but also decompose lignin (Hammel, 1997); their enzymatic degradative mechanisms are not yet well characterized.

Cellulose decomposition

Cellulose decomposition is an extracellular process that releases the glucose subunits as carbon and energy sources to the active microbial (largely fungal) populations. The fungal hyphae deliver three functionally distinctive, yet highly cooperative, cellulase enzymes to an eroding cellulosic surface: (i) endoglucanases, which randomly hydrolyze 1,4-bonds internal to the long cellulose molecules; (ii) exoglucanases, which sequentially cleave two-unit (cellobiose) molecules from one end of the chains freed by the endoglucanases; and (iii) β-glucosidases, which hydrolyze cellobiose (and other low molecular weight sugar fragments) into glucose molecules (Leschine, 1995; Hammel, 1997; Lengeler et al., 1999).

Cellulosomes

While the basic three-step mechanism of cellulose hydrolysis holds for all active microorganisms (fungi and bacteria), each possesses a refined cluster of cellulase enzymes that vary in the *endo-* versus *exo-* mode of attack and recognition of crystalline versus amorphous regions in cellulose molecules (see cellulose structure, in Figure 1). One unique metabolic adaptation in anaerobic cellulolytic bacteria and fungi that operate in submerged aquatic habitats is the *cellulosome* (Leschine, 1995; Bayer et al., 2004; Doi and Kosugi, 2004). Cellulosomes are organelles (high molecular weight, e.g., 65,000 kDa) – multienzyme complexes situated on the exterior of cells. Electron micrographs and other procedures reveal them as "extracellular protuberances" (Doi and Kosugi, 2004) that consist of a three-dimensional framework (scaffold) which arranges as many as 35 different enzymes that both bind to and hydrolyze the cellulose and related polysaccharides in plant cell walls. The genes that encode cellulosome proteins and the intricacies of gene resolution have begun to be explored (Doi and Kosugi, 2004).

Lignin

Lignin is not metabolized as a carbon and energy source – instead lignin is degraded by ligninolytic fungi to expose the more digestible polysaccharides of wood so that these can be cleaved by fungal cellulases and hemicellulases (Hammel, 1997). Ligninolysis is not yet fully understood (Hammel, 1997; Leonowicz et al., 1999; Hofrichter, 2002), though many rudimentary aspects have been described. The random, free-radical catalyzed synthesis of lignin renders it resistant to hydrolytic cleavage. Given the chemical recalcitrance of lignin, white rot fungi have evolved a novel biochemical approach to degrade it: random free-radical oxidation (combustion) reactions involving molecular oxygen. Lignin peroxidase, manganese peroxidase, and copper-containing laccase enzyme systems utilize very strong oxidants (ferric heme, Mn^{3+}, and Cu^{2+}, respectively) that attack C-C bonds in the lignin, removing electrons (Figure 3). The destabilized intermediates undergo ring-cleavage and ether-cleavage reactions that release low molecular weight lignin "structural monomers". In order for the delignification process to persist,

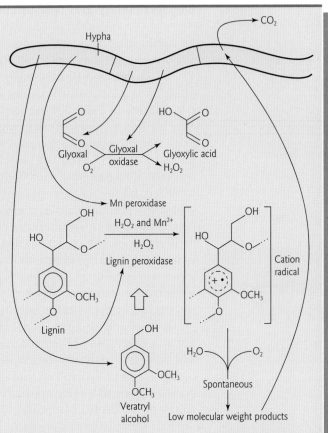

Figure 3 The ligninolytic system of the white rot fungus, *Phanerochaete chrysosporium*. See above box text for details. (From Lengeler, J.W., G. Drews, and H.G. Schlegel. 1999. *Biology of Prokaryotes*, fig. 9.10. Blackwell Science, Stuttgart. With permission from Blackwell Science, Stuttgart.)

laccase, ligninase peroxidase and manganese peroxide enzymes that have stolen electrons from the lignin structure must be reoxidized. Extracellular H_2O_2 (a strong oxidant) fulfills this role: H_2O_2 is generated by the reaction between O_2 gas and a commonly excreted fungal compound, glyoxal (Figure 3). Overall then, the three enzymes responsible for destroying lignin shuttle electrons from lignin, itself, to H_2O_2.

Biodegradability and recalcitrance of organic compounds

Prevailing wisdom in environmental microbiology deems that all naturally occurring biosynthetic compounds are biodegradable (Alexander, 1973, 1999; Wackett and Herschberger, 2001). Indeed, the majority of organic materials (e.g., carbon-containing compounds excluding graphite, CO, and CO_2) shown in Table 7.4 are useful carbon and energy sources or nutrients for heterotrophic prokaryotes and fungi. Some of the biopolymers (e.g., starch, protein) are useful to higher eukaryotes as well. The compounds in Table 7.4 that are resistant to enzymatic attack and metabolism are: coal, keragen, peat, and humic substances. Both *keragen* and *coal* are fossilized, highly metamorphosed, mineral-like forms of carbon found in geologic deposits (coal is in massive deposits, while keragen is highly dispersed throughout sedimentary rocks, such as shales). Peat, the geologic predecessor of coal, is accumulated plant biomass (especially *Sphagnum*-type mosses) in wetlands that are often acidic. The combination of anaerobic conditions, cellulosic/ligninic material, accrual of organic-acid fermentation products, and low pH leads to preservation of peat materials (including the remains of humans retrieved from ancient peat bogs!).

Like coal and keragen, *humic substances* (Table 7.4) resist microbiological attack because of their complex molecular structure. But, unlike coal and keragen, humic substances do not form in the presence of high temperature and pressure. The traditional view of humic substances has, for decades, been that they are high molecular weight, "polymer-like" networks of molecules composed of plant-derived and microbially derived random subunits linked by extracellular enzymatic and free-radical reactions into random covalent linkages (Figure 7.8; Stevenson, 1994). Sutton and Sposito (2005) have recently developed a new argument that depicts humic substances as "collections of diverse, relatively low-molecular-mass compounds forming dynamic associations stabilized by hydrophobic interactions and hydrogen bonds". Evidence for the new view was derived from physical, molecular, and spectroscopic data showing: (i) that humic substances are aggregates of micelle-like particles (400–800 nm in size) held together by hydrophic associations; (ii) the chemical moieties in these particles are diverse and derived from recognizable low molecular weight biomolecules such as lipids, lignin, carbohydrates, and protein; and (iii) that the major functional organic moieties (amide groups), thought to participate in forming the traditional polymer-like model of humic substances, are largely unoccupied. There are still many unanswered questions about humic substances: the dark, refractory, heterogeneous organic compounds that are crucial for creating agriculturally desirable soil properties and that are among the most widely distributed organic materials in terrestrial habitats.

Petroleum reservoirs (or crude oil, see Table 7.4) represent another form of metamorphosed organic matter (plants and algal lipids). Time, heat,

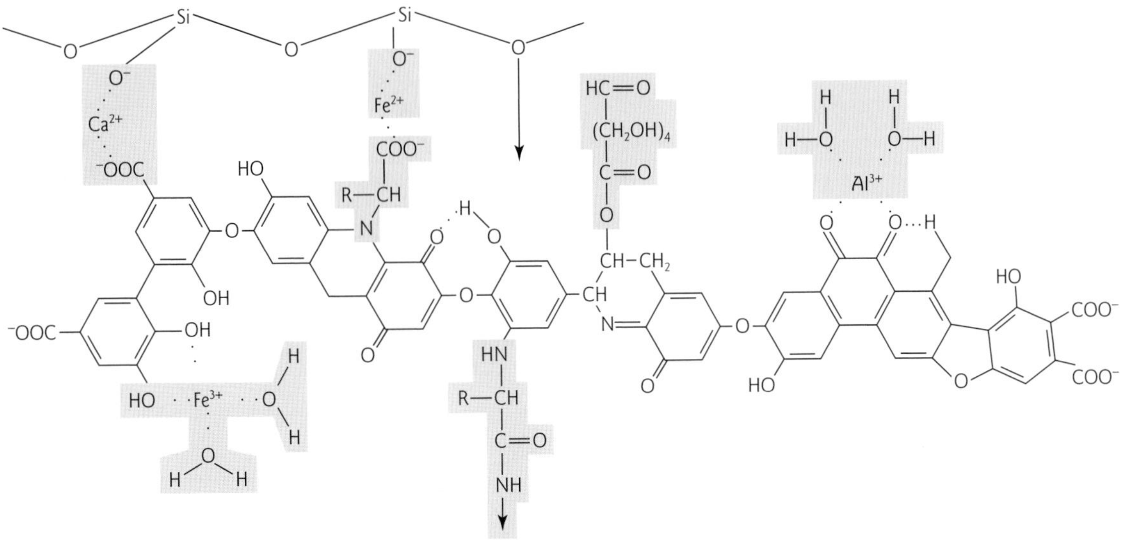

Figure 7.8 Schematic (traditional) model of humic acid. Included in the structure are molecular remnants of amino acids and sugars, showing possible associations with inorganic compounds including silica (Si-O) moieties at the surface of clay and other minerals. See text for details and an alternative model. (From Lengeler, J.W., G. Drews, and H.G. Schlegel. 1999. *Biology of Prokaryotes*, fig. 31.11. Blackwell Science, Stuttgart. With permission from Blackwell Science, Stuttgart.)

and pressure caused diagenetic alterations in the molecular structures of the original biomass so that liquid and gaseous hydrocarbons migrated until they accumulated in porous rocks, forming an oil field. While petroleum hydrocarbons are both "naturally occurring" and "biodegradable", the optimal conditions (enzymes, microorganisms, physiological reactions) for growth of heterotrophic hydrocarbon-degrading prokaryotes and some fungi are restricted. If this were not the case, geological petroleum reservoirs would not exist.

Two major classes of hydrocarbons are straight-chain alkanes and benzene-containing aromatic compounds (see also Section 8.3 and Boxes 8.5, 8.7). By definition, "hydrocarbons" consist almost exclusively of carbon and hydrogen atoms. Oxygen was eliminated from the organic precursor materials during oil diagenesis; only small amounts of nitrogen, sulfur, and trace metals remain (and are released during industrial combustion of petroleum fuels). Though highly reduced and thermodynamically unstable in the presence of potential electron-accepting oxidants (e.g., CO_2, SO_4^{2-}, Fe^{3+}, Mn^{4+}, NO_3^-, and O_2; see Section 3.8 and Figure 7.9), breakage of C-C bonds in petroleum molecules is a specialized metabolic task. When oxygen is a reactant, petroleum biodegradation is most rapid. Oxygenase enzymes add molecular oxygen to the molecule. This process can be catalyzed by a monooxygenase enzyme (forming one

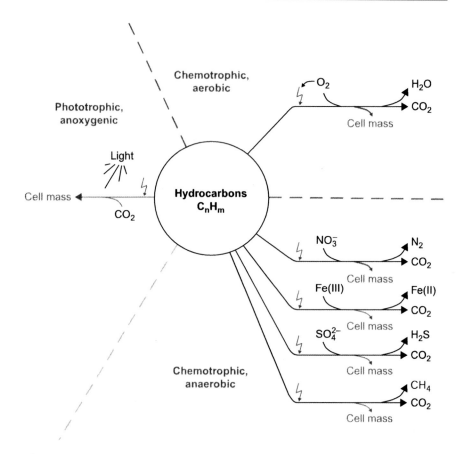

Figure 7.9 Experimentally verified possibilities for the microbial utilization of hydrocarbons. In all chemotrophic reactions, a part of the hydrocarbon is oxidized for energy conservation (catabolism) and another part is assimilated into cell mass. In the long-established aerobic oxidation of hydrocarbons (upper right), oxygen is not only the terminal electron acceptor, but is also needed for substrate activation (oxygenase reactions). The anaerobic pathways involve novel hydrocarbon activation mechanisms that differ completely from the aerobic mechanisms. Jagged arrows indicate hydrocarbon activation. (Reprinted from Widdel, F. and R. Rabus. 2001. Anaerobic biodegradation of saturated and aromatic hydrocarbons. *Curr. Opin. Biotechnol.* **12**:259–276. Copyright 2001, with permission from Elsevier.)

C-OH bond) or a dioxygenase enzyme (forming two adjacent C-OH bonds). Molecules with C-OH bonds are alcohols. Oxygen incorporation into hydrocarbon molecules requires reducing power [e.g., often in the form of NAD(P)H – a molecule related to electron carriers used in respiration; see Section 3.9 and Box 3.8] that converts oxygen from the zero oxidation state to −2 (Alexander, 1999; Lengeler et al., 1999; Van Hamme et al., 2003).

If the newly hydroxylated molecule was an alkane, it is enroute to conversion into a carboxylic acid and reactions, termed β-oxidation (Figure 7.10). Beta-oxidation is a six-step series that: (1) links the carboxylic acid to coenzyme A (CoA, composed of pantothenic acid and β-mercaptoethylamine) via a thioester bond; (2) dehydrogenates a C-C bond in the "beta position", two carbons in from the original terminus; (3) adds water to that bond, resulting in hydroxylation; (4) oxidizes the new hydroxyl to a keto group; (5) cleaves off a two-carbon moiety, acetyl-CoA, from the hydrocarbon chain while generating an SCoA-linked hydrocarbon molecule two carbons shorter than the original; and (6) repeats steps 1 to 5 until enzymatic digestion is complete (Lengeler et al., 1999). If the newly hydyroxylated molecule was originally an aromatic (benzene) ring, then aerobic metabolism proceeds via: (i) oxidative ring cleavage by dioxygenases; and (ii) further cleavage of C-C bonds, creating noncyclic, nonaromatic ring fission products, such as pyruvate, that feed into central metabolic pathways such as the Krebs citric acid cycle (bottom of Figure 7.11; Lengeler et al., 1999; see also Section 3.9 and Box 3.8). Though anaerobic metabolism of petroleum hydrocarbons is kinetically and energetically far less favorable than in the presence of oxygen, some aspects of

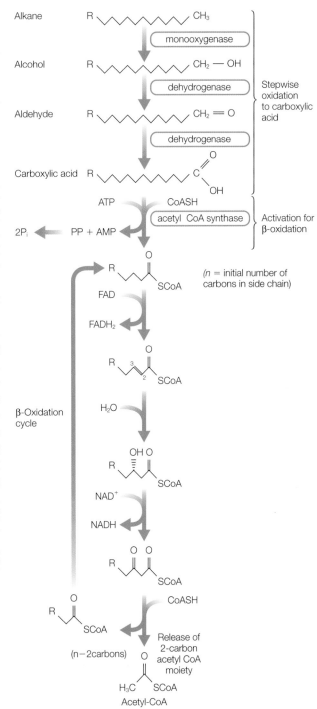

Figure 7.10 Pathway for aerobic metabolism of alkane petroleum hydrocarbons, featuring stepwise terminal oxidation to a carboxylic acid and subsequent β-oxidation. See text for details. Pi, inorganic phosphate; PP, pyrophosphate; SCoA, thioester-linked acetyl CoA. (Modified from Lengeler, J.W., G. Drews, and H.G. Schlegel. 1999. *Biology of Prokaryotes*, fig. 9.37. Blackwell Science, Stuttgart. With permission from Blackwell Science. Stuttgart.)

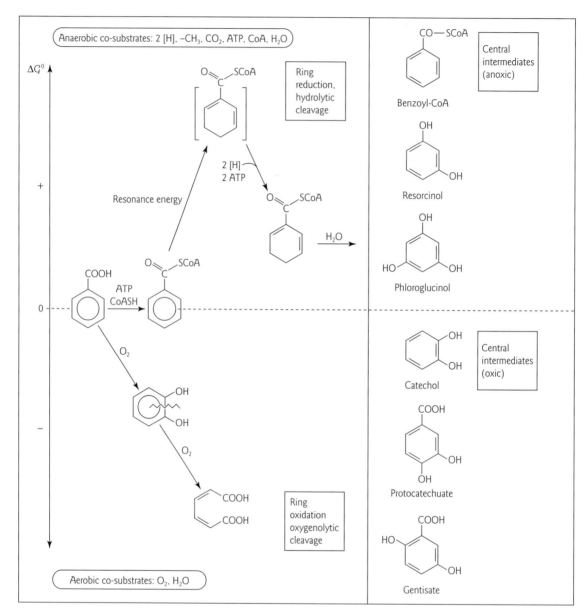

Figure 7.11 Comparison of aerobic (oxygen-dependent) and anaerobic (anoxic) metabolism of aromatic compounds. Shown are the co-substrates used and several key intermediary metabolites. Under oxic conditions, benzoate is converted, for example, to catechol, and the aromatic ring is oxygenolytically cleaved. Under anoxic conditions, the high resonance energy does not allow direct reduction of the aromatic ring. Rather, ring destabilization is achieved by forming the coenzyme A (CoA) thioester, which requires ATP. Then, the ring is reduced at the cost of another ATP, and finally the ring is opened hydrolytically. See text for additional details. (From Lengeler, J.W., G. Drews, and H.G. Schlegel. 1999. *Biology of Prokaryotes*, fig. 9.29. Blackwell Science, Stuttgart. With permission from Blackwell Science. Stuttgart.)

the process have been well characterized (see Figure 7.9; Spormann and Widdel, 2000; Widdel and Rabus, 2001; So et al., 2003; Heider, 2007). Alkanes are suspected to be initially anaerobically attacked by enzymes that convert alkanes to organic acids via addition of the four-carbon dicarboxylic acid, fumarate. Though these and subsequent biochemical steps are still being actively researched, it is likely that β-oxidation-type reactions then proceed (as described above, Lengeler et al., 1999). Without molecular oxygen as a reactant to destabilize the benzene ring of aromatic compounds (top of Figure 7.11), anaerobic microorganisms, instead, appear to use water, bicarbonate, ATP, CoA thioesters, methyl donors, and reduced and oxidized coenzymes to sequentially reduce and then hydrolytically cleave aromatic structures to low molecular weight intermediary metabolites that funnel into central biosynthetic and catabolic (respiratory) pathways (Lengeler et al., 1999). The current weight of evidence indicates that unsubstituted aromatic hydrocarbons (e.g., benzene, naphthalene) are initially methylated (to toluene and 2-methyl-naphthalene, respectively), before undergoing oxidation (of $-CH_3$ to $-COOH$) and ring cleavage reactions (Coates et al., 2002; Safinowski and Meckenstock, 2006; Heider, 2007).

Oxidation states of elements predict compound reactivity

Biosphere compounds (e.g., Table 7.4), particularly those of carbon, sulfur, and nitrogen, exhibit a wide range of oxidation states. The importance of elemental oxidation state, as a predictor of thermodynamic reactivity, was previously emphasized in Chapter 3 (Sections 3.6–3.9) and computation of elemental oxidation states was described in Box 3.4. Table 7.5 presents information that systematically declares the range of oxidation states of some of the carbon, nitrogen, and sulfur compounds listed in Table 7.4. The oxidation–reduction scale in Table 7.5 reminds the reader that the most oxidized forms of the elements (C as CO_2; S as SO_4^{2-}; and N as NO_3^-) are excellent electron acceptors for ATP generation in electron transport and other processes carried out by anaerobic microorganisms. At the opposite extreme, the most reduced forms of the elements (C as CH_4; S as H_2S; and N as NH_3) are excellent electron donors. Mid-range on the redox scale of Table 7.5 are compounds that can be oxidized *or* reduced. Examples of compounds exhibiting bidirectional (up and down) changes in oxidation states include: (i) carbohydrate (sugar) fermentation, in which a portion of the substrate pool is oxidized to CO_2 while the remainder of the pool is reduced to organic alcohols or acids (see Sections 3.8 and 3.10 and Figure 3.11); (ii) the disproportionation of thiosulfate by some anaerobic bacteria – generating both H_2S and SO_4^{2-}; and (iii) nitrite oxidation during the second major step of nitrification to nitrate, or nitrite reduction to NO in denitrification.

The cyclic nature of compound generation and destruction is apparent if information in columns 4 and 5 of Table 7.4 is scrutinized. To illustrate

Table 7.5

Average oxidation states of carbon, nitrogen, and sulfur atoms in compounds found widely in soils, sediments, waters, and the atmosphere

Oxidation–reduction scale	Carbon		Nitrogen		Sulfur	
	Compound	Oxidation state	Compound	Oxidation state	Compound	Oxidation state
Oxidized (good electron acceptor)	CO_2 (carbon dioxide)	+4	NO_3^- (nitrate)	+5	SO_4^{2-} (sulfate)	+6
			NO_2^- (nitrite)	+3	$S_2O_3^{2-}$ (thiosulfate)	+2
			NO (nitric oxide)	+2		
			N_2O (nitrous oxide)	+1		
					$(CH_3)_2SO$ (dimethyl sulfoxide)	0
	CH_2O (carbohydrate, biomass tissue)	0	N_2 (dinitrogen gas)	0		
					S^0 (elemental sulfur)	0
Fermentable (disproportionation-type reactions)	$C_2H_4O_2$ (acetic acid)	0				
	$C_3H_6O_2$ (propionic acid)	−0.66				
	$C_4H_8O_2$ (butyric acid)	−1			$(CH_3)_2S_2$ (dimethyl sulfide)	−2
	C_2H_6O (ethanol)	−2				
	C_8H_{18} (octane)	−2.25			$HSCH_2CHNH2COOH$ (cysteine)	−2
Reduced (good electron donor)	CH_4 (methane)	−4	NH_3 (ammonia)	−3	H_2S (hydrogen sulfide)	−2

this key concept, the reader's attention should be focused on the elemental sulfur (S^0) entries in Table 7.4. Elemental sulfur is microbiologically produced via partial oxidation of the electron donor, sulfide, by chemolithoautotrophs. Remarkably, elemental sulfur has three potential metabolic pathways of destruction: (i) continued oxidation to sulfate by chemolithoautotrophs using S^0 as an electron donor and oxidants like O_2 as electron acceptors; (ii) reduction to sulfide in a process known as "sulfur respiration" by anaerobic *Archaea* and/or heterotrophic bacteria using S^0 as an electron acceptor; and (iii) oxidation of S^0 to sulfate by phototrophic sulfur bacteria requiring reducing power to fix CO_2.

The main lesson that emerges from information in Table 7.4 is that all biogeochemically produced substances cycle from one form to another. However, the multiple processes influencing the metabolic fate of any given biosphere compound are separated in space and time. Furthermore, *the metabolic process that prevails in any given ecological context depends upon the dominant terminal electron-accepting process (see Sections 3.7 and 3.8), ambient nutrient limitations, and ambient geochemical conditions.*

In the next several sections we will tackle biogeochemical cycles of individual elements. Our goal is to use the physiological logic presented in Chapter 3 (Sections 3.6–3.11), to establish a rigorous foundation for systematically interpreting the catalog of biogeochemical reactions presented in Table 7.4.

7.4 ELEMENTAL BIOGEOCHEMICAL CYCLES: CONCEPTS AND PHYSIOLOGICAL PROCESSES

Complex mixtures of elements are cycled through food chains in nature

Chemically pure elements (O as O_2 gas, N as N_2, S as elemental sulfur, C as graphite, H as H_2 gas, Fe as the metal, etc.) can be found in the biosphere. But, in real-world habitats, such as soils, sediments, waters, and geologic formations, it is relatively rare to encounter any element in its pure state. Mostly, elements have a propensity to react with one another, either chemically or biochemically, thereby forming "compounds". As an example addressing inorganic compounds, O_2 (a strong oxidant) bonds readily with many elements – C, N, S, Fe, As, and H, creating, respectively, CO_2, NO_3^-, SO_4^{2-}, $Fe(OH)_3$, AsO_4^{3-}, and H_2O. Regarding organic compounds, the biomolecules that are the "stuff of life" are proteins, nucleic acids, carbohydrates, lipids, ATP, metalloenzymes – all of which are created through biosynthetic pathways that assemble compounds composed of the many nutrient elements.

At the level of ecological processes and trophic interactions (Figure 7.12), it is clear that the units of biomass transferred between primary, secondary,

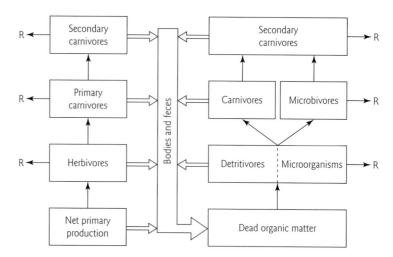

Figure 7.12 Trophic levels of ecosystems through two parallel food chains: one based on primary production and the other on dead biomass (detritus). The thin arrows show the flow of energy up the food chain (through living biomass) and the broad arrows show the complementary flow of detritus. R, respiration. (Reprinted from Staley, J.T. and G.H. Orians 2000. Evolution and the biosphere. *In*: M.C. Jacobson, R.J. Charlson, H. Rodhe, and G.H. Orians (eds) *Earth System Science: From biogeochemical cycles to global change*, pp. 29–61. Academic Press, San Diego, CA. Copyright 2000, with permission from Elsevier.)

and tertiary consumers are the elementally heterogeneous tissues or entire bodies of creatures that dwell in the biosphere. Ecologically, it is important to note that two food chains operate in parallel in virtually every habitat. One food chain (left side, Figure 7.12) is based directly on net primary productivity (NPP; photosynthesis or chemolithoautotrophy). The other (right side, Figure 7.12) is based on the detritus food chain in which deceased biomass and fecal materials from the extant creatures are transformed into new microbial (decomposer) biomass, that, itself, is the base of a food chain.

Thus, the real world is complicated and integrated. You cannot go out into a forest or lake or ocean or desert and find the isolated cycling of major elements (C, S, N, etc.). Nonetheless, the notion of discrete cycles of the elements is a useful exercise in biogeochemistry (see below).

Cyclic physiological processes: carbon, sulfur, and nitrogen

Thermodynamics (see Sections 3.6–3.8) gives us predictive power about microbially mediated reactions. Furthermore, the partial catalog of biosphere compounds (see Table 7.4 and Section 7.3) provides a glimpse of the pools of materials and their sources and sinks. It is conceptually insightful to split elements in real-world compounds away from one another. The exercise allows us to systematically assemble, focus, and

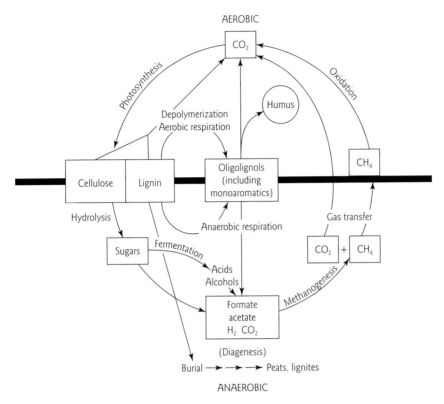

Figure 7.13 The carbon cycle, with emphasis on the processing of photosynthetically derived cellulose and lignin in aerobic and anaerobic habitats. (From Colberg, P.J. 1988. Anaerobic microbial degradation of cellulose, lignin, oligolignols, and monoaromatic lignin derivatives. *In*: A.J.B. Zehnder (ed.) *Biology of Anaerobic Microorganisms*, pp. 333–372. Wiley and Sons, New York. Reprinted with permission from John Wiley and Sons, Inc., New York.)

integrate information describing how, when, and why biogeochemical processes occur.

Carbon cycle

Figure 7.13 provides a generic overview of how carbon cycles in the biosphere and the responsible physiological processes. The division in the diagram between aerobic (top half) and anaerobic (bottom half) processes remind us of two important factors: (i) physiological conditions change in time and space; and (ii) those changes have major implications for predicting thermodynamic instabilities – hence the processes that microorganisms can and do catalyze. Let us begin with CH_4 (right side, Figure 7.13). As mentioned in Table 7.4, methane is a thermodynamically stable end-product of methanogenesis (dissimilatory reduction of CO_2) in anaerobic habitats lacking non-CO_2 electron acceptors. But methane is the most reduced form of carbon (see Table 7.5) and serves as an electron donor for methanotrophic microorganisms that link electron flow generated during methane oxidation to the reduction of O_2 (Mancinelli, 1995), NO_3^- (Raghoebarsing et al., 2006), and SO_4^{2-} (Strous and Jetten, 2004). These oxidations of CH_4 generate CO_2 (oxidation state of +4, Table 7.5), which is a carbon source for autotrophs. Although there are four distinctive biochemical pathways for autotrophic CO_2 fixation (Lengeler et al., 1999), Calvin cycle-carrying autotrophs (photosynthetic and chemosynthetic) predominate in biosphere habitats. Furthermore, *ribulose bisphosphate carboxylase* (RuBisCO) is the Calvin cycle's key enzyme. As indicated by the enzyme's title, ribulose bisphosphate carboxylase adds a CO_2 molecule to a phosphorylated form of the five-carbon sugar, ribose. This critical enzymatic step creates a six-carbon molecule, which is cleaved to two molecules of the three-carbon compound, 3-phosphoglycerate. Three turns of the Calvin cycle synthesizes the equivalent of one new 3-phosphoglycerate molecule – an essential biosynthetic building block that can lead to glucose, starch, and cellulose in plants and to gluconeogenesis in prokaryotes (via reversal of the glycolysis pathway; see Box 3.8).

Fixed CO_2 in plant or microbial biomass (shown as cellulose and lignin in Figure 7.13) has an intermediate oxidation state of ~0 (Table 7.5). Thus, biomass is susceptible to two potential physiological reaction pathways: (i) oxidation back to CO_2 by microorganisms using various electron acceptors (e.g., O_2, Fe^{3+}, Mn^{4+}, SO_4^{2-}); or (ii) reduction (often via fermentation) to an anaerobic intermediary metabolite pool of organic acids and hydrogen gas. The final step in Figure 7.13 completes the cycle: predominant mechanisms of methanogenesis use either H_2 as an electron donor and CO_2 as the acceptor creating methane, or ferments (disproportionates) methyl-containing or C_1 compounds (e.g., formate, carbon monoxide, acetate, methanol, methylamine, or dimethylsulfide) to CH_4 and CO_2. Acetogenesis (discussed in Section 3.8) may also occur. Information in Box 7.4 illustrates a direct relationship between sources and sinks of methane and the atmospheric concentration of this greenhouse gas.

Sulfur cycle

A generalized view of the physiological sulfur cycle is shown in Figure 7.14. As shown in Table 7.5, the fully reduced form of sulfur is hydrogen sulfide [H_2S; −2 oxidation state (−II); oxidation states are depicted in roman numerals on the left-hand axis of Figure 7.14]. Fully reduced sulfur may take the form of inorganic minerals or organic compounds such as amino acids (Table 7.4) and *dimethyl sulfide* (DMS). Box 7.5 elaborates on the properties and environmental role of DMS in marine systems. At the opposite extreme, fully oxidized sulfur exhibits a +6 (+VI) oxidation state as the sulfate anion (top of Figure 7.14). Other forms of sulfur exhibit

Box 7.4

Microbial controls on fluxes of methane to the atmosphere

In Section 7.2 it was pointed out that small changes in atmospheric methane concentrations have substantial impacts on global warming and climate change. There is a flux of methane from methane reservoirs and anaerobic sites of methanogenesis, which include soils, sediments, and the digestive tracts of termites, cows, and other ruminant animals. As methane diffuses up and away from its site of production (where it is a stable metabolic waste product), it migrates toward the atmosphere. In so doing, the methane passes through more oxidized zones where it is thermodynamically unstable – enriching methanotrophs that use SO_4^{2-}, NO_3^-, and O_2 as final electron acceptors. In these neighboring habitats, methane becomes a useful physiological electron donor. Thus, the flux of methane that reaches the atmosphere is regulated (and substantially diminished) by methanotrophic microbial populations (see Sections 3.11, 7.3 and 7.5, and Box 7.9).

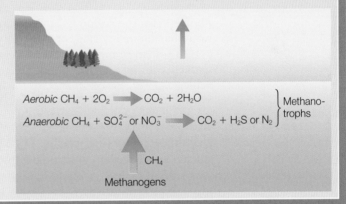

Lesson
Net methane flux to the atmosphere represents a balance between rates of production versus destruction. Clearly, understanding the microbial biogeochemical controls on these relative rates is of major scientific and biogeochemical concern.

Aerobic $CH_4 + 2O_2 \longrightarrow CO_2 + 2H_2O$

Anaerobic $CH_4 + SO_4^{2-}$ or $NO_3^- \longrightarrow CO_2 + H_2S$ or N_2 } Methanotrophs

CH_4

Methanogens

intermediate oxidation states, such as elemental S (zero oxidation state; see also Table 7.4). In the presence of oxidizing agents (especially O_2 and nitrate), reduced forms of sulfur, especially H_2S and S^0 are thermodynamically unstable – they are used as electron donors for ATP production by chemolithotrophic microorganisms (see Section 3.11).

Anaerobic, sulfur-rich surface waters, such as those in hot springs and Solar Lake in the Sinai Peninsula, harbor and release S^0 and H_2S. In these sites, bathed daily in sunlight, various groups of phototrophic sulfur bacteria utilize H_2S (an analog to the H_2O used in oxygenic photosynthesis) and S^0 as sources of electrons in photosynthetic dark reactions that fix CO_2, via RuBisCO-type reactions (see above). Arrows appear in Figure 7.14 representing this anaerobic phototrophic oxidation of reduced sulfur compounds (from H_2S to SO_4^{2-}, from H_2S to S^0, and from S^0 to SO_4^{2-}).

After the various oxidation reactions have generated the sulfate anion, this material is stable in the presence of oxygen (e.g., most oceanic waters) but in carbon-rich sediments where microbial metabolic demand for

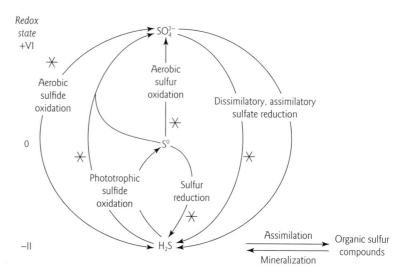

Figure 7.14 The sulfur cycle, showing key reactions and sulfur species. Note that the redox state of each sulfur compound is shown on the left axis in roman numerals. Reactions catalyzed exclusively by prokaryotes are marked with an asterisk. (From Lengeler, J.W., G. Drews, and H.G. Schlegel. 1999. *Biology of Prokaryotes*, fig. 32.3. Blackwell Science, Stuttgart. With permission from Blackwell Science. Stuttgart.)

electron acceptors (especially, O_2, NO_3^-, Fe^{3+}, and Mn^{4+}) exceeds supply, sulfate is an excellent electron acceptor. Sulfate-reducing prokaryotes convert sulfate to H_2S (a dissimilatory waste product); thus completing the physiological sulfur cycle.

Nitrogen cycle

An overview of the physiological basis for nitrogen cycling is shown in Figure 7.15. The intricacies of microbial nitrogen transformations are suggested by the variety of nitrogen compounds and their respective oxidation states shown in Tables 7.4 and 7.5. The basic principles used to organize carbon and sulfur cycles apply also to nitrogen: the oxidized forms are useful final electron acceptors and the reduced forms are useful electron donors. Furthermore, a small portion of the vast atmospheric pool of N_2 ($N \equiv N$; zero oxidation state) can be brought into the biologically useful nitrogen pool via the uniquely prokaryotic process of nitrogen fixation, which creates intracellular NH_3 (-3 oxidation state). For a description of N_2 fixation, see Box 7.6.

Focusing on the upper left-hand portion of Figure 7.15, N_2 gas has been reduced to the NH_3 state (e.g., as amino acids) within a microbial cell represented as particulate organic nitrogen (PON). Cell lysis, hydrolysis of protein, and metabolism by heterotrophs (a three-step process known

Box 7.5

Dimethyl sulfide: an unusual sulfur compound whose oceanic sources and transformations influence climate change

Marine algae produce dimethyl sulfonio-proprionate (DMSP) to maintain cellular osmotic balance in seawater. During decomposition of deceased algae, DMSP is used as a carbon and energy source by heterotrophic bacteria. The enzyme, DMSP lyase, splits the substrate into acrylate (related to proprionate, a useful carbon and energy source) and dimethyl sulfide (DMS) – a volatile gas. DMS escapes to the atmosphere where it can be oxidized by hydroxyl radicals to sulfate. Air-borne sulfate aerosol particles (originating from DMS) over oceanic waters are thought to contribute to cloud formation, whose reflectance may contribute to global cooling (Schlesinger, 1997; Yock, 2002).

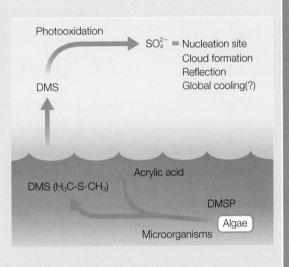

Recent studies have made it clear that the global cooling effect of DMS is regulated by two additional marine microbial processes. The DMSP precursor for DMS can be assimilated directly from seawater by phytoplankton (both prokaryotic *Prochlorococcus* and eukaryotic algae such as diatoms; Vila-Costa et al., 2006). Also, widespread marine prokaryotic heterotrophs (such as *Silicibacter* and *Pelagibacter ubique*) carry genes that encode a demethylase enzyme that diverts DMSP down a metabolic pathway that eliminates DMS as a possible endproduct (Howard et al., 2006). Both of the above processes may diminish the size and flux of the DMS pool in the oceans.

When DMS is produced (as above), it can also be transported to anaerobic habitats. There are three likely anaerobic metabolic fates of DMS: (i) during methanogenesis it may be converted to CH_4 and H_2S; (ii) as an electron donor for photosynthetic purple bacteria, DMS can be oxidized to dimethyl sulfoxide (DMSO); and (iii) a variety of chemoorganotrophs and chemolithotrophs oxidize DMS to DMSO to generate ATP using sulfate or nitrate as electron acceptors.

Clearly, the biogeochemistry of DMS represents an intricate balance between biological and chemical processes involving production, consumption, and transport.

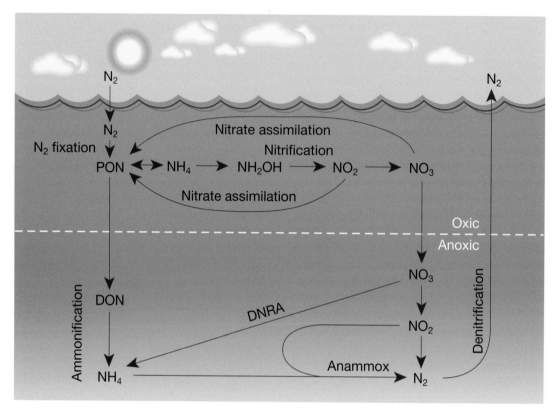

Figure 7.15 The nitrogen cycle. See text for details. DNRA, dissimilatory nitrate reduction to ammonium; DON, dissolved organic nitrogen; PON, particulate organic nitrogen, including phytoplankton. (Reprinted by permission from Macmillan Publishers Ltd: Nature, from Arrigo, K.R. 2005. Marine microorganisms and global nutrient cycles. *Nature* **437**:349–355. Copyright 2005.)

Box 7.6

Nitrogen fixation: an amazing and uniquely prokaryotic feat

Industrial and microbial processes fix N_2

The industrial source of commercial fertilizer nitrogen, the Haber–Bosch process, was invented in Germany around 1910. Using high pressure (200 atm or ~20,000 kPa) and temperature (450°C), atmospheric nitrogen (N_2; N≡N) is combined with H_2 gas to create NH_3. The prokaryotic enzyme complex, termed *nitrogenase*, carries out the same process at ambient temperatures and pressures.

Biology and biochemistry of nitrogen fixation

The overall reaction for biological nitrogen fixation is:

$$8H^+ + 8\ e^- + N_2 \rightarrow 2NH_3 + H_2,$$

requiring an input of ~16 to 24 ATP.

The enzyme system also fortuitously acts on the triple bonded molecule acetylene $HC\equiv CH$, creating ethene $H_2C=CH_2$.

Two distinctive nitrogenase enzyme systems have been described in prokaryotes.

1 The most commonly distributed system (in free-living aerobic and anaerobic prokaryotes, some associated as symbionts with root structures in plants) delivers ATP and reducing power (as a reduced flavodoxin protein) to a multistep electron transport chain that ends with a Fe-Mo protein (dinitrogenase) that converts a single molecule of N_2 to two molecules of NH_3. Hydrogen gas is also produced (Figure 1). At least two additional structural variants in the dinitrogenase enzyme have been characterized – one with a Fe-V metal cluster and another with only Fe.

  The reaction is poisoned by molecular oxygen, and aerobic prokaryotes (e.g., *Cyanobacteria*, *Azotobacter*, *Rhizobium*) have evolved clever and effective ways to shield their nitrogenase systems from atmospheric oxygen.

2 Another enzymatic strategy for nitrogen fixation has been recently described in *Streptomyces thermoautotrophicus* (Ribbe et al., 1997). Based on phylogenetic analysis of component proteins, this system appears to be an evolutionary hybrid – created by merging enzymes from the above (classic) nitrogenase system with other enzymes derived from the superoxide–dismutase oxygen-detoxification system. Remarkably, molecular oxygen is a required reactant in the electron-transport chain that delivers reducing power to a molybdenum-containing dinitrogenase enzyme.

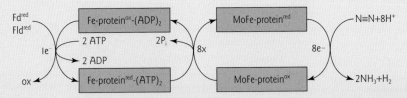

Figure 1 Model of the nitrogenase enzyme complex showing the transfer of electrons between protein carriers, ending in the release of ammonia and hydrogen. Fd, ferredoxin; Fld, flavodoxin. See box text for details. (From Lengeler, J.W., G. Drews, and H.G. Schlegel. 1999. *Biology of Prokaryotes*, fig. 8.14. Blackwell Science, Stuttgart. With permission from Blackwell Science, Stuttgart.)

as *ammonification*) release the N as NH_4^+. PON can also be transported to anoxic waters where heterotrophic activity may convert particulate organic materials to the dissolved form (dissolved organic nitrogen, DON) and eventually to NH_4^+ (Figure 7.15).

In the oxic upper layers of ocean water NH_4^+ is thermodynamically unstable – it is used as an electron donor by chemolithoautotrophs in the multistep process known as *nitrification*. The microorganisms responsible for nitrification use O_2 as the final electron acceptor in their respiratory chains to generate a proton motive force and ATP. Some populations (e.g., *Nitrosomonas* sp.) oxidize NH_4^+ to NO_2^-. Others (e.g., *Nitrobacter* sp.) complete the process by oxidizing NO_2^+ to NO_3^-. Additional details of nitrification and denitrification are provided in Box 7.7.

Box 7.7

Ways that denitrification and nitrification influence greenhouse gases and climate change

Section 7.2 presented the crucial roles that N_2O and ozone play in tropospheric global climate change. Surveys of gas fluxes into the atmosphere from terrestrial soils have proven that two nitrogen-containing gases, N_2O (nitric oxide) and NO (nitrous oxide) are routinely released. A sample data set from Amazon basin soils is shown in Table 1.

Table 1

Summary of annual N_2O and NO soil emissions from the Amazon basin (from Neill, C., P.A. Steudler, D.C. Garcia-Montiel et al., 2005. Rates and controls of nitrous oxide and nitric oxide emissions following conversion of forest to pasture in Rondonia. *Nutr. Cycling Agroecosyst.* **71**:1–15, table 2. With kind permission of Springer Science and Business Media)

Location	N_2O (kg N/ha/year)		NO (kg N/ha/year)	
	Forests	Pastures	Forests	Pastures
Nova Vida, Rondonia	1.7–2.0	3.1–5.1		
		1.2–1.7		
Nova Vida, Rondonia	4.3	0–1.3	1.4	0.1–0.4
Noval Rondonia	3.2			
Nossa Senhora, Rondonia				0.17
Paragominas, Parú	2.4	1.5	1.5	1.0
		0.2		0.6
Manaus, Amazonas	1.4			
Manaus, Amazonas	1.9	6		
Tapajos, Pará	2.3		1.7	

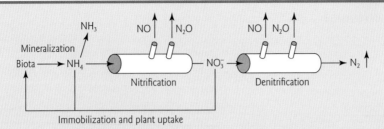

Figure 1 Microbial processes that yield nitrogen gases during nitrification and denitrification in soil. (From Schlesinger, W.H. 1997. *Biogeochemistry: An analysis of global change*, 2nd edn. Academic Press, San Diego, CA. Copyright 1997, with permission from Elsevier.)

It has been documented that both nitrification and denitrification biochemical pathways leak both NO and N_2O to the atmosphere. A broad scheme depicting N processing in terrestrial systems is shown in Figure 1. Special emphasis is given to gaseous release of NO and NO_2.

Biochemical basis of gas production

According to Conrad (Remde and Conrad, 1991; Conrad, 1996), production of NO and N_2O from soil are a composite of several different processes. In denitrification, NO and N_2O are directly produced during stepwise reduction of nitrate to N_2 (Figure 2). But intracellular retention of NO and N_2O is not 100% efficient: a portion of these gases escape from the metabolic pathways and reach the atmosphere (Davidson et al., 2000, 2004).

In nitrification, N_2O is formed during chemical decomposition of two intermediary metabolites (NH_2OH and NO_2^-; Figure 3). In addition, within a single bacterium capable of both nitrification and denitrification, the intermediary nitrification metabolite, NO_2^-, can be intercepted by the denitrification enzyme, nitrite reductase, yielding NO.

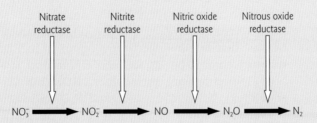

Figure 2 Summary of denitrification: outline of the pathway and enzymes involved. (Reprinted from Wrage, N., G.L. Velthof, M.L. van Beusichem, and O. Oenema. 2001. Role of nitrifier denitrification in the production of nitrous oxide. *Soil Biol. Biochem.* **33**:1723–1732. Copyright 2001, with permission from Elsevier.)

Box 7.7 *Continued*

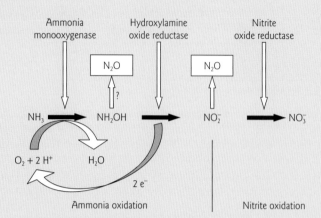

Figure 3 Summary of nitrification: outline of the pathway and enzymes involved. (Reprinted from Wrage, N., G.L. Velthof, M.L. van Beusichem, and O. Oenema. 2001. Role of nitrifier denitrification in the production of nitrous oxide. *Soil Biol. Biochem.* **33**:1723–1732. Copyright 2001, with permission from Elsevier.)

Atmospheric effects of N₂O and NO

N_2O is a greenhouse gas that contributes to global warming in the lower atmosphere. Furthermore, in the upper atmosphere, N_2O decomposes to a reactive species that consumes the ozone layer that shields the biosphere from DNA-damaging UV radiation. NO is not directly a greenhouse gas, but its reactions in the lower atmosphere lead to the formation of ozone, a pollutant and a greenhouse gas.

In an aerobic habitat, NO_3^- is thermodynamically stable. But when NO_3^- is transported to an anaerobic habitat it becomes an excellent final electron acceptor for anaerobes. As shown in the lower right region of Figure 7.15, there are three interacting processes that influence nitrate and its metabolites in anaerobic habitats: denitrification, dissimilatory nitrate reduction to ammonia (DNRA), and anammox.

1 In the (classic) *denitrification process*, single anaerobic microorganisms use NO_3^- as their terminal electron acceptor to fuel an ATP-generating respiratory chain. The intracellular endproduct is N_2 gas [produced via several intermediary metabolites: NO_2^- (nitrite, a key reactant for anammox), NO (nitric oxide), and N_2O (nitrous oxide); see Box 7.7].

2 *Dissimilatory nitrate reduction to ammonia* (DNRA) has been well described in enteric bacteria (like *Escherichia coli*; these inhabit digestive tracts of animals). DNRA is thought to be a physiological strategy that allows cells to avoid an intracellular excess of reducing equivalents and/or reduced toxic metabolites. Note that reduction of nitrate to ammonia consumes eight electrons, while reduction of nitrate to N_2

consumes five electrons. The extent that DNRA occurs in natural habitats (Figure 7.15; Tiedje, 1988; Welsh et al., 2001) is uncertain.

3 *Anammox* stands for "anaerobic ammonia oxidation". In this reaction, fully reduced NH_4^+ is used as an electron donor by chemolithoautotrophs that utilize NO_2^- as the electron acceptor. The recently discovered anammox reaction (see Section 7.5 and Boxes 7.8, 7.9) is carried out by members of the Planctomyces phylum of *Bacteria* that exhibit unique ("ladderane") membrane lipids that form an intracellular "anammoxysome" organelle – the site of membrane-bound enzymes that combine nitrite and ammonia to create hydrazine (N_2H_4) which is oxidized to N_2 gas. Electron flow in this process generates ATP (Kuypers et al., 2003; Strous and Jetten, 2004; Strous et al., 2006). Current estimates are that anammox may account globally for 30–50% of N_2 production in the ocean (Arrigo, 2005), though the proportion could actually be much higher (Kuypers et al., 2005).

Box 7.7 describes the impacts of two key nitrogen cycle processes (denitrification and nitrification) on global warming.

Science and the citizen

Arsenic-contaminated drinking water and the role of biogeochemistry

Headline news from the World Health Organization (WHO): arsenic-contaminated drinking water threatens the lives and well being of tens of millions of people worldwide, particularly in Bangladesh and West Bengal, India Arsenic is a toxic element capable of existing in five different valence states [As (–III), (0), (II), (III), and (V)]. Forms of arsenic range from sulfide minerals (e.g., $As_2S_3^-$) to elemental As to arsenic acid to arsenite (AsO_2^-) to arsenate (AsO_4^{3-}) to various organic forms that include methylated arsenates and trimethyl arsine. The dominant form of inorganic arsenic in aqueous, aerobic habitats is arsinate [oxidation state of +5 (V)]. In contrast, arsenite [oxidation state of +3 (III)] occurs widely in anaerobic aqueous environments (Oremland and Stolz, 2003).

Arsenic has been used by humans as a medicinal agent, a pigment, a pesticide, a poison, and in the production of glass and semiconductors. Inorganic forms of arsenic are more toxic than organic forms. In arsenite, the mobile trivalent (III) form is highly toxic and reacts with thiol groups of proteins within living cells; this rapidly disrupts many types of metabolism, particularly energy generation in the Krebs citric acid cycle. Though arsenate, the pentavalent (V) form of arsenic, exhibits a somewhat diminished toxicity and mobility, its resemblance to phosphate allows arsenate to interfere with respiratory processes – acting as an uncoupling agent in oxidative phosphorylation. Very few human organ systems escape the toxic effects of arsenic.

Arsenic is listed as a presumed carcinogenic substance based on the increased prevalence of lung and skin cancer observed in human populations with multiple exposures. Chronic

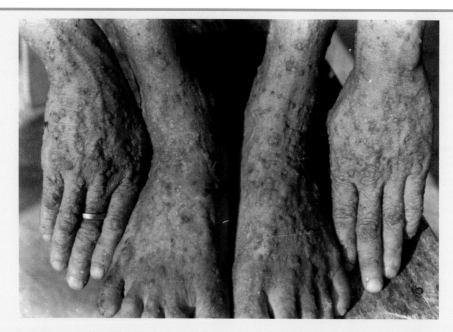

Figure 1 Subject from a village in West Bengal, India with the full panoply of arsenical skin lesions, including hyperkeratosis, suspected Bowen's disease, and nonhealing ulcers (suspected cancer). (Photo courtesy of Dr. Dipankar Chakraborti, Jadavpur University, India, with permission.)

exposure through drinking water leads to a variety of characteristic skin lesions (Figure 1). Globally, an estimated 100 million people are at risk of exposure to unacceptable arsenic levels in drinking well-water supplies (Islam et al., 2004).

SCIENCE: microbial processes that govern arsenic biogeochemistry

By studying Mono Lake in California, Oremland and Stolz (2003) have established the fundamental physiological and biogeochemical principles that apply to microbially mediated oxidation/reduction reactions of arsenic (Figure 2). The water column in Mono Lake, about 30 m deep, supports a gradient from oxic to anoxic conditions (left side, Figure 2). In a mid-gradient transition zone, arsenite [As(III)] is used as an electron donor by aerobic chemolithoautotrophic arsenite-oxidizing microorganisms (CAOs, center of Figure 2), producing arsenate [As(V)]. Farther down in the water column in the absence of oxygen, heterotrophic microorganisms are limited by the availability of final electron acceptors. Fueled by reduced carbon (CH_2O) raining down from the upper water column, dissimilatory arsenate-reducing prokaryotes (DARPs, right side of Figure 2) use arsenate as an electron acceptor; thus completing the local arsenic redox cycle.

In Bangladesh and India, arsenic is thought to originate from naturally occurring minerals in the highlands of the source rivers (Chakraborti et al., 2003). Though the same

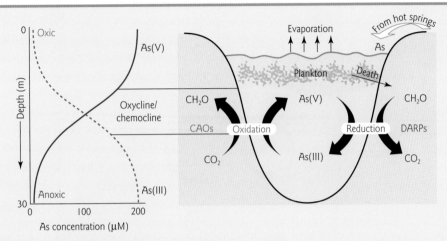

Figure 2 The chemical speciation of arsenic in the stratified water column of Mono Lake, California (left) as explained by the metabolism of arsenic by microbial populations present in the water column (right). Arsenic cycling occurs in the region of the chemocline. Arsenate reduction is mediated by DARPs that use released organic matter from dying plankton to fuel their respiration. Arsenite oxidation (aerobic and anaerobic) is mediated by CAOs that also contribute to secondary production by "fixing" CO_2 into organic matter. Arsenic first enters this alkaline (pH = 9.8), saline (290 g/L) lake as a dissolved component contained in the discharge from hydrothermal springs. Arsenic, as well as other dissolved constituents, reaches high concentrations because of the predominance of evaporation over precipitation in this arid region. (From Oremland R.S. and J.F. Stolz. 2003. The ecology of arsenic. *Science* **300**:939–944. Reprinted with permission from AAAS.)

microbiological principles of arsenic apply to both Mono Lake and Asia, the delta regions of India and Bangladesh are distinctive in several ways: (i) there is a high human population density; (ii) the use of agricultural activities that include irrigation and the use of nitrate fertilizers; (iii) the installation of tube wells (Figure 3) for extracting drinking water; and (iv) the presence of iron minerals in the sediments. The iron minerals interact both chemically and physically with the arsenic. New FeAs minerals can form. Also, mineral surfaces act as sorption sites that can bind and influence the mobility of arsenate and arsenite. In addition, Fe(III) may be used in preference to arsenate as an electron acceptor by anaerobic microorganisms (Islam et al., 2004).

The overall biogeochemistry of the tube-well systems (Figure 3) is complex. Interactions between multiple electron donors [organic carbon, sulfide, arsenite, Fe(II)] and acceptors [oxygen, nitrate, arsenate, Fe(III)] govern microbial processes. In turn, the microbial processes are influenced by geochemical reactions. Moreover, the setting is dynamic in space and time – reflecting both climate-related hydrologic events and human-managed agricultural practices. There is a pressing need to understand and manage this biogeochemical system because of its impact on the well being of millions of people.

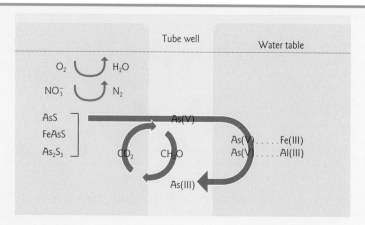

Figure 3 A conceptual model of how arsenic-metabolizing prokaryotes may contribute to the mobilization of arsenic from the solid phase into the aqueous phase in a subsurface drinking water aquifer. Arsenic is originally present primarily in the form of chemically reduced minerals, like realgar (AsS), orpiment (As$_2$S$_3$), and arsenopyrite (FeAsS). These minerals are attacked by chemolithoautotrophic arsenite-oxidizing microorganisms, which results in the oxidation of As(III), as well as iron and sulfide, with the concurrent fixation of CO$_2$ into organic matter. Construction of wells by human activity accelerates this process by providing the necessary oxidants like molecular oxygen or, in the case of agricultural regions, nitrate. The As(V) can subsequently be adsorbed onto oxidized mineral surfaces like ferrihydrite or alumina. The influx of substrate organic materials derived either from buried peat deposits, recharge of surface waters, or the microbial mats themselves promotes microbial respiration and the onset of anoxia, hence the conversion of As(V) to mobile As(III). DARPs then respire adsorbed As(V), resulting in the release of As(III) into the aqueous phase. (From Oremland R.S. and J.F. Stolz. 2003. The ecology of arsenic. *Science* **300**:939–944. Reprinted with permission from AAAS.)

Research essay assignment

Select an element other than carbon, sulfur, nitrogen, or arsenic and prepare an essay on its biogeochemistry. At the beginning of the essay, declare what motivated your selection of the element. Are you concerned with public health issues? Environmental pollution issues? Evolutionary issues? Or physiological or other issues?

Use the published literature (books, peer-reviewed publications) to develop a presentation that includes: (i) the element's biogeochemical cycle; (ii) key physiological mechanisms of the cycle; and (iii) an application of the cycle to everyday life of people.

7.5 CELLULAR MECHANISMS OF MICROBIAL BIOGEOCHEMICAL PATHWAYS

Learning the biochemical and genetic mechanisms of infectious disease provides myriad opportunities for medicine-based disease prevention. So too, mastering the intricacies of biogeochemical reactions can lead to biotechnology developments (see Section 8.6) and to wise ecological management practices. In the classic approaches to discovering biochemical pathways (for instance, photosynthesis in algae – the Calvin cycle of CO_2 fixation), a model organism that carries out the biochemical process needs to be grown in large quantities in the laboratory. Then, labeled substrates ("tracers", composed of radioactive atoms or rare stable isotopes) can be added to the active cells and the metabolic pathway can be discovered using analytical chemistry procedures to identify sequential metabolites. The structures of the metabolic intermediate compounds, themselves, are the basis for hypotheses about biochemical conversions and rates. Next, extracted enzymes from the dense cell preparations can be separated from one another and the individual enzyme activities can be tested and proven to carry out individual, sequential steps in the metabolic pathways. Furthermore, the genes encoding each enzyme can be identified in the test organisms [e.g., via site-directed mutagenesis (see Section 6.9 and Table 6.4), or by determining the amino acid sequence of the protein, inferring the DNA template encoding the protein, then probing genomic DNA extracts and sequencing the identified genes]. Typically, next examined are intra- and extracellular signaling networks (regulatory proteins and their biochemical cues) that control and modulate expression of the genes that encode the processes of interest. Topics explored by structural biological investigations add to understanding by revealing the three-dimensional arrangement of atoms in enzymes and of enzymes in membranes.

The types of biochemical and genetic studies described above have been extended to each of the biogeochemical processes shown in Figures 7.13–7.15 and listed in Table 7.4. Consequently there is an enormous amount of scientific information available about physiological, biochemical, and genetic mechanisms of microbially mediated reactions. As an example, Table 7.6 provides a listing of many microbial systems and their enzymes that catalyze nitrogen cycling.

The anatomy of discovery: anaerobic oxidation of ammonia and methane

As was discussed in Chapter 6 (Sections 6.6 and 6.10), selecting an ecologically important model organism as a source of biochemical and genetic information is a challenge. That challenge was met exceptionally well in two recent and remarkably successful cases that each uncovered and characterized microbial processes of global significance: anaerobic

Table 7.6

Types of microorganisms active in the nitrogen cycle and their enzyme systems (from Stein, L.Y. and Yung, Y.L. 2003. Production, isotopic composition, and atmospheric fate of biologically produced nitrous oxide. *Annu. Rev. Earth Planet. Sci.* **31**:329–356. Reprinted with permission from *Annual Reviews of Earth and Planetary Sciences*, Vol. 31. Copyright 2003 by Annual Reviews, www.annualreviews.org)

Bacterial group	Process	Main enzyme	Metabolism
Chemolithotrophic ammonia oxidizers	NH_3 oxidation to NO_2^- via NH_2OH	AMO HAO	Dissimilatory
	NH_2OH oxidation to NO and N_2O	HAO or nonenzyme	?
	NO_2^- reduction to NO	NIR	?
	NO reduction to N_2O	NOR	?
	NO_2^- reduction to N_2	?	?
Methane oxidizers	NH_3 oxidation to NO_2^- via NH_2OH	MMO P460	Fortuitous*
	NH_2OH oxidation to NO and N_2O	P460 or nonenzymatic	?
	NO_3^- reduction to NO_2^-	?	Assimilatory
	NO_2^- reduction to NH_3	?	Assimilatory
	N_2 fixation to NH_3	Nitrogenase	Assimilatory
	NO_2^- reduction to NO	NIR	?
	NO reduction to N_2O	NOR	
Heterotrophic ammonia oxidizers	NH_3 oxidation to NO_2^- via NH_2OH	AMO HAO	Fortuitous
Nitrite oxidizers	NO_2^- oxidation to NO_3^-	Nitrite oxidase	Dissimilatory
	NO_3^- reduction to NO_2^-, NH_3, and N_2O	Nitrite oxidase ?	
	NO oxidation to NO_2 and NO_3^-	?	?
Dissimilatory denitrifiers	NO_3^- reduction to NO_2^-	NAR, NAP	Dissimilatory
	NO_2^- reduction to NO	NIR	Dissimilatory
	NO reduction to N_2O	NOR NOS	Dissimilatory
	N_2O reduction to N_2	N_2OR	Dissimilatory
	NO oxidation to NO_3^-	?	?
	NO_3^- reduction to NH_3^- via NO_2^-	NAR NAS	Dissimilatory
Assimilatory denitrifiers	NO_3^- reduction to NO_2^-	NAS	Assimilatory
	NO_2^- reduction to NH_3	Siroheme NIR Hexaheme NIR	Assimilatory
Fungal denitrifiers	NO_2^- reduction to NO	Cu-NIR	?
	NO reduction to N_2O	P450nor	?
Nitrogen fixers	N_2 to NH_3	Nitrogenase	Assimilatory
Anammox	NH_3 and NO_2^- to N_2 via N_2H_4 and NH_2OH	Hydrazine hydrolase Hydroxylamine/hydrazine oxidoreductase	Dissimilatory

AMO, ammonium monooxygenase; Cu-NIR, copper-based nitrite reductase; HAO, hydroxlyamine oxygenase; MMO, methane monooxygenase; NAP, nitrate reductase (in periplasm of cell); NAR, nitrate reductase (membrane bound); NIR, nitrite reductase; NOR, nitric oxide reductase; N_2OR, nitrous oxide reductase; NAS, nitrate assimilation; NOS, nitrous oxide reductase; P450nor, cytochrome P450 nitric oxide reductase; P460, cytochrome P-460.
* Fortuitous here means of no direct physiological benefit; caused by nonspecific enzymes active in an unrelated process (co-metabolism).

Box 7.8

ammonia oxidation (anammox, Box 7.8) and anaerobic methane oxidation (AMO, Box 7.9).

Perusal of the information in Boxes 7.8 and 7.9 reveals interesting commonalities and contrasts in the progression of experiments that established anaerobic oxidation of ammonia and methane within biogeochemistry. As noted by Strous and Jetten (2004), both ammonium and methane are biochemically difficult to attack. Their activation under aerobic conditions

---

Box 7.8

Timeline and events in the discovery of anaerobic ammonia oxidation (anammox): from a denitrifying bioreactor to global cycling of nitrogen

1 *1965–1977. Chemical oceanographers speculate.* Thermodynamic calculations and ratios of nutrients in both ocean waters and phytoplankton ("Redfield ratios") suggest anoxic metabolism of ammonia (Strous et al., 1999).

2 *1990. Experimental verification of physiological process in engineered bioreactor systems.* Mass balances and stoichiometric relationships between compounds entering and exiting flow-through vessels colonized by microorganisms show that nitrite and ammonia are simultaneously consumed (Strous and Jetten, 2004).

3 *1995. Laboratory growth conditions of an enrichment culture show nitrate and ammonia are essential reactants for microbial growth of enriched, transferable microorganisms.* Sequencing of 16S rRNA genes identify the dominant organisms as a member of the *Planctomyces* phylum (Strous et al., 2002; Strous and Jetten, 2004).

4 *1999–2002. Dominant organisms in enrichment culture are physically purified – allowing detailed physiological studies.* Intermediary metabolites are confirmed – especially one unique to biology, hydrazine (H_2NNH_2). Autotrophy (CO_2 fixation) is shown to be linked to NH_4 oxidation. Purification of key enzyme: hydroxylamine/hydrazine oxidoreductase is housed in a membrane-bound organelle (the anammoxysome) composed of unique ladderane lipids (Jetten et al., 2003; Strous and Jetten, 2004). (For a metabolic scheme, see Figure 1 in Box 7.9.)

5 *2003–2005. Widespread geographic significance (Black Sea and coasts of Africa and Costa Rica).* In situ geochemical profiles show NH_4 depletion in nitrite-rich zones. Microbial communities incubated with $^{15}NH_4^+$ form N_2 with mixed isotopic composition ($^{14}N\ ^{15}N$), which distinguishes anammox from the denitrifying pathway of anaerobic N_2 formation (Kuypers et al., 2003, 2005).

6 *2006. Community genomics (metagenome) approach develops blueprint for biochemical and genetic basis of anammox process.* DNA sequences from shotgun, fosmid, and bacterial artificial chromosome (BAC) clone libraries (see Section 6.9) are assembled from the mixed microbial community in an anammox bioreactor. Efforts to assemble the genome of the active bacterium, named *Kuenenia stuttgartiensis*, are incomplete; they stall at five large contigs (see Section 6.9). But information does include many insights into the overall metabolism and evolution of the organisms (Strous et al., 2006).

Box 7.8 *Continued*

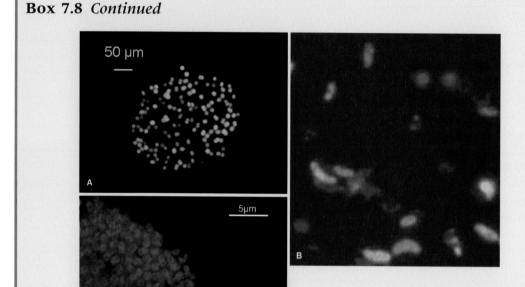

Figure 1 Microscopic images using fluorescent in situ hybridization (FISH) showing anammox microorganims in water samples from (A) coastal Africa, (B) the Black Sea, and (C) a bioreactor. (From Jetten, M.S.M., O. Sliekers, M. Kuypers et al., 2003. Anaerobic ammonium oxidation by marine and freshwater planctomycete-like bacteria. *Appl. Microbiol. Biotechnol.* **63**:107–114, fig. 2. With kind permission of Springer Science and Business Media.)

by monooxygenase enzymes makes use of oxygen's tendency to steal electrons via free-radical reactions. Notions about anaerobic metabolism of ammonium and methane originated with chemical oceanographers (Boxes 7.8, 7.9). Physiological documentation of the processes and elucidation of biochemical reaction mechanisms followed decades later. Both the paths of discovery began more than 50 years ago and both were hampered by scientific dogma that ammonium and methane were chemically inert under anaerobic conditions (Strous and Jetten, 2004).

As mentioned above, being able to cultivate (see Section 5.1) and use enrichment cultures (see Sections 6.2 and 6.10) are the gateways to discovering physiological and biochemical mechanisms in microorganisms. Pure cultures have not yet been isolated that are capable of anaerobically

Box 7.9

Timeline and events in the discovery of anaerobic methane oxidation (AMO): from field geochemistry to metagenomics

1   *~1957. Chemical oceanographers speculate.* Data sets from vertical sediment profiles show that methane and sulfate are simultaneously depleted in discrete zones in vertical sediment profiles.

2   *1979. Recognition of thermodynamic potential for CH_4 serving as e^- donor.* Early laboratory incubations led to the hypothesis "methanogenesis working in reverse in conjunction with a sulfate reducer". However, robust enrichment cultures are unsuccessful (Zehnder and Brock, 1979).

3   *1999. Use of field biogeochemistry and gene cloning.* Sediments above a methane hydrate deposit in the coastal Pacific Ocean have a lipid biomarker with a $^{13}C/^{12}C$ value that proves it was derived from the methane deposit on the ocean floor. The methane must have been the carbon source for the microorganisms. Clone libraries of 16S rRNA genes reveal several unusual sequences that suggest the identity of organisms involved in AMO (Hinrichs et al., 1999).

4   *2000. Fluorescent in situ hybridization (FISH) microscopy and radiotracer based physiology.* Microscopy (FISH) of field samples proves robust physical associations between cooperating methanogenic *Archaea* and sulfate-reducing *Bacteria*. Laboratory assays show that $^{35}SO_4$ is reduced to $H_2^{35}S$ (Boetius et al., 2000).

5   *2001. Confirmation of $^{13}C/^{12}C$ biomarker and FISH.* Spatially resolved mass spectrometry confirms that the field-derived associations of methanogens and sulfate reducers are composed of carbon whose $^{13}C/^{12}C$ ratio links them to in situ methane oxidation (Orphan et al., 2001).

6   *2002. Confirmation of biogeochemical significance: AMO creates large geochemical deposits in Black Sea.* Large reefs (deposits of $Ca/MgCO_3$) at the bottom of the Black Sea were created from the CO_2 produced by AMO; evidence is supported by FISH, conversion of $^{14}CH_4$ to $^{14}CO_2$, and stable isotope ratios in both lipid biomarkers and the carbonate minerals (Michaelis et al., 2002).

7   *2003. Field enzymology.* Extracted protein from Black Sea reefs reveals an abundant nickel-containing enzyme that is closely related to enzymes in methanogenic *Archaea* (Kruger et al., 2004).

8   *2004. Community genomics (metagenomic) approach develops preliminary genetic blueprint for biochemical and genetic basis of process.* DNA sequences from shotgun and fosmid libraries (see Section 6.9) are assembled from a partially purified microbial community from the coastal Pacific Ocean. Partial assembly of genome fragments from two archaeal genomes and three sulfate reducers result. Identification of most of the genes associated with methanogenesis provides strong support for the "reverse methanogenesis" hypothesis (Hallam et al., 2004).

Box 7.9 *Continued*

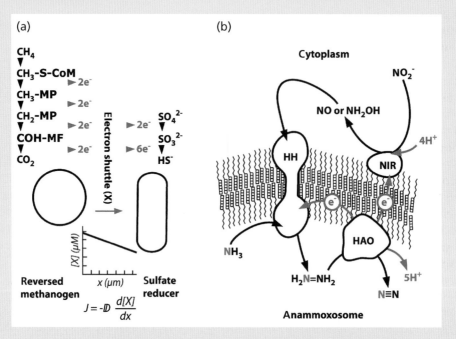

Figure 1 Metabolic schemes for (a) anaerobic methane oxidation, and (b) anammox.
(a) Possible biochemical pathway of methane oxidation. The reversed methanogen donates
electrons to the electron shuttle, which is scavenged by the sulfate reducer. Fick's first law
of diffusion dictates that a concentration gradient of the shuttle must exist between the two
syntrophic bacterial partners. This gradient limits the thermodynamic and kinetic efficiency
of the overall process. (b) Possible biochemical pathway of anammox. Black and gray nitrogen
atoms highlight the production of $^{29}N_2$ from one of the ^{15}N-labeled substrates, ammonium
or nitrite. In this model, the proton motive force is generated over the anammoxosome
membrane via separation of charges. CoM, coenzyme M; HAO, hydroxylamine/hydrazine
oxidoreductase; HH, hydrazine hydrolase; MF, methanofuran; MP, methanopterin; NIR, nitrite
reductase. (From Strous, M. and M.S.M. Jetten. 2004. Anaerobic oxidation of methane and
ammonium. *Annu. Rev. Microbiol.* **58**:99–117. Reprinted with permission from *Annual Reviews of
Microbiology*, Volume 58, copyright 2004 by Annual Reviews, www.annualreviews.org.)

oxidizing either ammonium or methane; however, enrichment cultures
are available for anammox (but not yet for AMO). This makes a big difference in rates of discovery and scientific progress. The six-step
sequence of investigation for anammox benefited significantly from the
availability of an engineered bioreactor system in which the microbially
mediated reaction between nitrite and ammonium could be carefully controlled and modified. The catalytic anammox activity could be transferred
from one culture vessel to another. The growing biomass fixed CO_2 (was

autotrophic), produced intermediary metabolites that could be traced and identified, and produced enzymes that mediated steps in the metabolic reaction pathway. Moreover, microscopic and biomarker (unique ladderane lipids and 16S rRNA sequence) based characterizations were completed. Armed with biochemical fundamentals, investigators were able to explore the ecological significance of anammox. As mentioned in Section 7.4, anammox is currently thought to account for *at least* 30–50% of all marine ammonium oxidation.

The path of AMO discovery is shown in Box 7.9. Because of the absence of laboratory-based enrichment cultures, approximately half of the eight steps shown in Box 7.9 were devoted to documenting AMO. All of the procedures used to document and biochemically characterize AMO began in anaerobic field-study sites that were highly enriched in methane (e.g., methane hydrate deposits in cold, deep Pacific Ocean sediments or in the Black Sea). A key factor in the logic of all the biomarker-based studies of AMO is the fortuitous fact that microbially produced methane has a unique and extremely negative $\delta^{13}C$ value (see Box 2.2). No other pools of carbon in nature are as depleted in ^{13}C as biogenic methane. Thus, a stable isotopic tracer experiment was built into the field-study sites! When isotope ratio mass spectrometery was used to analyze archaeal lipids extracted from sediments adjacent to the methane source, the $\delta^{13}C$ value of the lipids was extremely negative – this meant that the organisms in the sediments had to be using methane-derived carbon as their carbon source. Then, microscopy, fluorescent in situ hybridization (FISH), laboratory radiotracer studies (for both sulfate reduction and conversion of CH_4 to CO_2), and a sediment-derived enzyme all provided additional information about the AMO process (Box 7.9). AMO is carried out by a pair of physiologically cooperating microorganisms: a "methanogen operating in reverse" which delivers electrons to a sulfate-reducing bacterial partner. It is estimated that 75% of all marine methane oxidation is attributable to AMO (Strous and Jetten, 2004).

7.6 MASS BALANCE APPROACHES TO ELEMENTAL CYCLES

In Section 7.4 we acknowledged that, for physiological and biochemical reasons, it can be insightful to conceptually split the elements away from one another. If we examine the "stuff of life" (the live and dead biotic components of the biosphere plus nutrients; Section 7.3) and sort through these materials by elemental content, then budgets can be assembled. We have seen (Section 7.2) that such budgets help assess the status and health of Earth. These scientific exercises tally the sizes of nutrient pools and the fluxes of materials between them. Figures 7.16–7.18 portray global elemental cycles of carbon, sulfur, and nitrogen for terrestrial, atmospheric, and marine compartments of the globe.

Annual global carbon budget

The annual global carbon budget (Figure 7.16) provides insight into the function of the biosphere via the contrasting size and turnover times of the carbon pools. The current estimate of annual photosynthetic NPP (CO_2 fixation in excess of respiration) is between 50 and 60×10^{15} g C for both terrestrial and marine habitats (Reeburgh, 1997; Schlesinger, 1997). With the exception of prokaryotic biomass (see Section 4.5 and Table 4.10), the biomass of land plants ($550–680 \times 10^{15}$ g C) far exceeds that of any other biota and on a dry-weight basis plant biomass is ~50% cellulose and 15–36% lignin (see Section 7.3 and Box 7.3; Deobald and Crawford, 2002; Lynd et al., 2002). It is noteworthy that 20% of oceanic primary productivity is concentrated along continental coastlines (see Section 4.3). The largest active pools of carbon in terrestrial and oceanic compartments are soil organic matter (viable organisms plus recognizable dead and decaying biomass plus humic substances) and dissolved organic carbon (DOC), respectively. A small increase in the rate of microbial carbon mineralization from either of these two pools could significantly

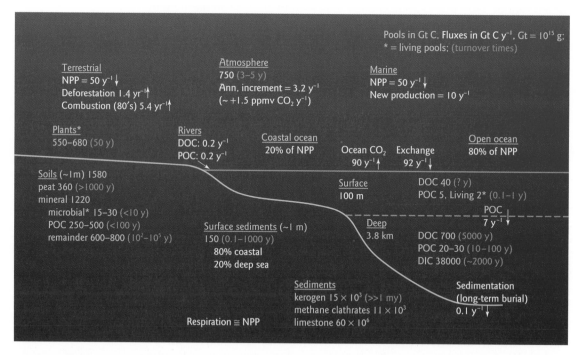

Figure 7.16 Global carbon reservoirs, fluxes, and turnover times. DIC, dissolved inorganic carbon; DOC, dissolved organic carbon; NPP, net primary productivity; POC, particulate organic carbon; ppmv, parts per million on a volume basis. (From Reeburgh, W.S. 1997. Figures summarizing the global cycles of biogeochemically important elements. *Bull. Ecol. Soc. Am.* **78**:260–267. Reprinted with permission from the Ecological Society of America, http://www.ess.uci.edu/~reeburgh/fig1.html.)

influence the concentration of atmospheric CO_2 (though this seems unlikely for deep oceanic DOC because of its slow turnover time of ~5000 years). Conversely, the stored soil and oceanic organic pools are so large that a small relative shift toward carbon storage (also termed sequestration) could possibly compensate for excesses in anthropogenic CO_2 emissions to the atmosphere.

Enormous additional pools of carbon, remote from the biosphere, occur in deep terrestrial and oceanic sites as keragen and methane hydrates (termed clathrates in Figure 7.16) and as carbonate minerals (limestone and dissolved inorganic carbon, DIC). Note that, in preparing Figure 7.16, Reeburgh (1997) explicitly showed an annual increase in atmospheric CO_2, as well as net absorption of CO_2 by ocean waters. Clearly the global carbon budget reveals a planet that is not in steady state. Moreover, the added CO_2 to ocean water has, via bicarbonate equilibrium reactions, the potential to cause gradual oceanic acidification.

Annual global sulfur budget

Highlighted in the global sulfur budget (Figure 7.17) are atmospheric transfers between terrestrial, oceanic, and atmospheric pools. As shown

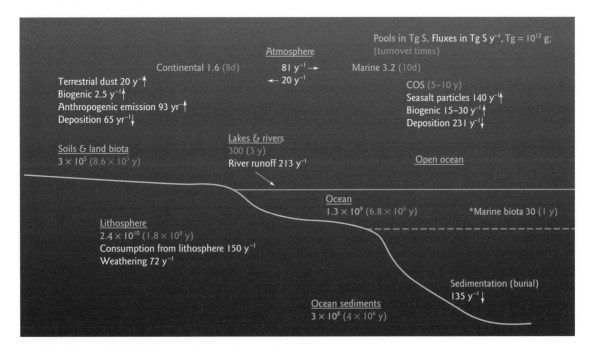

Figure 7.17 Global sulfur reservoirs, fluxes, and turnover times. (From Reeburgh, W.S. 1997. Figures summarizing the global cycles of biogeochemically important elements. *Bull. Ecol. Soc. Am.* **78**:260–267. Reprinted with permission from the Ecological Society of America, http;//www.ess.uci.edu/~reeburgh/fig6.html.)

in Figure 7.17, terrestrial anthropogenic emission of sulfur to the atmosphere exceeds biogenic emissions by nearly a factor of four. The main sources of anthropogenic sulfur releases (largely as SO_2 gas) are coal burning and sulfide-ore smelting (Charlson et al., 2000). When SO_2 is emitted, it forms sulfuric acid by reactions with water in the atmosphere. Acid rain and its geochemical and biological effects are central concerns for understanding the global sulfur cycle.

Owing to its relatively long turnover time (5–10 years), carbonyl sulfide (COS, alternatively OCS) is the most abundant sulfur gas in the atmosphere (~500 parts per trillion). According to Schlesinger (1997), the major sources of atmospheric COS is a photochemical reaction with dissolved organic matter in ocean water. Other COS sources include biomass burning, fossil fuel combustion, and atmospheric oxidation of industrially released carbon disulfide (CS_2). Schlesinger (1997) states that "our understanding of COS biogeochemistry is primitive". Like other sulfur-containing gases, COS is chemically oxidized to SO_4^{2-}, which participates in aerosol formation. The role of aerosols and cloud-induced changes in atmospheric reflectance of sunlight was previously discussed in Box 7.5. Atmospheric scientists (e.g., Wigley, 2006) have recently suggested large-scale engineered release of sulfate aerosol precursors into the atmosphere (simulating volcanic eruptions) as a means to offset global warming and to provide additional time to reduce human dependence on fossil fuels for energy. The major masses of sulfur in the global cycle are found in the lithosphere and oceanic compartments – largely as the crustal minerals gypsum ($CaSO_4$) and pyrite (FeS_2).

Annual global nitrogen budget

The annual global nitrogen budget (Figure. 7.18) reveals an excess of ~100 $\times 10^{12}$ g nitrogen brought into the biosphere from the atmosphere, relative to the amount returned to the atmosphere via denitrification-related processes. The increases in nitrogen made available to terrestrial and aquatic biota are derived largely from industrial fertilizer production, enhanced biological nitrogen fixation in agriculture, and the burning of nitrogen-containing fossil fuels. These three sources boost background N_2 fixation via naturally occurring free-living and symbiotic prokaryotes by approximately two-thirds (Galloway et al., 2004). The environmental impacts of this imbalance in nitrogen cycling (summarized by Jaffe, 2000) include: (i) radiative forcing of climate change by N_2O and O_3 in the troposphere (see Section 7.2 and Box 7.7); (ii) photochemical smog; (iii) acid precipitation (deposition of nitric acid, HNO_3); (iv) stratospheric depletion of the ozone shield by N_2O reactions; (v) groundwater contamination by nitrate; (vi) fertilization of the global carbon cycle; and (vii) potential shifts in plant, animal, and microbial biodiversity in fertilized habitats.

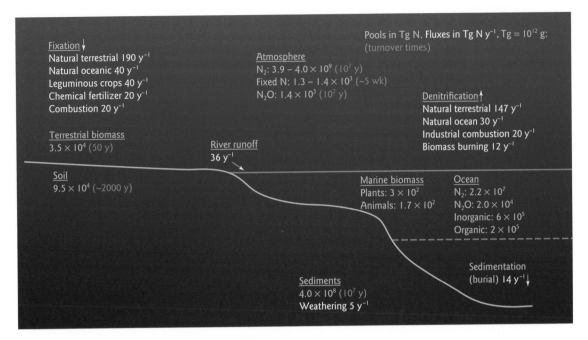

Figure 7.18 Global nitogen reservoirs, fluxes, and turnover times. (From Reeburgh, W.S. 1997. Figures summarizing the global cycles of biogeochemically important elements. *Bull. Ecol. Soc. Am.* **78**:260–267. Reprinted with permission from the Ecological Society of America http://www.ess.uci.edu/~reeburgh/fig3.htm.)

The cycles of carbon, sulfur, and nitrogen have been disturbed by human activities. All are strongly influenced by transformations mediated by prokaryotes. Furthermore, advanced understanding of all three elemental cycles (and others) will be achieved via close collaboration between microbiologists, geochemists, atmospheric chemists, engineers, and modelers.

STUDY QUESTIONS

1 Data in Table 7.1 show the mass of mineral nutrients needed to produce 100 g dry mass of microbial cells. Table 7.2 and Box 7.6 both mention vanadium (V) as a potential nutrient required for nitrogen fixation. If you wanted to implement an assay showing that V is required as a nutrient, what three main considerations would be essential before implementing the experiment? (Hint: consider practical issues about the organism tested, its physiology, and background levels of V.)

2 Table 7.3, entry 3 lists a *cadA* gene (encoding the cadA protein) that detoxifies both Cd^{2+} and Zn^{2+} in bacterial cells. Venture a simple educated guess why the protein recognizes both Cd^{2+} and Zn^{2+}. (Hint: inspect the periodic table of the elements.)

3 CFC-11 and CFC-12 are mentioned in Section 7.2 and appear in Box 7.1 because of their influence on climate change as greenhouse gases in the troposphere. What other environmental process (not discussed here in this book) makes CFCs a major environmental concern? Where in the atmosphere does this process occur?

4 Figure 7.4 shows historical records of the rise in greenhouse gases in the troposphere. Carbon dioxide and methane fluctuate sinusoidally each year (Section 7.2). CFC-12 and CFC-11 do not fluctuate annually. Why?

5 What is your personal opinion of global warming? In preparing your answer, consider historical, geologic, and biological observations and evidence near your home and abroad. Which four of the 12 factors influencing global warming (Figure 7.5) do you think are most important? Why?

6 Construct three sketches of carbon budgets for the place you live. Carry out this exercise at the simplest level (your body) and two progressively higher levels, such as your home, your apartment complex, your neighborhood, your village, region, and/or state. Draw boxes and label the carbon pools. Draw arrows and label the processes. Note your major uncertainties in each model. What happens as the size of the system of interest increases?

7 Table 7.4 and Box 7.3 convey information about high molecular weight carbon compounds and how they are metabolized by microorganisms.
   (A) Name two major structural differences between starch molecules (that *you* digest) and cellulose molecules (that bacteria and fungi digest).
   (B) In both humans and fungi, what is the endproduct of hydrolytic digestion of starch and how is it further metabolized?
   (C) What sets lignin apart from the other biopolymers shown in Table 7.4 and Box 7.3? Briefly describe lignin formation and biodegradation.

8 Forest trees are fully biodegradable. Is there a forest or woodland or park near your home?
   (A) Visit the forest and take notes on the degree of decay in the dead wood.
   (B) Can you see fungal hyphae? Mycelia? Fruiting bodies?
   (C) What is the evidence for decay?
   (D) How long does it take for decay to be complete? Venture a guess. Also, use information in Chapter 6 to design one or more experiments to answer the question.

9 Information in Section 7.3 (including Table 7.4 and Figures 7.9–7.11) describes microbial metabolism of petroleum hydrocarbons.
   (A) Summarize the biochemical "logic" (the physiological strategy) for destabilizing the aromatic ring of a molecule under both aerobic and anaerobic conditions.
   (B) There are structural similarities between aromatic compounds in petroleum and in lignin. In light of your answer to question 8D, how long do you think it takes naturally occurring microorganisms to biodegrade the aromatic (and other) components in spilled crude oil? Please answer this by consulting library sources describing the Exxon Valdez or other oil spills.

10 The vertical scale in Table 7.5 associates being reduced with serving as an electron donor and being oxidized with serving as an electron acceptor. Tetrachloroethene ($Cl_2C=CCl_2$) is a widely used industrial solvent in the metal machining and electronics industries. Tetrachloroethene (also known as perchloroethene, or PCE) is also a widespread groundwater pollutant.

(A) What is the oxidation state of carbon in PCE?

(B) What is the physiological role you expect PCE to play in microbial metabolism?

(C) If you were planning biodegradation tests for PCE, what experimental design would you use? What would you measure? Under what conditions would biodegradation occur?

11 Focus on the DMS and DMSO entries in Table 7.4:

(A) Diagram the DMS–DMSO redox cycle by placing DMS on the right-hand side, DMSO on the left-hand side, and adding semicircular arrows to connect them (one arrow above, and one arrow below). Label the top of the diagram "aerobic" and the bottom "anaerobic".

(B) In a marine setting, what would be the likely electron donors and acceptors to drive the cycle?

(C) Consider where DMSO falls on the redox scale in Table 7.5. What physiological or ecological circumstances would favor DMSO utilization as an electron *donor* rather than an electron *acceptor*?

12 Scrutinize Figure 7.12.

(A) Were you aware of the detritus-based food chains prior to reading Section 7.4?

(B) Choose a habitat of interest within your favorite ecosystem. Next, state the resources and organisms likely to be participating in the detritus food chain in that habitat.

(C) As a related library research task, search for information on the "microbial loop". What did you find? Research papers using this term attest to its scientific significance, especially in aquatic systems.

13 In depicting the carbon cycle, Figure 7.13 includes a pathway "burial → (diagenesis) → peats, lignites". Likewise, Figure 7.16, in depicting the global carbon budget, includes "sedimentation".

(A) Does this mean that fossil fuels are renewable?

(B) Could global warming be corrected by "sequestering" or removing atmospheric CO_2 from the biosphere? Please develop microbiological, geochemical, and/or engineering arguments to support your answer.

14 Consider the nitrogen cycle as described in Sections 7.3, 7.4, and 7.6 (as supported by information in Tables 7.4–7.6, Figures 7.15, 7.18, and Boxes 7.6, 7.7).

(A) If you were writing a science fiction novel, would you have invented a biogeochemical cycle more complex than the nitrogen cycle? Does it seem "stranger than fiction" to you?

(B) What does the complexity of pathways and microbial metabolic adaptations say about the role of nitrogen compounds in evolution (consider, especially, N_2 fixation)?

(C) The text of Box 7.6 provides a fortuitous physiological fact that has led to a widespread assay for determining nitrogenase activity in both microbial cultures and in environmental samples. What is that assay? Why is it useful and important?

(D) Summarize the multiple connections between nitrogen biogeochemistry and climate change.

(E) Prepare a diagram of the nitrogen cycle that is analogous to the one shown for the carbon cycle (Figure 7.13) and to the diagram you prepared in response to question 11A. Place ammonia on the far right and nitrate on the far left of the central horizontal line that divides processes into aerobic (top) and anaerobic (bottom). Now use this framework to create a unified conceptual summary of the nitrogen cycle – specifying key nitrogen compounds and the processes (e.g., nitrification, denitrification, nitrogen fixation, etc.) that connect them.

15 In reflecting on material presented in Section 7.5, consider parallels between discovering an AIDS vaccine and understanding mechanistic details of reactions that microorganisms carry out on the seafloor. Anammox (Box 7.8) and AMO (Box 7.9) are key biogeochemical processes that influence the "health" of the biosphere.

(A) If environmental and physiological conditions diminished AMO activity on the seafloor, what climate change scenario might develop?

(B) Likewise, if anammox activity on the seafloor diminished, what climate change scenario might develop?

In developing answers, please consider the intricacies of direct effects, indirect effects, elemental mass balances (e.g., Figures 7.16–7.18), interactions between elemental cycles, and potential compensatory processes.

REFERENCES

Alexander, M. 1973. Nonbiodegradable and other recalcitrant molecules. *Biotechnol. Bioeng.* **15**:611–647.

Alexander, M. 1999. *Biodegradation and Bioremediation*, 2nd edn. Academic Press. San Diego, CA.

Arrigo, K.R. 2005. Marine microorganisms and global nutrient cycles. *Nature* **437**:349–355.

Bayer, E.A., J.-P. Belaich, Y. Snoham, and R. Lamed. 2004. The cellulosomes; multienzyme machines for degradation of plant wall polysaccharides. *Annu. Rev. Microbiol.* **58**:521–554.

Boetius, A., K. Ravenschlag, C.J. Schubert, et al. 2000. A marine microbial consortium apparently mediating anaerobic oxidation of methane. *Nature* **407**:623–626.

Buist, G., J. Kok, K.J. Leenhouts, M. Pabrowska, G. Venema, and A.J. Haandrikman. 1995. Molecular cloning and nucleotide sequence of the gene encoding the major peptidoglycan hydrolase of *Lactococcus lactis*, a muramidase needed for cell separation. *J. Bacteriol.* **177**:1541–1563.

Chakraborti, D.S, C. Mukherjee, S. Pati, et al. 2003. Arsenic groundwater contamination in Middle Ganga Plain, Bihar, India: a future danger? *Environ. Health Persp.* **111**:1194–1201.

Charlson, R.J. 2000. The coupling of biogeochemical cycles and climate: forcings, feedbacks, and responses. *In*: M.C. Jacobson, R.J. Charlson, H. Rodhe, and G.H. Orians (eds) *Earth System Science: From biogeochemical cycles to global change*, pp. 439–448. Academic Press. San Diego, CA.

Charlson, R.J., T.L. Anderson, and R.E. McDubb. 2000. The sulfur cycle. *In*: M.C. Jacobson, R.J. Charlson, H. Rodhe, and G.H. Orians (eds) *Earth System Science: From biogeochemical cycles to global change*, pp. 343–359. Academic Press, San Diego, CA.

Coates, J.D., R. Chakraborty, and M.J. McInerney. 2002. Anaeorobic benzene biodegradation – a new era. *Res. Microbiol.* **153**:621–628.

Colberg, P.J. 1988. Anaerobic microbial degradation of cellulose, lignin, oligolignols, and monoaromatic lignin derivatives. *In*: A.J.B. Zehnder (ed.) *Biology of Anaerobic Microorganisms*, pp. 333–372. Wiley and Sons, New York.

Conrad, R. 1996. Soil microorganisms as controllers of atmospheric trace gases (H_2, CO, CH_4, OCS, N_2O, and NO). *Microbiol. Rev.* **60**:609–640.

Davidson, E.A., F.Y. Ishida, and D.D. Nepstead. 2004. Effects of an experimental draught on soil emissions of carbon dioxide, methane, nitrous oxide, and nitric oxide in a moist tropical forest. *Global Change Biol.* **10**:718–730.

Davidson, E.A., M. Keller, H.E. Erickson, L.V. Verchot, and E. Veldkamp. 2000. Testing a conceptual model of soil emissions of nitrous and nitric oxides. *Bioscience* **50**:667–680.

Deobald, L.A. and D.L. Crawford. 2002. Lignocellulose biodegradation. *In*: C.J. Hurst, R.L. Crawford, G.R. Knudsen, M.J. McInerney, and L.D. Stetzenbach (eds) *Manual of Environmental Microbiology*, 2nd edn, pp. 925–933. American Society for Microbiology, Washington, DC.

Devlin, T.M. (ed.) 2001. *Textbook of Biochemistry with Clinical Correlations*, 5th edn. Wiley and Sons, New York.

Doi, R.H. and A. Kosugi. 2004. Cellulosomes: plant-cell-wall degrading enzyme complexes. *Nature Rev. Microbiol.* **2**:541–551.

Ehrlich, H.L. 2002. *Geomicrobiology*, 4th edn. Marcel Dekker, New York.

Galloway, J.N., F.J. Dentener, D.G. Capone, et al. 2004. Nitrogen cycles: post, present and future. *Biogeochemistry* **70**:153–226.

Hallam, S.J., N. Putnam, C.M. Preston, et al. 2004. Reverse methanogenesis: testing the hypothesis with environmental genomics. *Science* **305**:1457–1462.

Hammel, K.E. 1997. Fungal degradation of lignin. *In*: G. Cadisch and K.E. Giller (eds) *Driven by Nature: Plant litter quality and decomposition*, pp. 33–45. CAB International Publishers, Cambridge, MA.

Heider, J. 2007. Adding handles to unhandy substrates anaerobic hydrocarbon activation mechanisms. *Curr. Opin. Chem. Biol.* **11**:188–194.

Henshaw, P.C., R.J. Charlson, and S.J. Burges. 2000. Water and the hydrosphere. *In*: M.C. Jacobson, R.J. Charlson, H. Rodhe, and G.H. Orians (eds) *Earth System Science: From biogeochemical cycles to global change*, pp. 109–131. Academic Press, San Diego, CA.

Hinrichs, H.-U., J.M. Hayes, S.P. Sylva, P.G. Brewer, and E.F. DeLong. 1999. Methane-consuming archaebacteria in marine sediments. *Nature* **398**: 802–805.

Hofrichter, M. 2002. Review: lignin conversion by manganese peroxidase (MnP). *Enzyme Microbial Technol.* **30**:454–466.

Houghton, J.T., Y. Ding, D.J. Griggs, et al. 2001. *Climate Change 2001: The science basis.* Cambridge University Press, Cambridge, UK.

Howard, E.C., J.R. Henriksen, A. Buchan, et al. 2006. Bacterial taxa that limit sulfur flux from the ocean. *Science* **314**:649–652.

Hughes, M.N. and R.K. Pool. 1989. *Metals and Microorganisms*. Chapman and Hall, London.

Islam, F.S., A.G. Gault, C. Boothman, et al. 2004. Role of metal-reducing bacteria in arsenic release from Bengal delta sediments. *Nature* **430**:68–71.

Jaffe, P.A. 2000. The nitrogen cycle. *In*: M.C. Jacobson, R.J. Charlson, H. Rodhe, and G.H. Orians (eds) *Earth System Science: From biogeochemical cycles to global change*, pp. 322–359. Academic Press, San Diego, CA.

Jetten, M.S.M., O. Sliekers, M. Kuypers, et al. 2003. Anaerobic ammonium oxidation by marine and freshwater planctomycete-like bacteria. *Appl. Microbiol. Biotechnol.* **63**:107–114.

Kruger, M., A. Meyerdierks, F.O. Glöckner, et al. 2004. A conspicuous nickel protein in microbial mats that oxidized methane anaerobically. *Nature* **426**:878–881.

Kuypers, M.M.M., G. Lovik, D. Woeblan, et al. 2005. Massive nitrogen loss from Benguela upwelling system through anaerobic ammonia oxidation. *Proc. Natl. Acad. Sci. USA* **102**:6478–6483.

Kuypers, M.M.M., A.O. Sliekers, G. Lavik, et al. 2003. Anaerobic ammonium oxidation by anammox bacteria in the Black Sea. *Nature* **422**:608–611.

Lengeler, J.W., G. Drews, and H.G. Schlegel. 1999. *Biology of Prokaryotes*. Blackwell Science, Stuttgart.

Leonowicz, A., A. Matuszewska, J. Luterek, et al. 1999. Biodegradation of lignin by white rot fungi. *Fungal Genet. Biol.* **27**:175–185.

Leschine, S.B. 1995. Cellulose degradation in anaerobic environments. *Annu. Rev. Microbiol.* **49**:399–426.

Lynd, L.R., P.J. Weimer, W.H. van Zyl, and I.S. Pretorius. 2002. Microbial cellulose utilization: fundamentals and biotechnology. *Microbiol. Molec. Biol. Rev.* **66**:506–577.

Madigan, M.T. and J.M. Martinko. 2006. *Brock Biology of Microorganisms*, 11th edn. Prentice Hall, Upper Saddle River, NJ.

Mancinelli, R.L. 1995. The regulation of methane oxidation in soil. *Annu. Rev. Microbiol.* **49**:581–605.

Michaelis, W., R. Seifert, K. Hauhaus, et al. 2002. Microbial reefs in the Black Sea fueled by anaerobic oxidation of methane. *Science* **297**:1013–1015.

Neill, C., P.A. Steudler, D.C. Garcia-Montiel, et al. 2005. Rates and controls of nitrous oxide and nitric oxide emissions following conversion of forest to pasture in Rondonia. *Nutr. Cycling Agroecosyst.* **71**:1–15.

Niedhardt, F.C., J.L Ingraham, and M. Schaechter. 1990. *Physiology of the Bacterial Cell*. Sinauer Associates, Sunderland, MA.

Oremland, R.S. and J.F. Stolz. 2003. The ecology of arsenic. *Science* **300**:939–944.

Orphan, V.J., C.H. House, K.-U. Hinrich, K.D. McKeegan, and E.F. DeLong. 2001. Methane-consuming archaea revealed by directly coupled isotopic and phylogenetic analysis. *Science* **293**: 484–487.

Raghoebarsing, A.A., A. Pol, K.T. van de Pas-Schoonen, et al. 2006. A microbial consortium couples anaerobic methane oxidation to denitrification. *Nature* **440**:918–921.

Reeburgh, W.S. 1997. Figures summarizing the global cycles of biogeochemically important elements. *Bull. Ecol. Soc. Am.* **78**:260–267.

Remde, A. and R. Conrad. 1991. Role of nitrification and denitrification for NO metabolism in soil. *Biogeochemistry* **12**:189–205.

Ribbe, M., D. Gadkari, and O. Meyer. 1997. N_2 fixation by *Streptomyces thermautotrophicus* involves a molybdenum-dinitrogenase and a manganese-superoxide oxido-reductase that couples N_2 reduction to the oxidation of superoxide produced from O_2 by a molybdenum-CO dehydrogenase. *J. Biol. Chem.* **272**:26627–26633.

Rodhe, H. 1990. A comparison of the contribution of various gases to the greenhouse effect. *Science* **248**:1217–1219.

Rodhe, H. 2000. Modeling biogeochemical cycles. *In*: M.C. Jacobson, R.J. Charlson, H. Rodhe, and G.H. Orians (eds) *Earth System Science: From biogeochemical cycles to global change*, pp. 62–84. Academic Press, San Diego, CA.

Safinowski, M. and R.U. Mechenstock. 2006. Methylation is the initial reaction in anaerobic naphthalene degradation by a sulfate-reducing enrichment culture. *Environ. Microbiol.* **8**:347–352.

Schaechter, M., J.L. Ingraham, and F.C. Niedhardt. 2006. *Microbe.* American Society for Microbiology Press, Washington, DC.

Schlesinger, W.H. 1997. *Biogeochemistry: An analysis of global change*, 2nd edn. Academic Press, San Diego, CA.

Silver, S. 1998. Genes for all metals – a bacterial view of the periodic table. *J. Indust. Microbiol. Biotechnol.* **20**:1–12.

So, C.M., C.D. Phelps, and L.Y. Young. 2003. Anaerobic transformation of alkanes to fatty acids by sulfate reducing bacterium strain Hxd3. *Appl. Environ. Microbiol.* **69**:3892–3900.

Spormann, A.M. and F. Widdel. 2000. Metabolism of alkylbenzenes, alkanes, and other hydrocarbons in anaerobic bacteria. *Biodegradation* **11**:85–105.

Staley, J.T. and G.H. Orians. 2000. Evolution and the biosphere. *In*: M.C. Jacobson, R.J. Charlson, H. Rodhe, and G.H. Orians (ed.) *Earth System Science: From biogeochemical cycles to global change*, pp. 29–61. Academic Press, San Diego, CA.

Stanier, R.Y., J.C. Ingraham, M.L Wheelis, and P.R. Pointer. 1986. *The Microbial World*, 5th edn. Prentice Hall, Englewood Cliffs, NJ.

Stein, L.Y. and Y.L. Yung. 2003. Production, isotopic composition, and atmospheric fate of biologically produced nitrous oxide. *Annu. Rev. Earth Planet. Sci.* **31**:329–356.

Stevenson, F.J. 1994. *Humus: Chemistry, genesis, composition, reactions*, 2nd edn. Wiley and Sons, New York.

Strous, M., J.A. Fuerst, E.H. Kramer, et al. 1999. Missing lithotroph identified as a new planctomycete. *Nature* **400**:446–449.

Strous, M. and M.S.M. Jetten. 2004. Anaerobic oxidation of methane and ammonium. *Annu. Rev. Microbiol.* **58**:99–117.

Strous, M., J.G. Kuenen, J.A. Fuerst, M. Wagner, and M.S.M. Jetten. 2002. The anammox case: a new experimental manifesto for microbiological ecophysiology. *Antonie van Leeuwenhoek* **81**:693–702.

Strous, M., E. Pelletier, S. Mongenot, et al. 2006. Deciphering the evolution and metabolism of an anammox bacterium from a community genome. *Nature* **440**:790–794.

Sutton, R. and G. Sposito. 2005. Molecular structure in soil humic substances: the new view. *Environ. Sci. Technol.* **39**:9009–9015.

Tiedje, J.M. 1988. Ecology of denitrification and dissimilatory nitrate reduction to ammonia. *In*: A.J.B. Zehnder (ed.) *Biology of Anaerobic Microorganisms*, pp. 179–244. Wiley and Sons, New York.

Van Hamme, J.D., A. Singh, and O.P. Ward. 2003. Recent advances in petroleum microbiology. *Microbiol. Molec. Biol. Rev.* **67**:503–549.

Vila-Costa, M., R. Simo, H. Harada, J.M. Gasol, D. Slezak, and R.P. Kiene. 2006. Dimethylsulfoniopropionate uptake by marine phytoplankton. *Science* **314**:652–654.

Wackett, L.P., A.G. Dodge, and L.B.M. Ellis. 2004. Microbial genomics and periodic table. *Appl. Environ. Microbiol.* **70**:647–655.

Wackett, L.P. and C.D. Herschberger. 2001.

Biocatalysis and Biodegradation. American Society for Microbiology Press, Washington, DC.

Wackett, L.P., W.H. Orme-Johnson, and C.T. Walsh. 1989. Transition metal enzymes in bacterial metabolism. *In*: T.J. Beveridge and R.J. Doyle (eds) *Metal Ions and Bacteria*, pp. 165–246. Wiley and Sons, New York.

Welsh, D.T., G. Castadelli, M. Bartoli, et al. 2001. Denitrification in an interstitial seagrass meadow, a comparison of ^{15}N-isotope and acetylene-block techniques: dissimilatory nitrate reduction to ammonia as a source of N_2O? *Marine Biol.* **139**: 1029–1036.

Widdel, F. and R. Rabus. 2001. Anaerobic biodegradation of saturated and aromatic hydrocarbons. *Curr. Opin. Biotechnol.* **12**:259–276.

Wigley, T.M.L. 2006. A combined mitigation/geoengineering approach to climate stabilization. *Science* **314**:452–454.

Wrage, N., G.L. Velthof, M.L. van Beusichem, and O. Oenema. 2001. Role of nitrifier denitrification in the production of nitrous oxide. *Soil Biol. Biochem.* **33**:1723–1732.

Xiao, S., D.W. Kicklighter, J.M. Melillo, A.D. McGuire, P.H. Stone, and A.P. Sokolov. 1997. Linking a global terrestrial biogeochemical model and a 2-dimensional climate model: implications for the global carbon budget. *Tellus* **49B**:18–37.

Yock, D.C. 2002. Dimethylsulfonioproprionate: its sources, role in the marine food web, and biological degradation to dimethylsulfide. *Appl. Environ. Microbiol.* **68**:5804–5815.

Zabel, R.A. and J.J. Morrell. 1992. *Wood Microbiology: Decay and its prevention*. Academic Press, San Diego, CA.

Zahrl, D., M. Wagner, K. Bischof, et al. 2005. Peptidoglycan degradation by specialized lytic transglycosylases associated with type III and type IV secretion systems. *Microbiology* **151**:3455–3467.

Zehnder, A.J.B. and T.D. Brock. 1979. Anaerobic methane oxidation: occurrence and ecology. *Appl. Environ. Microbiol.* **39**:194–204.

FURTHER READING

IPCC (Intergovernmental Panel on Climate Change). 2007. *Climate Change* 2007. Fourth assessment report of the IPCC. IPCC Secretariat, Geneva. http://www.ipcc.ch

8

Special and Applied Topics in Environmental Microbiology

As stated in Chapter 1, the core discipline of environmental microbiology centers on a "house-like" structure in which "evolution" is the base, "thermodynamics" and "habitat diversity" are the walls, and in which "physiology" and "ecology" serve as the two roof pieces. The house has gradually been built in Chapters 1–7. Here, we reach from the discipline's core foundations to special and applied topics.

First and foremost in extending environmental microbiology's core is the need to recognize other organisms (both microorganisms and higher eukaryotes) as significant biogeochemical and evolutionary forces for life on Earth. Striking evolutionary adaptations reflecting relationships between organisms have developed. In this chapter, we cover fundamental principles of ecological relationships and then focus on microorganisms' relationships to plants and humans. Furthermore, because the major environmental microbiological questions (who? what? when? where? how? and why?) have been asked (see Chapters 1–7), we can now use the answers and the methods of inquiry to address important topics in engineering (biodegradation and bioremediation), natural history (biofilms, evolution of metabolic pathways), biotechnology (energy production, genetically engineered crops), and medicine (emerging human pathogens, antibiotic resistance). Case studies are featured prominently in the sections focusing on the evolution of metabolic pathways and environmental biotechnology.

8.1 OTHER ORGANISMS AS MICROBIAL HABITATS: ECOLOGICAL RELATIONSHIPS

Resource exploitation has been a major theme of this book. By understanding the materials and conditions that constitute selective pressures in a given habitat, we can predict and discover novel microbial processes and the organisms that carry them out. During early planetary history, prokaryotes were the sole forms of life (see Chapter 2) and the key selective pressures were geochemical ones found in ancient oceans (Figure 8.1, bottom).

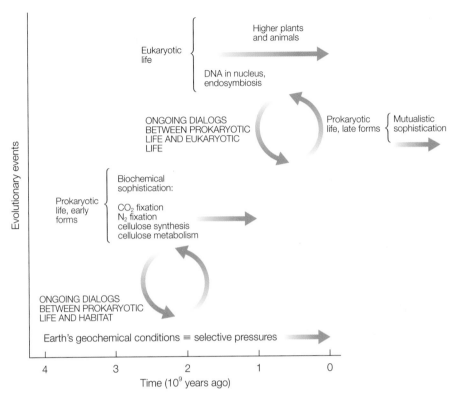

Figure 8.1 Sequence of evolutionary events leading to key mutualistic interactions between prokarytoic and eukaryotic forms of life.

As large, complex eukaryotic creatures (especially plants and animals) developed, these life forms, themselves, became new habitats that offered exploitable resources (Figure 8.1, top). It is clear, then, that *inner and outer compartments of other organisms* have played important roles in the evolution of members of the microbial world. In Chapter 7 (especially Section 7.3, Table 7.4), we focused upon resources available in *deceased biomass*. Here in this section, we develop an overview of the many types of mutual adaptations that can arise between microorganisms and their *viable partners*.

The spectrum of ecological relationships between organisms

Because prokaryotes preceded eukaryotes as Earth's inhabitants, eukaryotes have always had prokaryotic partners (Figure 8.1). The relationships between organisms span a broad range and the impacts on one partner can be very different from those on the other partner. Figure 8.2 provides an overview of the types of effects that organisms can have upon one another (from neutral to positive and neutral to negative). Ecologists have developed specific terminologies for describing the degree of positive or negative interaction and which partner is impacted.

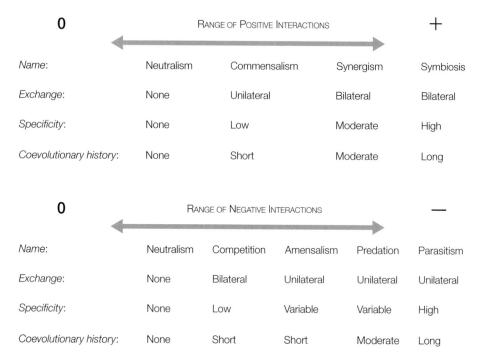

Figure 8.2 Spectrum of ecological interactions between two organisms. Shown are eight key categories of interaction and the degree of evolutionary dialog for each.

At the mid-point between the extremes of positive and negative interactions is "neutralism" (listed twice in Figure 8.2, on the left side). Neutralism represents the state in which organisms fail to interact with one another. From a microbial ecology perspective, neutralism is a rare situation that occurs in new, barely colonized habitats (such as recently emerged volcanic islands), or in habitats whose nutrient contents are so low that slow metabolism or dormancy predominates as a lifestyle. Box 8.1 provides information that elaborates upon the eight categories of ecological relationships that extend from symbiosis (also known as mutualism, at the positive extreme) to parasitism (at the negative extreme). Competition, amensalism, and predation are the intermediary categories of negative interaction. Commensalism and synergism (also known as protocooperation) are the intermediary categories of positive interaction. Key aspects of all eight categories include the type of exchange (one way or two way), the specificity of the relationship, and the extent of shared coevolutionary history of the organisms (Figure 8.2). In general, stable, nutrient-rich habitats are the sites where diverse, highly specialized biota are likely to evolve. These stable habitats are also sites where coevolution of symbiotic and parasitic ecological relationships are likely to occur.

Box 8.1

Eight categories of ecological relationships ranging from positive (symbiosis) to negative (parasitism) with characteristics and examples

Type of interaction	Category	Characteristics	Examples/habitats
None	Neutralism	Lack of interactions; low population densities; low growth rates; aliens in foreign habitats	Deserts, atmosphere, snow, ice, glaciers
Positive	Commensalism (mesa = table = table scraps; low specificity)	One-way positive exchange; one organism benefits from another's "table scraps"	Algae and plants leak carbon (photosynthate) to heterotrophs; methanogens provide methane to methanotrophs
	Synergism (also "proto-cooperation"; moderate specificity)	Two-way positive exchange; mutually beneficial to both partners	An alga provides carbon to epiphytic heterotrophs that synthesize a B vitamin required by the alga; hydrogen consumption by methanogens drives fermentation reactions in anaerobic food chains
	Symbiosis (also "mutualism"; high specificity)	Two-way positive exchange that allows a new biological trait to be expressed; this may be an obligate relationship; symbionts live singly in a different way than when together	*Rhizobium*–legume symbiosis, lichens, corals, ruminant animals, hydrothermal vent animals
Negative	Competition (low specificity)	Organisms compete for nutrient-limiting resources	In soil, sediment, and waters; many nutrients (e.g., C, N, P, Fe), electron donors, and electron acceptors may be limiting
	Amensalism (variable specificity)	One-way negative exchanges; one organism is antagonistic to another, often by releasing inhibitory chemicals	Antibiotic compounds from one microbe may prevent growth of a neighbor; in wine, ethanol produced by yeast inhibits bacterial growth; organic acids released by some anaerobes inhibit growth of adjacent populations
	Predation (variable specificity)	Big eats small; prey is engulfed; feeding is abrupt	Density-dependent feeding by protozoa on bacteria in soil and sediments; nematode-trapping fungi in soil; *Bdellovibrio* bacterium that attacks, penetrates, and consumes other bacteria
	Parasitism (high specificity)	Little debilitates big over prolonged period; parasite may require host (obligate) or be able to survive in absence of host (facultative); relationship may be balanced or destructive	Virus infections of microorganisms and animals

A survey of symbioses and their characteristics

There are at least three main questions to answer in attempting to understand symbioses:

1 What are their key features in terms of participants and resources exchanged?
2 How did symbioses develop over evolutionary time?
3 What are the detailed genetic and biochemical signaling pathways that allow contemporary symbioses to form and be maintained?

Only the first of these three questions will be addressed extensively in this section. Regarding the evolutionary development of symbioses, it is reasonable to speculate that an initially commensalistic relationship might later become synergistic and then symbiotic (see Figure 8.2 and Box 8.1). This progression from casual exchange to highly specialized, even obligate, metabolic function is likely to have evolved over time via a series of genetic and biochemical adaptations of one organism to the other. Contemporary symbiotic systems are endpoints of evolution that exhibit sophisticated biochemical signaling pathways that lead to the establishment of successfully functioning symbioses (Hoffmeister and Martin, 2003). Analogous to our own immune system, a host of a symbiotic relationship must recognize the partner – distinguishing the partner from other microorganisms that may be invaders, pathogens, or parasites. A glimpse into the biochemical and genetic intricacies of the early steps of one symbiosis (*Rhizobium*–legume, described below) is presented in Box 8.2. Signaling molecules, known as "nod factors", are produced by the symbiotic bacteria. These interact specifically with membrane protein receptors at the surface of legume roots and the interaction causes structural alterations of root hairs, eventually leading to the formation of nitrogen-fixing root nodules.

At least three key physiological traits make some prokaryotes desirable symbiotic partners (Figure 8.1):

1 Autotrophy – especially CO_2 fixation linked to chemolithotrophic metabolism, such as sulfide oxidation (for background, see Sections 3.3 and 7.4).
2 Nitrogen fixation – a uniquely prokaryotic trait that is predominantly expressed at very low oxygen concentration and converts atmospheric N_2 to ammonia and amino acids (for background, see Sections 7.4 and 7.5).
3 Digestion of complex high molecular weight organic compounds – particularly cellulose and other plant polysaccharides (for background, see Section 7.3).

One or more of these traits contributes to the (often obligate) symbiotic relationships formed among microorganisms and between microorganisms and animals. By merging their genetic and biochemical traits, the two symbiotic partners achieve physiological abilities and remarkable lifestyles that neither can achieve alone.

Box 8.2

Early biochemical and genetic details of recognition and root-hair curling essential for the successful establishment of *Rhizobium*–legume symbiosis

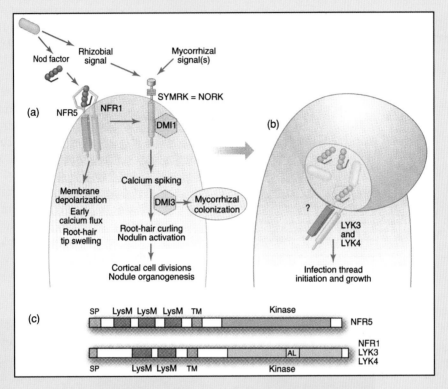

Figure 1 (a) Symbiotic rhizobial bacteria secrete Nod factors that are perceived by two linked transmembrane protein receptors (referred to as NFR1 and NFR5). The activated NFR1–NFR5 receptor initiates rapid calcium influx and swelling of root-hair tips, which are early events in the plant symbiotic response that may be specific for bacterial symbionts. Activation of the receptor is also required to activate another protein receptor complex (NORK–DMI1), which results in plant symbiotic responses to both bacterial and fungal symbionts. The NORK–DMI1 complex may also be involved in direct recognition of rhizobial and mycorrhizal signals. (b) Rhizobial bacteria entrapped in a curling root hair. In order for rhizobia to enter root hairs and to initiate the formation of infection threads and nodulation, the Nod factors they release must be recognized by highly specific plant receptors. Another set of membrane protein receptors (LYK3 and LYK4) of the plant host are involved in Nod factor recognition. (c) Schematic map of the many functional domains within the protein receptors (NFR5, NFR1, LYK3, and LYK4) in the nitrogen-fixing plant hosts. NFR5 has three LysM domains and its kinase domain does not contain an activation loop. NFR1, LYK3, and LYK4 have two LysM domains, and their kinase domains contain an activation loop. AL, activation loop in the kinase domain; SP, signal peptide; TM, transmembrane domain. (From Cullimore, J. and J. Denarie. 2003. Plant sciences: how legumes select their sweet talking symbionts. *Science* **302**:575–578. Reprinted with permission from AAAS.)

Table 8.1 provides an overview of many of the known types of symbiotic relationships, and their essential characteristics. Entries 1 and 2 of Table 8.1 remind the reader that the ability to hydrolyze cellulose is lacking in all animals higher on the evolutionary scale than mollusks [with the exception of silverfish (*Lepisma lineata*); Schlegel and Jannasch, 2006]. Cellulose synthesis evolved in prokaryotes (Nobles and Brown, 2004), therefore they are the key agents of cellulose biodegradation (see Figure 8.1). Our world would be a very different place without the cellulose metabolism carried out by anaerobic microbial communities in specialized organs in the gastrointestinal tract of animals. These communities are essential for the animals' normal nutrition which absorbs organic acid fermentation products through intestinal walls. If these microbial communities vanished, the global biogeochemical cycling of carbon would be drastically inhibited and many animals critical for human nutrition and commerce (cows, sheep, goats, horses, camels) would starve.

The next four entries in Table 8.1 focus on the soil habitat. For ~50 million years, tropical ants of the genus *Atta* have delivered leaf fragments to fungal gardens (entry 3; Figure 8.3). In these gardens, the fungi digest the leaf matter and the growing fungal biomass serves as food for the ants. An additional dimension of the symbiosis is the cultivation, in cavities within the ants, of bacteria that produce antibiotics used by the ants to keep parasites of the fungal garden at bay (Currie et al., 2006). The

Figure 8.3 Photograph of an ant of the genus *Atta* in its fungal garden. (Courtesy of Alex Wild Photography, www.myrmecos.net, with permission.)

Table 8.1

Examples of symbioses and their key characteristics

Entry	Habitat	Type of symbiosis	Partner A	Partner B	Beneficial resources exchanged between partners	References
1	Gastrointestinal tracts of animals	Ruminants (cows, sheep, goats, camels, giraffes) whose stomachs are extended at the entry point (foregut)	Herbivorous animal host with large stomach compartment known as the rumen. The animal is unable to digest cellulose and many other plant polysaccharides	Microbial community of anaerobic bacteria and protozoa	The animal provides a warm, well-mixed habitat periodically replenished with plant material. The microorganisms ferment the plant material into fatty acids absorbed through the animal stomach wall	Schlegel and Jannasch, 1992; Russell and Rychlik, 2001
2	Gastrointestinal tracts of animals	Animals whose stomachs are modified near the exit point (hindgut). These include vertebrates (horses, pigs, rats, rabbits) and invertebrates (termites, wood roaches)	Herbivorous animal host whose intestinal tract features extended hindgut compartments (blind sacs, "ceca")	Microbial community of anaerobic bacteria and protozoa	The animal provides a warm, well-mixed habitat periodically replenished with plant material. The microorganisms ferment the plant material into fatty acids absorbed through the animal stomach wall	Schlegel and Jannasch, 1992
3	Soil	Ant–fungus	Attine leaf-cutting ants of the tropics (e.g., genus *Atta*)	Fungi in the family *Lepiotaceae*	The ants carefully create, maintain, and disseminate fungal gardens that serve as the ant's main food source. The fungi are fed and digest leaf material delivered by the ants; they also protect the ants from competitors and parasites	Currie, 2001; Currie et al., 2003, 2006

Table 8.1 *Continued*

Entry	Habitat	Type of symbiosis	Partner A	Partner B	Beneficial resources exchanged between partners	References
4	Soil	*Rhizobium*–legume	Leguminous plants, ranging from arctic annuals to tropical trees	Bacteria of the genus *Rhizobium*, *Bradyrhizobium*, and *Azorhizobium*	Specialized organs (nodules) form on roots and stems of plants. These are composed of plant and bacterial tissue. Photosynthetic carbon from the plant is exchanged for fixed nitrogen from the bacterium	Van Rhijn and Vanderleyden, 1995; Perrett et al., 2000
5	Soil	Mycorrhizae (arbuscular; endo)	Most vascular plants, including representatives of angiosperms, gymnosperms, and bryophytes	Many soil fungi (especially *Glomus*)	Fungi inhabit root cortical cells and obtain carbon from the plant. Fungi deliver mineral nutrients from soil to the root cortical cells	Harrison, 2005
6	Soil	Mycorrhizae (ecto)	Many trees and shrubs of temperate regions	Many soil fungi, including *Basidiomycetes* and *Ascomycetes*	Resource exchange is like that of arbuscular mycorrhizae; however, in ectomycorrhizae, the root is enveloped in a fungal network; root tissue cells are surrounded, not penetrated	Martin et al., 2001
7	Plant, rock, and soil surfaces	Lichens	Fungi	Algae (often cyanobacteria)	Close interweaving of fungal and algal tissues. The fungi provide a physical anchor, mineral nutrients, and desiccation resistance. The algae provide carbon (through photosynthesis) and nitrogen (if a cyanobacterium)	Nash, 1996; Adams et al., 2006

8	Tropical wetlands	Azolla–Anabaena	Azolla (a fern)	N$_2$-fixing cyanobacterium, Anabaena	The small floating fern provides cavities within its leaves that serve as a habitat for the cyanobacteria. The cyanobacterium fixes nitrogen in this intracellular association	Adams et al., 2006
9	Rice plants	Fungus–bacterium in rice-blight disease	Rhizopus, plant pathogenic fungus that attacks rice	Intracellular bacterium	Fungus long-thought to be responsible for the plant disease is pathogenic because endosymbiotic bacteria synthesize rhizoxin	Partida-Martinez and Hertwick, 2005
10	Ocean hydrothermal vents	Riftia worms	Riftia worms that lack a gastrointestinal tract	Sulfide-oxidizing chemolithoautotrophic bacteria	Specialized organ in worm is colonized by the bacteria. Bacteria receive shelter and H$_2$S; the worm receives fixed carbon	Jeanthon, 2000; Cavanaugh et al., 2006
11	Ocean hydrothermal vents	Large clam (Calyptogena)	Bivalve clam with gills colonized by H$_2$S-oxidizing bacteria	Sulfide-oxidizing chemolithoautotrophs	Endosymbiont bacteria provide carbon and energy. Host provides shelter and transport of resources	Jeanthon, 2000; Cavanaugh et al., 2006
12	Coral reefs	Coral reef-forming animals	Sea anemone-like anthozoans that secrete protective exoskeletons of calcium carbonate	Unicellular dinoflagellate marine algae, Symbiodinium	Anthozoans provide shelter for algal symbionts whose photosynthetic activity supports growth and maintenance of the host	Baker, 2003

Table 8.1 Continued

Entry	Habitat	Type of symbiosis	Partner A	Partner B	Beneficial resources exchanged between partners	References
13	Ocean waters	Luminous squid plus *Vibrio fischeri* bacteria	Squid, *Euprymna scolopes*, with bioluminescent light organ on head	*Vibrio fischeri*, a bioluminescent bacteria	Specialized cavity provides housing and nutrients for bacteria whose ability to produce light assists the squid in avoiding predators	Ruby, 1996, 1999; Nyholm and McFall-Ngai, 2004; Lupp and Ruby, 2005
14	Ocean floor, whale fall	*Osedax* bone-eating worm	Mouthless, gutless polychaete worm that decomposes whale bones	*Oceanospirillales*, bacteria	Adult worms possess elaborate posterior root-like extensions that penetrate bones. Bacteria are thought to digest collagen- and cholesterol-rich tissues in exchange for shelter	Goffredi et al., 2005
15	Marine invertebrates	Shipworms	Wood-boring marine bivalve worms of the family *Teredinidae*	Cellulytic, nitrogen-fixing bacteria, *Terredinibacter turnerae*	Worms burrow into and consume floating or submerged wood; bacteria that colonize worm gills receive cellulose and shelter in exchange for cellulose digestion and amino acids	Felbeck and Distel, 1992; Distel et al., 2002

16	Insects	Intracellular obligate endosymbioses of internal structures, termed "bacteriocytes" associated with gut, ovary, and other insect organs	Several large orders of insects including aphids, psyllids, whiteflies, mealy-bugs, tsetse flies, weevils, carpenter ants, and termites	Many are members of γ- and β-Proteobacteria	The bacteria receive food and shelter; they dwell within clusters of structures, termed "bacteriocytes" inside the insect. Specific functions of insect–endosymbiont partnerships remain unknown in most cases. Main benefit to hosts is probably correction of dietary imbalances by synthesis of growth factors such as amino acids. Other benefits may include resistance to parasitism and broader food source utilization	Wernegreen, 2002; Scarborough et al., 2005; Baumann et al., 2006
17	Protozoa	Protozoa–intracellular symbiont	Many types of amoebae, flagellates, and ciliates	Many types of bacteria	By colonizing the interior of protozoa, bacteria generally receive food and shelter. If the bacteria give nothing back in return, they are parasites. However, positive effects range from synthesis of growth factors to providing electron sinks for anaerobic energy metabolism, to conferring selective advantages such as defense against predators	Horn and Wagner, 2004; Görtz, 2006

Figure 8.4 Photograph of the root of a pine seedling showing an ectomycorrhizal fungal sheath formed by *Amanita muscaria*, a Basidiomycete. Magnification approximately ×24. (Courtesy of R. Molina, US Forest Service, with permission.)

Rhizobium–legume symbiosis (entry 4 of Table 8.1, see also Box 8.2) is a remarkably intricate cooperative effort between bacteria and plants: within specialized nodules, oxygen tension is buffered at very low levels, allowing expression of bacterial nitrogen-fixation genes and delivery of fixed nitrogen to the plant in exchange for refuge and carbon.

Mycorrhiza is a term literally meaning "fungus root". This commingling of tissues from two organisms, the mycorrhizal symbiosis, is an essential feature of the biology and ecology of the majority (>80%) of terrestrial plants. In this mutually beneficial relationship, the propensity of fungal hyphae to explore and exploit soil nutrients augments the plant root's own role in nutrient and water uptake. Furthermore, the fungal endosymbiont renders the plant less susceptible to infection by pathogenic fungi. In return for these services, the fungus receives plant photosynthate as a carbon source. When examined under a microscope, mycorrhizae exhibit two major morphological types. In the arbuscular form (also known as "endo", entry 5 of Table 8.1), the fungal filaments occur as inter- and intracellular coils and tree-like "arbuscules", largely within the body of the root. By contrast, the ectomycorrhizal morphology found on roots of many coniferous trees (entry 6; Figure 8.4) features an extensive hyphal sheath that coats the exterior of the root surface. Other subclasses of the symbiosis

Figure 8.5 Photograph of the small water fern, *Azolla*, that houses within its tissue the nitrogen-fixing endosymbiont *Anabaena*. (Courtesy of Kurt Snyder and Wikipedia, with permission.)

are recognized based largely upon the taxonomy of the host plant – especially orchyd, ericoid, arbutoid, and monotropoid mycorrhiza (Martin et al., 2001).

Fungi also play a vital role in lichen symbiosis (Table 8.1, entry 7), which relies upon intimate physical and nutritional linkages between the fungal heterotroph and a photosynthetic autotroph (either eukaryotic algae or prokaryotic cyanobacteria). Together, the two lichen partners are able to colonize harsh, dry, low-nutrient habitats (rocks, tree surfaces, facades of buildings) – cooperatively growing and reproducing by scavenging water, light, and minerals. The symbiosis between the tropical fern (*Azolla*) and cyanobacterium (*Anabaena*) (Table 8.1, entry 8; Figure 8.5) has been exploited by rice farmers for centuries as a way to boost the nitrogen status of rice paddies. The final plant-related entry in Table 8.1 (entry 9), is the recently discovered relationship between an intracellular bacterium that colonizes the plant-pathogenic fungus, *Rhizopus*, responsible for rice blight disease. Remarkably, the key toxin that confers virulence on the fungal pathogen is synthesized by the endosymbiotic bacterium. This new information invites new strategies for controlling the plant disease: inhibit the bacterium that inhabits the fungus.

Figure 8.6 Photograph of the squid, *Euprymna scolopes*, that features a light organ colonized by the bacterium *Vibrio fischeri*. (Courtesy of C. Frazee and M. McFall-Ngai via E. G. Ruby, University of Wisconsin, with permission.)

The next six entries in Table 8.1 make it clear that marine habitats support a wide diversity of symbiotic relationships. On the dark ocean floor where hydrothermal vents release hot sulfide-rich water, there are oases of life supported by sulfide-oxidizing chemo-lithotroph bacteria. Two important animal species that flourish in hydrothermal vent regions only do so because they have specialized organs (Table 8.1, entries 10 and 11) that deliver shelter and sulfide to sulfide-oxidizing, autotrophic bacterial endosymbionts – primarily in exchange for carbon. In the shallow photic zones of tropical oceans, photosynthetic algae (dinoflagellates of the genus *Symbiodinium*) play the role for coral-forming animals that sulfide autotrophs do for hydrothermal vent animals. Carbon fixed by coral-inhabiting algae (at a density as high as 10^{10} algal symbionts per m^2 of coral surface) allows the coral animals (sea anemone-like Anthozoans) to grow and secrete calcium carbonate exoskeletons that accrue and provide crucial habitats for other forms of marine life (entry 12 of Table 8.1). In shallow tropical waters near Hawaii, a squid (*Euprymna scolopes*, Figure 8.6) supports an entirely different type of symbiosis: the exchange of a sheltered habit within the squid's body for an endosymbiotic bacterium's ability to glow (entry 13). The luminescent marine bacterium *Vibrio fischeri* is specifically recognized by the host and is adapted to colonize the squid's light organ, which can eliminate shadows cast by the squid, enabling it to evade nocturnal predatory attack from below.

The next two entries in Table 8.1 (entries 14 and 15) provide additional examples of how the prokaryotic digestive traits can be exploited by animal hosts. On the ocean floor, deceased whale biomass ("whale falls") provide a banquet of resources for heterotrophic life, both microbial and animal. In these whale falls, bone marrow is an unusual resource – rich in cholesterol and collagen and encased in the bone mineral, hydroxyapatite. A worm of the genus *Osedax* (Figure 8.7), adapted to burrow into the bone marrow, relies upon bacterial endosymbionts to digest marrow-derived carbon compounds. Similarly, shipworms (entry 15) bore into submerged wooden materials, but without their nitrogen-fixing, cellulose-degrading bacterial endosymbionts, ship-worms would not be successful.

Figure 8.7 Photograph of the bone-eating marine worm, *Osedax frankpressi*, that relies on bacterial endosymbionts for digestion of bone marrow constituents. The image shows two red-and-pink-tufted worms in a whale vertebra with their green roots and white ovisacs exposed. (Courtesy of G. Rouse, Scripps Institute of Oceanoagraphy, with permission.)

The penultimate entry in Table 8.1 lists insects as the habitat for symbioses. *Bacteria*-containing "bacteriocyte" structures within insect tissues have been microscopically detected and catalogued for more than four decades (Baumann et al., 2006). As many as 10% of some insect families are inhabited by obligate symbiotic bacteria that have not yet been cultivated in the absence of their insect hosts. The relationships between aphids (plant sap-ingesting insects; Figure 8.8) and their endosymbiotic bacteria (*Buchnera*) have been particularly well characterized using non-culture-based molecular procedures, including cloning and sequencing of 16S rRNA genes. The phylogeny of the endosymbiotic aphid bacteria perfectly matches that of the aphid hosts. This provides strong evidence for strict transmission of the endosymbionts from generation to generation of aphid progeny with no influx of genes from other bacteria throughout the 150–250 million years of the evolution of the symbiosis. The implication is that all contemporary aphid–bacterial symbioses are derived from what may have been a single infection by a free-living bacterium long

Figure 8.8 Photograph of plant sap-ingesting aphids, which harbor the bacterial endosymbiont *Buchnera*. (Courtesy of Wikipedia, GNU Free Documentation License.)

ago. As discussed in Sections 3.2 and 5.9, endosymbionts and intracellular parasites are likely to shed genes not required in their new habitats. Indeed, the genome of *Buchnera aphicola* APS [655,725 base pairs (bp)] is one-tenth the size of many free-living bacteria. Physiological, biochemical, developmental, and genetic details of the many types of insect–bacterium relationships (entry 16) await elucidation by ongoing research. The general consensus is that the insect endosymbionts benefit their hosts nutritionally (by synthesizing amino acids and growth factors) and also by conferring resistance to parasites and extending the insect's range of habitats by extending the insect's utilizable food sources.

The final entry in Table 8.1 presents protozoa as habitats that have been colonized by bacteria over evolutionary time. Bacterial endosymbionts are very common in amoebae, flagellates, and ciliates (Figure 8.9). Such associations were detected microscopically more than a century ago. Interest in the bacterial endosymbionts of protozoa is spurred for a variety of reasons that include both evolutionary and medical issues. Because protozoa reside in the lower trunk of the *Eukarya* in the tree of life (see Sections 5.5 and 5.6), our understanding of the development of cellular organelles (e.g., mitochondria, chloroplasts) can be advanced by studying protozoa. Furthermore, there is a continuum of relationships exhibited by intracellular residents of protozoa (from pathogens and parasites to symbionts and organelles) that may provide important insights into the emergence of new microbial agents of human disease. The physiological impact of many intracellular protozoan residents have not yet been discovered; while others have definitively been shown to benefit their host through mechanisms that include: the synthesis of growth factors (e.g., heme, lysine), serving as an electron sink in the hosts' anaerobic metabolism, and conferring resistance to predators.

8.2 MICROBIAL RESIDENTS OF PLANTS AND HUMANS

In an insightful early essay entitled, "Eukaryotes as habitats for bacteria", Schlegel and Jannasch (2006) pointed out that "eukaryotes present a multitude of habitats for bacteria . . . the surfaces, cavities, crevices, and intercellular spaces open to the air, as well as intestinal tracts, exudates, and excretory substances offer opportunities for the growth of many bacteria". Earlier in this book, we obtained a preview of how important other life forms can be as microbial habitats: Table 4.9 (Section 4.5) assessed microbial biomass associated with gastrointestinal tracts of animals. Here we briefly survey plants and humans as habitats for microorganisms.

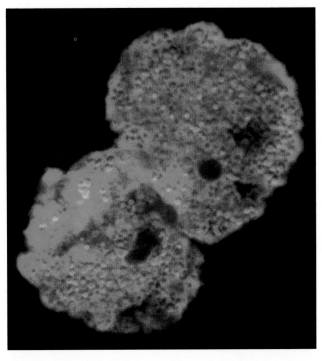

Figure 8.9 Photograph of a protozoan and its intracellular endosymbionts visualized using fluorescent in situ hybridization (FISH) and confocal laser scanning microscopy. The fluorescent probes were designed to specifically target the protozoan host, *Acanthamoeba* (green signals), and the intracellular bacteria, *Protochlamydia amoebophila* (red signals), respectively. Scale of protozoan cell, ~30 μm. (From Horn, M., A. Collingro, S. Schmitz-Esser, et al., 2004. Illuminating the evolutionary history of Chlamydiae. Science **304**:728–730. Copyright 2004, AAAS/Science. Courtesy of M. Horn, Universitat Wien, Austria.)

Plants

Plants consist of two primary microbial habitats: the phyllosphere and the rhizosphere (Figure 8.10). The phyllosphere is defined as the aerial portions of plants (trunk, branches, stems, buds, flowers, leaves); phyllosphere inhabitants are termed epiphytes. The majority of information about microbial epiphytes has focused upon the leaf habitat where bacterial

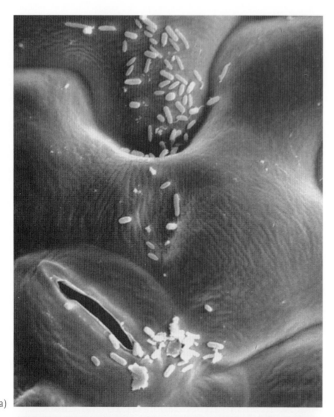

(a)

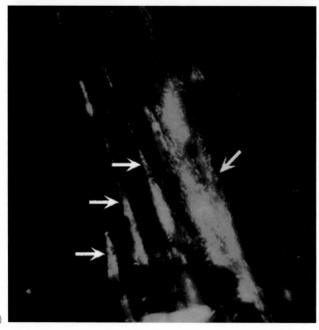

(b)

densities often average 10^6–10^7 per cm^2 (Lindow and Brandl, 2003). Given the extensive standing global plant biomass (see Sections 4.1 and 7.6), planetary phyllosphere bacterial populations are impressive – having been estimated at ~10^{26} cells (Lindow and Brandl, 2003). Thus, the microbial phyllosphere has the potential to influence many processes globally – as well as individual plants, locally.

Like all other microbial habitats, understanding the phyllosphere requires that we pose several basic questions:

1 What conditions and resources prevail in the phyllosphere?
2 What organisms occur?
3 What processes occur?
4 What is the motivation for microbiological inquiry?

These four questions are addressed by information in Table 8.2.

Figure 8.10 (a) Microscopic image of microorganisms found on the leaves of a bean plant. (b) Biofilm of fluorescently labeled bacteria on the root of a tomato plant. A large microcolony of bacteria is apparent on the root surface and is indicated by the yellow arrow. The white arrows highlight three smaller colonies that have formed at plant root cell boundaries, which may be the site of release of root exudates used by bacteria as nutrient sources. ((a) Courtesy of G.A. Beatttie, Iowa State Univeristy, with permission. (b) From Davey, M.E. and G.A. O'Toole. 2000. Microbial biofilms: from ecology to molecular genetics. *Microbiol. Molec. Biol. Rev.* **64**:847–867. With permission from the American Society for Microbiology.)

Table 8.2

Four fundamental ecological questions and answers about the plant phyllosphere (compiled from Lindow and Brandl, 2003)

Question	Answer
What conditions and resources prevail in the phyllosphere?	• Physical conditions: fluctuations in temperature, humidity, free-standing rainwater, ultraviolet light • Plant surfaces: waxy surface cuticle, subcuticle voids, intercellular spaces within plant stomata • Carbon sources: waxes, sloughed cells, sugars (glucose, fructose, sucrose) • Spatially heterogenous (microscale) distributions of nutrients (sugars) and micronutrients, such as iron
What organisms occur?	• Patchy, spatially heterogenous distributions of heterotrophic microorganisms and plant pathogens • Nonculture-based investigations of epiphytic microorganisms indicate higher community complexity than has been indicated from culture-based studies
What processes occur?	• Commensalistic relationships between plant and epiphytic microorganisms • Competition among heterotrophs for limited nutrients • Plant immune-like defenses against pathogenic bacteria include localized cell death (the hypersensitive response) • Plant pathogens may release virulence factors that assist in their colonization and infection • Microorganisms may also release surfactants and auxins that assist them colonizing the plant
What is the motivation for microbiological inquiry?	• Understanding the ecology of an important biosphere habitat • To use that understanding to control microbial agents of plant disease, food-transmitted human disease, and crop damage • Microbial ecological strategies can establish microorganisms that are antagonistic to plant pathogens (biological control) • Microbial ecological strategies can use competitive exclusion to: (i) curtail establishment by plant pathogens; and (ii) curtail frost damage instigated by phyllosphere bacteria that catalyze ice nucleation

The term "rhizosphere" was first used by Hiltner (1904) to indicate the zone of soil where exudates released from plant roots have the potential to influence soil microorganisms (Jones, 1998; Bowen and Rovira, 1999; Pinton et al., 2001a, 2001b; Bais et al., 2006; Cardon and Whitbeck, 2007). The physical, chemical, and microbiological complexity of the soil habitat have been discussed at length in Sections 4.2, 4.6, 5.1, 5.4, and 6.10. Soil

Table 8.3

Key types of influence and processes that occur in the plant rhizosphere (compiled from Pinton et al., 2001a; Uren, 2001)

Type of influence from root or microbe	Examples
Physical change of habitat	Growing root surface extending through soil Moisture flux toward root
Chemical change of habitat	Nutrient (e.g., N, P, K) depletion near root surface Root products: • diffusates (sugars, amino acids, organic acids, inorganic ions, oxygen, growth factors, water) • excretions: CO_2, protons, bicarbonate, ethylene • secretions: mucilage, enzymes, iron-binding siderophores, allelochemicals that inhibit other organisms • debris: root cap cells, sloughed tissues
Microbiological processes in habitat	Mycorrhizal infection Nitrogen fixation Pathogen infection Competition for nutrients among heterotrophs Interactions with soil-fauna (e.g., nematodes, insects)
Plant processes in habitat	Acquisition of nutrients (e.g., by iron uptake or phosphorus solubilization) Acquisition of water via transpiration and as modulated by mucilage release Protection against toxic agents (e.g., complexation of Al^{3+}) Protection against competition and plant pathogens (e.g., allelochemicals that inhibit other organisms) Establishment of symbiotic relationships (root exudates may guide chemotaxis by *Rhizobia* and mycorrhizae)

contains a yet-to-be understood assortment of ~10^9 microorganisms per gram, composed of tens of thousands of species. The extension and growth of a root within the soil provides plant-tissue surfaces for microbial colonization and creates microscale biogeochemical gradients that are far different from those in bulk soil (Table 8.3). Because soil is a carbon-limited habitat, it is conceptually obvious that the portion of soil colonized by plant roots (composed largely of metabolically active, growing, leaky, sometimes senescing, carbonaceous tissue) would harbor microbial populations distinctive from those in bulk soil. The following are crucial for developing an understanding of rhizosphere ecology, microbiology, and biochemistry:

- knowing the identity of materials released by the plants; and
- knowing the impact of released materials on the soil ecosystem – other plants, microorganisms, and soil fauna (such as invertebrates).

According to Uren (2001), "root products" are all of the substances produced by roots that are released into the rhizosphere (Table 8.3). All variables (e.g., plant type, soil type, plant stresses) aside, approximately 50% of the fixed carbon in photosynthate is committed to roots – half of this is retained as root tissue (25%) and the remainder is relegated to respiration (~15%), debris (~10%), diffusates (~1%), and secretions (~1%). Within Table 8.3 is a summary of the possible impacts and functional roles of root secretions for plants. In assessing the intricacies of rhizosphere processes, it is important to be aware that most compounds do not persist in soil in the free and active form for very long.

> A secretion must be free to diffuse through a portion of the rhizosphere, but a sort of tyranny of distance exists. The longer it takes or the further it must travel, the greater is the chance that it will be rendered ineffective by microbial degradation/assimilation, chemical degradation/reaction, sorption, or a combination of these processes (Uren, 2001).

Clearly, rhizosphere biology and microbiology offer fascinating challenges for future research.

Humans

Humans present several habitats for microbial colonization. Together these habitats support the "human microbiome", which consists of 10^{13}–10^{14} microorganisms. While each prokaryotic cell is small, the tally of microbial residents amounts to ~10 times the number of human cells (somatic plus germ cells) in the human body. Thus, we humans are "walking minorities" – outnumbered by the (largely) prokaryotic cells that we support externally, and, especially, internally (see below). We are born 100% human but live and die 90% microbial. If we count the number of genes, the human contribution to the total (22,287 genes in the human genome) is even smaller – amounting to ~1% of the aggregate gene pool (Gill et al., 2006). Here we focus upon the skin and gastrointestinal tract of humans.

Skin

Fundamental ecological characteristics of the human skin and its microbial inhabitants are presented in Table 8.4. The epidermal layer of our bodies serves as a protective barrier that is replaced every 28 days. The outer-surface stratum corneum consists of 25–30 layers of flattened,

Table 8.4

Four fundamental ecological questions and answers about the human skin (compiled from Roth and James, 1988; Taylor et al., 2003; Brüggemann et al., 2004; http://www.nuskin.com/corp/science/skinscience/skin_anatomy.shtml)

Question	Answer
What conditions and resources prevail in the skin habitat?	• Physical conditions: fluctuations in ventilation, desiccation, light, humidity, pH, and releases from glands [pores, sweat, sebaceous (oil production associated with hair follicles), lymph] • *Epidermis*: the outermost surface (~1 mm thick): – 50 to 100 layers of cells that (from top to bottom) include: stratum corneum, stratum lucidum, stratum granulosum, stratum spinosum, and stratum basale – key cell types are keratinocytes and melanocytes – key biomolecules are keratin and lipids – renewed every 28 days • *Dermis*: located between the epidermis and hypodermis: – fibrous network ~2 mm thick – cell types include fibroblasts, mast cells, blood vessels, and lymph vessels – key biomolecules are collagen (major protein), elastin protein, and glycoaminoglycan "ground" substances • *Hypodermis*: the deepest layer and composed largely of fat cells (for insulation)
What microorganisms?	• Major methodologies have been limited to cultivation-based procedures • Two ecological types: stable residents (endogenous) and transient residents (exogenous) • Dominant cultivated taxa include *Staphylococcus*, *Propionibacterium*, coryneforms, *Micrococcus*, and yeasts • *Propionibacterium acnes*, a normal skin inhabitant, carries several immunogenic factors thought to trigger acne skin disease
What processes occur?	• Commensalistic relationships between host and dermal inhabitants • In the underarm area (axilla) and other occluded sites, microbial metabolism of skin secretions is thought to create volatile, odorous products • Constant shedding of surface cells and keratin by host • Sebaceous glands, at the base of hair follicles, release oils and lipids that can be metabolized by microorganisms to fatty acids. At pH 5.5, these fatty acids have an antimicrobial effect on many microorganisms • If the epidermis is breached, pathogenic infections may develop. Bacterial pathogens include *Staphylococcus*, *Corynebacterium*, *Propionibacterium*, *Micrococcus*. Fungal pathogens include *Tinea versicolor*, *Tinea pedis*, and *Candida*
What is the motivation for inquiry?	• Medical treatment of dermatological disease • Commercial development of deodorant products • Ecological aspects of disease and disease transmission

dead keratinocyte cells that are continually shed by friction and are replaced by cells formed in deeper layers. Spaces between the keratinocytes are filled with epidermally produced proteins, lipids, and fatty acids. Together, the cells and filler serve as a barrier for moisture loss and against entry of foreign matter (allergens, microorganisms, chemical irritants).

Across the geography of the human body, skin conditions vary markedly in pH, moisture, density of hair follicles, gland (sweat, sebaceous, lymph) secretions, temperature, ventilation, and exposure to light (Table 8.4). When ecological conditions are altered (by covering an exposed area of skin with a bandage, washing with soap, or applying topical antibiotics), new selective pressures alter the composition of the microbial community. One key trait that distinguishes transient (exogenous) skin populations from true (endogenous) residents is adherence. Normal microbial residents and pathogens of the human skin may feature specialized surface structures, termed "adhesins" that facilitate cell attachment to collagen-rich interstitial areas of the skin (Nitsche et al., 2006).

Based on the carbon sources available to skin residents, the major ecological niche is for commensal heterotrophs (several groups of Gram-positive bacteria and some yeasts) that are able to grow on proteinaceous keratin and collagen and/or secreted oils, lipids, and/or organic acids (Table 8.4). Current information on the identity of skin microflora still relies on cultivation-based procedures. Thus, analogous to other habitats, there is good reason to expect drastic revision of our view of the identity of "normal skin residents" after 16S rRNA-based and/or metagenomic studies have been completed (see Sections 5.1, 6.7, and 6.9). Shifts in host immunity and breaches in the epidermis can allow commensal organisms to act as pathogens – as is the case for many medically important infections from the bacterial genus *Staphylococcus*. In addition to obvious medical concerns caused by bacteria and fungi (dermatomycoses), the microbial ecology of human skin impacts how humans perceive one another via body odor. It is widely accepted that underarm zones of our bodies (the axilla) accumulate initially odorless natural skin secretions that are converted to volatile odorous products by our indigenous microbial skin populations (Taylor et al., 2003). Understanding the populations and altering their physiological activity have obvious commercial applications in the personal-care industry (Table 8.4). Nearly two decades ago, Roth and James (1988) completed a review article on the microbiology of the human skin whose concluding paragraph still applies today:

> Humans exist in an environment replete with microorganisms, yet only a few of these microorganisms become residents on the skin surface. These resident flora and the skin constitute a complex ecosystem in which organisms adapt to changes in the microenvironment and to coactions among microorganisms. The skin possesses an assortment of protective mechanisms to limit colonization, and the survival of organisms on the surface lies in part in the ability of the organisms to resist these mechanisms. Microbial

colonization on the skin adds to the skin's defense against potentially pathogenic organisms. Although microbes normally live in synergy with their hosts, at times colonization can lead to clinical infection. Common infections consist of superficial infections of the stratum corneum or appendages, which can respond dramatically to therapy but commonly relapse. In rare circumstances these infections can be severe, particularly in immuno-compromised patients or hospitalized patients with indwelling foreign devices.

Gastrointestinal tract

The gastrointestinal (GI) tract of an adult human is a tube about 10 m in length that begins with the mouth and esophagus and ends with the rectum and anus. In between are organs that process the food, absorb nutrients, absorb fluids, and allow microbial colonization of processed materials so that, in the end, fecal material consists primarily of bacterial biomass. The stomach is a temporary storage vessel for ingested material where gastric juices (especially HCl) are added and where churning action converts the contents to chyme. The three segments of the small intestine are the major sites for exposing materials from the stomach to digestive fluids (from the pancreas, liver, and intestinal walls themselves) and for absorbing digested food stuffs via high surface-area villus cells that line

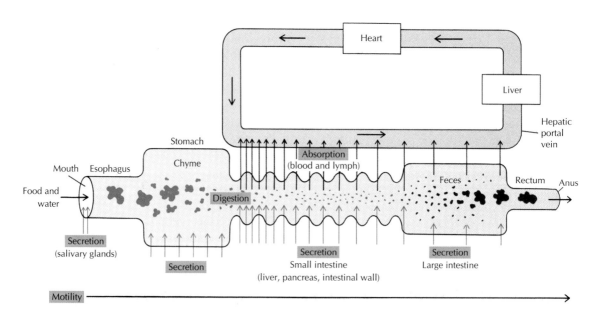

Figure 8.11 Model of the human gastrointestinal tract. (From Vander, A.J., J.H. Sherman, and D.S. Luciano. 1980. *Human Physiology: The mechanisms of body function*, 3rd edn. McGraw-Hill, New York. Reproduced with permission from the McGraw-Hill Companies.)

the small intestine walls. The absorbed nutrients are delivered to the circulatory system and liver (Figure 8.11). Undigested food residues, highly colonized by microorganisms, enter the large intestine whose bands of exterior muscles and inner-surface goblet cells dehydrate and absorb nutrients from the gut contents until they are released as feces.

It is in the large intestine (cecum, rising colon, transverse colon, sigmoid colon, plus descending colon) where a symbiosis (mutualism) analogous to those listed in Table 8.1 (entries 1 and 2, for herbivorous animals) develops for humans (Bäckhed et al., 2005). Box 8.3 provides a summary of the recently recognized benefits contributed to humans by their intestinal microflora. Also summarized in Box 8.3 are the results of two recent nonculture-based inquiries attempting to characterize the composition and metabolic function of the intestinal microflora. In terms of molecular sequencing effort (see Sections 6.7 and 6.9), the two studies were massive. In the first study, a 16S rRNA approach documented more than 13,000 taxonomic sequences from 21 intestinal or fecal samples from three healthy individuals. One striking result from the study was the fact that 1524 archaeal sequences belonged to a single group (*Methanobrevibacter smithii*). This apparently limited archaeal diversity could possibly be a result of polymerase chain reaction (PCR) bias (see Section 6.7) in the methodology. Alternatively, this result may reveal

---

Box 8.3

Recently recognized roles of the intestinal bacterial symbiosis (mutualism) in human physiology and related nonculture-based studies of intestinal microorganisms (compiled from Bäckhed et al., 2005; Eckburg et al., 2005; Gill et al., 2006)

Human benefits from intestinal microorganisms
- The distal human intestine is an anaerobic bioreactor programmed with an enormous population (10^{13} to 10^{14} cells).
- Resident microorganisms protect against injury of the human epithelial cells that line the gut.
- The resident microbial community synthesizes and releases nutrients (amino acids and vitamins) used by the human host.
- The microbial community regulates epithelial development within the gastrointestinal tract.
- The microorganisms digest and process otherwise indigestible plant polysaccharides.
- Differences in microbial community composition among healthy individuals may contribute to differences in individual human physiology and difference in susceptibility to disease.

Box 8.3 *Continued*

Synopses of innovative studies using nonculture-based procedures

Study 1
- *Goal*: Ask "Who is there"? A comparative study.
- *Strategy*: Sample several colonic mucosal sites and feces of three healthy humans (total number of samples processed = 21).
- *Method*: Extract nucleic acids. Use PCR to amplify 16S rRNA genes. Clone and sequence the biomarker genes (see Section 6.7).
- *Results*: 13,355 prokaryotic 16S rRNA gene sequences from two broad sources in the large intestine of three people.
- *Message*: 11,831 bacterial 16S rRNA gene sequences and 1524 archael 16S rRNA gene sequences; the most extensive survey to date.
  (A) The habitat is highly selective: (i) >99% of the bacterial sequences fell within only two taxonomic divisions (*Bacteriodetes* and *Firmicutes*); and (ii) all archaeal sequences belonged to a single species (*Methanobrevibacter smithii*).
  (B) The majority of sequences were novel, corresponding to uncultivated microorganisms.
  (C) Significant variability in community composition was found between human subjects and sampling sites within a given subject.

Study 2
- *Goal*: Ask "What genes are there?" by performing metagenomic functional analysis (see Sections 3.2 and 6.9).
- *Strategy*: Sample fecal DNA from two healthy individuals. Create metagenomic DNA libraries.
- *Method*: Use shotgun sequencing procedures to attempt to clone and sequence all genes. Also use PCR to amplify, clone, and sequence 16S rRNA genes.
- *Results*: ~78 × 10^6 bp of DNA were sequenced and 2062 PCR-amplified 16S rRNA genes were examined. COG (clusters of orthologous groups) and KEGG (Kyoto encyclopedia of genes and genomes) approaches were used to assess gene function (see Section 3.2).
- *Message*:
  (A) Both shotgun cloning and PCR amplification of 16S rRNA genes revealed a dominance of *Firmicutes*-related taxa. *Bacteriodetes* were largely absent.
  (B) Shotgun sequencing of functional genes revealed enriched genes for polysaccharide hydrolysis and uptake, amino acid synthesis, vitamin synthesis, and methanogenesis.

remarkable biological specificity. Another significant (and perhaps predictable) finding of study 1 in Box 8.3 was that microbial species diversity was greater than previously documented – consisting largely of novel sequences representing novel microorganisms.

In the second study described in Box 8.3, both shotgun metagenomic analysis (see Section 6.9) and PCR-based cloning and sequencing of 16S rRNA genes were utilized. One major surprise in the taxonomic survey was the absence of the *Bacteriodetes* division that was one of the two dominant groups of the *Bacteria* found in the first study (Box 8.3). By using clusters of orthologous groups (COG, see Section 3.2) and a related procedure for functionally characterizing genes (KEGG, Kyoto encyclopedia of genes and genomes), a glimpse was gained into the selective pressures that shape the composition of intestinal microbial communities. Compared to other known microbial genomes and to the human genome, the metagenome of the human intestine was enriched for polysaccharide uptake, for polysaccharide hydrolysis, for methanogenesis, and for synthesis of amino acids and vitamins. These trends in the microbial metagenome of the human intestine converge with other independent nutritional and medical investigations to support the idea that the relationship between humans and our microbiome is a symbiotic one. Certainly, the microorganisms benefit from the food and shelter we provide. In exchange, the human host benefits medically and nutritionally in ways listed at the top of Box 8.3.

8.3 BIODEGRADATION AND BIOREMEDIATION

This section is designed to familiarize a reader with definitions, principles, and facts about how microbial processes can be used to eliminate environmental pollution. We draw upon fundamentals of physiology (see Chapters 3 and 7) and apply them to technologies aimed at cleaning up both organic and inorganic environmental contaminants.

Background on environmental pollution, biodegradation, and biotransformation

There are a wide variety of industrial organic and inorganic compounds that humans manufacture, use, control, and mismanage. Many of these materials are environmental pollutants, intentionally or inadvertently released to soil, sediment, or aquatic habitats. Examples of environmental pollutants include fossil fuels such as gasoline that contains benzene, toluene, and ethylbenzenes, and xylenes (BTEX), and metals such as chromium and mercury, used in industry and commerce.

Obviously, we need to understand the environmental fate of released pollutants because they may threaten human health, ecosystem function, and/or environmental quality. Documenting the environmental fate of pollutant materials is a challenge. Not only is a mass balance-type accounting of pollutants difficult in open field sites, but many competing biotic and abiotic processes may influence pollutant behavior (see Sections 1.4 and 6.3–6.8).

The term *biodegradation* is largely synonymous with "microbial metabolism of *organic compounds*". By breaking carbon-carbon bonds in organic pollutant molecules, microorganisms are often uniquely capable of altering the chemical structure of the compounds – often rendering them harmless. For a glossary of terms pertinent to biodegradation and bioremediation, see Box 8.4. Biodegradation processes are governed by the fundamental thermodynamic and biochemical principles already discussed in Chapter 3 (Sections 3.6–3.11) and applied to the biogeochemical cycling of carbon, sulfur, and nitogen compounds in Chapter 7 (Sections 7.3–7.5). Biodegradation reactions are merely the subset of physiological reactions that act upon unusual, often synthetic, industrial organic compounds that are considered environmental pollutants.

Many *inorganic chemicals* (e.g., heavy metals, acid mine drainage, cyanide, mercury, radionuclides, and oxyanions including nitrate and perchlorate) are environmental pollutants whose chemical behavior (especially toxicity, solubility, and mobility) can be altered by microbial processes (see below). The goal of bioremediation technology is to manage biodegradation (organic materials) and biotransformation (both organic and inorganic materials) processes so that environmental pollution is attenuated or destroyed.

Environmental microbiologists frequently use *model systems* to gain physiological, biochemical, and genetic information about biodegradation and biotransformation reactions. Such information is routinely obtained

---

Box 8.4

Glossary of terms pertinent to biodegradation and bioremediation (reprinted and modified with permission from Madsen, E.L. 1991. Determining *in situ* biodegradation: facts and challenges. *Environ. Sci. Technol.* 25:1662–1673. Copyright 1991, American Chemical Society)

abiotic reactions These include all of the reactions not encompassed by biotic reactions. Included are inorganic, organic, photolytic, surface-catalzyed, sorptive, and transport processes.

biodegradation A subset of biotransformation which causes simplification of an organic compound's structure by breaking intramolecular bonds. The simplification may be subtle, involving merely a substituent functional group, or severe, resulting in mineralization.

bioremediation A managed or spontaneous process in which biological, especially microbiological, catalysis acts on pollutant compounds, thereby remedying or eliminating environmental contamination (*biorestoration* is a synonym).

biotic reactions For the purposes of this book, biotic reactions are synonymous with biotransformation reactions. However, depending on the system of interest, biotic reactions may include uptake and metabolism of organic compounds by plants and animals.

biotransformation A broad term signifying alteration of the molecular structure of a chemical by microbiological, usually enzymatic, catalysis. Alterations include those that increase (e.g., condensation reactions) and decrease (e.g., mineralization reactions) the number or complexity of intramolecular bonds. Reactions may be both intracellular and extracellular. Biotransformation of both organic and inorganic compounds may be catalyzed by microorganisms – especially oxidation/reduction reactions that may alter the mobility of inorganic compounds.

cometabolism Cometabolism is the fortuitous modification of one molecule by an enzyme which routinely acts on another (primary substrate) molecule. The primary substrate supports growth of microorganisms that produce one or more enzymes of low specificity that also act on the cometabolized substrate. The cometabolized substrate is usually altered only slightly and does not enter catabolic and anabolic pathways of the microbial cell. Therefore, the responsible organism does not benefit from cometabolic reactions. Microbial growth does not result and cometabolic reactions are not expected to accelerate. However, other organisms may be able to mineralize products of cometabolism.

in situ biodegradation In Latin, "in situ" means "in its original place". In situ biodegradation focuses on activating microbial processes for the destruction of environmental pollutants where they are found in the landscape. Mobilizing the pollutants away from the spill site for physical, chemical, or biological treatment in a reaction vessel (i.e., incinerator or bioreactor) is contrary to the definition of in situ biodegradation.

metabolic adaptation Metabolic adaptation is operationally defined as an enhancement of biodegradation potential that follows exposure of a microbial community to the organic compounds of interest. The specific mechanisms responsible for metabolic adaptation are seldom investigated, but they include enzyme induction, growth of biodegrading microorganisms, and genetic change.

microbial metabolism This term has many commonalities with "biotransformation", however the emphasis is upon an integration of physiological pathways and energy flow within an organism. Microbial metabolism is usually proven and explored using laboratory assays that contrast the alteration of organic compounds in the presence of live versus killed microbial cells. Metabolic pathways are discovered by identifying the sequential production of intermediary metabolites (and, ideally, the responsible enzymes and genes that encode those enzymes) produced as microorganisms act upon the compounds.

mineralization Conversion of an organic molecule into its inorganic constituents (e.g., CO_2, NO_3^-, SO_4^{2-}, PO_4^{3-}). Mineralization occurs when an organic compound is altered by central catabolic cellular mechanisms. The responsible organism(s) typically benefit from mineralization and reactions – thus, microbial growth is expected, and a substantial portion (~50%) of the carbon in the original organic molecule is usually incorporated into biomass.

natural attenuation Generally refers to the physical, chemical, or biological processes which, under favorable conditions, lead to the reduction of mass, toxicity, mobility, volume, or concentration of organic contaminants in soil, sediment, and/or groundwater. The reduction takes place as a result of processes such as biological or chemical degradation, sorption, and others.

by applying chemical, physiological, and genetic assays to laboratory incubations of flasks containing pure cultures of microorganisms, mixed cultures, or environmental samples (soil, water, sediment). Under controlled experimental conditions, an "abiotic control" treatment can be prepared that mimics a "live" treatment containing viable microorganisms. Contrasts in the behavior of the environmental pollutant under scrutiny are often crystal clear when comparing live and abiotic treatments: the pollutant is consumed only in the presence of viable microorganisms. In this way, the *roles of microorganisms in biodegradation and biotransformation* are readily documented (see Box 8.4).

Relationships between the technology of bioremediation and the physiological processes of biodegradation and biotransformation

Bioremediation is the intentional use of biodegradation and biotransformation processes to eliminate or attenuate environmental pollutants from sites where they have been released. Bioremediation technologies use the physiological potential of microorganisms, as documented most readily in laboratory assays (see above), to eliminate or reduce the concentration of environmental pollutants in field sites to levels that are acceptable to site owners and regulatory agencies that may be involved (Madsen, 1998).

The fundamental divisions in approaches to implementing bioremediation technology are based on two questions: "Where will the contaminants be metabolized?" and "How aggressively will site remediation be approached?" Regarding location, microbial processes destroy, immobilize, or detoxify environmental contaminants in situ, where they are found in the landscape, or ex situ, which requires that contaminants be mobilized into some type of containment vessel (a bioreactor) for treatment (Figure 8.12). Regarding aggressiveness, *intrinsic bioremediation* is passive – it relies on the innate capacity of microorganisms present in field sites to respond to and metabolize the contaminants. Because intrinsic bioremediation occurs in the landscape where both indigenous

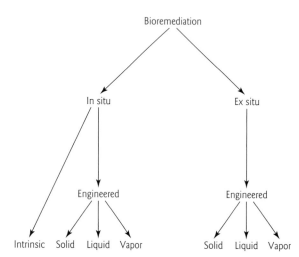

Figure 8.12 Overview of bioremediation approaches. Categories are based, respectively, on where remediation will occur (in situ versus ex situ), on how aggressively remediation is pursued (engineered versus intrinsic) and on the status of the treatment system: solid-, liquid-, or vapor-phase treatments. (From Madsen, E.L. 1998. Theoretical and applied aspects of bioremediation: the influence of microbiological processes on organic compounds in field sites. *In*: R. Burlage, R. Atlas, D. Stahl, G. Geesey, and G. Sayler (eds) *Techniques in Microbial Ecology*, pp. 354–407. Oxford University Press, New York. By permission of Oxford University Press, Inc.)

microorganisms and contaminants reside, this type of bioremediation necessarily occurs in situ. Alternatively, *engineered bioremediation* takes an active role in modifying a site to encourage and enhance the capabilities of microorganisms to transform and/or biodegrade pollutants. Each of the two major engineered bioremediation approaches may exploit solid-, slurry-, or vapor-phase systems for encouraging microorganisms to proliferate and metabolize the contaminant chemicals (Figure 8.12).

Selection of the most effective bioremediation strategy is based on characteristics of the contaminants (toxicity, molecular structure, solubility, volatility, susceptibility to microbial attack), the contaminated site (geology, hydrology, soil type, climate, and the legal, economic, and political pressures felt by the site owner), and the microbial process that will be exploited, such as pure culture, mixed cultures, and their respective growth conditions, and supplements. Engineered bioremediation relies on a variety of engineering procedures – control of water flow, aeration, chemical amendments, physical mixing, and the like – that influence both microbial populations and targeted contaminants (Madsen, 1998). Furthermore, the efficacy of the remediation processes must be documented by chemical analysis of water, air, and soil taken from the contaminated site.

Intrinsic bioremediation

Intrinsic bioremediation is the management of contaminant biodegradation without taking any engineering steps to enhance the process. It uses the innate capabilities of naturally occurring microbial communities to metabolize environmental pollutants. The capacity of native microorganisms to carry out intrinsic bioremediation must be documented in laboratory biodegradation tests performed on site-specific samples. Furthermore, the effectiveness of intrinsic bioremediation must be proven with a site monitoring regime that includes chemical analysis of contaminants, final electron acceptors, and other reactants or products indicative of biodegradation processes. This bioremediation strategy differs from the no-action alternative in that it requires adequate assessment of the existing biodegradation rates and potential and adequate monitoring of the process. It may be used alone or in conjunction with other remediation techniques. For intrinsic bioremediation to be effective, the rate of contaminant destruction must be faster than the rate of contaminant migration. These relative rates depend on the type of contaminant, the microbial community, and the hydrogeochemical conditions of the site. Intrinsic bioremediation has been documented for a variety of contaminants and habitats, including low molecular weight polycyclic aromatic compounds in groundwater, gasoline-related compounds in groundwater, crude oil in marine waters (Box 8.5), and low molecular weight chlorinated solvents in groundwater.

Box 8.5

Intrinsic bioremediation of the *Exxon Valdez* oil spill in Prince William Sound, Alaska

The accident

On March 24, 1989, at 12:03 a.m. the oil tanker *Exxon Valdez* hit Bligh Reef, Prince William Sound, Alaska. Eight of 11 cargo tanks ruptured, spilling 260,000 barrels (~40 million liters) of crude oil into Prince William Sound. There was immediate exposure of the intertidal and subtidal zones to oil. After 3 days of calm weather and little or no oil-spill-response effort, a major windstorm drove the oil slick southward towards Smith, Naked, and Knight Islands (Figure 1). The oil reached beaches in Chugach National Forest, Kenai Fjords National Park, and other parks and wildlife regions. The oil impacted sites reach as far as 1000 km away and ~2500 km of shoreline was contaminated.

Immediate effects

At the time of the accident, reproduction and migration was in progress for many fish, birds, and mammals. For example, Pacific herring were spawning in intertidal and subtidal eel grass, and juvenile salmon had begun migration towards ocean waters.

The oil contacted and contaminated thousands of seabirds, the salmon hatcheries, the commercial fisheries, and influenced the activities and food sources of many native Alaskans (whose culture is reliant upon fish, shellfish, and

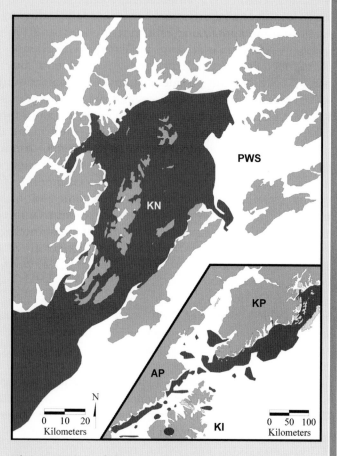

Figure 1 Map of the spread of oil and the shorelines contaminated after the grounding of the *Exxon Valdez* at Bligh Reef in northern Prince William Sound (PWS). Oil was transported to the southwest, striking Knight Island (KN) and other PWS islands, the Kenai Peninsula (KP), the Kodiak Island archipelago (KI), and the Alaska Peninsula (AP). Dark areas are those affected by oil; white areas are open water; and lightly shaded areas are land. (From Petersen, C.H., S.D. Rice, J.W. Short, et al. 20003. Long-term ecosystem response to the Exxon Valdez oil spill. *Science* **302**:2082–2086. Reprinted with permission from AAAS.)

wildlife). Salmon egg mortality was high (~67%) for several years. The tally of oil-induced wildlife casualties included 250,000 seabirds, 2800 sea otters, 260 bald eagles, 30 harbor seals, and 13 orca whales. The US Fish and Wildlife Service estimated recovery times of 20–70 years for some wildlife populations. Large animals (e.g., Sitka black-tailed deer, brown bear) largely escaped injury.

Cleanup efforts

The initial human, corporate, and governmental response to the spill was largely one of confusion. Organized, coordinated oil recovery efforts were delayed. Opportunities during the first 3 days of calm weather were missed. It was determined that responsibility for cleanup was Exxon's. Cleanup crews were uncertain how to tackle a spill of that magnitude. Technologies were unproven. Booms (used to corral the oil) and skimmers (used to remove oil from the ocean surface) were applied, but the windstorm on day 3 interfered. Chemical dispersants (that diminish the chance of coating the beaches but increase the toxicity of oil) were used to some degree. Due to the windstorm, large masses of oil washed onto coastlines and the oil was redistributed by tides. Attempts were made to "clean" oily shorelines with hot water at high pressure; subsequent evaluations of this approach were that it did more harm (to biota) than good.

Bioremediation

Intrinsic bioremediation (also termed natural attenuation) is the evolutionarily proven suite of processes that occur due to innate forces of nature (physical, chemical, and biological processes). Many populations of microorganisms native to ocean water possess enzymatic capabilities that allow them to utilize petroleum as a carbon and energy source. These naturally occurring marine microorganisms digest and grow on hydrocarbons. Thus, microorganisms were the key players in destroying the oil released by the *Exxon Valdez* tanker. Because ambient levels of nitrogen and phosphorus in ocean water are low, these essential inorganic nutrients limit the physiological activity of the hydrocarbon-degrading microorganisms as they digest large masses of oil. To meet the demand for nitrogen and phosphorus, the cleanup workers in Alaska proposed adding fertilizer to the oil-impacted sites. This fertilization strategy was approved by a US Environmental Protection Agency (USEPA) science advisory board (Table 1).

Documenting bioremediation

- **In an open, heterogeneous, dynamic habitat like Prince William Sound, how do you prove that the biodegradation processes you *hope* are occurring *truly are* occurring?**

  Answer: Information in Table 1 provides three examples of strategies that environmental microbiologists have developed for establishing field evidence of biodegradation.

The overarching principle addressed by the strategies in Table 1 is that microbial processes leave a "footprint" in the field that can be documented by drawing on environmental microbiology's toolbox (see Sections 5.2 and Section 6.5).

Box 8.5 *Continued*

Table 1

Strategies used to prove that biodegradation by microorganisms in field sites contributes to remediation of organic environmental contamination

Strategies for proving field biodegradation	Principles and examples
Replicated field plots with and without applied fertilizer	Some field sites are amenable to controlled treatments designed to stimulate microbiological activity. Comparing the loss of pollutants from plots with and without nutrients can effectively demonstrate the role of microorganisms in field biodegradation. An oil-based (oleophilic) N- and P-fertilizer that adhered to crude oil was dispersed to blocks of the Prince William Sound shoreline impacted by the oil. Oil disappeared more rapidly in treated, compared to untreated, blocks
Internal conservative tracers to document relative disappearance	The loss of certain compounds can be assessed relative to the persistence of less biodegradable, but similarly transported, compounds. Researchers analyzing water-borne hydrocarbons from the *Exxon Valdez* oil spill found that using ratios of straight- to branched-chain alkanes (C_{17}/pristane, C_{18}/phytane), and ratios of other compounds to hopanes in crude oil diminished over time
Molecular biological indicators can qualitatively verify that a suspected process is active	Based on prior knowledge of the biochemical pathways, metabolites, enzymes, and genes responsible for pollutant metabolism, a variety of assays can be applied to field samples. The goal is to find biodegradation-specific biomarkers (e.g., metabolites, mRNA) inside but not outside the contaminated habitat

Effectiveness of intrinsic bioremediation in Prince William Sound

There have been several surveys of Prince William Sound aimed at assessing ecosystem recovery. Outstanding questions address: (i) the status of the wildlife; and (ii) the petroleum residues remaining in the habitat.

1 Regarding wildlife, experts have concluded that current practices for assessing ecological risks of oil in the oceans should be improved to accommodate site-specific, food web-based toxicology. This approach is required to understand and ultimately predict chronic, delayed, and indirect long-term risks and impacts (Petersen et al., 2003).

2 Regarding petroleum residues, a recent survey of 32 shorelines for oil on the land surface and at 0.5 m depth (Short et al., 2004, 2006) has revealed that oil remains in 59 of 662 quadrants – largely in the subsurface zones.

Thus, low-level chronic exposure of marine wildlife continues.

• If bioremediation is effective, why does oil still persist in Prince William Sound?
Answer: Intrinsic bioremediation and natural attenuation have successfully removed vast quantities of oil in Prince William Sound. However, some habitats within the sound (especially the subtidal zone) offer suboptimal physiological conditions for oil-degrading microorganisms. The timeframes for nature and microorganisms need not match the impatient expectations of humans.

Engineered bioremediation

Engineered bioremediation either accelerates intrinsic bioremediation or replaces it completely through the use of site modification procedures, such as excavation, hydrologic manipulations, and the installation of bioreactors that allow concentrations of nutrients, electron acceptors, or other materials to be managed in a manner than hastens biodegradation reactions. Engineered bioremediation is especially well suited for treating nonvolatile, sparingly soluble contaminants whose properties impede successful treatment by other technologies. Engineered bioremediation may be chosen over intrinsic bioremediation because of considerations of

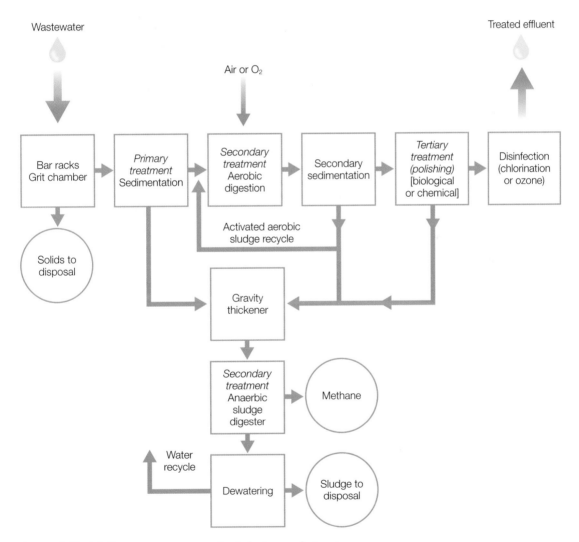

Figure 8.13 (a) Sewage treatment plant design and structure.

Figure 8.13 (b) City of Ithaca wastewater treatment plant, Ithaca, NY. (Photo courtesy of E. L. Madsen.)

time, cost, and liability. Because engineered bioremediation accelerates biodegradation reactions, this technology is appropriate for situations where time constraints for contaminant elimination are short or where transport processes are causing the contaminant plume to advance rapidly. The need for rapid pollutant removal may be driven by an impending property transfer or by the impact of the contamination on the local community. A shortened clean-up time means a correspondingly lower cost of maintaining the site.

Engineered ex situ bioremediation has been used in municipal sewage treatment systems (Figure 8.13) for over a century. In sewage treatment systems, wastewaters from municipalities are directed through an array of controlled environments that encourage microbial growth in filters, tanks, and digestors. Physical, chemical, and microbiological manipulations remove carbonaceous, nitrogenous, and other materials from water before it is discharged into rivers, lakes, or oceans.

The processes that occur in sewage treatment plants (Figure 8.13) are the biogeochemical reactions (especially respiration, nitrification, denitrification, ammonification, sulfate reduction, and methanogenesis) discussed at length in Sections 3.6–3.11 and 7.2–7.6. In essence, municipal wastewater treatment facilities *intervene* between human activities that load waters with undesirable materials and downstream natural bodies of water.

Box 8.6

The Sharon–anammox process: specialized microbiology and engineering for specialized needs

For many years, wastewater engineers have been confronted with a challenging task: "What is the best way to process ammonium-rich sludge-digester effluent?" In standard wastewater treatment schemes, elimination of ammonium requires two steps: (i) the ammonium must be aerobically oxidized to nitrate via nitrification; and (ii) this is followed by reduction of nitrate to nitrogen gas via denitrification. A significant amount of oxygen is required for the first step and a significant amount of electron donor (often, added methanol) is required for the second step.

Recently two innovative procedures have been combined to attain a new technology for treating high-ammonium wastewaters. As is clear from the name, the Sharon–anammox process takes advantage of anammox reactions (see Sections 7.4–7.5) that, under anaerobic conditions, use ammonium as an electron donor and nitrite as an electron acceptor to produce nitrogen gas.

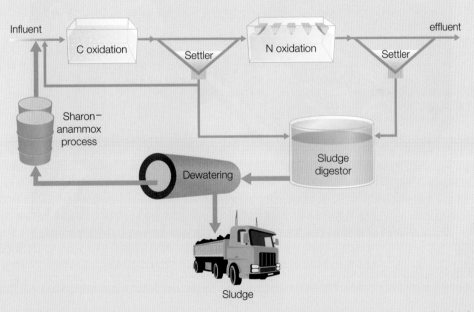

Figure 1 The Sharon–anammox process augments traditional sewage treatment. The high-ammonia effluent from the sludge digestor is freed of nitrogen via the two-step Sharon–anammox process (green tanks). (Reprinted from van Dongen, U., M.S.M. Jetten, and M.C.M. van Loosdrecht. 2001. The Sharon–anammox process for treatment of ammonium rich wastewater. *Water Sci. Technol.* **44**(1):153–160, with permission from the copyright holders, IWA.)

Box 8.6 *Continued*

In the new technology, which is patented and successfully operating on the industrial scale in many European cities, the Sharon process prepares the desired 1 : 1 (nitrite : ammonium) reactant mixture for anammox by delivering a limited amount of oxygen to the high-ammonium waste. High-flow conditions favor rapidly growing *Nitrosomonas* bacteria that convert ammonium to nitrite, while the slower growing nitrite oxidizers are flushed from the system.

A schematic diagram of the Sharon–anammox system in place in the Rotterdam (the Netherlands) wastewater treatment plant is shown in Figure 1. After dewatering, high-ammonium waste from a sludge digestor is delivered first to the Sharon bioreactor (Figure 2) and then to the anammox bioreactor. Nitrogen is effectively removed without any requirement of carbon amendments used routinely in denitrification-based treatment of ammonium. Furthermore, the overall oxygen requirement for nitrogen removal is reduced by 60% and CO_2 emissions drop.

Figure 2 The Sharon system bioreactor, Rotterdam, the Netherlands. (Photo courtesy of Grontmij NV, the Netherlands.)

Before the water is discharged from the treatment plant, specialized habitats within the plant create enrichment cultures of mixed, naturally occurring microbial communities. By making a living in these habitats, the microorganisms degrade organic environmental pollutants, release CO_2, oxidize ammonia, reduce NO_3^- to N_2, generate methane, and accumulate phosphorus. Sewage treatment plants are superb examples of ex situ bioremediation and biogeochemistry. The Sharon–anammox process (Box 8.6) provides an example of how the combination of specialized microorganisms and engineering can address a specialized industrial need: high-ammonium waste streams.

Engineered in situ bioremediation was implemented by R. Raymond and colleagues to clean up petroleum-contaminated groundwater over two decades ago. In the pioneering version of this technology, a groundwater circulation system was established that enhanced the mixing of contaminants, microbial cells, and nutrients designed to encourage aerobic catabolic reactions. Table 8.5 compares in situ and ex situ approaches to engineered bioremediation. In situ bioremediation approaches strive to engineer the landscape to mimic conditions known to foster biodegradation reactions readily demonstrable in laboratory flasks. Thus, the major barrier to successful in situ engineered bioremediation is the impossibility of fully controlling the variety of intractable processes and heterogeneities characteristic of open field sites. Engineered bioremediation must contend with variability in properties of the site, pollutant chemicals, microorganisms, and regulatory agencies overseeing cleanup efforts. Bioreactors central to ex situ strategies to engineered bioremediation offer better control over biodegradation processes, as pollutant metabolism within the bioreactors can be verified and enhanced.

When compared side by side it is clear that the two engineered bioremediation strategies share certain obstacles, but each also offers different advantages (Table 8.5). If contaminants are strongly sorbed onto soil or sediment solids, the ex situ approach may be less appropriate. In either case, qualitative evidence must be obtained to prove that microbiological processes are responsible for pollutant loss. Ideally, quantitative mass balances should be assembled that confirm stoichiometric relationships between physiological reactants (e.g., electron donors and acceptors) and metabolic endproducts (e.g., CO_2, H_2S, CH_4; National Research Council, 2000). Computer modeling of site-specific biogeochemical reactions is one promising approach for quantifying the proportion of pollutant loss attributable to biotic versus abiotic processes when site remediation technologies are implemented.

Susceptibility of organic compounds to bioremediation

Organic compounds that are most susceptible to microbial metabolism are naturally occurring, have a simple molecular structure, are water soluble, exhibit no sorptive tendencies, are nontoxic, and serve as a growth substrate for microorganisms. By contrast, compounds that are resistant to microbial metabolism exhibit properties such as complex molecular structure, low water solubility, strong sorptive interactions, toxicity, and/or do not support growth of microorganisms. Table 8.6 provides an

Table 8.5

Comparison of in situ and ex situ strategies for engineered bioremediation systems (modified from Madsen, E.L. 1998. Theoretical and applied aspects of bioremediation: the influence of microbiological processes on organic compounds in field sites. *In*: R. Burlage, R. Atlas, D. Stahl, G. Geesey, and G. Sayler (eds) *Techniques in Microbial Ecology*, pp. 354–407. Oxford University Press. New York. By permission of Oxford University Press, Inc.)

Feature	In situ strategy	Ex situ strategy
Location	In the landscape	In a controlled bioreactor
Requirements	Engineer the landscape to resemble a laboratory flask	Move contaminants from landscape to on-site bioreactors
Characteristics	Relatively poor control of biodegradation process	Greater control
Obstacles	Complexities of landscape that may prevent success	Complexities of landscape partially overcome
	Pollutant mixtures	Pollutant mixtures
	Unknown site histories	Unknown site histories
	Mass balances uncertain	Decent bioreactor mass balances
	Biotic versus abiotic processes	Biotic processes defined in bioreactor
	Incompatibility between site characteristics and microbiological processes	Incompatibility of site characteristics and microbiological processes can be overcome
	Production of pollutants by microorganisms	Production of pollutants by microorganisms can be minimized
	How clean is clean?	How clean is clean?

overview of the categories of contaminants, the mechanisms of microbe–contaminant interaction, the type of contaminant alteration, and the compounds' overall susceptibility to bioremediation measures. Table 8.6 presents a broad perspective on how chemical and microbiological properties jointly affect prospects for bioremediation. Box 8.7 describes the chemical structures for a selection of common organic environmental contaminants.

Petroleum hydrocarbons are naturally occurring chemicals featuring a variety of molecular weights and functional groups (see Section 7.3). BTEX components of gasoline are widely distributed and have been the focus of substantial biodegradation and bioremediation research. BTEX, the low molecular weight fuels, and structurally simple alcohols and ketones are readily mineralizable aerobically, hence, these compounds have been successfully removed from contaminated sites via established bioremediation procedures (see also Table 8.6 and Section 7.3). One prominent fuel component, notably resistant to biodegradation, is methyl *tert*-butyl ether (MTBE). MTBE has been added to gasoline at up to 15% by volume and features three traits [a stable ether linkage, a tertiary carbon structure,

Table 8.6

Overview of biodegradation and bioremediation potential for particular classes of environmental contaminants (modified from National Research Council. 2000. *Natural Attenuation for Groundwater Remediation*. National Academies Press, Washington, DC. Reprinted with permission from the National Academies Press. Copyright 2000, National Academy of Sciences)

Chemical class	Mechanisms of microbe–contaminant interactions	Type(s) of contaminant alteration	Susceptibility to microbiological transformation	
			Aerobic	Anaerobic
Organic				
Petroleum hydrocarbons				
Low molecular weight				
BTEX	Carbon and electron-donor source	Mineralized to CO_2	1	2
Gasoline, fuel oil	Carbon and electron-donor source	Mineralized to CO_2	1	2
High molecular weight				
Oils, PAHs	Carbon and electron-donor source	Mineralized to CO_2 or partially degraded	1, 2	2, 4
Creosote	Carbon and electron-donor source	Mineralized to CO_2 or partially degraded	1, 2	2, 4
Oxygenated hydrocarbons				
Low molecular weight				
Alcohols, ketones, esters, ethers	Carbon and electron-donor source	Mineralized to CO_2	1, 2	2
MTBE	Cometabolized; occasionally used as carbon and electron-donor source	Partially degraded, sometimes mineralized to CO_2	2–5	4, 5
Halogenated aliphatics				
Highly chlorinated	Electron acceptor under anaerobic conditions; cometabolized	Partially degraded, dechlorinated	2–5	2–5
Less chlorinated	Electron acceptor under anaerobic conditions; carbon and electron-donor source; cometabolized	Partially degraded, dechlorinated	2–5	2–5

Table 8.6 Continued

Chemical class	Mechanisms of microbe–contaminant interactions	Type(s) of contaminant alteration	Susceptibility to microbiological transformation*	
			Aerobic	Anaerobic
Halogenated aromatics				
Highly chlorinated	Electron acceptor under anaerobic conditions; carbon and electron-donor source; cometabolized	Partially degraded, dechlorinated	2–5	2, 3
Less chlorinated	Electron acceptor under anaerobic conditions; carbon and electron-donor source; cometabolized	Partially degraded, mineralized to CO_2	1, 2	2
PCBs				
Highly chlorinated	Electron acceptor under anaerobic conditions	Partially degraded, dechlorinated	4	2, 3
Less chlorinated	Electron acceptor under anaerobic conditions; carbon and electron-donor source	Partially degraded or fully mineralized to CO_2	1, 2	2, 4
Dioxins	Electron acceptor under anaerobic conditions	Partially degraded	4	4
Nitrogen-containing explosives (TNT, RDX)	Cometabolized	Partially degraded; immobilized by precipitation or polymerization	2	2
Inorganic				
Metals				
Copper, nickel, zinc	Sorbs to extracellular polymers and biomass	Immobilized by sorption	2	2
Cadmium, lead	Sorbs to extracellular polymers and biomass	Immobilized by sorption; methylation possible	2	2
Iron, manganese	Electron acceptor under anaerobic conditions; oxidized to form insoluble hydroxides; sorbs to extracellular polymers and biomass	Mobility (solubilization) increased by reduction; immobilized by precipitation and sorption	1	1

Chromium	Enzymatically oxidized or reduced to promote detoxification; cometabolized; sorbs to extracellular polymers and biomass	Immobilized by precipitation	2	2
Mercury	Enzymatically oxidized or reduced or methylated to promote detoxification; sorbs to extracellular polymers and biomass	Volatilized or immobilized by sorption, methylation, and precipitation	2	2
Nonmetals				
Arsenic	Enzymatically oxidized or reduced or methylated; electron acceptor under anaerobic conditions; oxidation of reduced forms linked to microbial growth; sorbs to extracellular polymers and biomass	Volatilized or immobilized by precipitation and sorption	2	2
Selenium	Enzymatically oxidized or reduced and methylated; electron acceptor under anaerobic conditions; cometabolized; sorbs to extracellular polymers and biomass	Volatilized or immobilized by precipitation of elemental Se or sorption	1	2
Oxyanions				
Nitrate	Electron acceptor under anaerobic conditions	Reduced to nontoxic nitrogen	4	1
Perchlorate	Electron acceptor under anaerobic conditions	Reduced to nontoxic chloride ion	4	2, 5
Radionuclides				
Uranium	Electron acceptor under anaerobic conditions; sorbs to extracellular polymers and biomass	Immobilized by precipitation	4	2
Plutonium	Cometabolized; sorbs to extracellular polymers and biomass	Mobility increased by reduction to soluble Pu(III); immobilized by precipitation and sorption	4	2
Technetium	Enzymatically oxidized or reduced; cometabolized; sorbs to extracellular polymers and biomass	Immobilized by precipitation	4	2

BTEX, benzene, toluene, ethylbenzenes, and xylenes; MTBE, methyl *tert*-butyl ether; PAHs, polycyclic aromatic hydrocarbons; PCBs, polychlorinated biphenyls; RDX, Royal Dutch Explosive; TNT, trinitrotoluene.

* The numeric entries for each compound class provide a rating of susceptibility to microbial transformation under anaerobic conditions (in the presence of oxygen) and anaerobic conditions (when oxygen is absent): 1, readily mineralized or transformed; 2, degraded or transformed under a narrow range of conditions; 3, metabolized partially when second substrate is present (cometabolized); 4, resistant; 5, insufficient information.

Box 8.7

A selection of widespread organic environmental contaminants and their chemical structures

Compound	Chemical structure
Alcohols	R—OH Hydroxyl group Methanol H_3C—OH
Alkanes	Saturated with H atoms, no double bonds Octane H_3C—$(CH_2)_6$—CH_3
Alkenes	Sites of unsaturation, at least one C double bond Ethene 2HC=CH2
Alkynes	Triple-bonded C R—C $\equiv$ C—R Acetylene HC$\equiv$CH
Aromatic hydrocarbons	Hydrocarbons containing benzene rings
Atrazine	
Benzene	 Aromatic ring C_6H_6
Chlorobenzene	
Creosote	Oily yellow-to-black substance rich in phenols, PAHs, and cresols. Used as a wood preservative
Cresol(s) (three different isomers)	
2,4-Dinitrotoluene	
Dioxins (e.g., TCDD)	 2.3.7.8-Tetrachloro-dibenzo-*p*-dioxin (TCDD)

Esthers

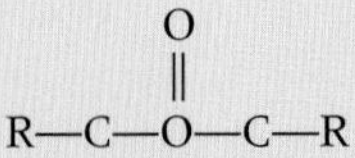

$$R-C-O-C-R$$
(with O double-bonded to the first C)

Ethers

$$R-C-O-C-R$$

Ethyl benzene

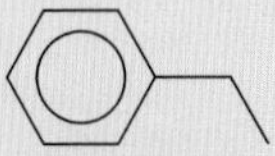

$$(C_6H_6)-CH_2-CH_3$$

Ketones

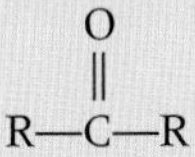

$$R-C-R$$
(with O double-bonded to C)

Melamine

$$HN_2 \quad N \quad NH_2$$
$$N \quad N$$
$$NH_2$$

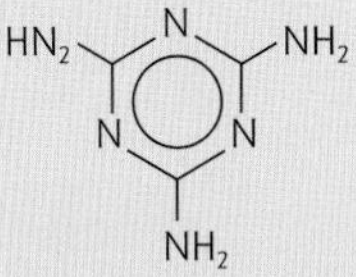

MTBE (methyl-tert-butyl ether)

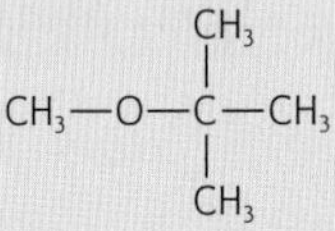

$$CH_3-O-\underset{\underset{CH_3}{|}}{\overset{\overset{CH_3}{|}}{C}}-CH_3$$

Nitrobenzene

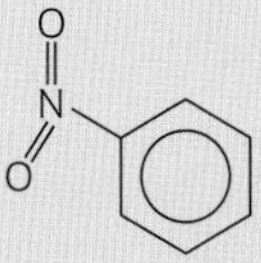

Pentachlorophenol (PCP)

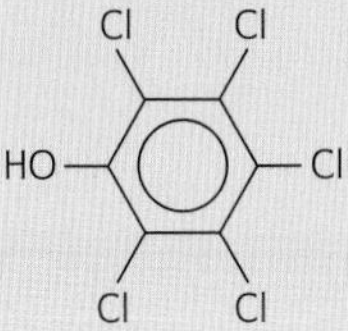

Perchloroethene (PCE)

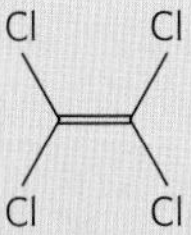

Phenol

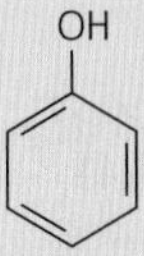

Polychlorinated biphenyls (there are 209 different PCB molecules; shown at right is 2,3,4,5,6-2', 3',4',5',6-decachlorobiphenyl)

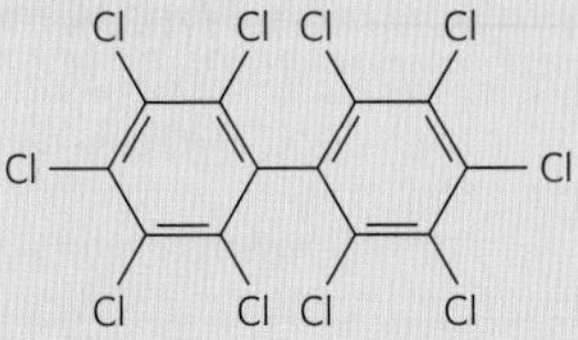

Box 8.7 *Continued*

Polycyclic aromatic
hydrocarbons (PAHs)

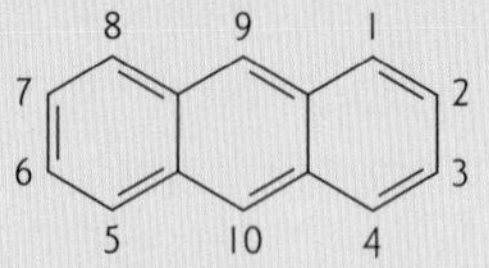

Anthracene, a three-ring PAH

Styrene

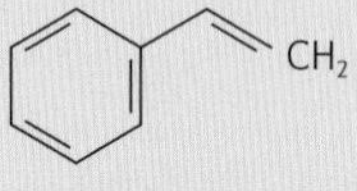

RDX (Royal Dutch Explosive)

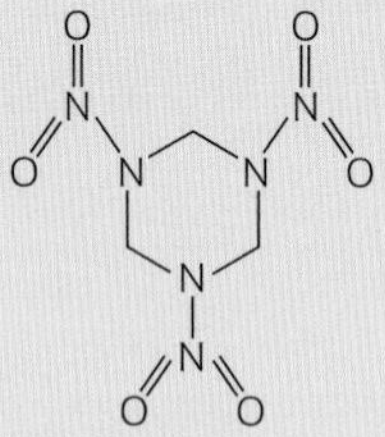

Toluene

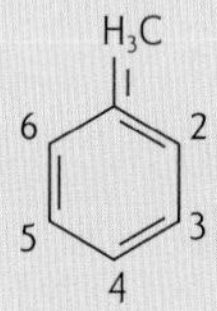

TNT (trinitrotoluene)

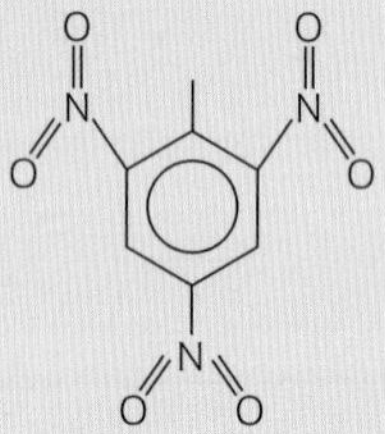

Trichloroethene

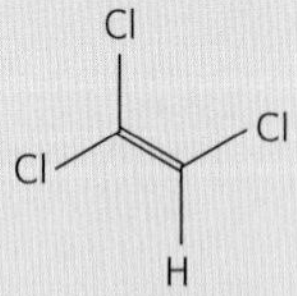

Xylene(s)
(three different isomers)

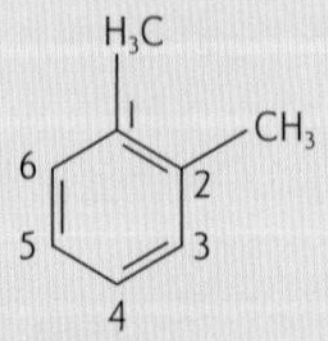

Xylene (o-shown)

Nomenclature: R, organic moiety; —, terminal methyl group (e.g., –CH$_3$).

and adenosine triphosphate (ATP) generation at or below cell maintenance demands] that prevent MTBE from readily serving as a carbon and
electron source for microbial growth. Consequently, MTBE is often a threat
to water quality. When microorganisms possess one of a variety of nonspecific oxygenase enzyme systems, oxygen atoms can be fortuitously
inserted into MTBE by a process known as a cometabolism (see Box 8.4).
Recently, many BTEX components, traditionally considered resistant to
anaerobic microbial attack, have also been found to be biodegradable under
a variety of anaerobic conditions (Table 8.6; see also Section 7.3). The
high molecular weight petroleum components and creosotes are only slowly
metabolized – partly as a result of their structural complexity, low solubility, and strong sorptive characteristics. Thus, bioremediation techniques
for the latter classes of petroleum hydrocarbons listed in Table 8.6 are
still emerging.

Though halogenated organic compounds have been found in nature,
these are not of commercial significance compared to the synthetic halogenated chemicals listed in Table 8.6. When halogen atoms (chlorine,
bromine, fluorine) are introduced into organic molecules, many properties, such as solubility, volatility, density, hydrophobicity, and toxicity
change markedly. These changes confer improvements that are valuable
for commercial products, but also have serious implications for microbial
metabolism. The susceptibility of the chemicals to enzymatic attack is
sometimes drastically altered by halogenation and persistent compounds
often result.

Halogenated aliphatic compounds are straight-chain hydrocarbons in
which varying numbers of hydrogen atoms have been replaced by halogen atoms. Halogenated aliphatics are effective solvents and degreasers
that have been widely used in many manufacturing and service industries. Some highly chlorinated representatives of this class, such as
tetrachloroethene, are completely resistant to aerobic microbial attack,
while being susceptible to anaerobic reductive dehalogenation under
methanogenic and other anaerobic conditions (Table 8.6). In fact, recent
laboratory and field evidence shows that complete reductive dechlorination from tetrachloroethene to the nontoxic, plant growth hormone, ethene,
can occur (see also Chapter 3, Science and the citizen box). Furthermore,
as the degree of halogenation in aliphatics diminishes, susceptibility to
other metabolic reactions increases. For instance, vinyl chloride (a carcinogen created by reductive halogenation of three of the four chlorines
on the tetrachloroethene molecule) can be used as an electron donor by
both iron-reducing and aerobic microorganisms that oxidize vinyl chloride to CO_2. Aerobic cometabolism of partially halogenated aliphatics (e.g.,
trichloroethene) has been tested as a bioremediation strategy: microorganisms are supplied with substrates such as methane, toluene, or
phenol, which induce expression of nonspecific oxygenase enzymes that
insert oxygen atoms across carbon–carbon bonds. This radically alters the

toxicity, stability, and biodegradability of the halogenated ethenes. Thus, one treatment approach for chlorinated aliphatics is to remove chlorine atoms anaerobically (via reductive dehalogenation) and then complete the biodegradation process using aerobic cometabolism. Procedures for bioremediating sites contaminated with chlorinated aliphatic are developing rapidly. Both anaerobic treatment (driven by supplementing field-site waters with an electron source for reductive dechlorination) and aerobic cometabolic treatment (driven by additions of methane or aromatic substrates) have been field tested.

Halogenated aromatics, such as phenoxyacetic acid pesticides, pentachlorophenol, polychlorinated biphenyls (PCBs), and others (Table 8.6), consist of one or more benzene rings, which bear halogens as well as other chemical functional groups (i.e., hydroxyls, carboxyls, etc.). The aromatic benzene nucleus is susceptible to both aerobic and anaerobic metabolism, though the latter occurs relatively slowly (see Section 7.3). Overall, however, the presence of halogen atoms on the aromatic ring, their position, and their interaction with functional groups is what governs biodegradability. A high degree of halogenation may prevent aromatic compounds from being oxidatively (aerobically) metabolized, as is the case for PCBs. However, as discussed above for the aliphatic compounds, the highly halogenated aromatics are subject to anaerobic reductive dehalogenation. As the halogen atoms are replaced by hydrogens, the molecules become susceptible to aerobic attack. Thus a common bioremediation scenario for treating soils, sediments, or water contaminated with halogenated aromatic chemicals is anaerobic dehalogenation followed by aerobic mineralization of the residual compounds. It should be noted, however, that when a proper substituent group accompanies the halogens on the aromatic ring, aerobic metabolism may proceed rapidly, as is the case for pentachlorophenol (Section 8.5).

Other prominent chlorinated aromatic contaminants are dioxins (e.g., tetrachlorodibenzo-*p*-dioxin, TCDD) and polychlorinated dibenzofurans (PCDF). TCDD and PCDF are potent teratogens and carcinogens whose complex chemical structure and strong sorptive properties render them nearly nonbiodegradable – although dechlorination reactions have been reported (Table 8.6). Microorganisms can also metabolize some nitro-aromatic compounds, common components of explosives and pesticides (Section 8.5). Dinitrotoluene, produced during the manufacture of polyurethane foams and explosives, is biodegradable (Spain et al., 2000; Section 8.5). However, despite decades of study, the explosives, trinitro-toluene (TNT) and Royal Dutch explosive (RDX), have not been shown to serve as readily utilizable carbon or energy sources for microbial growth. Under both aerobic and anaerobic conditions, the nitrate moieties on explosives are reduced to amino groups that can cause the compounds to be toxic, undergo polymerization reactions, and/or strongly sorb onto soil solids. Recent reports have shown that aerobically grown bacteria can use TNT as a nutritional nitrogen source. Nonetheless, the known microbial transformations of dioxins and nitroaromatics are limited in extent (Table 8.6). Bioremediation technology for these compounds is still emerging.

Susceptibility of inorganic materials, including metals, nonmetals, and radionuclides, to bioremediation

Microbiological processes also affect inorganic environmental contaminants – many of which are listed in the lower portion of Table 8.6. Unlike organic compounds that are often susceptible to partial structural alteration or complete detoxification to carbon dioxide by microorganisms, the majority of inorganic contaminant compounds are subject only to changes in speciation that may influence contaminant mobility. Microorganisms may cause precipitation, volatilization, sorption, and solubilization of inorganic compounds. Mechanistically, these reactions can be the result of direct enzymatic processes such as oxidation, reduction, methylation, or uptake – many of which were discussed in Section 7.1. Reaction mechanisms can also be indirect (nonenzymatic), resulting from microbiological production of metabolites or biomass that can strongly influence the behavior of inorganic contaminants via redox, acid/base, coprecipitation, sorption, and other geochemical means. As is evident in the lower half of Table 8.6, microorganisms influence metals, nonmetals, and radionucleotides in diverse ways. Bioremediation technologies for all of the toxic inorganic materials listed in Table 8.6 except nitrate are still emerging.

One nearly universal means by which microorganisms have been shown to reduce concentrations of inorganic contaminants in water (e.g., Cu, Ni, Zn, Cd, Pb; Table 8.6) is by immobilizing aqueous-phase inorganics in microbial biomass and/or microbial exopolymers. The mechanisms involved range from nonspecific electrostatic sorptive interaction between cationic metals and anionic extracellular polysaccharides to highly specific active transport systems that cause metals to accumulate in high concentration within microbial cells. The utility of these sequestration reactions has been proven primarily in engineered wastewater treatment systems where metal-laden water flows over fixed biofilms, which later can be removed from the treatment system so that the toxic inorganics can be recovered. Biomass-mediated sorption reactions clearly influence the behavior of inorganic contaminants. However, any bioremediation technology based on sorption requires periodic harvesting of biomass – which may not be feasible in some settings (e.g., groundwater).

Many inorganic contaminants, especially metals, become relatively soluble, hence mobile, at low pH. In contrast to the various bioremediation approaches that rely on immobilization reactions, the opposite (washing inorganic contaminants out of a habitat) can theoretically be achieved by directing low pH waters through contaminated sites. The acidification step can be mediated by a variety of microbial processes, which includes the oxidation of elemental sulfur.

The often highly abundant nontoxic metals, iron (Fe) and manganese (Mn), exist in reduced and oxidized states. The oxidized states [Fe(III), Mn(IV)] react chemically to form oxyhydroxide precipitations that serve as physiological electron acceptors for anaerobic microbial food chains (see Sections 3.7, 3.8, 3.10 and 7.3). The endproducts of Fe and Mn reduction [Fe(II) and Mn(II)] are relatively soluble and may migrate to aerobic habitats, where reoxidation and precipitation can also be catalyzed by microorganisms. The behavior of many of the toxic metals discussed below is intimately tied to the microbially mediated cycling of Fe and Mn because the toxic metals may be immobilized (through coprecipitation and sorptive reactions with many Fe and Mn oxides) or solubilized [by being reduced via chemical reactions with Fe(II) and Mn(II)]. Thus, most of the inorganic compounds in Table 8.6 are shown to undergo immobilization reactions via sorption and precipitation.

Chromium (Cr) is a metal whose key oxidation states are Cr(VI) and Cr(III). In aqueous environments Cr(VI) predominates as the mobile and highly toxic anions, chromate (CrO_4^{2-}) and dichromate ($Cr_2O_7^{2-}$). Reduced Cr(III) is less toxic and less mobile because it precipitates at pH 5 and above. A variety of both aerobic and anaerobic microorganisms have been shown to enzymatically reduce Cr(VI) to Cr(III), but the physiological reason for this ability has not been adequately investigated. Among the hypotheses explaining the reduction reactions are: survival (i.e., detoxification), cometabolism (i.e., fortuitous enzymatic reactions), and the use of Cr(VI) as a physiological electron acceptor (to date, only equivocal evidence for the latter hypothesis has been obtained). Direct microbial detoxification (reduction) of Cr(VI) is unlikely to be a useful remediation technology in anaerobic habitats because the reduction occurs spontaneously in the presence of sulfide, Fe(II), and some organic compounds. Although microbial production of sulfide, Fe(II), and reduced organic compounds is generally reliable, additional research is required before judging if Cr(VI) reduction has the potential to serve as a useful bioremediation tool (Table 8.6).

Mercury (Hg) is a toxic metal whose predominant forms include mercuric ion [Hg(II)], elemental mercury [Hg(0)], and the biomagnification-prone organic mercury compounds, monomethyl and dimethyl mercury. All transformations of mercury by microorganisms are considered detoxification reactions that are intended to mobilize mercury away from microbial cells (see also Sections 7.1 and 7.3). Most reactions are enzymatic, carried out by both aerobes and anaerobes, and involve the uptake of Hg(II) followed by its reduction to volatile forms [elemental Hg(0), methyl, and dimethyl mercury; Table 8.6). Hg(II) also forms highly insoluble precipitates with sulfide; thus, one indirect microbial detoxification strategy involves the stimulation of sulfate-reducing microorganisms. Engineered systems that first reduce the mercuric ions and then purge the volatile mercury from water have been designed and implemented (Section 8.6).

In addition to mercury, microorganisms are capable of methylating other metals (Cd, Pb, Sn, Te, Se; Se methylation is discussed in detail below). Additional methylation reactions may occur as a result of nonbiological transmethylation by microbially produced methylated donor compounds such as trimethyl tin. These donors may react with ionic forms of palladium, thorium, platinum, and gold, but the resultant reduced metals may not be chemically stable. The significance of these unusual metal methylation reactions for bioremediation is uncertain.

As mentioned in Chapter 7 (Science and the citizen box), arsenic (As) is a toxic element capable of existing in five different valence states [As(III), As(0), As(II), As(III), and As(V)]. Forms of arsenic range from sulfide minerals (e.g., As^2S^{3-}) to elemental arsenic to arsenic acid to arsenite (AsO_2^-) to arsenate (AsO_4^{3-}) to various organic forms that include methylated arsenates and trimethyl arsine. Clearly, the chemical and microbiological reactions of arsenic are complex. Both anionic forms (arsenite and arsenate) are highly soluble and highly toxic – interfering with various enzyme functions and oxidative phosphorylation, respectively. However, no form of arsenic is nontoxic. Microorganisms transform arsenic for three fundamental physiological reasons: (i) under anaerobic conditions, arsenate [As(V)] can be used as a final electron acceptor; (ii) under aerobic conditions, reduced arsenic (e.g., arsenite) has been shown to serve as an electron donor, generating energy (ATP); and (iii) under both anaerobic and aerobic conditions, arsenic can be detoxified by methylation, oxidation, or reduction mechanisms that mobilize arsenic away from microbial cells. Engineered bioremediation strategies that rely on mobilizing methylated arsenic from water have been implemented. All in situ bioremediation strategies for arsenic cleanup need to contend with both the complexities of arsenic microbial transformations and site-specific geochemical conditions. Arsenic bioremediation technology is still emerging (Table 8.6).

Although selenium (Se) is an important and beneficial micronutrient for plants, animals, humans, and some microorganisms (largely because of its role in some key amino acids), this element can be toxic at greater than trace concentrations. In natural environments, selenium has four predominant inorganic species: Se(VI) (selenate, SeO_4^{2-}), Se(IV) (selenite, SeO_3^{2-}), Se(0) (elemental selenium), and Se(II) (selenide). Like arsenic, selenium also has many volatile organic forms; these include: dimethyl selenide, dimethyl diselenide, methane selanone, methane selenol, and dimethyl selenyl sulfide. Each of these compounds exhibits its own chemical and biochemical behavior, mobility, and toxicity. The various forms of selenium are transformed by microorganisms according to their physiological needs and the ambient thermodynamic conditions (see Table 8.6). Reduced inorganic selenium compounds have been shown to be oxidized under aerobic conditions, though not linked to microbial growth. Oxidized selenium (selenate) can serve as a final electron acceptor for anaerobic microorganisms – resulting in production of selenide and/or elemental selenium. Methylation of the various selenium compounds is thought to be a protective detoxification mechanism that mobilizes selenium away from microbial cells. Thus, like arsenic, the environmental fate of selenium is governed by complex interactions between chemical and physiological processes. For instance, anaerobic microbial reduction of selenate and selenite to insoluble elemental selenium represents an important mechanism for immobilizing and removing selenium from aqueous solution. Furthermore, the various volatile, methylated forms of selenium are sufficiently mobile that aerobic deselenification (largely via dimethyl selenide formation) of highly contaminated Californian soils has been demonstrated in field experiments. Given the complex chemical and biological processes that influence the fate of selenium, effective bioremediation practices in the subsurface are still emerging (see Table 8.6).

Oxyanions are water soluble, negatively charged chemical species in which a central atom is surrounded by oxygen. Nitrate (NO_3^-) is a naturally occurring oxyanion commonly found at low concentrations in soils, sediments, and both surface water and groundwater as a result of the biogeochemical cycling of organic matter (see Sections 7.3 and 7.4). Nitrate is a readily utilizable form of nitrogen that can be assimilated into amino acids by plants and microorganisms. Although serving to supply nitrogen, an essential nutrient, nitrate is also a serious health concern for at least two reasons: (i) it can be a chemical or microbiological precursor for nitrite, which spontaneously combines with amino compounds to form highly carcinogenic nitrosamines; and (ii) nitrate can be reduced to nitrite in the stomachs of infants, which can cause the respiratory stress diseases, methemoglobinemia.

Nitrate is produced from ammonia by nitrifying microorganisms under aerobic conditions. The major microbial process that destroys nitrate is dissimilatory reduction to dinitrogen gas (denitrification; see Sections 7.3 and 7.4). Genetic, biochemical, physiological, ecological, environmental, and engineering aspects of denitrification have been examined for decades. As shown in Table 8.6, nitrate is used as a physiological electron acceptor under oxygen-free (anaerobic) conditions. The denitrification process is widespread among microorganisms, and occurs reliably in every anaerobic habitat with abundant carbon and electron sources. Denitrification is a well-established bioremediation process and is used routinely in sewage treatment plants to curb eutrophication.

The oxyanions chlorate (ClO_3^-) and perchlorate (ClO_4^-) or their precursors (chlorine dioxide, hypochlorite, chlorite) are produced by a variety of paper manufacturing, water disinfection, aerospace, and defense industries. Many of these oxygenated chlorine compounds have recently been found to be naturally occurring at trace concentrations. These highly oxidized forms of chlorine are energetically very favorable physiological electron acceptors for microorganisms. Compared to denitrification, knowledge of chlorate and perchlorate biodegradation reactions is limited. However, laboratory studies using both pure bacterial culture and environmental samples (soil, freshwater sediments, sewage) have shown that in the presence of common electron donors (carbohydrates, carboxylic acids, amino acids, even H_2 and H_2S), microorganisms can grow at the expense of perchlorate and chlorate; thus reducing them to the nontoxic chloride ion. Furthermore, bioreactors have recently been engineered to successfully convert chlorate and perchlorate to chloride. Thus, the prospects for (per)chlorate bioremediation are favorable.

Uranium (U) is a radioactive element that releases α-, β-, and γ-radiation that can be toxic. Uranium can exist in the oxidation states of U(0), U(III), U(IV), U(V), and U(VI), though in nature insoluble U(IV) and soluble U(VI) predominate. Recently U(VI) [in the form of the uranate oxygen complex $(UO_2)^{2+}$] has been shown to serve as a terminal electron acceptor for anaerobic microorganisms (see Table 6.9 and Section 6.10); thus, in anaerobic habitats, growth-linked reduction (hence immobilization) of uranium should be a reliable process. Although the reverse process, microbial oxidation of U(IV) to U(VI) under aerobic conditions has been demonstrated, this process has not been shown to be physiologically beneficial to the responsible microorganisms. Direct chemical oxidation of U(IV) by molecular oxygen [creating U(VI)] may also influence the robustness of uranium bioremediation strategies. Other radioactive elements, plutonium (Pu) and technecium (Tc), have also been shown to be susceptible to microbial transformation. Iron-reducing microorganisms were found to reduce insoluble Pu^{4+} to the more soluble Pu^{3+}; thus, oxidation/reduction reactions provide tools for emerging bioremediation strategies for U, Pu, Tc (Table 8.6), and other radionuclides.

8.4 BIOFILMS

Biofilms are matrix-enclosed microbial accretions that adhere to biological or nonbiological surfaces (Hall-Stoodley et al., 2004; Kolter and Greenberg, 2006). Typically, biofilms form in habitats such as submerged surfaces exposed to flowing water (Figure 8.14). However, knowledge of biofilms is also crucial for understanding the microbiology of other habitats – ranging from plant roots to sewage-treatment plants to the human lung and medical devices implanted in the human body. A summary of the major attributes of biofilms is presented in Table 8.7.

The fundamental lesson emerging from the study of biofilm microbial communities is that the suspended (planktonic) lifestyle is very different from the attached lifestyle: The behavior and physiological, biochemical, and genetic responses of microorganisms in biofilms are distinctive from responses of the same microorganisms when living as isolated suspended cells.

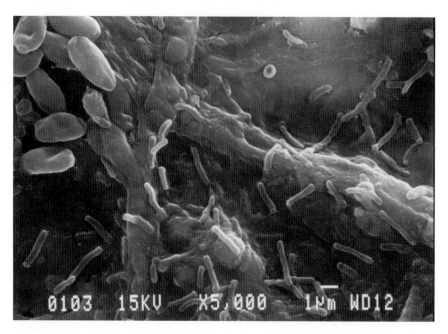

Figure 8.14 Microbial biofilm from the hull of an ocean-going ship. (Courtesy of R. Gubner, Korrosions- och Metallforskningsinstitutet AB, Stockholm, Sweden, with permission.)

Hall-Stoodley et al. (2004) have pointed out that remnants of 3.2 billion-year-old filamentous biofilms were found in the Kornberg formation in South Africa. The implication is that there have always been strong physiological and ecological advantages for microorganisms in nature to form aggregations of cells attached to solid surfaces. By adhering to one another and to solid surfaces, microorganisms are able to modify their immediate surroundings (their microhabitat) – buffering it against rapid physical and chemical fluctuations (Table 8.7). Such surface associations also take advantage of ionic, electrostatic, and hydrophobic interactions that occur between minerals and many other solid surfaces that tend to concentrate important nutrients in aqueous habitats.

Naturally occurring biofilms may be found in our mouths, in toilet bowls, on submerged aquatic plants, and on medical devices (such as catheters and prosthetic heart valves) (Hall-Stoodley et al., 2004). Biofilm development has also been examined in model systems of culturable bacteria – leading to detailed biochemical and genetic information about events that govern biofilm development. Box 8.8 provides a diagrammatic view of the broad theme of biofilm development (Davey and O'Toole, 2000; Watnick and Kolter, 2000) and three individual bacterial variations on the theme. The following are the main features of biofilm development:

Table 8.7

Attributes of biofilm microbial communities (compiled from Davey and O'Toole, 2000; Hall-Stoodley et al., 2004; Ramey et al., 2004; Kirisits and Parsek, 2006; Kolter and Greenberg, 2006)

Attribute	Details
Biofilm definition	Aggregates of microbial cells associated with solid surfaces
Biofilm composition	Adherent cells in a complex matrix of extracellular polymeric substances (EPSs) including exopolysaccharides, proteins, and DNA
Physical aspects of biofilm formation	At the nanometer and micrometer scale, chemical, nutritional, and physical properties of aqueous solutions change at the solid–liquid interface
Selective pressures for biofilm formation	Resistance to fluctuations in physical/chemical stressors: pH, temperature, light, O_2, hydrodynamic shear stress, dehydration Complementary nutritional and physiological associations between bacteria (e.g., anaerobic food webs)
Specialized microbiological aspects of biofilm formation	Cell–cell signaling often known as "quorum sensing" (see Box 8.9) Resistance to doses of chemotherapeutic agents (such as antibiotics) that are lethal to planktonic microorganisms Genetic exchange between densely packed microbial cells Other density-dependent ecological processes such as viral infections and predation by protozoa

1 Suspended planktonic cells have specific extracellular surface structures (e.g., flagella) essential for reaching and remaining on a surface.

2 The expression of specific genes (and their protein products) allows attached cells to proliferate, forming a monolayer.

3 Cell proliferation and the expression of additional sets of genes control the formation of microcolonies.

4 A "mature" biofilm often features a network of three-dimensional towers and channels whose formation is, again, linked to particular expressed genes.

5 The life of any particular biofilm is dynamic and subject to hydrodynamic stresses of flowing water. The fragmentation and release of cells back into the planktonic phase occurs during all biofilm developmental phases.

As is clear from details presented in the text of Box 8.8, variations on the five-part theme (above) are species-specific. Intracellular, genetic, and regulatory cues are distinctive for the three pure-culture bacterial systems described in Box 8.8. In the *Pseudomonas aeruginosa* system, development from the microcolony to the mature biofilm stage is dependent upon both the synthesis of alginate (an extracellular polysaccharide that augments the biofilm's intercellular matrix) and a density-dependent cell-cell signaling system known as "quorum sensing" (Box 8.9). Clearly biofilms play a very important role in many naturally occurring, engineered, and medical systems. Biofilm research will continue to be a major frontier in environmental microbiology far into the future.

Box 8.8

Biofilm development (compiled from Davey and O'Toole, 2000; Moorthy and Watnick, 2005; Ramsey and Wozniak, 2005; Van Houdt and Michiels, 2005; Kolter and Greenberg, 2006)

Figure 1 outlines the three models for the early stages in biofilm formation in three of the best-studied model organisms, *Pseudomonas aeruginosa*, *Escherichia coli*, and *Vibrio cholerae*.

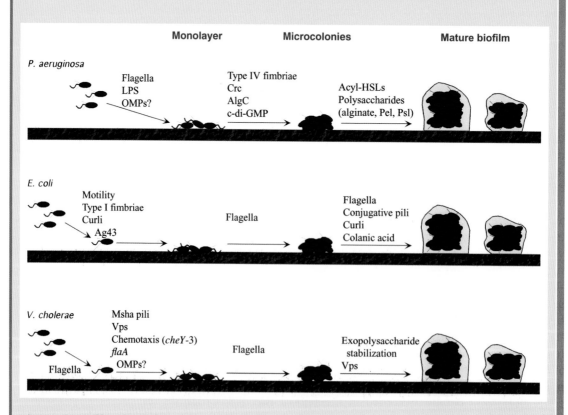

Figure 1 Biofilm development in three Gram-negative organisms. (Modified from Davey, M.E. and G.A. O'Toole. 2000. Microbial biofilms: from ecology to molecular genetics. *Microbiol. Molec. Biol. Rev.* **64**:847–867. With permission from the American Society for Microbiology.)

In *P. aeruginosa*, flagella are required to bring the bacterium into proximity with a solid surface, and lipopolysaccharide (LPS) mediates early interactions, with an additional possible role for outer membrane proteins (OMPs). Once bacteria are on the surface in a monolayer, type IV fimbriae-mediated twitching motility is required for the cells to aggregate into microcolonies. The production of fimbriae is regulated at least in part by nutritional signals via a protein known as Crc. Documented changes in gene expression at this early stage include the messenger molecule c-di-GMP, up-regulation of the polysaccharide biosynthesis genes (alginate, *pel*, and *psl*) and down-regulation of flagellar synthesis. The production of cell-to-cell signaling molecules (acyl-HSLs; see Box 8.9) is required for formation of the mature biofilm. Alginate (AlgC) may also play a structural role in this process.

In *E. coli*, flagellum-mediated swimming is required for both approaching and moving across a solid surface. Organism–surface interactions require surface appendages (type I fimbriae and curli) and the outer membrane protein Ag43. Finally, colanic acid, curli, and conjugative pili are required for development of the normal *E. coli* biofilm architecture.

V. cholerae, like *E. coli*, utilizes the flagella to approach and spread across a solid surface. Chemotaxis (CheY-3), Msha pili, and possibly one or more unidentified outer membrane proteins, are required for attachment to the surface and monolayer formation. This initial surface attachment appears to be stabilized by extracellular polysaccharide (EPS). Formation of the mature biofilm, with its associated three-dimensional structure, also requires production and stabilization (via Bap1 and LeuO proteins) of EPS. Vps refers to the EPS produced by *V. cholerae*. The *flaA* gene encodes the flagella subunit protein.

8.5 EVOLUTION OF CATABOLIC PATHWAYS FOR ORGANIC CONTAMINANTS

There are approximately 2×10^7 naturally occurring organic compounds in the biosphere globally and ~100,000 distinct natural products in a given freshwater sediment sample (Seffernick and Wackett, 2001; Wackett and Hershberger, 2001). Furthermore, several million organic compounds have been synthesized by organic chemists (Seffernick and Wackett, 2001). Many synthetic compounds have novel molecular structures [e.g., polymeric plastics and halogenated pesticides like DDT (dichloro-diphenyl trichloroethane)], though the uniqueness of synthetic ("xenobiotic") compounds is not easy to discern, given the immense and unknown diversity of naturally occurring compounds.

Current dogma in the biodegradation literature is that "all naturally occurring organic compounds are biodegradable". A corollary of this is that "truly novel molecular structures should not be biodegradable because there is no evolutionary precedent, no selective pressure, for their metabolism". After all, genetic mutation involves alteration of an existing DNA

Box 8.9

Quorum sensing is a general cell–cell communication system important for some biofilms and in a variety of other microbial habitats (compiled from Miller and Bassler, 2001; Winans and Bassler, 2002)

Quorum sensing is a chemical communication system that is widespread in prokaryotes. It is used by bacteria to sense one another and to coordinate gene expression among neighboring microbial populations. In *Pseudomonas aeruginosa*, the small excreted organic molecule, known as acyl-homoserine lactone (acyl-HSL in Box 8.8), has the potential to bind to a specific cellular receptor. Once bound together, the acyl-HSL–receptor complex activates the transcription of genes crucial for mature biofilm development by *P. aeruginosa*. Gene activation (and translation of the encoded proteins) will only occur when the concentration of acyl-HSL is above a threshold that is reached when a critical mass of *P. aeruginosa* cells are adjacent to one another. A generalized scheme for quorum-sensing-type regulatory circuits is shown in the figure.

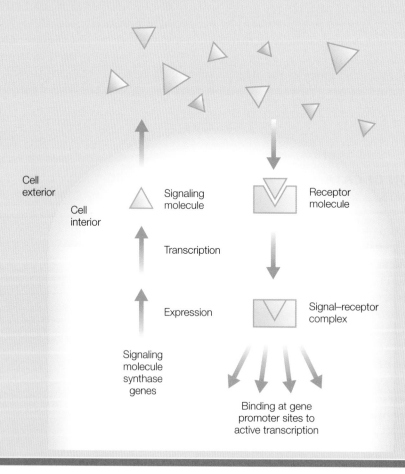

Cell exterior

Cell interior

Signaling molecule

Receptor molecule

Transcription

Expression

Signal–receptor complex

Signaling molecule synthase genes

Binding at gene promoter sites to active transcription

Quorum sensing has many manifestations in coordinated stimulus response systems of the microbial world (Miller and Bassler, 2001; Winans and Bassler, 2002). This cell–cell communication phenomenon was originally discovered in *Vibrio fischeri*, where quorum sensing regulates expression of bioluminescence (Nealson and Hastings, 1979). Quorum sensing is also a regulatory mechanism for coordinating populations of bacteria involved in the following:

- The production of antibiotics.
- The activation of pathogenesis factors.
- Triggering intracellular transfer of plasmid DNA.
- The morphological development of some microorganisms.

A diversity of chemical signaling compounds has been discovered. The compounds are all produced intracellularly, diffuse outside the cells, and only reach the critical intracellular threshold for gene activation when a high density (a "quorum") of cells are present. What emerges is a sophisticated and widespread network of cell–cell signaling system that modulate microbial behavior and interactions in nature.

sequence in a way that varies the catalytic or binding site of a protein. However, if there is no selective pressure for the genetic change, it is unlikely that a newly evolved protein function will persist and proliferate in the gene pool. Selective pressures are crucial for identifying new changes in protein function that are beneficial to the host.

In the past several decades, we know of a handful of novel synthetic organic compounds that were initially considered nonbiodegradable and later deemed biodegradable (often under restricted physiological conditions; Table 8.8). Among the compounds listed in Table 8.8 are monomers used in pesticides, chemical feed stocks, PCBs, chlorinated solvents, and polymeric products. An extremely insightful source of evidence for the evolution of new biodegradation pathways in microorganisms is the molecular biology and biochemistry of the metabolic pathways, themselves (Johnson and Spain, 2003). Table 8.9 provides a listing of key types of evidence for recent and ongoing evolution of catabolic pathways for pollutant compounds: patterns in sequenced biodegradation genes, their regulation, and the low specificity of encoded enzymes. The characteristics indicative of recent catabolic pathway evolution are consistent with and predictable from the four proposed mechanistic steps of pathway evolution shown in the lower half of Table 8.9 (Johnson and Spain, 2003).

Concrete examples of the principles of catabolic pathway evolution shown in Table 8.9 are detailed below for the biodegradation of atrazine and chlorobenzene.

Table 8.8

Selected examples of organic environmental pollutants that were initially considered nonbiodegradable and later reclassified as biodegradable (from J. Spain, personal communication; Seffernick and Wackett, 2001; Wackett et al., 2002; Johnson and Spain, 2003)

Compound	Use	Physiological role of compound	Genetic change conferring biodegradability
Melamine	Building block for early industrial polymers	Electron donor and carbon source	New enzymatic activity, melamine deaminase, in new combination with other genes
Atrazine	Chlorinated triazine herbicide	Electron donor and carbon source	New enzymatic activity, especially atrazine chlorohydrolase, in new combination with other genes
Styrene	Monomer for industrial polymer, polystyrene	Electron donor and carbon source	New enzymatic activity, styrene epoxide isomerase, in new combination with other genes
2,4-dinitrotoluene (2,4-DNT)	Manufacture of polyurethane foam; production of explosives	Electron donor and carbon source	New enzymatic activity, 2,4-DNT dioxygenase, in new combination with other genes
Chlorobenzene	Solvent for pesticide formulations; degreasing agent	Electron donor and carbon source	New enzymatic activity, chlorobenzene dioxygenase, in new combination with other genes
Nitrobenzene	Production of the industrial compound, aniline; ingredient in shoe and floor polishes	Electron donor and carbon source	New enzymatic activity, nitrobenzene nitroreductase, in new combination with other genes
Pentachlorophenol (PCP)	Antifungal wood preservative	Electron donor and carbon source	New enzymatic activity, PCP 4-monooxygenase, in new combination with other genes
Polychlorinated biphenyls (PCBs)	Dielectric fluid used in electrical transformers	If highly chlorinated, final electron acceptor. If lightly chlorinated, electron donor and carbon source	[There are 209 different forms (congeners) of PCB molecules; explanations for metabolism resist generalization]
Tetrachloroethene and trichloroethene	Industrial solvents and degreasing agents	Final electron acceptors	New enzymatic activity, tetrachloroethene reductive dehalogenase

Several of the listed compounds (especially styrene) may be naturally occurring. Thus, some metabolic pathways may have evolved prior to widespread manufacture and environmental release of the compounds.

Table 8.9

Evidence for the recent evolution of catabolic pathways for pollutant compounds and proposed mechanistic steps that produce new pathways (compiled from Johnson and Spain, 2003)

Biochemical and genetic evidence for early stage of catabolic pathway evolution
- Scattered organization within the host cell of the genes encoding the catabolic enzymes
- Primitive or inefficient regulation of enzyme synthesis
- Poorly adapted degradative enzymes (low affinity for the substrate and low catalytic specificity)
- Remnants of mobile genetic elements (such as transposons) that may have transferred the genes from other hosts

Proposed evolutionary steps that lead to new metabolic pathways
- Gene duplication and mutation allow for alteration or broadening of the substrate preference of the enzymes
- Genes and gene clusters are recruited and assembled via various horizontal-gene transfer mechanisms
- Variant strains of microorganisms able to use the new substrate can proliferate and selective pressure allows strains carrying the best adapted enzymes to thrive
- Further advantage is granted when organisms develop mechanisms that regulate the synthesis and activity of the enzymes

Case studies for the evolution of catabolic pathways: atrazine and chlorobenzene

Microorganisms evolved to be able to utilize atrazine and chlorobenzene as growth substrates. Discovering how this evolution occurred elegantly illustrates environmental microbiological inquiry (Box 8.10). Information in Box 8.10 focuses primarily upon the natural history and genetic basis of atrazine and chlorobenzene metabolism. The early industrial monomer, melamine, is the first entry in Box 8.10 because its pre-existing biodegradation pathway set the stage for atrazine.

Atrazine is a member of the s-triazine class of herbicides, developed in the 1950s, that kill susceptible plants by attaching to the quinone-binding protein in photosystem II (Wackett et al., 2002). The atrazine structure is similar to the diazine structures of DNA bases, cytosine and adenine, but the atrazine molecule features three major alterations (a Cl group, a third N within the ring structure, and *N*-alkyl substitutions; see Box 8.7) that impede microbial metabolism. Until the 1990s, all biodegradation studies showed that atrazine could not be used as a carbon and energy source by soil microorganisms: at best, a nonbeneficial (cometabolic) minor change in the atrazine molecule was documented. However, after 1993 several independent investigations were able to detect, isolate, and characterize bacteria that used atrazine as a carbon and energy source for growth. Extensive biochemical and genetic characterization of the *Atz* gene cluster, its related genes and enzymes, and the mobile genetic

Box 8.10

Three case studies demonstrating the evolution of metabolic pathways for the biodegradation of organic environmental contaminants (compiled from van der Meer et al., 1998; Seffernick and Wackett, 2001; Wackett et al., 2002; Müller et al., 2003)

Compound and its use	Details documenting evolution
Melamine: triazine synthetic polymer intermediate; introduced in the early 1900s	Considered nonbiodegradable in the 1930s, but reclassified as slightly biodegradable in the 1960s. Melamine is metabolized by five enzymes encoded by five genes (*trzABCDE*) to urea, which is hydrolyzed to ammonia and CO_2. The middle intermediary metabolite is cyanuric acid. The gene *trzA* encodes hydrolytic deamination reactions and is a member of the amidohydrolase family of enzymes
Atrazine: major triazine agricultural herbicide used to protect corn; introduced in the 1960s	Considered nonbiodegradable until the mid-1990s when bacteria able to grow on atrazine were isolated. Extensive biochemical and genetic characterization of atrazine biodegradation show that: • the metabolic pathway converts atrazine to cyanuric acid in three steps • the genes coding for enzymes responsible for the three steps (*atzABC*) all belong to the same family of enzymes (amidohydrolase), closely related to *trzA* (for melamine) • nearly identical *atzABC* gene sequences have been found in atrazine-metabolizing bacteria in North America, Europe, Australia, and Asia • although the genes have been found on plasmids (extrachromosomal, often self-transmissible, independently replicating, genetic elements), no single type of plasmid is common to all atrazine-degrading bacteria • sequence analysis of DNA flanking the *atz* genes has discovered evidence for transposons. Thus transposons may be involved in horizontal gene transfer of atrazine biodegradation genes between different hosts
Chlorobenzene: solvent in pesticide formulations, also used as a degreasing agent; produced by chlorination of benzene in the presence of a catalyst; first synthesized in 1851; used in high volume during World War I	Chlorobenzene contamination persisted in a shallow aquifer at Kelly Air Force base for ~30 years. In the 1990s the levels of contamination began to decline Chlorobenzene-degrading bacteria were not in the contaminated zone prior to the chlorobenzene spill, nor could they be isolated outside the contaminated zone. However, bacteria able to grow on toluene and chlorocatechol were present in the indigenous microflora across the site A bacterium able to grow on chlorobenzene, *Ralstonia* sp. JS705, was isolated from the contaminated aquifer and was characterized biochemically and genetically

Compound and its use	Details documenting evolution
	The chlorobenzene degradation pathway and the gene fragments encoding the enzymes represent a merger of two biodegradation pathways: toluene (*todCBA*) and chlorocatechol (*clcAB*). Thus, the chlorobenzene genes in the strain JS705 were assembled in situ within the contaminated aquifer
	Analysis of the DNA sequence of the new combination of genes allowed the investigators to reconstruct the genetic history of strain JS705. It was originally a toluene-degrading bacterium and received the chlorocatechol gene on a self-transmissible mobile genetic element, termed a "genomic island" (see Science and the citizen box, Chapter 5). Sequence analysis also implicated involvement of insertion (IS) elements, hence transposons (other agents of gene rearrangement), in achieving the final configuration of genes conferring chlorobenzene metabolism to strain JS705

elements that carry them (Box 8.10) makes it clear that a combination of gene duplication, mutational refinement, recombination, and proliferation was the broad mechanism that made atrazine biodegradable around the globe today.

The last case study in Box 8.10 describes microbial metabolism of chlorobenzene. The scenario portrayed for chlorobenzene is very different from that of atrazine because it is site-specific. Chlorobenzene is a colorless organic liquid with a faint, almond-like odor whose primary industrial use is the manufacture of dyestuffs and insecticides. It is also a solvent for adhesives, drugs, rubber, paints, dry cleaning, and textile manufacture. Long-term exposure to chlorobenzene may damage the liver, kidney, and central nervous system of humans. Information presented in Box 8.10 describes how chlorobenzene spilled into the subsurface at Kelly Air Force Base (near San Antonio, Texas), persisted for several decades, and then the concentrations began to decline. Culturable bacteria capable of growing on two compounds related to chlorobenzene (toluene, chlorocatechol) were found throughout the study site. Only after natural attenuation (see Section 8.3) of chlorobenzene had begun, was it possible to isolate (from the contaminated portion of the study site) bacteria that utilized chlorobenzene as a carbon and energy source. Biochemical and genetic analysis of the genes conferring chlorobenzene metabolism on the newly isolated bacterium revealed that the new gene cluster was a hybrid of those from toluene and chlorocatechol metabolic pathways and that two mobile genetic elements (a "genomic island" and transposons) were the likely causes of site-specific evolution of chlorobenzene catabolism (Box 8.10).

8.6 ENVIRONMENTAL BIOTECHNOLOGY: OVERVIEW AND EIGHT CASE STUDIES

Definitions and scope

According to Glick and Pasternak (2003), the term *biotechnology* was created in 1917 by a Hungarian engineer, Karl Ereky, to describe an integrated process for the large-scale production of pigs using sugar beets as their source of food. In Ereky's view, biotechnology was "all lines of work by which products are produced from raw materials with the aid of living things". In 1961, biotechnology was redefined by Carl Göran, as "the industrial production of goods and services by processes using biological organisms, systems, and processes" (Glick and Pasternak, 2003). A twenty-first century definition of biotechnology might be "the integrated use of biochemistry, molecular, biology, genetics, microbiology, plant and animal science, and chemical engineering to achieve industrial goods and services". An example is production of human insulin by cloning the gene into the bacterium, *Escherichia coli*.

It follows that *environmental biotechnology* is a subset of biotechnology that addresses environmental needs and problems, environmental goods and services. So, what are environmental needs, problems, goods, and services?

Figure 8.15a highlights several of the major global environmental problems faced by humans in the twenty-first century. Figure 8.15b illustrates the intricate connections between biotechnology and environmental biotechnology, between human needs, commercial needs, cultural needs, and between goods and services, economy and ecology.

Many boundaries between disciplines and human endeavors are crossed by environmental biotechnology. It extends from pollution control to food production and to medicine. Throughout this book, we have found that connections are both implicit and explicit in the science and practice of environmental microbiology and biogeochemistry. At the level of an individual bacterial cell, the cycles of carbon, nitrogen, sulfur, iron, and other elements are intimately linked (see Chapters 3 and 7). Similarly, at the levels of global circulation in the atmosphere and the oceans (see Chapter 4), human activities (e.g., emissions of CO_2, chlorofluorocarbon refrigerants, or the pesticide, DDT) in one place influence conditions in other places.

Traditional environmental biotechnology includes: agriculture (tilling the soil, sowing seeds, harvesting crops), improved agriculture (plant breeding, seed storage, *Rhizobium* inoculation for enhanced nitrogen fixation), composting, forestry, aquaculture, wastewater treatment, constructed wetlands, and microbial leaching of mining ores for metal recovery. More recent environmental biotechnology developments have utilized

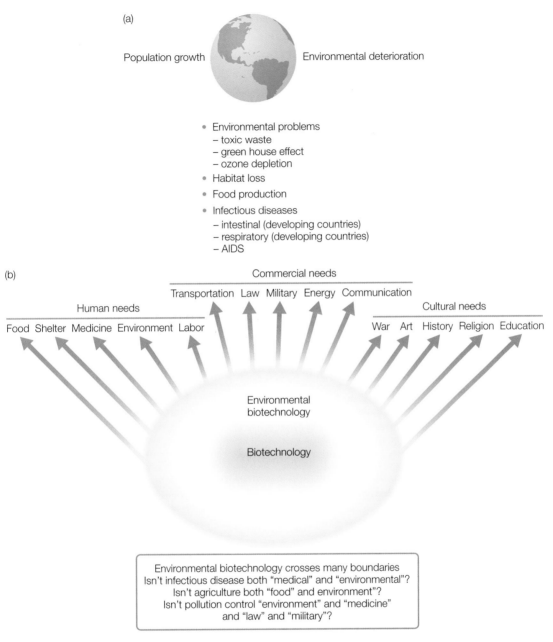

Figure 8.15 Overview of global environmental problems in the 21st century and the cross-disciplinary nature of environmental biotechnology. (a) The relationships between human population growth, environmental deterioration, and health. (b) The "goods and services" needed by humans are provided, in a cross-disciplinary manner, by biotechnology and environmental biotechnology.

genetic engineering approaches to address goods and services related to environmental and related concerns. The immense breadth of environmental technological issues cannot be treated here. However, eight case studies that sample the issues and approaches to environmental biotechnology appear below.

Our current civilization is fueled largely by petroleum, natural gas, and coal. There is a growing awareness that the economic, political, and environmental costs of fossil fuels are high (Logan, 2004). Therefore considerable intellectual and technological efforts are being directed toward discovering renewable energy sources for the twenty-first century. The first three case studies address energy biotechnology. The next four case studies illustrate how genetic engineering of microorganisms and plants has the potential to improve sustainability in agriculture, manufacturing, and environmental clean up. The final case study examines severe acute respiratory syndrome (SARS) as an example of how molecular detective work can discover the origin of a newly emerging human disease.

Background to genetic engineering

Genetic engineering (heralded as the "eighth day of creation", Judson, 1996) has transformed biology from a descriptive science into a powerful technological tool (Glick and Pasternak, 2003; Primrose and Twyman, 2006). In a nutshell, DNA technology allows DNA fragments (genes) to be identified, characterized, modified, and used to create novel products (chemicals or organisms) or processes, or (in the case of gene therapy) to cure diseases by correcting its genetic cause in vivo. A major success story in genetic engineering is the production of the human hormone, insulin (critical for treating diabetes). Diabetes affects millions of people who, due to a deficiency in insulin production, cannot regulate their own blood-sugar levels. Prior to genetic engineering, insulin was extracted from the pancreas of cows and pigs. The human genes for insulin were cloned and expressed in the bacterium, *E. coli*. Now genetically engineered insulin is an improvement over cow and pig insulin, plus it is more abundant and less expensive.

Case study I: harvesting electricity from anaerobic bacteria – microbial fuel cells

Sections 3.8–3.10 emphasized that a major theme of the "anaerobic way of life" is contending with final-electron-acceptor-limited habitats. Electron sinks are critical for ATP generation during electron transport in microorganisms. In essence, electrons in reduced substrates (e.g., carbohydrates, organic acids, H_2, S^{2-}, CH_4) begin an "electric circuit" that ends when the electrons reach the electron sink furnished by a final electron acceptor. In 2002, Bond et al. reasoned that it might be possible to establish an electrical connection between electron-acceptor-limited microorganisms in anaerobic sediments and the best electron sink of all, molecular oxygen (see Section 3.8). An electrode (anode) was inserted into highly reducing marine sediments rich in carbon and active anaerobic microorganisms. The anode was connected in an electric circuit to a cathode positioned in adjacent aerobic (O_2 rich) water. Bond et al. (2002) were able to measure an electric current in the circuit. The milliamp-level electric current was able to power small scientific devices that might be deployed in remote locations (Figure 8.16). Furthermore, the output of electric current from model fuel cell systems could be boosted simply by delivering a pulse of the reduced-carbon electron donor (Figure 8. 16). In essence, electricity was generated via oxidation of reduced carbon. Confirmation and refinement of these anaerobic electric fuel cells was obtained by Chaudhuri and Lovley (2003) who found that a pure culture of the iron-reducing bacterium *Rhodoferax ferroreducens* was able to carry out the same type of electric-current generation described above at ~80% efficiency, fueled with dissolved sugars, including glucose, fructose, sucrose, and xylose. The science and technology of microbial fuel cells have recently been reviewed by Logan and Regan (2006) and Logan et al. (2006). It is possible that the above types of microbial fuel cells could be the basis for specialized electricity generation in the future.

Case study 2: hydrogen gas from microbial fermentation

Hydrogen gas has been recognized as a very promising alternative to petroleum-based fuels – especially in automobiles (Logan, 2004). Hydrogen can be produced from renewable material such as biomass, and hydrogen can be transported and used in internal combustion engines. Furthermore, combustion of H_2 does not contribute to global warming by adding CO_2 to the atmosphere.

As mentioned above and in Section 3.8, anaerobic bacteria (e.g., *Clostridium*) and communities of anaerobic microbial populations are routinely limited by available electron acceptors. Hydrogen gas (H_2) generated by proton-reducing reactions is a common fermentation byproduct generated during electron-acceptor-limited microbial processes.

Ordinarily excess carbonaceous waste products from the food industry are managed in ways that consume, rather than generate, energy. To investigate the feasibility of generating H_2 from food industry waste, Oh and Logan (2005) monitored H_2 production in an anaerobic bioreactor supplied with heat-treated wastewater from a cereal food-processing plant. Hydrogen gas concentration reached a maximum of 32–39% in the bioreactor headspace. If properly engineered and refined, this "biogas" process could become the basis of a practical energy-generation technology. Assuming the organic waste being processed in the fermentation technology is glucose, the maximum theoretical yield of hydrogen gas is 12 moles per mole of glucose:

(a) • Harvesting electricity

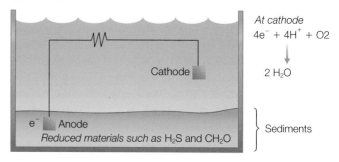

• ─\/\/\─ = Resistor ≡ Appliance   (bulb) ≡ Voltmeter, ammeter

(b) • Power generation in W/m^2 of electrode surface

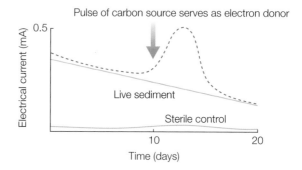

• *Contemporary application*: Power for marine oceanographic instruments (sensors, hydrophones, etc.)

• *Future application*: Highly engineered, highly efficient microbial fuel cells

Figure 8.16 Generation of electricity by microbial physiological processes in sediment. The electric circuit is created by connecting an anode in the electron-rich anaerobic sediment to a cathode in aerobic water. The figure shows an electric circuit (a), power output (b), and applications of the biogeochemical battery (c). Note that electrons flow from reduced materials at the anode to an electron acceptor (O_2) at the cathode (a) and that the addition of an organic carbon electron source causes a pulse of electric power (b). (Adapted from Bond et al., 2002.)

$$C_6H_{12}O_6 + 6H_2O \rightarrow 12\ H_2 + 6CO_2$$

Very real physiological constraints currently limit the theoretical yield, by a factor of three, to 4 moles of H_2 per mole of glucose. In the future, it is conceivable that genetic engineering of the fermenting microorganisms may boost the hydrogen production yield.

Case study 3: genetic engineering of photosynthesis-based production of hydrogen gas

Solar energy is unquestionably the most abundant source of energy on Earth. Each year across the void of space, the sun delivers to our planet ~5.7×10^{24} J of radiation (Miyake et al., 1999). This is ~10,000 times more than the total energy consumed by humans. From an evolutionary standpoint, the development of oxygenic photosynthesis (see Sections 2.8–2.11) was one of the most innovative of biological events. In searching for sources of clean, safe, renewable energy, genetic engineers have ideas of new light-harvesting innovations.

As presented in Chapters 2 and 3 (Sections 2.10, 3.7, and 3.9), oxygenic photosynthesis's key advance was to use energy in light to split water into oxygen and protons – liberating high-energy electrons that can flow through electron transport chains to create the cell's energy currency, ATP. A goal pursued by some genetic and biochemical engineers is to modify electron flow in photosynthesis so that it is directed towards enzymes that synthesize hydrogen gas. Hydrogen gas production (in small quantities) was discovered in cyanobacteria in the 1890s and in eukaryotic green algae in the 1940s (Prince and Kheshgi, 2005). Though the physiological reactions are a reality, selective evolutionary forces have not led to process optimization. Figure 8.17 illustrates the basic bioengineering concept that fuses two well-known systems: the photosystem that generates high-energy electrons and ATP and the nitrogenase system which uses electrons and ATP to synthesize H_2 gas (see Section 7.4, Box 7.6). Also shown in Figure 8.17 is a photoheterotroph-based strategy (e.g., *Rhodopseudomonas palustris*) in which the bacterial-type anoxygenic photosynthetic apparatus generates ATP and organic substrates serve as electron donors. Prince and Kheshgi (2005) have stated that hydrogen production via hydrogenase, instead of nitrogenise, offers advantages that include higher efficiency and elimination of ATP consumption. Although the basic concept of coupling photosynthetic apparatus to proton-reducing apparatus is highly meritorious, there are many details that need to be addressed before the idea becomes a technology. Such details include:

- Which photosystem in which host? (Oxygenic photosynthesis in algae cyanobacteria or anoxygenic photosynthesis in a bacterium?)
- Which proton-reducing system? (Nitrogenase or hydrogenase?)
- How can the oxygen-sensitive proton-reducing systems (hydrogenase and nitrogenase) be protected from O_2?
- How can photosystem efficiency be boosted?
- How can hydrogen-consuming pathways in host cells be eliminated?
- Will hybrid protein systems assemble and mature and retain their functionality?

Biological H_2 production in microorganisms has recently been reviewed by Rupprecht et al. (2006) and by Prince and Kheshgi (2005).

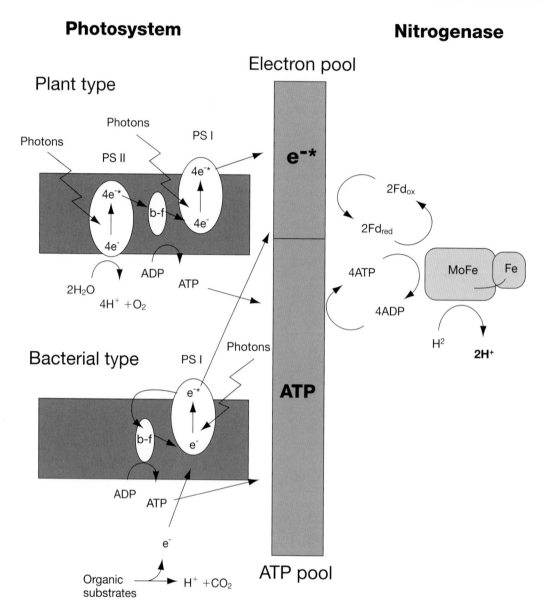

Figure 8.17 Prototype mechanism of light-to-hydrogen conversion using a combination of photosystems (PSI and PSII) from plants or bacteria combined with nitrogenase system. The figure shows electron flow originating from both light energy (used to split water, middle left) and the oxidation of organic substances (lower left). Two critical physiological pools result: electrons and adenosine triphosphate (ATP). Both nitrogenase and hydrogenase enzyme systems have the potential to use electrons to convert protons to hydrogen gas. Substituting hydrogenase enzymes for nitrogenase (shown) may offer advantages such as higher efficiency and no ATP demand (see text). ADP, adenosine diphosphate; Fd, ferredoxin. (Modified from Miyake, J., M. Miyake, and Y. Asada. 1999. Biotechnological hydrogen production: research for efficient light energy conversion. *J. Biotechnol.* 70:89–101. Copyright 1999, with permission from Elsevier.)

Case study 4: genetically engineering the Bt insecticide in crop plants

Bacillus thuringiensis (Bt) is the ubiquitous spore-forming bacterium that produces insecticidal protein crystals called "Bt toxin". Though the bacterium was discovered in 1901, spores and protein crystals from this bacterium were not used commercially as an insecticide until the 1950s (Glick and Pasternak, 2003). Within the bacterial spore is a "parasporal body" that houses what is known as the "pro-toxin", an inactive form of the toxin. When the parasporal crystal is ingested by the target insect, proteolytic cleavage releases the active toxin. The toxin inserts itself into the membrane of the gut epithelial cells of the insect, causing excessive loss of ATP through an ion channel. Cellular metabolism ceases, the insect stops feeding, becomes dehydrated, and eventually dies (Glick and Pasternak, 2003).

The genes for Bt toxin have been cloned and transferred to the chloroplasts of many crop plants (e.g., cotton, potatoes). Vast plantings of these crops no longer need to be sprayed with chemical insecticides that may be inefficiently applied and have unanticipated environmental impacts. Though the potential negative environmental impacts of genetically engineered organisms are highly controversial, it is undeniable that Bt crops have had dramatic environmental and economic benefits.

Case study 5: engineering crop plants for herbicide resistance

The herbicide glyphosate is a synthetic compound similar to the amino acid, glycine (Figure 8.18). Glyphosate inhibits a key enzyme involved in the biosynthesis of aromatic amino acids in both plants and bacteria. A gene encoding a glyphosate-resistant version of the biosynthetic enzyme was discovered in *E. coli*. That gene was cloned and introduced into plant cells in tobacco, petunia, tomato, potato, and cotton crops (among others). The transgenic crops produce a sufficient amount of the enzyme to allow amino acid synthesis to proceed when their own enzyme is inactivated by glyphosate application. The result is a "package deal" from the manufacturer: when glyphosate is applied to farm land, only planted transgenic crops resist the applied herbicide.

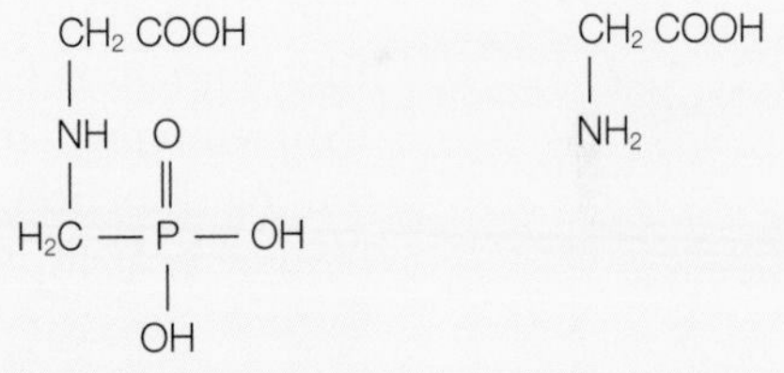

Figure 8.18
Structures of the herbicide, glyphosate, and the amino acid, glycine.

Case study 6: genetic engineering of plants to detoxify mercury

In Chapter 7 (Sections 7.1 and 7.3), we discussed interactions between microorganisms and mercury. Some of the interactions were also mentioned as mercury bioremediation strategies in Section 8.3. An elaboration of the mercury biogeochemical cycle is depicted in Figure 8.19.

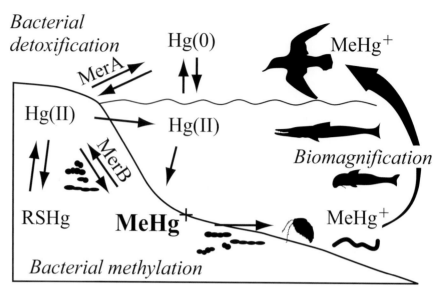

Figure 8.19 The mercury biogeochemical cycle. MeHg$^+$, methyl mercury; RSHg, mercury bound to organic molecules via sulfur (Reprinted by permission from Macmillan Publishers Ltd: *Nature Biotechnology*, from Bizily, S.P., C.L. Rugh, and R.B. Meagher. 2000. Phytodetoxification of hazardous organomercurials by genetically engineered plants. *Nature Biotechnol.* **18**:213–217. Copyright 2000.)

Mercury is found in the naturally occurring mineral, cinnabar, and in other minerals. Mercury has many industrial uses – in, for example, thermometers, biocides, industrial catalysts, and gold-mining operations. Unfortunately, mercury is also a highly toxic neurotoxin and impairs kidney function. Methyl mercury impedes biochemical reactions by binding to DNA, RNA, and sulfhydryl groups in proteins. While the cationic form of mercury Hg(II) is likely the most environmentally abundant, it is converted to the highly toxic methyl mercury (CH$_3$Hg$^+$), largely by sulfate-reducing populations in aquatic sediments. As mentioned in Sections 7.1 and 7.3, pure cultures of bacteria capable of detoxifying (mobilizing) Hg(II) and CH$_3$Hg$^+$ have been characterized, and genes encoding the key reactions (Figure 8.20) have been cloned and sequenced. *merB* encodes the enzyme (organomercurial lyase) that converts methyl mercury to Hg(II). Furthermore, *merA* encodes the enzyme (mercuric reductase) that reacts with reduced nicotinamide adenine dinucleotide phosphate (NADPH) to convert Hg(II) to elemental mercury Hg(0).

$$R\text{–}CH_2\text{–}Hg^+ + H^+ \xrightarrow{\text{MerB}} R\text{–}CH_3 + Hg(II)$$

$$Hg(II) + NADPH \xrightarrow{\text{MerA}} Hg(0) + NADP^+ + H^+$$

Figure 8.20 The bacterial mercury-processing enzymes, organomercurial lyase (MerB) and mercuric reductase (MerA), catalyze the reactions shown.

In order to investigate the feasibility of using plants to detoxify mercury-contaminated soils and sediments, Bizily et al. (2000) transferred and expressed *merAB* in *Arabidopsis thaliana* (a small flowering plant related to cabbage that is easily amenable to genetic engineering). When grown hydroponically, not only did the transgenic plants resist toxic concentrations of CH_3Hg^+ and Hg(II), but they converted these compounds to Hg(0). These data establish the principle that transgenic plants have the potential to grow in mercury-contaminated soils and mobilize mercury out of that habitat. Presumably, once removed from soil, the vapor-phase Hg(0) would be scrubbed and recovered from the air.

Case study 7: biodegradable plastic

Carbonaceous, cellular reserve storage products that guard against starvation in bacteria (see Section 3.5) were described as early as 1926 (Lenz and Marchessault, 2005). Polyhydroxyalkanoates (PHAs; Figure 8.21) occur as high molecular weight (mass of 50–100 kDa) storage bodies in many types of Gram-positive and Gram-negative bacteria (Figure 8.22). At least 75 different genera of bacteria have been shown to produce PHAs and

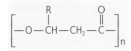

Figure 8.21 Molecular structure of polyhydroxyalkanoate molecules that constitute biodegradable plastics. (Reprinted from Reddy, C.S.K., R. Ghai, Rashmi, and V.C. Kalia. 2003. Polyhydroxyalkanoates: an overview. *Bioresource Technol.* **87**:89–101. Copyright 2003, with permission from Elsevier.)

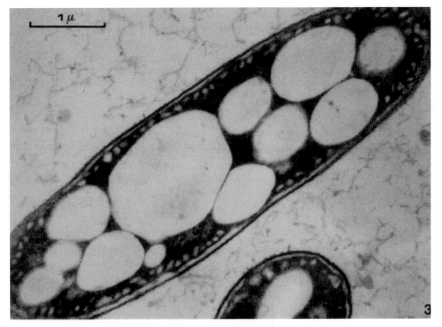

Figure 8.22 Photomicrograph of polyhyroxyalkanoate bodies within bacterial cells. This transmission electron micrograph of an ultrathin section shows *Azotobacter chroococcum*. (Reprinted with permission from Lenz, R.W. and R.H. Marchessault. 2005. Bacterial polyesters: biosynthesis, biodegradable plastics and biotechnology. *BioMacromolecules* **6**:1–8. Copyright 2005, American Chemical Society.)

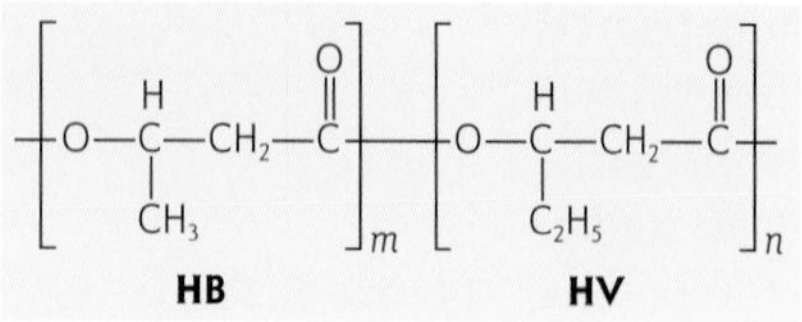

Figure 8.23 Structure of the microbially synthesized plastic co-polymer, hydroxybutyrate (HB) hydroxyvalerate (HV). (Reprinted with permission from Lenz, R.W. and R.H. Marchessault. 2005. Bacterial polyesters: biosynthesis, biodegradable plastics and biotechnology. *BioMacromolecules* **6**:1–8. Copyright 2005, American Chemical Society.)

these exhibit many structural variations. The R group shown in Figure 8.21 symbolizes an akyl side chain that varies in length and branching mode so that more than 100 different monomer units have been identified as constituents of PHA (Reddy et al., 2003).

PHAs are nontoxic polymers that are biocompatible (nonallergenic, nontoxic) and biodegradable. They can be produced from renewable resources and feature many traits similar to modern synthetic plastics: a high degree of polymerization, highly crystalline, optically active, and insoluble in water. One remarkable bacterium, *Ralstonia eutrophus*, is capable of channeling as much as 90% of its dry weight into PHA. Moreover, the chemical composition (hence physical traits) of PHA polymers can be manipulated by varying the organic substrates fed to the PHA-producing bacteria. For instance, the PHA co-polymer formed by *R. eutrophus* grown on a mixture of glucose and proprionic acid is a random polyester of hydroxybutyrate (HB) and hydroxyvalerate (HV; Figure 8.23). The HB–HV co-polymer has a lower melting point than a HB homopolymer – conferring better mechanical and processing characteristics.

As early as 1982, Imperial Chemical Industries (ICI) in the UK announced a product development program on PHA-based biodegradable plastics – creating the plastic containers dubbed "Biopol" (Figure 8.24). Over the past two decades, other corporations in the USA and Japan have created biodegradable plastic products ranging from packaging materials to agricultural mulch film to resins. Overall, three types of biodegradable plastics have been introduced: photodegradable, semibiodegradable (a hybrid of starch-based polymer and polyethylene), and completely biodegradable (Reddy et al., 2003).

One production strategy for PHA-based plastics is to use industrial bioreactors (fermentors) to grow PHA-synthesizing heterotrophic bacteria on carbon sources ranging from glucose to methanol to cane molasses to cheese whey. Clearly the economic viability of PHA-based plastic manufacturing is dependent upon many factors – including the costs of the bacterial growth substrates.

Another strategy for PHA-based plastics involves transgenic plants. The biochemical pathways for PHA biosynthesis begin with the intermediary metabolite, acetyl coenzyme A (CoA). Three enzymes encoded by the gene cluster *phbCAB* create the PHA polymer. The *phbCAB* gene cluster has been cloned and expressed in plastids of *Arabidopsis thaliana* – successfully producing the PHA polymer (Reddy et al., 2003). For potential commercialization in plants, oil seed crops (e.g., rape seed, sunflower) and starch crops (e.g., corn) represent good prospects for large-scale agricultural production of PHA plastics. In the end, marketplace issues, such as consumer

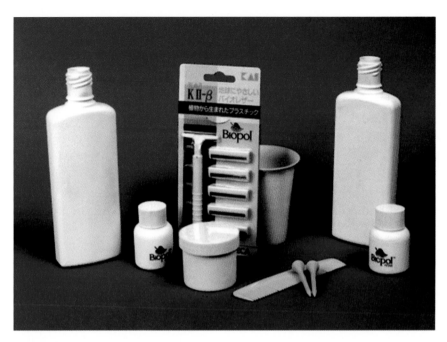

Figure 8.24 Commercially manufactured, biodegradable, molded plastic containers created from microbial polyesters. (Reprinted with permission from Lenz, R.W. and R.H. Marchessault. 2005. Bacterial polyesters: biosynthesis, biodegradable plastics and biotechnology. *BioMacromolecules* **6**:1–8. Copyright 2005, American Chemical Society.)

demand and the price of traditional plastics, will govern the destiny of biodegradable plastic products.

Case study 8: environmental epidemiology of a newly emerged disease, SARS

Severe acute respiratory syndrome (SARS) erupted late in 2002, eventually sickening thousands of people around the globe. The illness was first reported in the Guangdong Province of the People's Republic of China where several hundred people exhibited a severe and unusual form of pneumonia. After similar cases were detected in patients in Hong Kong, Viet Nam, and Canada during February and March 2003, the World Health Organization (WHO) issued a global alert for the illness designated "SARS" (Rota et al., 2003). Transmissibility of SARS became obvious when health-care workers and household members who had cared for SARS patients in Hong Kong and Viet Nam also exhibited symptoms. Many of the SARS cases could be traced to contacts with a single health-care worker from Guangdong Province, who visited Hong Kong, where he was hospitalized with pneumonia and died. The disease spread at an alarming rate and had an equally alarming fatality rate: by the end of April 2003, >4300 SARS cases and 250 SARS-related deaths had been reported to the WHO from over 24 countries. In some instances, death due to respiratory failure reached 10%. All told, 800 people died.

• **What was the origin of the new disease?**
Answer: It had "jumped" from animals to humans.
• **What tools are available to characterize the disease and its causative agent?**
Answer: Epidemiology, culture-based isolation of the infectious agent, genomics, and structural biology.

In an extraordinary effort, the WHO organized a network of 13 laboratories in 10 countries. These teams identified a virus (by repeated cultivation in human tissue-culture cells) associated with SARS during the third week of March 2003. Two weeks later, the entire genome of the causative agent, a carona virus, was sequenced (~30,000 bp; Marra et al., 2003).

It was no surprise that the type of virus associated with SARS was a carona virus. Carona viruses are RNA viruses with a long history of causing respiratory and intestinal diseases in humans and other animals. Typically, they have narrow host ranges.

• **So where did the new virus come from?**
• **What was present in the Guangdong Province of China that allowed the SARS pathogen to "emerge"?**

The current thought is that masked palm civet cats (Figure 8.25), traded in large, open-air markets of China, were the source of the carona virus. More than 13% of approximately 500 animal traders from Guangdong Province tested positive for the virus in blood samples taken during the initial SARS outbreak. (In control groups only 1–3% of animal traders tested positive.) Domesticated cats and ferrets are other animals in Guangdong Province that are easily infected by the virus and routinely contact humans. Some public health experts have recommended that China develop new strict regulations on commercial animal handling and trading that may curb emergence and spread of disease in the future.

• **But what allowed the host range of the SARS virus to broaden from civet cats to humans?**

Genomic sequence comparisons showed that viruses specific to animals and humans differed by only a handful of mutations. Li et al. (2005) analyzed the influence of a change in one protein in the SARS virus that seems critical for recognition of its human host. The crystal structure of

Figure 8.25 Photograph of the masked palm civet cat traded in open-air markets in China. These animals are thought to be a likely source of the SARS virus. (Courtesy of Robert Siegel, Stanford University, with permission.)

a spike-shaped viral recognition protein was analyzed while in physical contact with its attachment (and fusion) site on human cells. By examining the details of molecular bonding in the virus–human complex, Li et al. (2005) discovered that two amino acids differ from those in strains of the animal viruses unable to infect humans. It is thought that the conformational change effected by the new amino acids allowed the altered animal virus to recognize its (new) human host. New information about the structural mechanism for virus-host recognition may assist in the development of carona virus vaccines that prevent infection (Li et al., 2005).

Thus, information from a multitude of sources (natural history, epidemiology, structural biology, and genomics) converged to help explain the emergence of SARS.

8.7 ANTIBIOTIC RESISTANCE

Our shield against infections

A revolutionary advancement in modern medicine was the discovery of antibiotics – compounds that can be taken orally or intravenously by humans to destroy microbial agents of disease. Using the lexicon from werewolf folklore, antibiotics are "silver bullets" – able to act selectively on target organisms, while sparing the infected host. The antibiotic revolution was initiated by A. Fleming's 1929 discovery of penicillin (which interferes with cell wall synthesis primarily in Gram-positive bacteria) and by the 1944 discovery by A. Schatz and S. Waksman of streptomycin (which interferes with ribosomal protein synthesis in both Gram-negative and Gram-positive bacteria). Prior to antibiotics, chemotherapy for microbial infections relied on sulfa drugs (e.g., sulfonamides; often allergenic in humans) and arsenic-containing compounds (often toxic to humans).

An appreciation of the impact of antibiotics on contemporary human well being can be developed by imaging a world without antibiotics.

• **How safe would you and your doctor feel if pharmacies no longer filled prescriptions for antibiotics?**

Table 8.10 provides a listing of the many diseases and bacterial infectious agents that have the potential to threaten human health in the United

Table 8.10

Reported bacterial diseases and infectious agents in the United States (modified from MADIGAN, M. and J. MARTINKO. 2006. *Brock Biology of Microorganisms*, 11th edn, p. 833. Prentice Hall, Upper Saddle River, NJ. Copyright 2006. Reprinted by permission of Pearson Education, Inc., Upper Saddle River, NJ)

Anthrax	Pertussis
Botulism	Plague
Brucellosis	Psittacosis
Chancroid	Q fever
Chlamydia trachomatis	Rocky Mountain spotted fever
Cholera	Salmonellosis
Diphtheria	Shigellosis
Ehrlichioss	Streptococcal diseases, invasive, group A
Enterohemorrhagic *Escherichia coli*	Streptococcal toxic shock syndrome
Escherichia coli O147:H7	*Streptococcus pneumoniae*, drug-resistant and invasive disease
Gonorrhea	Syphilis
Haemophilus influenzae, invasive disease	Tetanus
Hansen's disease (leprosy)	Toxic shock syndrome
Hemolytic uremic syndrome	Tuberculosis
Legionellosis	Tularemia
Listeriosis	Typhoid fever
Lyme disease	Vancomycin intermediate *Staphylococcus aureus* (VISA)
Meningococcal disease	Vancomycin-resistant *Staphylococcus aureus* (VRSA)

Viral-, fungal-, protozoan- and helminth-caused infections were omitted from this list. These agents of disease pose special challenges for development of selective therapeutic agents because they rely on eukaryotic-type cellular metabolism.

States and other countries. Figure 8.26 illustrates the impact of antibiotics on the US population by contrasting the total rate of mortality (per 100,000 people per year; black line) with the mortality rates from noninfectious (blue line) and infectious (red line) causes from 1900 to 1996. Prominent features of the graph are:

- The death rate from noninfectious causes has remained nearly constant.
- A short-lived spike in the death rate occurred in 1918 that is attributable to the 1918 influenza pandemic.
- The death rate from infectious disease has plummeted dramatically.

The decline in infectious disease-caused death reflects several factors, such as improved sanitation, medical care, and standard of living (Armstrong et al., 1999). But a major contribution to the decline was the clinical use of sulfa drugs (1935), penicillin (1941), and streptomycin (1943). In the early years, demand for antibiotics in the treatment of infectious disease often far exceeded supply. For example, during World War II, penicillin was in such scarce supply that it was routinely recovered from the urine of patients in hospital wards, and then reused.

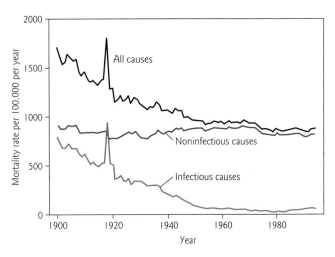

Figure 8.26 Mortality rates in the US population for all causes (black line), noninfectious causes (blue line), and infectious diseases (red line). (From Armstrong, G.L., L.A. Conn, and R.W. Pinner. 1999. Trends in infectious disease mortality in the United States during the 20th century. *J. Am. Med. Assoc.* **281**:61–66. Copyright 1999, American Medical Association. All rights reserved.)

A shield that needs constant renewal

Antibiotics are effective in thwarting disease-causing agents because the biochemical structure and function of the agents are distinctive from those of humans. In Chapter 2 (Section 2.12, Table 2.3) we reviewed some of the major biochemical distinctions between *Eukarya*, *Archaea*, and *Bacteria*. What arose during evolution translates into drug design strategies that are crucial for medicine and public health. Key cellular targets for antibiotics in *Bacteria* are cell wall synthesis, nucleic acid replication, protein synthesis, cytoplasmic membrane, and folic acid metabolism. Information in Figure 8.27 provides a summary of the modes of action of 30 antibiotics in widespread use today (Madigan and Martinko, 2006).

But there is a problem: antibiotics that were effective yesterday, become ineffective today. Table 8.11 displays the deployment dates of 10 antibiotics and the dates that microorganisms were found to be resistant

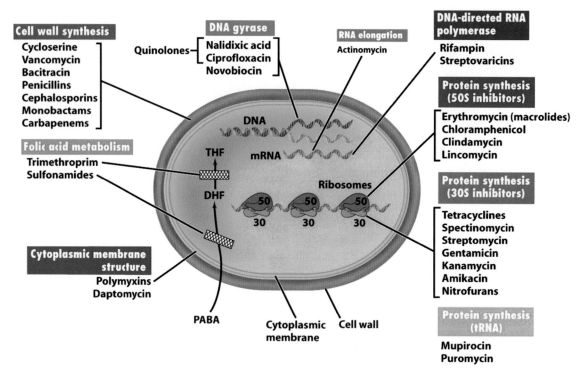

Figure 8.27 Mode of action of major antimicrobial agents. DHF, dihydrofolate; DNA, deoxyribonucleic acid; mRNA, messenger RNA; PABA, para-amino benzoic acid; THF, tetrahydrofolate; tRNA, transfer RNA. (From MADIGAN, M. and J. MARTINKO. 2006. *Brock Biology of Microorganisms*, 11th edn, p. 683. Prentice Hall, Upper Saddle River, NJ. Copyright 2006. Reprinted by permission of Pearson Education, Inc., Upper Saddle River, NJ.)

Table 8.11

Dates of deployment of representative antibiotics and the evolution of resistance (from Palumbi, S.R. 2001. Humans as the world's greatest evolutionary force. *Science* **293**:1786–1790. Reprinted with permission from AAAS)

Antibiotic	Evolution of resistance to Antibiotics	
	Year deployed	Year resistance observed
Sulfonamides	1930s	1940s
Penicillin	1943	1946
Streptomycin	1943	1959
Chloramphenicol	1947	1959
Tetracycline	1948	1953
Erythromycin	1952	1988
Vancomycin	1956	1988
Methicillin	1960	1961
Ampicillin	1961	1973
Cephalosporins	1960s	Late 1960s

to each (Palumbi, 2001). Antibiotic-resistant disease-causing microorganisms are a serious concern for us all. As one antibiotic loses effectiveness, it must be replaced by another. Thus, our shield against diseases is a slowly eroding one that needs constant renewal.

Mechanisms of antibiotic resistance

Each antibiotic targets a specific cellular function (such as inhibiting peptidoglycan biosynthesis or altering membrane sterols to increase cell permeability). In the face of selective pressure for survival, bacteria have responded with adaptations that resist antibiotics in eight different ways (Table 8.12; Davies, 1994). Key mechanisms of antibiotic resistance include: efflux pumps that reduce the intracellular concentration of the antibiotic, enzymes that inactive the antibiotic, and alteration of the structure of the cellular site targeted by the antibiotic (Table 8.12).

The eight mechanisms of antibiotic resistance are phenotypes. The way to thwart the development of such phenotypes is to understand their

Table 8.12

Mechanisms of antibiotic resistance in bacteria. The genes for each of these resistance traits can be transferred between bacteria (from Davies, J. 1994. Inactiviation of antibiotics and the dissemination of resistance genes. *Science* **264**:375–382. Reprinted with permission from AAAS)

Mechanism	Antibiotic
Reduced uptake into cell	Chloramphenicol
Active efflux from cell	Tetracycline
Modification of target to eliminate or reduce	β-lactams (e.g., penicillin G, amoxicillin)
binding of antibiotic	Erythromycin, lincomycin
Inactivation of antibiotic by enzymic modification:	
Hydrolysis	β-lactams
	Erythromycin
Derivatization	Aminoglycosides
	Chloramphenicol
	Fosfomycin
	Lincomycin
Sequestration of antibiotic by protein binding	β-lactams
Metabolic bypass of inhibited reaction	Fusidic acid
Binding of specific immunity protein to antibiotic	Sulfonamides
	Trimethoprim
	Bleomycin
Overproduction of antibiotic target (titration)	Sulfonamides
	Trimethoprim

underlying genetic basis. The critical questions that address the vicious cycle between new antibiotic deployment and microbial resistance are:

1 How do new *genotypes* for drug resistance arise?
2 How do the genotypes become widely distributed among pathogenic microorganisms?

The antibiotic resistance gene pool

As stated clearly by Davies (1994) and documented convincingly by D'Costa et al. (2006), antibiotic-producing microorganisms that are native to natural habits (such as soil) constitute a major reservoir for genes that encode antibiotic resistance. For decades, the filamentous Gram-positive streptomycetes have been used by pharmaceutical companies as a source of antibiotic discovery and production. These same organisms have self-protection mechanisms. D'Costa et al. (2006) isolated 480 *Streptomyces* bacteria and screened them for resistance to 21 different antibiotics or drugs (including all major types and targets of activity; some having seen clinical use for decades and some brand new). The results were chilling:

• Every *Streptomyces* strain was resistant to multiple drugs (the average number of resistances was seven or eight).
• Two strains were resistant to 15 of the 21 drugs.
• Resistances were found for both naturally occurring and for chemically synthesized antibiotics.
• Many of the resistances were linked to enzymatic inactivation of the antibiotics.

While self-protection mechanisms in streptomycetes have long been known (the producer *needs* to be resistant), the breadth of genes conferring multiple drug resistance and their effectiveness on antibiotics not yet in wide clinical use were sobering.

• **Do mutation and recombination also contribute to the genetic pool of antibiotic resistance?**
Answer: Yes. For example, J. Davies (1994) has stated that the role of mutation is especially important in the evolution of resistance to β-lactam antibiotics, such as penicillins and cephalosporins. The precise mechanisms by which antibiotic-resistance genes and gene clusters evolve have yet to be discovered. But the scenario described in Section 8.5 for the evolution of new metabolic pathways (gene duplication, mutation, recruitment, recombination, and selection) generally applies.

In summary, there are eight mechanisms of antibiotic resistance (Table 8.12) and their underlying genetic bases are derived from both short-term mutation/selection and a large long-standing reservoir of antibiotic-producing microorganisms in nature. The question, "how do the genotypes become widely distributed among pathogenic microorganisms", is addressed next.

Dissemination of resistant genes

It is one thing to be aware that genetic mechanisms of antibiotic resistance may exist in some soil-dwelling strain of *Streptomyces* in some remote location. It is quite another thing to know that the host of a mobile genetic element (e.g., plasmid or transposon) conferring resistance to vancomycin is here in the hospital in patients being treated for an infectious disease with vancomycin. The first scenario is cause for mild concern. The latter scenario is alarming.

Even though there are many barriers against facile exchange of genetic material between taxonomically unrelated microorganisms (e.g., low frequency of contact, surface exclusion, enzymes that digest foreign DNA; Thomas and Nielsen, 2005), retrospective studies prove that it does occur (see Section 5.9). Like mutations, horizontal gene transfer frequencies may be rare. But if the selective pressure for such rare events is strong, then the new genetic combinations [host plus transferred gene(s)] will flourish. *Thus, it is crucial to realize that proliferation of antibiotic resistance requires two on-going events*:

1 The transfer itself.
2 Selective pressures that allow the transferred genes to proliferate in the gene pool.

Davies (1994) has provided a model for the acquisition of genetic determinants for antibiotic resistance in pathogenic bacteria (Figure 8.28). The model begins (top of Figure 8.28) with an antibiotic-resistance gene pool (whose origins were addressed above). The four fundamental steps in the dissemination of resistance genes across taxonomic boundaries to pathogens are:

1 Uptake into cell cytoplasm of the resistance-conferring genetic elements; the best characterized hypothetical mechanisms are transformation (uptake of naked DNA), conjugation (intercellular delivery of plasmids between adjacent cells), and transduction (virus-mediated genetic exchange).
2 Formation of small multidrug-resistant mobile genetic elements. [The examples of mobile genetic elements show in Figure 8.28 are integrons. Integrons have been discovered to be widespread in nature. They are mobile genetic elements with a relatively conserved structure that includes an enzyme (integrase) that mediates insertion into host DNA, a "cassette" area that can carry and receive antibiotic-resistance genes, and several promoters that activate gene expression.]
3 Intracellular incorporation of the small multidrug-resistance cassettes into larger replicating elements such as plasmids.

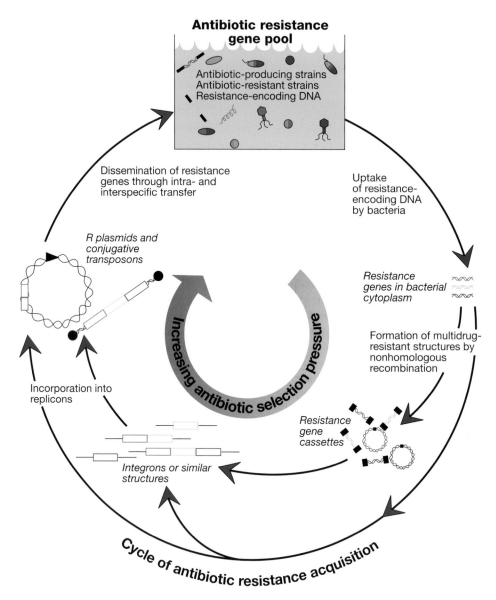

Figure 8.28 A scheme showing the route by which antibiotic resistance genes are acquired by bacteria in response to the selection pressure of antibiotic use. The resistance gene pool represents all potential sources of DNA encoding antibiotic resistance determinants in the environment. This includes hospitals, farms, or other places where antibiotics are used to control bacterial development. After uptake of single- or double-stranded DNA by the bacterial host, the incorporation of the resistance genes into stable replicons (DNA elements capable of autonomous replication) may take place by several pathways which have not yet been identified. The involvement of integrons, as shown here, has been demonstrated for a large class of transposable elements in a group of bacteria known as the Enterobacteriaceae. The resulting resistance plasmids could exist in linear or circular forms in bacterial hosts. The final step in the cycle – dissemination – is brought about by one or more of the gene transfer mechanisms that commonly include transduction, transformation, and/or conjugation (see text for details). (From Davies, J. 1994. Inactivation of antibiotics and the dissemination of resistance genes. *Science* **264**:375–382. Reprinted with permission from AAAS.)

4 Dissemination of the resistance-encoding DNA back to microbial communities that constitute the antibiotic-resistant gene pool.

An urgent need to curb development of antibiotic resistance

There is an overwhelming consensus among experts today that the current (and developing) arsenal of antimicrobial therapeutic agents may soon fail to keep pace with the microbial adaptation to antibiotics. Table 8.13

Table 8.13

Calls from the scientific and medical professions for reform in the use of antimicrobial compounds so that they remain effective (compiled from Bell, 1998; Levy 2005; Weber, 2005)

Source	Message
Journal of the American Medical Association, 2005. Editorial: "Appropriate use of antimicrobial drugs. A better perspective is needed", by J. T. Weber	• Doctors and patients may be inappropriately overusing antibiotics. This may hasten the emergence of antibiotic resistance in pathogenic bacteria. Efforts to curb inappropriate use of antibiotics should be strengthened
Center for Disease Control. *Emerging Infectious Diseases*, 1998. "Controversies in the prevention and control of antimicrobial drug resistance", by D. Bell	• In hospitals: antimicrobial resistance is spread via high rates of antimicrobial drug use and inadequate infection-control processes • In communities: antibiotics may be prescribed when not required – partly to meet patient demands and partly be "safe, rather than sorry" • In veterinary medicine: both medical and nonmedical use of antibiotics should be reduced • In developing countries: there are problem in drug resistance in treating diarrheal diseases, sexually transmitted diseases, pneumonia, tuberculosis, malaria, and hospital infections; resources for surveillance and control of antibiotic resistance are inadequate • In clinical laboratories: pathogenic bacteria isolated from diseased hospital patients need to be more carefully screened for low-level resistance to antibiotic therapy
Advanced Drug Delivery Reviews, 2005. "Antibiotic resistance – the problem intensifies", by S. B. Levy	• The frequency of antibiotic-resistant bacteria is increasing – especially worrisome is the emergence of new combinations of multidrug resistance in patients suffering from infectious disease

provides excerpts from three authoritative reports describing the improper use of antibiotics. Such calls for "restrained, proper use of antibiotics" are largely from the medical and public health professionals who rely on antimicrobial therapy to cure human disease. The concern is urgent because the ineffectiveness of antibiotic treatments seems to be accelerating. Resistance to antibiotics in microorganisms is caused by two simultaneous factors:

(Mobile resistance genes) × (Selective pressure from released antibiotics) = Widespread resistance

In the above equation, humans have little control over spontaneous horizontal gene transfer and evolution in nature. In contrast, the daily actions of people (individual citizens, doctors, health care workers, veterinarians, livestock managers, the poultry industry, the swine industry, farmers, marketers, pharmacists, the pharmaceutical industry, the agricultural industry) have a direct impact on releasing antibiotics to our bodies, animals, soils, sediments, and waters where such releases foster an antibiotic-resistance gene pool.

Vast amounts of antibiotics are released

If alive today, Fleming, Schatz, and Waksman would be shocked at the tonnage of antibiotics produced and disseminated globally. Antibiotic use has expanded from therapeutic cures to prophylactic disease prevention, to promotion of growth in the large-scale farming of animals (Table 8.14). Thus, commercial markets for antibiotics in human health care, veterinary

Table 8.14

Medical and commercial uses of antibiotics (compiled from Levy, 2002)

Type of antibiotic use	Reason	Type of industry
Therapeutic	Remedial treatment of disease	Human medicine, veterinary
Prophylaxis	Disease prevention	Human medicine, veterinary, aquaculture, honey bees, agriculture
Animal growth promotion	Subtherapeutic doses to promote growth in animals	Livestock and food production

medicine, and food production industries are immense. One source (Wenzel and Edmond, 2000) states that in the United States each year, 160 million prescriptions are written for humans and 23 million kg (23,000 metric tons) of antibiotics are used – approximately 50% used by patients and 50% used in animals, agriculture, and aquaculture. Thus, the US population of 300 million people received an average of 30 prescriptions per 100 persons per year and ~4 kg of antibiotics per 100 persons per year (Wenzel and Edmond, 2000). Another source (Levy, 2002) states that of the 8 billion animals (especially chickens, turkeys, cattle, and pigs) raised for human consumption in the United States, most receive some antibiotics during their short lifetimes. In the 1950s, use of sub-therapeutic levels of antibiotics on animals was found to improve animal growth. When considered in terms of nationwide use today, the amount of subtherapeutic antibiotic usage in animals is 4–5 times that used for treatment of animal diseases (Levy, 2002). Thus, there is no question about the enormity of antibiotic use and release.

Can and should antibiotic releases be curtailed?

Unequivocally, the answer to this question is "Yes". Organizations ranging from the World Health Organization (WHO), the Centers for Disease Control (CDC), to the Alliance for Prudent Use of Antibiotics (APUA) have mounted public campaigns to educate all components of society that control the use and release of antibiotics (institutions, industries, businesses, doctors, farmers, health care workers, individuals). Control of antibiotic resistance is a *social problem* of global proportion where major economic, cultural, scientific, and governmental forces meet. The two basic principles of wise antimicrobial usage are as follows:

1 Always administer antibiotics at a dosage and for a period of time that eliminates the pathogens.
2 If the compounds are administrated at inadequate potency, for too short a time, or for the wrong disease, there will be an increased likelihood that pathogens will not be killed. Instead, they may adapt and replicate, and add new traits to the growing pool of antibiotic-resistance mechanisms that erode our shield against infectious disease.

The issues surrounding the management of antibiotics are complex because there are more than 150 compounds used as antibiotics today. These compounds are partitioned between therapeutic, prophylactic, growth promotion, human, and animal usage. Figure 8.29 provides an insightful overview of the many industrial, economic, behavioral, and environmental connections between antibiotic use and human well being. Looking forward, there is an obvious need to reassess personal and public policies that influence the effectiveness of antibiotics used to cure and prevent infectious disease.

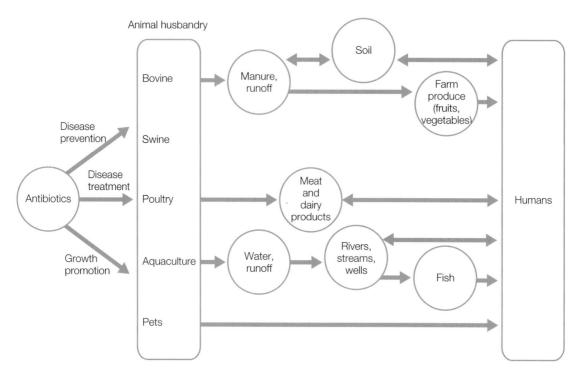

Figure 8.29 Following three kinds of antibiotic use in animals, antibiotics and antibiotic-resistant bacteria enter many habitats, and can eventually come in contact with and impact people. (Adapted from Levy, 2002.)

STUDY QUESTIONS

1 Imagine that for 10 weeks in a row, at 1-week intervals, you have scraped a film of microorganism from between your front teeth. After dispersing and plating the cells, 10 types of microbial colonies consistently grow up from agar plates with starch as the carbon source. For the same 10 weeks, you also sampled the food you eat and assembled cultured types of microorganisms from your food sources.

(A) Outline an experimental plan aimed at distinguishing endogenous (stable) residents of your mouth versus exogenous (transient) residents.

(B) Consider attempting to classify your teeth-inhabiting microflora according to the eight ecological relationships described in Box 8.1 and Figure 8.2. How difficult would it be to characterize just one of your isolated microorganisms according to the eight categories of ecological relationships shown?

(C) Briefly list and/or describe at least four types of experimental results and/or procedures that would allow you to define the organisms as "commensals".

(D) List four methodological challenges that you see in the human mouth microbial system that hamper you from achieving your goal of rigorously deciding if an isolated bacterium is exogenous, endogenous, or a commensal.

2 If you suspect that some of the isolates from your teeth (from question 1) are antagonistic to one another:
   (A) How would you demonstrate this in the laboratory?
   (B) How would you demonstrate this in your mouth?
   (C) Would amensalism have implications for dentistry and tooth decay?

3 Information in Box 8.3 states that the microbial residents of the human intestine are "symbiotic".
   (A) Do you find the arguments convincing?
   (B) Develop counterarguments that the organisms are commensals, not symbionts.
   (C) Experimentally, what do you need to do to prove a symbiotic relationship?
   (D) Why is the proof particularly challenging for the human ecosystem?
   (E) Would it be easier to identify symbionts in other ecosystems? Why?

4 Table 8.1 lists 17 different types of symbiotic relationships.
   (A) Which of these had you never heard about prior to reading the table entries?
   (B) Write a 2–3-page essay that compares two different symbiotic relationships. One of the relationships should be chosen from Table 8.1. The second should be chosen from another source such as the peer-reviewed published scientific literature or the world wide web. For the two symbioses, organize your essay around a new table prepared as follows: make three columns labeled "characteristics", "symbiosis 1", and "symbiosis 2". Under the "characteristics" heading, enter "habitat where relationship evolved", then "key nutritional and physiological features of the habitat", then "benefits derived by the cooperating partners", then "likely sequential developments in the symbiosis, beginning with neutralism", then "physiological and/or adaptive traits achieved through symbiosis", and then "known genetic adaptations by either partner". Next fill in most (if not all) of the details of the six characteristics for both symbioses that you have chosen. The text portion of the essay should describe patterns of contrasts in the table and critically evaluate why some of the information may be particularly strong, weak, or absent.

5 Early discoveries about the mycorrhizal symbiosis were made by plant pathologists. These experimentalists routinely grow plants in sterile and nonsterile potting soil with and without a fungal inoculum. In normal assays of pathogenesis, stunted plant growth occurs in the experimental treatments that receive the fungal inoculum.
   (A) In assays with mycorrhizal fungi, what types of plant responses were found?
   (B) With the (initially) unexpected results in your answer to part A, what do you guess were the experimenter's first reactions?
   (C) What would the next likely steps be for the experimenter?
   (D) After verification, what would the next series of scientific procedures be? (Hint: think "hypotheses".)
   (E) Given the complexity of the rhizosphere habitat (Section 8.2), do you think that progress in explaining the benefits of mycorrhizal infection was easy or not? Please explain.

6 Section 8.2 presents a perspective on humans as habitats for microorganisms.
   (A) Define the "human microbiome".
   (B) What is the habitat tunover time for the skin? For the intestine?

(C) What are the physiological and ecological implications of turnover time for the endogeonous populations? For the exogenous populations?

7 Critically evaluate the sampling plans and analysis procedures used in the two studies on human intestinal microorganisms described in Box 8.3. If you were in charge of designing new studies to extend the findings shown in Box 8.3, what would the new goals and experimental plan be? Please explain.

8 Choose an organic compound – preferably one whose key metabolic pathway has been previously characterized in model bacteria. Be sure that the involved enzymes and genes have also been characterized. Envision this compound as a pollutant in a groundwater plume moving toward a river that provides drinking water to your home town. Reconsider what you know about the fundamentals of microbiology, physiology, biochemistry, and genetics and the many tools used for exploring these (e.g., Chapter 6 and Chapter 7, particularly Section 7.5). Now, develop a proposal that designs and implements a *comprehensive* set of assays that can be applied to the contaminated field site. The goal of these assays should be to verify that the compound you are interested in is, in fact, being biodegraded. Please incorporate within your proposal experimental designs that include control treatments and sampling procedures (e.g., Section 6.5). State your rationale, goals, and the criteria you will be using to prove that the biodegradation reactions you hope are occurring, are truly occurring.

9 Consider microbial metabolism of tetrachloroethene.
(A) Draw the chemical structure for the molecule.
(B) What is the oxidation state of carbon in the molecule?
(C) Can the carbon be oxidized?
(D) Can the carbon be reduced?
(E) Reason from basic principles what the likely bioremediation strategy is for tetrachloroethene-contaminated groundwater.
(F) Use the logic of chemistry to prepare the stoichiometric reaction for the tetrachloroethene biodegradation. Verify this by searching the world wide web and/or peer-reviewed scientific literature.
(G) What is the electron donor in the reaction? Where does it come from in tetrachloroethene-contaminated field sites?

10 Consider microbial metabolism and bioremediation of one of the BTEX components, benzene.
(A) Use the world wide web or other sources to find benzene's aqueous solubility.
(B) Assume that benzene is contaminating aerobic groundwaters adjacent to a gas station and that aerobic microorganisms capable of growing on benzene are present. The rudiments of the metabolic pathway for metabolism of aromatic compounds were presented in Section 7.3 and Figure 7.11. Write out the stoichiometry of aerobic conversion of benzene to carbon dioxide.
(C) Based on your answer to part B, how many moles of oxygen are required to biodegrade the benzene in 1 L of benzene-saturated water?
(D) Now recall that spilled gasoline can occur as nonaqueous-phase liquids (NAPLs) that float on the water table adjacent to spill sites. The NAPL provides a benzene supply that constantly replenishes the dissolved-phase benzene. How likely is it that the water will remain aerobic?
(E) Explain the factors that control if and when anaerobic conditions develop in the contaminated site.
(G) If oxygen is depleted, what happens to the benzene and other BTEX compounds?

11 Consider the bioremediation of a toxic inorganic element of your choice, say uranium.
   (A) Define the predominant chemical species for that element and their solubility, mobility, oxidation states, and the key microbiological reactions that convert one to another.
   (B) Define the microbial process(es) that attenuates the compound (e.g., attenuation is often achieved by converting the element to a form that precipitates – becoming immobile).
   (C) Once "immobilized", what steps need to be taken to be sure the element remains in place?
   (D) Can you envision any additional physical, chemical, mixing, or other engineering concerns that might hamper successful implementation of bioremediation schemes for the toxic element of interest?

12 Consider Section 8.4 on biofilms.
   (A) Had you been aware of biofilms before reading the section?
   (B) How widespread do you think biofilms are in nature? Justify your answer using ecological/environmental facts and the logic of microbial physiology.

13 DDT is an insecticide used globally from ~1934 to 1972. Despite DDT's extreme effectiveness in curbing malaria and other tropical diseases, the use of the compound was severely curtailed because of DDT's tendency to biomagnify up the ecological food chain and impair reproduction in a variety of bird species. Consider Section 8.5 on the evolution of metabolic pathways.
   (A) What is the chemical structure of DDT? (To answer this, go to the peer-reviewed literature or the world wide web.)
   (B) Given the information in Tables 8.8, 8.9, and Box 8.10 venture a hypothesis that may explain why a catabolic pathway for DDT has not yet evolved in microorganisms.

14 Consider Section 8.6 on environmental biotechnology.
   (A) Scrutinize Figure 8.15. Name two or three additional issues, concerns, and/or technologies that you feel also should be represented in the figures. Please justify your additions.
   (B) Eight case studies were presented in Section 8.6. How many of the presented cases were new to you? Choose one of the topics you had never heard of before and another (from current news stories, the peer-reviewed scientific literature, or from the world wide web) and write an essay (2–3 pages in length) comparing the histories, technical backgrounds, and prospects for technological success of the two selected studies.

15 Consider Section 8.7 on antibiotic resistance.
   (A) Which of the diseases listed in Table 8.10 have you or a family member had?
   (B) Were you and your family members cured with the help of antibiotics?
   (C) Each time you took antibiotics did you *precisely* follow the dosage guidelines?
   (D) Do you feel uneasy if you visit the doctor for an infection and he/she fails to prescribe antibiotics?
   (E) Prior to reading Section 8.7, were you aware of the extent of antibiotic use in medicine and industry?
   (F) What is the "equation" that ensures that the mechanism of evolution to antibiotic resistance continues to successfully operate and erode our shield against antibiotics?
   (G) What is the title of S. B. Levy's 2002 book about antibiotic resistance? Briefly summarize: (i) why "paradox" was chosen as a key word in the title; and (ii) the current thinking on the genetic mechanisms that allow antibiotic resistance to evolve and be transferred between microorganisms.

REFERENCES

Adams, D.G., B. Bergman, S.A. Nierzwicki-Bauer, A.N. Rai, and A. Schubler. 2006. Cyanobacterial–plant symbioses. *In*: M. Dworkin, S. Falkow, E. Rosenberg, K.-H. Schleifer, and E. Stackebrandt (eds) *The Prokaryotes*, Vol. 1, 3rd edn, pp. 331–363. Springer-Verlag, New York.

Armstrong, G.L., L.A. Conn, and R.W. Pinner. 1999. Trends in infectious disease mortality in the United States during the 20th century. *J. Am. Med. Assoc.* **281**:61–66.

Bäckhed, R., R.E. Ley, J.L. Sonnenburg, D.A. Peterson, and J.I. Gordon. 2005. Host–bacterial mutualism in the human intestine. *Science* **307**:1915–1920.

Bais, H.P., T.L. Weir, L.G. Perry, S. Gilroy, and J.M. Vivanco. 2006. The role of root exudates in rhizosphere interactions with plant and other organisms. *Ann. Rev. Plant Biol.* **57**:233–266.

Baker, A.C. 2003. Flexibility and specificity in coral–algal symbiosis; diversity, ecology, and biogeography of *Symbiodinium*. *Annu. Rev. Ecol. Evol. Syst.* **34**:681–689.

Baumann, P., N.A. Moran, and L. Baumann. 2006. Bacteriocyte-associated endosymbionts of insects. *In*: M. Dworkin, S. Falkow, E. Rosenberg, K.-H. Schleifer, and E. Stackebrandt (eds) *The Prokaryotes*, 3rd edn, pp. 403–438. Springer-Verlag, New York.

Bell, D. 1998. Controversies in the prevention and control of antimicrobial drug resistance. *Emerg. Infec. Dis.* **4**:473–474.

Bizily, S.P., C.L. Rugh, and R.B. Meagher. 2000. Phytodetoxification of hazardous organomercurials by genetically engineered plants. *Nature Biotechnol.* **18**:213–217.

Bond, D.R., D.E. Holmes, L.M. Tender, and D.R. Lovley. 2002. Electrode reducing microorganisms that harvest electricity from marine sediments. *Science* **295**:483–485.

Bowen, G.D., and A.D. Rovira. 1999. The rhizosphere and its management to improve plant growth. *Adv. Agron.* **66**:1–102.

Brüggemann, H., A. Henne, F. Hoster, et al. 2004. The complete genome sequence of *Propionibacterium acnes*, a commensal of human skin. *Science* **305**:671–673.

Cardon, Z.G. and J.L. Whitbeck. 2007. *The Rhizosphere: An ecological approach*. Academic Press, San Diego, CA.

Cavanaugh, C.M., Z.P. McKiness, I.L.G. Newton, and F.J. Stewart. 2006. Marine chemosynthetic symbioses. *In*: M. Dworkin, S. Falkow, E. Rosenberg, K.-H. Schleifer, and E. Stackebrandt (eds) *The Prokaryotes*, Vol. 1, 3rd edn, pp. 475–510. Springer-Verlag, New York.

Chaudhuri, S.K. and D.R. Lovley. 2003. Electricity generation by direct oxidation of glucose in mediatorless microbial fuel cells. *Nature Biotechnol.* **21**:1229–1232.

Cullimore, J. and J. Denarie. 2003. Plant sciences: how legumes select their sweet talking symbionts. *Science* **302**:575–578.

Currie, C.R. 2001. A community of ants, fungi and bacteria: a multilateral approach to studying symbiosis. *Annu. Rev. Microbiol.* **55**:357–380.

Currie, C.R., M. Poulsen, J. Mendenhall, J.J. Boomsma, and J. Billen. 2006. Coevolved crypts and exocrine glands support mutualistic bacteria in fungus-growing ants. *Science* **311**:81–83.

Currie, C.R., B. Wong, A.E. Stuart, et al. 2003. Ancient tripartite coevolution in the Attine ant–microbe symbiosis. *Science* **299**:386–388.

D'Costa, V.M., K.M. McGrann, D.W. Hughes, and G.D. Wright. 2006. Sampling the antibiotic resistome. *Science* **311**:374–377.

Davey, M.E. and G.A. O'Toole. 2000. Microbial biofilms: from ecology to molecular genetics. *Microbiol. Molec. Biol. Rev.* **64**:847–867.

Davies, J. 1994. Inactivation of antibiotics and the dissemination of resistance genes. *Science* **264**:375–382.

Distel, D.L., D.J. Beaudoin, and W. Morrill. 2002. Coexistence of multiple proteobacterial endosymbionts in the gills of the wood-boring bivalve *Lyrodus pedicellatus* (Bivalvia: Teredinidae). *Appl. Environ. Microbiol.* **68**:6292–6299.

Eckburg, P.B., E.M. Bik, C.N. Bernstein, et al. 2005. Diversity of the human intestinal microbial flora. *Science* **308**:1635–1638.

Felbeck, H. and D.L. Distel. 1992. Prokaryotic symbionts of marine invertebrates. *In*: A. Balows, H.G. Trüper, M. Dworkin, W. Harder, and K.-H. Schleifer (eds) *The Prokaryotes*, 2nd edn, pp. 3891–906. Springer-Verlag, New York.

Gill, S.R., M. Pop, R.T. DeBoy, et al. 2006. Metagenomic analysis of the human distal gut microbiome. *Science* **312**:1355–1359.

Glick, B.R. and J.J. Pasternak. 2003. *Molecular Biotechnology: Principles and application of recombinant DNA*, 3rd edn. American Society of Microbiology Press, Washington, DC.

Goffredi, S.K., V.J. Orphan, G.W. Rouse, et al. 2005. Evolutionary innovation: a bone eating marine symbiosis. *Environ. Microbiol.* **7**:1369–1378.

Görtz, H.-D. 2006. Symbiotic-associations between ciliate and prokaryotes. *In*: M. Dworkin, S. Falkow, E. Rosenberg, K.-H. Schleifer, and E. Stackebrandt (eds) *The Prokaryotes*, 3rd edn, pp. 364–402. Springer-Verlag, New York.

Hall-Stoodley, L., J.W. Costerton, and P. Stoodley. 2004. Bacterial biofilms: from the natural environment to infectious diseases. *Nature Rev. Microbiol.* **2**:95–108.

Harrison, M.J. 2005. Signaling in the arbuscular mycorrhizal symbiosis. *Annu. Rev. Microbiol.* **59**: 19–42.

Hiltner, L. 1904. Uber neurer Erfahrungen and Problem auf dem Gebeit der Bodenbakteriologie und unter besonderer Berucksichtigung der Grundungung und Brache. *Arb. Dtisch. Lamwirt. Ges.* **98**:59.

Hoffmeister, M. and W. Martin. 2003. Interspecific evolution: microbial symbiosis, endosymbiosis and gene transfer. *Environ. Microbiol.* **5**:642–649.

Horn, M., A. Collingro, S. Schmitz-Esser, et al. 2004. Illuminating the evolutionary history of Chlamydiae. Science **304**:728–730.

Horn, M. and M. Wagner. 2004. Bacterial endosymbionts of free-living amoebae. *J. Eukaryot. Microbiol.* **51**:509–514.

Jeanthon, C. 2000. Molecular ecology of hydrothermal vent microbial communities. *Antonie van Leeuwenhoek* **77**:117–133.

Johnson, G.R. and J.C. Spain. 2003. Evolution of catabolic pathways for synthetic compounds: bacterial pathways for degradation of 2,4-dinitrotoluene and nitrobenzene. *Appl. Microbiol. Biotechnol.* **62**:110–123.

Jones, D.L. 1998. Organic acids in the rhizosphere – a critical review. *Plant Soil* **205**:25–44.

Judson, H.F. 1996. *The Eighth Day of Creation: Makers of the revolution in biology*. Cold Spring Harbor Press, Cold Spring Harbor, NY.

Kirisits, M.J. and M.R. Parsek. 2006. Does *Pseudomonas aeroginosa* use intercellular signalling to build biofilm communities? *Cell. Microbiol.* **8**: 1841–1849.

Kolter, R. and E.P. Greenberg. 2006. Microbial

sciences: the superficial life of microbes. *Nature* **441**:300–302.

Lenz, R.W. and R.H. Marchessault. 2005. Bacterial polyesters: biosynthesis, biodegradable plastics and biotechnology. *BioMacromolecules* **6**:1–8.

Levy, S.B. 2002. *The Antibiotic Paradox: How the misuse of antibiotics destroys their curative powers*, 2nd edn. Perseus Publishing, Cambridge, MA.

Levy, S.B. 2005. Antibiotic resistance – the problem intensifies. *Adv. Drug Del. Rev.* **57**:1446–1450.

Li, F., W. Li, M. Forzan, and S.C. Harrison. 2005. Structure of SARS caronovirus spike receptor-binding domain complexed with receptor. *Science* **309**:1864–1868.

Lindow, S.E. and M.T. Brandl. 2003. Microbiology of the phyllosphere. *Appl. Environ. Microbiol.* **69**: 1875–1883.

Logan, B.E. 2004. Extracting hydrogen and electricity from renewable resources. *Environ. Sci. Technol.* **38**:161A–167A.

Logan, B.E. and J.M. Regan. 2006. Microbial fuel cells – challenges and applications. *Environ. Sci. Technol.* **40**:5172–5180.

Logan, B.E., B. Hamelers, R. Rozendal, et al. 2006. Microbial fuel cells – methodology and technology. *Environ. Sci. Technol.* **40**:5181–5192.

Lupp, C. and E.G. Ruby. 2005. *Vibrio fischeri* uses two quorum-sensing systems for the regulation of early and late colonization factors. *J. Bacteriol.* **187**:3620–3629.

Madigan, M.T. and J.M. Martinko. 2006. *Brock Biology of Microorganisms*, 11th edn. Prentice Hall, Upper Saddle River, NJ.

Madsen, E.L. 1998. Theoretical and applied aspects of bioremediation: the influence of microbiological processes on organic compounds in field sites. *In*: R. Burlage, R. Atlas, D. Stahl, G. Geesey, and G. Sayler (eds) *Techniques in Microbial Ecology*, pp. 354–407. Oxford University Press, New York.

Madsen, E.L. 1991. Determining *in situ* biodegradation: facts and challenges. *Environ. Sci. Technol.* **25**:1662–1673.

Marra, M.A., S.J.M. Jones, C.R. Astell, et al. 2003. The genome sequence of the SARS-assoicated coronavirus. *Science* **300**:1399–1404.

Martin, F.M., S. Perotlo, and P. Bonfante. 2001. Mycorrhizal fungi: a fungal community at the interface between soils and roots. *In*: R. Pinton, Z. Varanini, and P. Nannipieri (eds) *The Rhizosphere:*

Biochemistry and organic substances at the soil–plant interface, pp. 263–296. Marcel Dekker, New York.

Miller, M.B. and B.L. Bassler. 2001. Quorum sensing in bacteria. *Annu. Rev. Microbiol.* **55**:165–199.

Miyake, J., M. Miyake, and Y. Asada. 1999. Biotechnological hydrogen production: research for efficient light energy conversion. *J. Biotechnol.* **70**:89–101.

Moorthy, S. and P.I. Watnick. 2005. Identification of novel stage-specific genetic requirements through whole genome transcription profiling of *Vibrio cholerae* biofilm development. *Molec. Microbiol.* **57**:1623–1635.

Müller, T.A., C. Werlen, J. Spain, and J.R. van der Meer. 2003. Evolution of a chlorobenzene degradative pathway among bacteria in a contaminated groundwater mediated by a genomic island in *Ralstonia. Environ. Microbiol.* **5**:163–173.

Nash, T.H., III (ed.) 1996. *Lichen Biology.* Cambridge University Press, New York.

National Research Council. 2000. *Natural Attenuation for Groundwater Remediation.* National Academic Press, Washington, DC.

Nealson, K.H. and J.W. Hastings, 1979. Bacterial luminescence: its control and ecological significance. *Microbiol. Rev.* **43**:496–518.

Nitsche, D.-P., H.M. Johansson, I.M. Frick, and M. Morgelin. 2006. Streptococcal protein FOG, a novel matrix adhesin interacting with collagen I *in vivo. J. Biol. Chem.* **281**:1670–1679.

Nobles, D.R., Jr. and R.M. Brown Jr. 2004. The pivotal role of cyanobacteria in the evolution of cellulose synthases and cellulose synthase-like proteins. *Cellulose* **11**:437–448.

Nyholm, S.V. and M.J. McFall-Ngai. 2004. The winnowing: establishing the squid–*Vibrio* symbiosis. *Nature Rev. Microbiol.* **2**:632–642.

Oh, S.E. and B.E. Logan. 2005. Hydrogen and electricity production from a food processing wastewater using fermentation and microbial fuel cell technologies. *Water Res.* **39**:4673–4682.

Palumbi, S.R. 2001. Humans as the world's greatest evolutionary force. *Science* **293**:1786–1790.

Partida-Martinez, L.P. and C. Hertwick. 2005. Pathogenic fungus harbors endosymbiotic bacteria for toxin production. *Nature* **437**:884–888.

Perret, X., C. Staehelin, and W.J. Broughton. 2000. Molecular basis of symbiotic promiscuity. *Microbiol. Molec. Biol. Rev.* **64**:180–201.

Petersen, C.H. S.D. Rice, J.W. Short, et al. 20003. Long-term ecosystem response to the Exxon Valdez oil spill. *Science* **302**:2082–2086.

Pinton, R., Z. Varanini, and P. Nannipieri. 2001a. The rhizosphere as a site of biochemical interactions among soil components, plants, and microorganisms. *In*: R. Pinton, Z. Varanini, P. Nannipieri (eds) *The Rhizosphere: Biochemistry and organic substances at the soil–plant interface*, pp. 1–17. Marcel Dekker, New York.

Pinton, R., Z. Varanini, P., and Nannipieri (eds) 2001b. *The Rhizosphere: Biochemistry and organic substances at the soil–plant interface.* Marcel Dekker, New York.

Primrose, S.B. and R.M. Twyman. 2006. *Principles of Gene Manipulation and Genomics*, 6th edn. Blackwell Science, Oxford, UK.

Prince, R.C. and H.S. Kheshgi. 2005. The photobiological production of hydrogen: potential efficiency and effectiveness as a renewable fuel. CRC *Crit. Rev. Microbiol.* **31**:19–31.

Ramey, B.E., M. Koutsoudis, S.B. van Bodman, and C. Fuqua. 2004. Biofilm formation in plant–microbe associations. *Curr. Opin. Microbiol.* **7**:602–609.

Ramsey, D.M. and D.J. Wozniak. 2005. Understanding the control of *Pseudomonas aeroginosa* alginate synthesis and the prospects for management of chronic infections in cystic fibrosis. *Molec. Microbiol.* **56**:309–322.

Reddy, C.S.K., R. Ghai, Rashmi, and V.C. Kalia. 2003. Polyhydroxyalkanoates: an overview. *Bioresource Technol.* **87**:137–146.

Rota, P.A., M.S. Oberste, S.S. Monroe, et al. 2003. Characterization of a novel coronavirus associated with severe acute respiratory syndrome. *Science* **300**:1394–1399.

Roth, R.R. and W.D. James. 1988. Microbial ecology of the skin. *Annu. Rev. Microbiol.* **42**:441–464.

Ruby, E.G. 1996. Lessons from a cooperative bacterial–animal association: The *Vibrio fischeri–Euprymna scolopes* light organ symbiosis. *Annu. Rev. Microbiol.* **50**:591–624.

Ruby, E.G. 1999. The *Euprymna scolopes–Vibrio fischeri* symbiosis: a biomedical model for the study of bacterial colonization of animal tissue. *J. Molec. Microbiol. Biotechnol.* **1**:13–21.

Rupprecht, J., B. Hankamer, J.H. Mussnug, G. Ananyev, C. Dismukes, and O. Kraus. 2006. Perspectives and advances in biological H_2 production in microorganisms. *Appl. Microbiol. Biotechnol.* **72**:442–449.

Russell, J.B. and R.L. Rychlik. 2001. Factors that alter rumen microbial ecology. *Science* **292**:1119–1122.

Scarborough, C.L., J. Ferrari, and H.C.J. Godfray. 2005. Aphid protected from pathogen by endosymbiont. *Science* **310**:1781.

Schlegel, H.G. and H.W. Jannasch. 2006. Prokaryotes and their habitats. *In*: M. Dworkin, S. Falkow, E. Rosenberg, K.-H. Schleifer, and E. Stackebrandt (eds) *The Prokaryotes*, Vol. 1, 3rd edn, pp. 137–184. Springer-Verlag, New York.

Seffernick, J.L. and L.P. Wackett. 2001. Rapid evolution of bacterial catabolic enzymes: a case study with atrazine chlorohydrolase. *Biochemistry* **40**: 12747–12753.

Short, J.W., M.R. Lindeberg, P.M. Harris, J.M. Maselko, J.J. Pela, and S.D. Rice. 2004. Estimate of oil persisting on the beaches of Price William Sound 12 years after the Exxon Valdex oil spill. *Environ. Sci. Technol.* **38**:19–25.

Short, J.W., J.M. Maselko, M.R. Lindeberg, P.M. Harris, and S.D. Rice. 2006. Vertical distribution and probability of encountering intertidal Exxon Valdez oil on shorelines of three embayments within Prince William Sound, Alaska. *Environ. Sci. Technol.* **40**:3723–3729.

Spain, J.C., J.B. Hughes, and H.-J. Knackmuss (eds) 2000. *Biodegradation of Nitroaromatic Compounds and Explosives.* Lewis Publishers, Boca Raton, FL.

Taylor, D., A. Daulby, S. Grimshaw, G. James, J. Mercer, and S. Vaziri. 2003. Characterization of the microflora of the human axilla. *Internatl. J. Cosmet. Sci.* **25**:137–145.

Thomas, C.M. and K.M. Nielsen. 2005. Mechanisms of, and barriers to, horizontal gene transfer between bacteria. *Nature Rev. Microbiol.* **3**:711–721.

Uren, N.C. 2001. Types, amounts, and possible functions of compounds released into the rhizosphere by soil-grown plants. *In*: R. Pinton, Z. Varanini, P. Nannipieri (eds) *The Rhizosphere: Biochemistry and organic substances at the soil–plant interface*, pp. 19–40. Marcel Dekker, New York.

van der Meer, J., R. Werlen, S. Nishino, and J.C. Spain. 1998. Evolution of a pathway for chlorobenzene metabolism leads to natural attenuation in contaminated groundwater. *Appl. Environ. Microbiol.* **64**:4185–4193.

van Dongen, U., M.S.M. Jetten, and M.C.M. van Loosdrecht. 2001 The Sharon–anammox process for treatment of ammonium rich wastewater. *Water Sci. Technol.* **44**(1):153–160.

Van Houdt, R. and C.W. Michiels. 2005. Role of bacterial cell surface structures in *Escherichia coli* biofilm formation. *Res. Microbiol.* **156**:626–633.

Van Rhijn, P. and J. Vanderleyden. 1995. The *Rhizobium*–plant symbiosis. *Microbiol. Rev.* **59**:124–142.

Vander, A.J., J.H. Sherman, and D.S. Luciano. 1980. *Human Physiology: The mechanisms of body function*, 3rd edn. McGraw-Hill, New York.

Wackett, L.P. and C.D. Hershberger. 2000. *Biodegradation and Biocatalysis*. American Society for Microbiol Press, Washington, DC.

Wackett, L.P., M.J. Sadowsky, B. Martinez, and N. Shapir. 2002. Biodegradation of atrazine and related s-triazine compounds: from enzymes to field studies. *Appl. Microbiol. Biotechnol.* **58**:39–45.

Watnick, P. and R. Kolter. 2000. Biofilm, city of microbes. *J. Bacteriol.* **182**:2675–2679.

Weber, J.T. 2005. Appropriate use of antimicrobial drugs. *J. Am. Med. Assoc.* **294**:2354–2356.

Wenzel, R.P. and M.B. Edmond. 2000. Managing antibiotic resistance. *New Engl. J. Med.* **343**:1961–1963.

Wernegreen, J.J. 2002. Genome evolution in bacterial endosymbionts of insects. *Nature Rev. Microbiol.* **3**:850–858.

Winans, S.C. and B.L. Bassler. 2002. Mob psychology. *J. Bacteriol.* **184**:873–883.

9

Future Frontiers in Environmental Microbiology

Like beauty, humor, and gemstones, environmental microbiology is multifaceted.
- *Is it microbial ecology or environmental science?*
- *Is it physiology or civil engineering?*
- *Is it natural history or ecosystem science?*
- *Is it medical epidemiology or habitat colonization?*
- *Is it limnology or biogeochemistry?*
- *Is it genomics or biotechnology?*
- *Is it biodegradation or evolution?*

The answer, of course, is that environmental microbiology is all these and more. Clearly, the boundaries of environmental microbiology are blurry. Thus, a truly comprehensive text on environmental microbiology is very difficult to assemble. This present book could be 10 times its current size and still be considered incomplete. Chapters 1 and 2 of this text set the stage or background for Chapters 3 through 7, which established environmental microbiology's core set of principles, facts, logic, and methodologies. Chapter 8 provided a sampling of environmental microbiology's extensions and applications. The goal of this chapter is to succinctly examine some of environmental microbiology's potential future directions and frontiers.

Chapter 9 Outline

9.1 The influence of systems biology on environmental microbiology
9.2 Ecological niches and their genetic basis
9.3 Concepts help define future progress in environmental microbiology

9.1 THE INFLUENCE OF SYSTEMS BIOLOGY ON ENVIRONMENTAL MICROBIOLOGY

The growing discipline of systems biology will inevitably influence biology as a whole, including human medicine and environmental microbiology. Systems biology is the logical "next step" in the ongoing "omics" revolution. According to Kitano (2002), "to understand biology at the systems level, we must examine the structure and dynamics of cellular and organismal function, rather than the characteristics of the isolated parts of a cell or organism". The goal of systems biology is to combine molecular information of various types in models that describe and predict function at the cellular, tissue, organ, and even whole organism level. Systems biology is further defined in Box 9.1. Systems biology follows directly from the nested sciences of genomics, transcriptomics, proteomics, and metabolomics (see Sections 3.2, 6.9, and 6.10). Now that we have ways to generate vast

---

Box 9.1

What is systems biology?

Systems biology is a way to make sense of the information flooding in from the "omics"-based procedures in biology. The four main "omics" approaches appear in the table below.

Omics system	Information
Genome: the genetic code (DNA) of an organism; determined by DNA sequencing	Genetic blueprint for cells and organisms Genome sizes range from ~500,000 bp (intracellular bacterial endosymbionts) to ~3,000,000,000 bp (humans) to 34,000,000,000 bp (the Easter lily plant) The genes encode structural proteins and regulatory networks for all cellular functions [e.g., the 25 clusters of orthologous groups (COG) categories described in Section 3.2 and Table 3.2]
Transciptome: the mRNA pool of cells, tissues, or an organism; determined by microarray analyses; mapped onto the genome of an organism	The subset of the genome expressed as mRNA The transcriptome is dynamic in time – responding to factors in an organism's environment that range from nutrients to organism age to disease onset
Proteome: the protein pool of cells, tissues, or an organism; determined by mass spectrometric analysis of proteins extracted and digested from whole cells; mapped onto the genome of an organism	Abundances of individual proteins in cell extracts or organ tissues at a given time under specified conditions
Metabolome: the metabolite pool of cells, tissues, or an organism; determined by liquid chromatography/mass spectrometry; mapped onto the proteome, transcriptome, and genome of an organism	Abundances of biochemical intermediary metabolites in cell extracts or organ tissues at a given time under specific conditions

Box 9.1 *Continued*

When "omics" procedures are applied to a given organism (e.g., a bacterium or a human), under particular conditions, at a given time, enormous amounts of data are generated. The best way to digest, assimilate, and interpret the web-like mechanistic linkages in "omics"-based data is to use computational models (Figure 1; Kitano, 2002; Price et al., 2004; Kell et al., 2005; Arnaud, 2006). Such models of "omics"-derived data are currently research tools used to advance fundamental understanding of cell function. The integrative, systems biology approach to fundamental biological questions is predicted to have applications in personalized human medicine in the future.

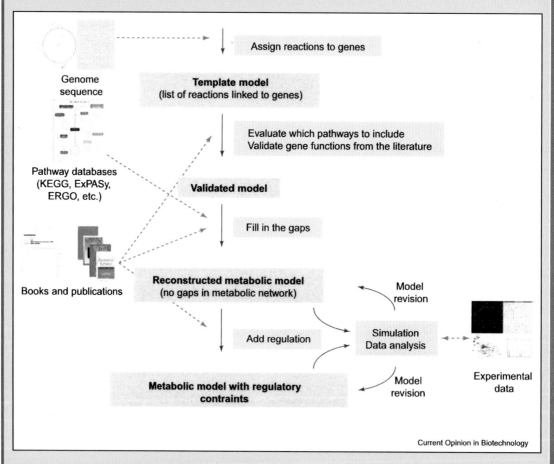

Current Opinion in Biotechnology

Figure 1 The process of systems biology-based metabolic model construction. Information flow is depicted as red dashed arrows and actions as blue arrows. (Reprinted from Borodina, I. and J. Nielsen. 2005. From genomes to *in silico* cells via metabolic networks. *Curr. Opin. Biotechnol.* **16**:350–355. Copyright 2005, with permission from Elsevier.)

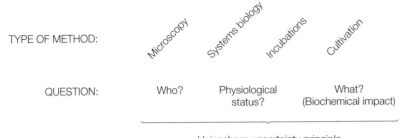

Figure 9.1 Future model of environmental microbiology. "Biomarkers" in Figure 6.5 have been replaced by "systems biology".

data sets, we need computational tools to receive and process the data (on DNA, mRNA, proteins, and metabolites) so that the data can be interpreted. Data, alone, have no meaning.

In Chapter 3 (Section 3.2), the rudiments of genomic analyses were presented. Genomic technology cascades into the transcriptome, proteome, and metabolome. These amount to pieces of information about an organism that can be systematically assembled into hypothetical "roadmaps" of biochemical reactions and gene regulatory networks. The roadmap is a first step. Systems biology seeks "to know the traffic patterns, why such traffic patterns emerge, and how we can control them" (Kitano, 2002).

> • **Where does systems biology fit in the discipline of environmental microbiology?**
> **Answer**: Systems biology replaces "biomarkers" as one of the four basic environmental microbiological tools. Figure 9.1 presents a model of environmental microbiology in which systems biology appears instead of "biomarkers".

The ever increasing sophistication of systems biology is likely to contribute significantly toward future scientific progress in environmental microbiology. Of course, methodological improvements in microscopy, physiological incubations, and cultivation procedures (Figure 9.1) will also contribute to future progress.

> • *When* and *where* will systems biology first be applied in environmental microbiology?
> **Answer**: Very soon, in simple habitats with low microbial genetic diversity.

There are two prerequisites that pave the way for systems biology approaches: (i) a relatively simple biotic system (e.g., a single organism); and (ii) a genomic map for the system. In applying systems biology to environmental microbiology, early headway can best be made using the simplest of naturally occurring microbial communities and by relying upon metagenomics as a genetic and functional blueprint for the system (see Section 6.9). Of all the microbial systems examined to date, the one that is perhaps most environmentally extreme, hence microbiologically simple, is the Iron Mountain acid mine site in California (Tyson et al., 2004). As described in Section 6.9, this site features a pH of <1, very low nutrient status, and very low microbial diversity; both metagenomic and proteomic analyses have already begun (Ram et al., 2005; Allen et al., 2007). Thus, the Iron Mountain microbial community and habitat constitute an excellent prototype study for using systems biology to advance environmental microbiology.

Another ideal prototype system for applying systems biology to environmental microbiology is the ocean. Marine microbial ecologists have an immense advantage over other microbial ecologists because their study habitats (ocean waters) are: (i) globally distributed; (ii) well mixed; (iii) accessible at a human scale; (iv) amenable to thorough geochemical habitat characterization methods; and (v) populated by relatively low-density, low-diversity microbial communities. In contrast to marine microbiology, many other subdisciplines in environmental microbiology (e.g., soil microbiology) focus on a habitat with immense spatial heterogeneity at micron and other scales – where microbial inhabitants may never be fully sampled and where the environmental resources that exert selective pressures have never been fully characterized (see Section 4.6). Marine microbiologists can sample and map geochemical gradients that occur over tens of meters. The waters are relatively uniform and sampling does not leave a hole behind in the habitat. Then, flow cytometers can detect *all* the cells in a liter of water and genetic markers can distinguish and quantify previously defined, fully sequenced, genomic "ecotypes" of microorganisms native to the ocean and known to function there (such as *Prochlorococcus*, *Pelagibacter*, *Silicibacter*, and *Trichodesmium*). The combination of habitat characterization and genomic tools allow basic ecological questions about selective pressures and adaptations to be posed. Hypotheses naturally follow. Issues such as mechanisms controlling biogeographic patterns (growth, nutrients, other resources, predation, viruses, adverse geochemistry, competition, light, etc.) can be addressed.

Box 9.2 provides an example of a genomics-assisted investigation, an "ecological genomics" study, in marine microbiology. Ecological genomics is an important step towards using systems biology in environmental microbiology. As described in Box 9.2, Coleman et al. (2006) have the right tools (e.g., ocean waters as habitat, *Prochlorococcus* as a key photosynthetic player in the habitat, cultivated representatives of *Prochlorococcus* with sequenced genomes, metagenomic techniques for community DNA, distinctive physiological characteristics of *Prochlorococcus* strains from shallow and deep water) for asking basic questions about evolution. The report by Coleman et al. (2006) makes a convincing

Box 9.2

Moving toward systems biology: ecological genomics of the ocean (Coleman et al., 2006)

Background

* *Prochloroccus* is a numerically dominant, photo-synthetic bacterium (Figure 1) in oligotrophic ocean waters – sometimes accounting for up to half of the photosynthetic biomass of some regions.
* The ocean-water habitat of *Prochlorococcus* presents gradients of light, temperature, and nutrients.
* Several strains of *Prochlorococcus* have been isolated and cultured in the laboratory and their genomes have been sequenced. Thus, a genetic blueprint for these organisms has been established.

What Coleman et al. (2006) did

* The genome sequences of two different strains of *Prochlorococcus* were compared and five large pieces of foreign DNA, termed "genomic islands" were discovered. Based on a variety of criteria, the foreign pieces of DNA did not resemble the DNA of the host – they stood out like islands surrounded by water (see Science and the citizen box, Chapter 5). "Signatures of mobility" in the DNA sequences indicated that viruses were the agents that imported the foreign DNA.

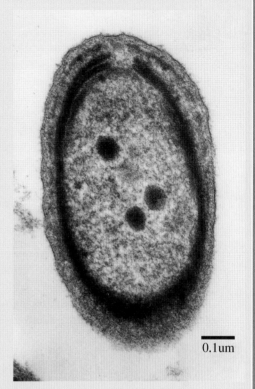

0.1um

Figure 1 Photomicrograph of *Prochlorococcus*. (Courtesy of Claire Ting, Department of Biology, Williams College, with permission.)

* By analyzing the sequences of genes encoded in the genomic islands, Coleman et al. (2006) were able to discern that the islands con-ferred specific physiological functions such as nutrient uptake (amino acids, manganese, and cyanate), stress response (to light and low phosphate), and susceptibility to virus infec-tion (via cell surface molecules).
* In laboratory assays of gene expression under high- and low-light conditions, differen-tial expression of genomic island genes was observed. This confirmed their physiological importance.
* Genomic fragments from the naturally occurring microbial community were retrieved from the Sargasso Sea and from Hawaiian waters. Many of these genomic fragments

Box 9.2 *Continued*

contained gene sequences that could be mapped onto genomic island-containing regions of *Prochlorococcus* chromosomes in the cultivated strains.

* Clear patterns in variations of the genomic islands in wild populations emerged.

What the findings mean

The genomes of wild populations of *Prochlorococcus* seem to be composed of a consistent "core", augmented by islands of high genetic variability. Functional genes carried within the islands play an important role in allowing distinctive ecotypes of *Prochlorococcus* to adapt to selective pressures such as high-light conditions in shallow waters, and low-nutrient conditions. Via these genomic islands, different niches in the oceans seem to be exploited by distinctive *Prochlorococcus* populations.

and insightful case for *Prochlorococcus* adaptation and niche differentiation (see Section 9.2) caused by "genomic islands" mobilized by viruses, within naturally occurring *Prochlorococcus* populations.

DeLong and Karl (2005) have established an international center for microbial oceanography. A systematic plan will be implemented that links activities and information from a dozen oceanography-related subdisciplines (Figure 9.2). Note that core activities at the center include genome libraries, bioinformatics, DNA arrays, proteomics (see Section 6.9 and Table 6.4), and modeling – all reflecting future influences of systems biology.

9.2 ECOLOGICAL NICHES AND THEIR GENETIC BASIS

Genome as evolutionary record

It is a truism that the current state of a microorganism's genome reflects its heritage. That heritage is a series of evolutionary events that combine vertical transmission of genes, lateral gene transfer, gene loss, mutations, rearrangements, and selective pressures (Kunin and Ouzounis, 2003; Kazazian, 2004; Ochman and Davalos, 2006). The legacy of such events, the genome itself, can be viewed as a collection of robust, highly refined genetic networks. But, mixed in with the evolutionary "successes" (the sophisticated regulatory networks) are evolutionary "failures" and "works in progress" that include random insertions, pseudogenes, and remnants of traits no longer useful (Kunin and Ouzounis, 2003; Ochman and Davalos, 2006). Thus, even in the postgenomic age, the detailed evolutionary histories of microorganisms are obscure. There is no doubt, however, that selection, ecological fitness, and niche have been dominant forces shaping the outcomes of genome evolution.

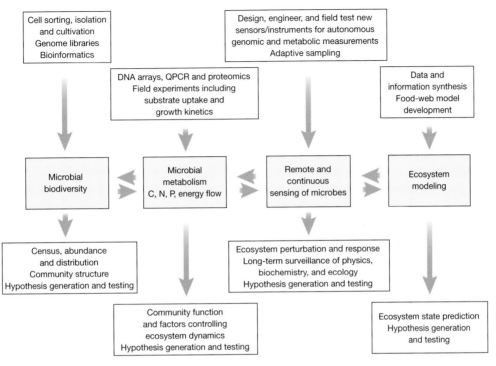

Figure 9.2 Organizational plan for coordinating the dozen subdisciplines necessary to advance understanding of microbial oceanography. QPCR, quantitative polymerase chain reaction. (Reprinted by permission from Macmillan Publishers Ltd: *Nature*, from DeLong, E.F., and D.M. Karl. 2005. Genomic perspectives in microbial oceanography. *Nature* **437**:336–342. Copyright 2005.)

Ecological niches

It is also a truism that every organism has its own ecological niche. Niche is an idea, a concept that humans have created about the ecological role of a species in its ecological community (Ricklefs and Miller, 2000). Hutchinson (1958) originated the concepts of "niche space" and "niche hyper volume". These ideas are based on the notion that environmental factors (such as temperature, pH, nutrient availability, food types, and food size) occur as gradients that can be plotted as axes in n-dimensional space. For each environmental factor (each dimension in niche space), a species can survive over a defined range. When all of the ranges for each of n environmental factors are integrated together, they can be plotted in n-dimensional space to represent a "Hutchinsonian hypervolume" that defines niche. The *fundamental niche* (Figure 9.3) is the largest ecological hypervolume that an organism or species can possibly occupy and is based mainly on interactions with the physical environment, in the

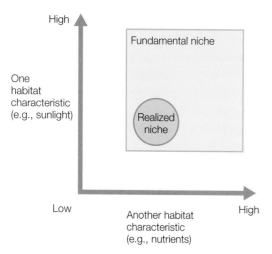

Figure 9.3 A view of fundamental and realized ecological niches in two dimensions. (Based on Hutchinson, 1958.)

absence of competition. In contrast, the *realized niche* (Figure 9.3) is the portion of the fundamental niche that the organism occupies after interacting, especially competing, with other organisms. Competition between organisms for resources is considered a major evolutionary and ecological force. Competitive exclusion of one organism by another occurs because no two species can occupy the same niche. In nutrient-rich, stable, physically diverse habitats (such as tropical rain forests), high species diversity is thought to be the result of niche differentiation. Species evolve toward difference in niche. In so doing, competition between species is reduced (Whittaker, 1970).

Methodological strategies for discovering the genetic basis for ecological fitness

Discovering the ecological niches of microorganisms within their native habitats is a major goal of environmental microbiology (see Section 6.10). *An even more fundamental issue is discovering the genetic basis for ecological fitness and niche.* This area represents a major frontier for future research (Rediers et al., 2005; Saleh-Lakha et al., 2005).

In medical microbiology, a variety of elegant molecular biological strategies have been employed to discover the genetic basis for infectious diseases (Box 9.3). Disease prevention strategies rely upon a mechanistic understanding of biochemical and signaling pathways in pathogenic microorganisms and their hosts. The molecular approaches described in Box 9.3 have successfully identified, in pathogenic bacteria, virulence factors that are expressed only in an infected host. There is an obvious parallel between medical microbiology and environmental microbiology: the infected host is equivalent to colonized soil, sediment, or body of water. Both disciplines advance by identifying genes for ecologically important processes that are expressed differentially in habitats of interest.

As described in Box 9.3, there are two basic strategies for finding ecologically significant, fitness-conferring genes: (i) screening of mRNA pools extracted from the habitat of interest (versus a control habitat); and (ii) screening a large library of mutant strains for genes that are specifically induced in the native habitat and confer survival to the host. Both of these strategies have begun to be implemented in environmental microbiology. The strategies and examples are listed in Box 9.4.

Based on information shown in Box 9.4, it is clear that the promise of applying medical gene-discovery procedures to environmental habitats has not yet been realized. The first three studies listed in Box 9.4 examined

Box 9.3

Discovering the genetic basis for ecological fitness in medical microbiology

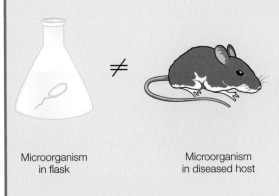

Microorganism
in flask

Microorganism
in diseased host

When a microorganism is grown in a laboratory flask, a small subset of its entire genome is activated (transcribed into mRNA) and manifest as the proteins that catalyze the catabolic and anabolic reactions that create new cells. Medical microbiologists have persistently sought ways to identify virulence genes that are crucial for causing diseases in humans and animals. Such genes are ordinarily intractable (unexpressed) during routine laboratory growth of bacterial pathogens. Once discovered and understood, virulence genes serve as gateways for battling diseases.

• **How can genes for pathogenesis be discovered?**
Answer: Devise a way to identify genes expressed by pathogens exclusively while they are inside a diseased host. The chances are good that these "habitat-specific" genes are necessary for disease development and/or survival in the host.

Many different techniques to find the host-induced genes for pathogenesis have been devised. The procedures include: "in vivo expression technology", "differential fluorescence induction", "signature-tagged mutagenesis", "differential display using arbitrarily primed PCR", "subtractive and differential hybridization", and "selective capture of transcribed sequences" (Chiang et al., 1999; Rediers et al., 2005).

Broadly, these techniques adopt either of two basic strategies:

1 Screening mRNA pools extracted from pathogens within their hosts. Contrasts between genes expressed in the host versus in a control (nonhost) environment reveal host-specific genes that are then cloned and sequenced.

2 Screening mutant strains of a pathogen for rare genes that are induced specifically and confer survival while within the host [this involves careful creation of a large library of mutation-carrying strains, comparing their survival inside versus outside a host, and then characterizing (sequencing) the genes of interest].

mRNA expression in microbial communities (soil), whose complexity impaired data interpretation. The studies by Botero et al. (2005) and Groh et al. (2005) moved forward, technically, but were "feasibility studies" that refined methodological procedures as a prelude for discoveries. On

Box 9.4

Promising approaches for discovering the genetic basis for ecological fitness in environmental microbiology

Microorganism in flask $\neq$ Microorganism in nature

Strategy	Examples	References
Screening of mRNA pools extracted from habitat	• mRNA differential display technique applied to induction of toluene biodegradation genes in soil	Fleming et al., 1998
	• mRNA differential display technique applied to cyclohexanone monooxygenase activity in soil	Brzostowicz et al., 2003
	• Arbitrarily primed polymerase chain reaction (PCR) examining plant decomposition genes in soil	Aneja et al., 2004
	• mRNA recovery using polyA-tailing procedures, followed by formation of cDNA libraries. This was a "feasibility study" that should apply to mRNA, but it examined rRNA recovery from hot springs in Yellowstone National Park	Botero et al., 2005
	• mRNA recovery via random priming, to produce cDNA libraries. rRNA in marine and freshwater microbial communities was selectively removed prior to analysis. Many gene transcripts were linked to environmentally important processes such as sulfur oxidation (*SoxA*), assimilation of C1 compounds (*fdh1B*), and acquisition of nitrogen via polyamine degradation (*aphA*)	Poretsky et al., 2005
Screening of mutant strains for selective expression of genes in the environment	• Signature tagged mutagenesis (STM) was used on *Desulfovibrio desulfuricans* and *Shewanella oneidensis* to study genes contributing to survival in sediment. The procedure successfully identified mutants in chemotaxis genes in both test bacteria. STM-identified genes conferring fitness for the sediment habit were reserved for future publication	Groh et al., 2005

the other hand, Poretsky et al. (2005) were actually able to "listen" to the "messages" (the mRNA pool) of planktonic bacterial communities in marine and freshwater. The genes expressed in situ by naturally occurring microorganisms provided glimpses of the identity of active populations, their physiological status, and the real-world resources the populations confront (see Section 6.10). In the future, these types of information need to be evaluated, confirmed, compiled, and extended to additional habitats.

Understanding the genetic basis for niche, fitness, and ecological success is a meritorious goal and a significant challenge for future environmental microbiologists. To quote Rediers et al. (2005):

> Analysis of ecological success is far from straight-forward: it is a complex and multidimensional phenotype determined by interconnected regulatory pathways involving both individual genes and gene networks. Natural selection, which is largely responsible for shaping the determinants of ecological success, does so by operating on interacting systems (more so than on single genes) to generate specific morphologies, physiologies and behaviors.

9.3 CONCEPTS HELP DEFINE FUTURE PROGRESS IN ENVIRONMENTAL MICROBIOLOGY

Figure 9.4 provides a summary of six ways to conceptualize the essence of environmental microbiology and its goals:

1 Figure 9.4a (from Section 1.1) reminds the reader of the five basic components (evolution, habitat diversity, thermodynamics, physiology, and ecology) that build the "house of environmental microbiology".

2 Figure 9.4b (from Figure 1.5) emphasizes the dynamic interfacial nature of environmental microbiology. Frontiers emerge as progress is made simultaneously in environmental science and microbial ecology.

3 Figure 9.4c (from Figure 6.11) portrays environmental microbiology as the iterative (cyclic) application of methodological tools from many biological disciplines to field sites. The tools are applied in ways that confirm and validate the identities of environmentally relevant microorganisms and their genes.

4 Figure 9.4d (the concentric circle model) emphasizes that Chapter 6's fundamental questions (Who? What? When? Where? How? Why?) are surrounded (enabled and limited) by the methodological tools used by environmental microbiologists.

5 Figure 9.4e (from Figure 6.5) divides the many environmental microbiological tools into four basic types (microscopy, biomarkers, incubations, and cultivation) and emphasizes two inescapable experimental facts: (i) the validity of all data generated by experimental approaches is threatened by environmental microbiology's Heinsenburg uncertainty principle (see Section 6.4); and (ii) when sampling real-world microbial

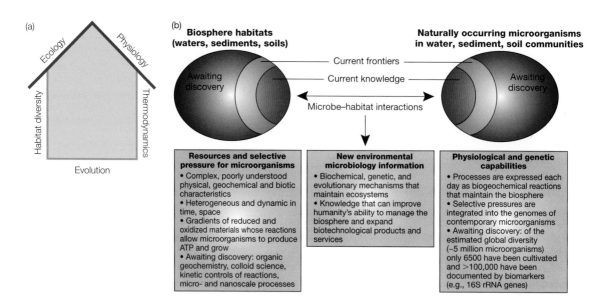

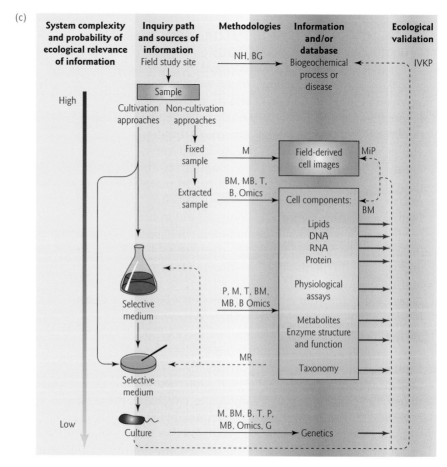

Figure 9.4 Six diagrams that conceptually define environmental microbiology. See text for details.

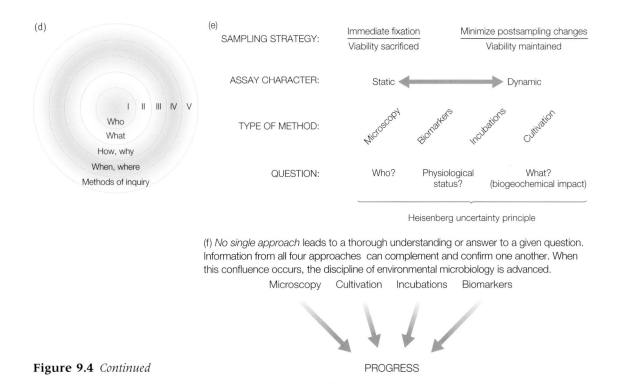

Figure 9.4 *Continued*

habitats, each of the four types of experimental tools demands an optimal sampling procedure.

6 The message from Figure 9.4f (from Box 5.1) is that no single type of environmental microbiological tool is sufficient to deliver robust, new information. Instead, complementary information from independent, convergent sources is the path toward progress.

- **This begs the question: What is progress?**

Information in Box 9.5 attempts to define "scientific progress" in environmental microbiology and predicts that accelerated progress is inevitable. The metaphor for advancing the discipline as environmental microbiology is not one of a mountain that, once climbed, is conquered. Instead, a more appropriate metaphor for environmental microbiology is the human mind. Humans never acquire complete, true, and accurate knowledge of the world that surrounds them. The constant accrual of sensory and other information throughout our lifetimes allows us to navigate the world – even if the full complexity (chaos? beauty? reality?) of our surroundings is never fully revealed. The "mind" of environmental microbiology is an accruing synthesis of facts, principles, and relationships

Box 9.5

Defining progress in environmental microbiology

- Scientific progress in environmental microbiology can be defined as "adding new information to the advancing frontiers of Who?, What?, When?, Where?, How?, and Why?"
- This book is simultaneously a synopsis of, and testimony to, progress in environmental microbiology.
- The path forward in environmental microbiology is depicted by two arrows on the graph to the right.

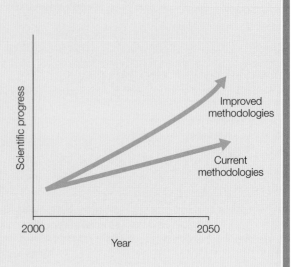

Current methodologies (low arrow scenario)

If current methods remained static, environmental microbiologists could march onward for ~50 years describing new cultivated microorganisms (Chapters 5 and 6), discovering new uncultivated microorganisms (Chapters 5 and 6), describing new habitats, symbioses, evolutionary adaptations (Chapters 4 and 8), and their biochemical, genetic, genomic, and structural biological bases (Chapter 7). These discoveries appear weekly and/or monthly in peer-reviewed scientific journals such as *Science, Nature, Proceedings of the National Academy of Sciences, Applied and Environmental Microbiology, Environmental Microbiology, Microbial Ecology, FEMS Microbial Ecology, Geomicrobiology*, and others.

Improved methodologies (high arrow scenario)

Environmental microbiology is a methods-limited, multidisciplinary science (see Chapters 1 and 6). Methodological advancements are constantly being made in the sets of tools used by environmental microbiologists (e.g., microscopy, analytical chemistry, DNA sequencing, proteomics, nanotechnology, microbial cultivation procedures, systems biology). Therefore, the rate of progress in environmental microbiology over the next several decades will certainly accelerate.

derived from environmental microbiology's many contributing disciplines (from microbiology and soil science, to oceanography and ecology, to physiology and biochemistry, to genomics and systems biology; see Section 1.5 and Table 1.5). We will never know it all because there is always more to discover – new habitats, new cultured microorganisms, physiological adaptations, ecological relationships, selective pressures, enzymatic structures, even new worlds (e.g., Mars). To quote C. Woese (2004):

Science is an endless search for truth. Any representation of reality we develop can be only partial. There is no finality, sometimes no single best representation. There is only deeper understanding, more revealing and enveloping representations. Scientific advance, then, is a succession of newer representations superseding older ones, either because an older one has run its course and is no longer a reliable guide for a field or because the newer one is more powerful, encompassing, and productive than its predecessor(s).

Thus, predicting the details of future progress in environmental microbiology is not facile. As shown in Box 9.5, accelerating progress is certain. This progress must be a balance between an ever-deepening understanding of environmental microbiology's many subdisciplines and the emergent properties of complex systems (Woese, 2004).

STUDY QUESTIONS

1 On systems biology:
   (A) If you were an instructor of an introductory biology course, how would you define systems biology?
   (B) What means and examples would you use to convey what systems biology seeks to accomplish?
   (C) Does it make sense to apply systems biology to environmental microbiology now?
2 On the genetic basis of fitness:
   (A) What ecological system do you find particularly fascinating? (Name one, for example coral reef, redwood forest, salt marsh.)
   (B) Why?
   (C) What microbiologically mediated biogeochemical processes occur there? (Name several.)
   (D) Venture several guesses about specific fitness traits that microorganisms need to be successful in your chosen ecosystem.
   (E) How would you investigate the genetic basis of these fitness traits?
3 Which of the conceptual models for the organization of environmental microbiology (see Box 9.5) do you find to be most insightful in encompassing the many elements of the discipline? Why?

REFERENCES

Allen, E.E., G.W. Tyson, R.J. Whittaker et al. 2007. Genome dynamics in a natural archaeal population. *Proc. Natl. Acad. Sci. USA* **104**:1883–1888.

Aneja, M.K., S. Sharma, J.C. Munch, and M. Schloter. 2004. RNA fingerprinting – a new method to screen for differences in plant litter degrading microbial communities. *J. Microbiol. Meth.* **59**:223–231.

Arnaud, C.H. 2006. Systems biology's clinical future. *Chem. Eng. News* **84**:17–26.

Borodina, I. and J. Nielsen. 2005. From genomes to *in silico* cells via metabolic networks. *Curr. Opin. Biotechnol.* **16**:350–355.

Botero, L.M., S. D'Imperio, M. Burr, T.R. McDermott, M. Young, and D.J. Hassett. 2005. Poly(A) polymerase modification and reverse transcriptase PCR amplification of environmental RNA. *Appl. Environ. Microbiol.* **71**:1267–1275.

Brzostowicz, P.D., D.M. Walters, S.M. Thomas, V. Nafarajan, and P.E. Rouviere. 2003. mRNA differential display in a microbial enrichment culture: simultaneous identification of three cyclohexanone monooxygenases from three species. *Appl. Environ. Microbiol.* **69**:334–342.

Chiang, S.L., J.J. Mekalanos, and D.W. Holden. 1999. In vivo genetic analysis of bacterial virulence. *Annu. Rev. Microbiol.* **53**:129–154.

Coleman, M.I., M.B. Sullivan, A.C. Martiny, et al. 2006. Genomic islands and the ecology and evolution of *Prochlorococcus*. *Science* **311**:1768–1770.

DeLong, E.F. and D.M. Karl. 2005. Genomic perspectives in microbial oceanography. Nature **437**: 336–342.

Fleming, J.T., W.H. Yao, and G.S. Sayler. 1998. Optimization of differential display prokaryotic mRNA: application to pure cultures and soil microcosms. *Appl. Environ. Microbiol.* **64**:3698–3706.

Groh, J.L., Q. Luo, J.D. Ballard, and L.R. Krumholz. 2005. A method adapting microarray technology for signature-tagged mutageness of *Desulfovibrio desulfuricans* G20 and *Shewanella oneidensis* MR-1 in anaerobic sediment survival experiments. *Appl. Environ. Microbiol.* **71**:7064–7074.

Hutchinson, G.E. 1958. Concluding remarks. *Cold Spring Harbor Symp. Quant. Biol.* **22**:415–427.

Kazazian, H.H. 2004. Mobile elements: drivers of genome evolution. *Science* **303**:1626–1632.

Kell, D.B., M. Brown, H.M. Davey, W.B. Dunn, I. Spasic, and S.G. Oliver. 2005. Metabolic footprinting and systems biology: the medium is the message. *Nature Rev. Microbiol.* **3**:557–563.

Kitano, H. 2002. Systems biology: a brief overview. *Science* **295**:1662–1664.

Kunin, V. and C.A. Ouzounis. 2003. The balance of driving forces during genome evolution in prokaryotes. *Genome Res.* **13**:1589–1594.

Ochman, H. and L.M. Davalos. 2006. The nature and dynamics of bacterial genomes. *Science* **311**:1730–1733.

Poretsky, R.S., N. Bano, A. Buchan, et al. 2005. Analysis of microbial gene transcripts in environmental samples. *Appl. Environ. Microbiol.* **71**: 4121–4126.

Price, N.D., J.L. Reed, and B.O. Palsson. 2004. Genome-scale models of microbial cells: evaluating the consequences of constraints. *Nature Rev. Microbiol.* **2**:886–897.

Ram, R.J., N.C. Ver Berkmoes, M.P. Thelan, et al. 2005. Community proteomics of natural microbial biofilm. *Science* **308**:1915–1920.

Rediers, H., P.B. Rainey, J. Vanderleyden, and R. De Mot. 2005. Unraveling the secrete lives of bacteria: use of in vivo expression technology and differential fluorescence induction promoter traps as tools for exploring niche-specific gene expression. *Microbiol. Molec. Biol. Rev.* **69**:217–261.

Ricklefs, R.E. and G.L. Miller. 2000. *Ecology*, 4th edn. W.H. Freeman, New York.

Saleh-Lakha, S., M. Miller, R.G. Campbell, et al. 2005. Microbial gene expression in soil: methods, applications and challenges. *J. Microbiol. Meth.* **63**:1–19.

Tyson, G.W., J. Chapman, P. Hugenholtz, et al. 2004. Community structure and metabolism through reconstruction of microbial genomes from the environment. *Nature* **428**:37–43.

Whittaker, R.H. 1970. *Communities and Ecosystems*. MacMillan Publishing Co., New York.

Woese, C.R. 2004. A new biology for a new century. *Microbiol. Molec. Biol. Rev.* **68**:173–186.

Glossary

agar A vegetable gelatin made from various kinds of algae or seaweed. Used as a gelling agent in solid culture media for isolating and cultivating microorganisms.

agar plate An agar plate is a sterile Petri dish that contains agar plus culture media, and is used to cultivate microorganisms.

algae Chiefly aquatic, eukaryotic one-celled or multicellular plants without true stems, roots, or leaves, that are typically autotrophic, photosynthetic, and contain chlorophyll.

allochthonous Transported in from elsewhere; materials found in a place other than where they and their constituents were formed.

amoeba Any of various one-celled aquatic or parasitic protozoans, having no definite form and consisting of a mass of protoplasm containing one or more nuclei surrounded by a flexible outer membrane. Amoebae move by means of pseudopods.

anabolism Metabolic reactions that build cellular constituents, consuming energy (ATP) in the process.

Archaea One of the three major trunks (domains) in the tree of life based on small subunit rRNA phylogeny.

assimilation, assimilatory Physiology. The conversion of nutrients into living tissue via biosynthetic metabolic reactions.

autochthonous Originating or formed in the place where found.

ATP Adenosine triphosphate, the energy currency of cellular metabolism. Catabolism generates ATP, anabolism consumes it.

autotroph An autotroph (or primary producer) is an organism that uses CO_2 as a carbon source. CO_2 fixation is driven by energy (ATP) derived from light or chemicals.

Bacteria One of the three major trunks (domains) in the tree of life based on small subunit rRNA phylogeny.

bacteria A generic term referring to prokaryotic forms of life (*Bacteria* and *Archaea*). The use of the term "bacteria" is historical baggage from early microbiologists' inability to distinguish between *Bacteria* and *Archaea*. In contemporary biology, the term is inherently inaccurate and often deemed obsolete. Because *Bacteria* and *Archaea* are vastly different (phylogenetically and phenotypically), referring to these two domains with a single term may cause erroneous thought.

basalt The most common type of solidified lava; a dense, dark gray, fine-grained igneous rock that is composed chiefly of plagioclase, feldspar, and pyroxene minerals.

biomarker A biochemical substances whose presence and structure serves as a signature of life. Examples of biomarkers include 16S rRNA, membrane lipids, and chlorophyll.

catabolism The metabolic breakdown of large molecules in living organisms, with accompanying release of energy, captured as ATP.

chemosynthesis, chemosynthetic Generation of ATP by reactions that oxidize organic and/or inorganic substances.

chromatin Condensed chromosomal DNA.

chromosome Major reservoir of genetic information in a cell; haploid cells have a single copy of their chromosome, while diploid cells have two copies.

ciliate, ciliated protozoa Any of various protozoans of the class Ciliata, characterized by an exterior covered with numerous short, whip-like appendages, called cilia (*see* flagellum).

clade A group of organisms, such as a genus, whose members share homologous features derived from a common ancestor. On a phylogenetic tree, a clade is a coherent cluster of related sequences.

conjugation Genetic transfer of plasmid DNA between cells; cell-to-cell contact is required.

culturable (*adj.*) Capable of being cultured. A microorganism is culturable if a suitable liquid or solid medium, supporting propagation, has been devised.

cultured (*adj.*) A microorganism that can be and has been grown under defined, controlled conditions. Domesticated.

cultivated *See* cultured.

cyanobacteria Free-living prokaryotic organisms without organized chloroplasts but having chlorophyll *a* and oxygen-evolving photosynthesis; capable of fixing nitrogen in heterocysts. Occurring in lichens and other symbiotic relationships. Commonly called "blue-green algae".

detritus Disintegrated or eroded matter: the remnants of prior life, "the detritus of the past".

dissimilation, dissimilatory Physiology. Participation in cellular processes without incorporation into biomass. Final electron acceptors are essential for physiological reactions but their reduced forms are waste products (e.g., H_2O, H_2S, CH_4) that are not assimilated.

electron microscopy Any of a class of microscopic procedures that use electrons rather than visible light to produce magnified images.

elective enrichment A method of isolating microorganisms capable of utilizing a specific substrate by incubating an inoculum in a medium containing the substrate. The medium may contain substances or have characteristics that inhibit the growth of unwanted microorganisms.

endolithic life Literally, "within rock". Microorganisms that have colonized and live within the pore space of geologic formations, from Antartica to the ocean floor.

endospore A small asexual spore, produced within a cell, that serves as a resistant resting stage in some prokaryotes.

enrichment culture A medium of known composition, providing specific physiological conditions to favor the growth of a particular type of microorganism.

epifluorescent microscopy A microscopic technique in which the impinging short-wavelength light causes target fluorescent molecules to emit longer wavelength light back to the viewer.

Eukarya One of the three major trunks (domains) in the tree of life based on small subunit rRNA phylogeny. *Eukarya* include complex nucleated life forms such as protozoa, plants, and animals.

eutroph A physiological type of microorganism that is suited to growth on nutrients at high concentration.

eutrophic (adj.) A habitat having waters rich in mineral and organic nutrients that promote a proliferation of microorganisms. If large blooms of photosynthetic algae occur in eutrophic water, their decay by heterotrophic prokaryotes may deplete oxygen in the water.

facultative Versatile; capable of functioning under varying environmental conditions. The term is often applied to microorganisms that can live with or without oxygen (facultative anaerobe).

flagellated protozoa Any of a large group of single-celled, microscopic, eukaryotic organisms having one or more large whip-like appendages, known as flagella (*see* flagellum)

flagellum In prokaryotes, a whip-like motility appendage present on the surface of some species. Flagella are composed of a protein called flagellin. Prokaryotes can have a single flagellum, a tuft at one pole, or multiple flagella covering the entire surface. In eukaryotes, flagella are thread-like protoplasmic extensions used to propel flagellates. Flagella have the same basic structure as cilia but are longer in proportion to the cell bearing them and present in much smaller numbers.

fungi Major group of heterotrophic, eukaryotic, single-celled, multinucleated, or multicellular organisms, including yeasts, molds, and mushrooms.

genotype Genetic makeup, as distinguished from physical and biochemical traits, of an organism or a group of organisms. Genotype is a set of fixed genetic characteristics, while phenotype is the subset of expressed characteristics. (Contrast with phenotype.)

granite A common, coarse-grained, light-colored, hard igneous rock consisting chiefly of quartz, orthoclase or microcline, and mica, and occurring as massive geologic formations, especially in mountain ranges.

genome The complete set of DNA carried by an organism. In prokaryotes, this means chromosomal plus plasmid DNA.

heterotroph Organism requiring an organic form of carbon as a carbon source. Heterotrophs do not fix CO_2.

horizontal gene transfer Horizontal gene transfer is any process in which an organism transfers genetic material (i.e., DNA) to another cell that resides in another branch of the evolutionary tree (lines of descent).

humic substances A series of relatively high molecular weight, yellow- to black-colored organic substances formed by secondary synthesis reactions in soils. The term is used in a generic sense to describe the colored material or its fractions obtained on the basis of solubility characteristics. These materials are distinctive to soil environments in that they are dissimilar to the biopolymers of microorganisms and higher plants (including lignin).

humus A brown or black organic substance consisting of partially or wholly decayed vegetable or animal matter that provides nutrients for plants and increases the ability of soil to retain water (*see* humic substances).

hypha, hyphae Tubular structures that constitute the basic cellular unit of most fungi.

hypoxia An environmental condition featuring reduced oxygen concentration.

igneous Geology. Resulting from, or produced by, the action of fire. Lavas and basalt are igneous rocks.

intrinsic bioremediation Elimination of pollutants at a given site by the naturally occurring microbial populations functioning in the naturally occurring chemical, biological, and geologic conditions. Also known as natural attenuation, when dominated by biological processes.

Koch's postulates A set of criteria, devised by Robert Koch in 1884, to establish causality between a pathogenic microorganism and its impact on a diseased host.

lateral gene transfer *See* horizontal gene transfer.

ligand An ion, a molecule, or a molecular group that binds to another chemical entity to form a larger complex.

lithotroph Physiology. Term used in reference to an organism's energy source (electron donor); an organism that uses an inorganic substrate (such as ammonia, hydrogen) as an electron donor in energy metabolism. Literally, lithothroph means "rock eater".

lysis To break open a cell.

lysogeny The condition of a host bacterium that has incorporated a virus into its own genetic material. When a virus infects a bacterium it can either destroy its host (lytic cycle) or be incorporated in the host genome in a state of lysogeny.

metabolism The integration of cellular biochemical reactions; often subdivided into catabolism (generation of energy as ATP) and anabolism (biosynthetic reactions that build and maintain new cells at the expense of ATP).

metamorphic Geology. Pertaining to, produced by, or exhibiting certain changes which minerals or rocks may have undergone since their original deposition; especially applied to the recrystallization that sedimentary rocks have undergone through the influence of heat and pressure, after which they are called metamorphic rocks.

micron (μ) A unit of length equal to one-millionth (10^{-6}) of a meter.

mineralization Conversion of organic substances to inorganic ones. For example, the production of CO_2 or NO_3^- from decaying organic matter.

mRNA Messenger RNA, the key intermediary in gene expression. mRNA serves as template for translation of the genetic code into chains of amino acids that constitute proteins.

mycelium, mycelia Fungal biomass, a network of hyphae.

natural attenuation Generally refers to the physical, chemical, or biological processes which, under favorable conditions, lead to the reduction of mass, toxicity, mobility, volume, or concentration of organic and/or inorganic contaminants in soil, sediment, and/or groundwater. The reduction takes place as a result of processes such as biological or chemical degradation, sorption, and others.

obligate The ability to live in only a set parameter of conditions; for example, an obligate anaerobe can survive only in the absence of oxygen.

oligocarbophile Able to grow only on highly diluted carbon substrates (*see* oligotroph).

oligotroph A physiological type of microorganism (or phase in the life of marine microorganisms) that is suited to growth on nutrients at low concentration.

oligotroph (adj) A habitat low in mineral and organic nutrients.

organotroph Microbial nutrition dependent upon organic compounds.

osmotroph An organism that obtains nutrients through the active uptake of soluble materials across the cell membrane. This class of organism, which includes the prokaryotes and fungi, cannot directly utilize particulate material as nutrients. (Contrast with phagotroph.)

OTU Operational taxonomic unit; this often refers to distinctive 16S rRNA gene sequences retrieved from environmental samples. OTU is the safe terminology in environmental microbiology because it avoids the use of the term "species" (which is very difficult to define).

oxidative phosphorylation Use of electron transport to produce ATP, via proton motive force.

oxidative stress A state of metabolic imbalance caused by increased levels of free radicals and other oxidation-promoting molecules that may result in cell membrane damage, cell death, and damage to genetic material.

PCR Polymerase chain reaction; a technique for amplifying DNA, making it easier to isolate, clone, and sequence.

Petri dish A shallow glass or plastic cylindrical dish that microbiologists use to culture microorganisms. It was named after the German bacteriologist Julius Richard Petri (1852–1921) who invented it in 1877 when working as an assistant to Robert Koch.

phage Viruses that infect prokaryotes.

phagotroph An organism that engulfs food. (Contrast with osmotroph.)

phenotype The observable physical or biochemical characteristics of an organism, as determined by both genetic makeup and environmental influences. (Contrast with genotype.)

photosynthesis A biochemical process in which plants, algae, and some prokaryotes harness the energy of light. The "light reactions" generate ATP and reducing power; the dark reactions fix CO_2 into biomass.

phylogeny Phylogeny documents the evolutionary relationship between organisms. The phylogeny of a particular organism reflects its ancestry, its own evolutionary developments.

phylum, phyla Used in classification and taxonomy. A phylum is a broad group of related microorganisms and contains one or more classes. A group of similar phyla forms a domain.

planetisemal Astronomy. Any of innumerable small bodies thought to have orbited the Sun during the formation of the planets.

plasmid Extrachromosomal DNA that replicates independently of the chromosome. The metabolic functions encoded by plasmid DNA typically are peripheral (e.g., catabolism of unusual carbon compounds), and not essential for cell function.

plates Microbiology. Plates refer to Petri dishes containing a microbial growth medium, usually solidified with agar.

prokaryote A unicellular organism lacking a nuclear membrane, a discrete nucleus, and other specialized compartments within the cell. *Bacteria* and *Archaea* are prokaryotes. The use of the term "prokaryote" is historical baggage from early microbiologists' inability to distinguish between *Bacteria* and *Archaea*. In contemporary biology, the term is inherently inaccurate and often deemed obsolete. Because *Bacteria* and *Archaea* are vastly different (phylogenetically and phenotypically), referring to these two domains with a single term may cause erroneous thought.

proton motive force Generated by electron transport, this is the separation of H+ and OH-icons across biological membranes; used by the ATP synthase enzyme to generate ATP.

quorum sensing The ability of populations of prokaryotes to communicate and coordinate behavior via signaling molecules.

16S rRNA and 18S rRNA The "1S" designation refers to "Svedberg units", which are sedimentation coefficients derived from the physical means of separating components of ribosomes from one another. Prokaryotes have 16S, while eukaryotes have 18S ribosomal RNA subunits. (*See* rRNA; small subunit rRNA.)

rDNA DNA sequences coding for rRNA.

rRNA Ribosomal RNA; major structural component of ribosomes. Ribosomes are the site of protein synthesis in cellular life. (*See* small subunit rRNA.)

regulatory gene A DNA sequence whose encoded protein (after transcription and translation) controls the expression of other genes. (Contrast with structural gene.)

resource A metabolic asset, present in the habitat of a microorganism, that can be used in biochemical reactions to generate ATP and/or be used as cellular building blocks to create new cells.

respiration Membrane-based coupling of electron flow between a reduced energy source (organic or inorganic) and an oxidized terminal-electron acceptor (e.g., oxygen, nitrate) to produce ATP.

scanning electron microscopy (SEM) Any of a class of microscopic procedures that use electrons rather than visible light to produce magnified images, especially of objects having dimensions smaller than the wavelengths of visible light, with linear magnification approaching or exceeding a million (10^6).

sedimentary Rocks formed from material, including debris of organic origin, deposited as sediment by water, wind, or ice and then compressed and cemented together by pressure.

senescence Decline or degeneration of cellular function, as with maturation, age, or disease stress.

slime mold Any of various single-celled eukaryotes that grow on dung and decaying vegetation and have a life cycle characterized by a slime-like amoeboid stage and a multicellular reproductive stage.

small subunit rRNA Ribosomal RNA from the 30S ribosomal subunit of prokaryotes or the 40S ribosomal subunit of eukaryotes. Small subunit rRNA plays a crucial cellular role in the translation of mRNA into protein. Used in molecular phylogeny to establish the three domains of life. (*See* rRNA.)

spore A small, usually single-celled, reproductive body that is highly resistant to desiccation and heat and is capable of growing into a new organism, produced especially by certain prokaryotes, fungi, algae, and nonflowering plants.

structural gene A DNA sequence whose encoded protein (after transcription and translation) has a definite catalytic or structural role in cell metabolism. (Contrast with regulatory gene.)

sulfatara A volcanic area or vent, characterized by high temperature, sulfur vapors, and steam.

taxonomy The science of identification, classification, and nomenclature.

thallus Fungal biomass, a network of mycelia.

tomography Any of several techniques for making detailed X-rays of a predetermined plane section of a solid object while blurring out the images of other planes. Multiple planar images can then be assembled.

transduction The transfer of microbial DNA by viruses from one infected microorganism to another.

transformation A process by which the genetic material carried by an individual microbial cell is altered by incorporation of exogenous DNA into its genome. The source DNA taken up by the recipient cell is free of any vector (no virus, no plasmid).

transmission electron microscopy (TEM) Procedure using a beam of highly energetic electrons to examine objects very closely, on a fine scale. The microscope shines a beam of electrons through an object and the transmitted result is projected onto a phosphor screen.

tree of life Graphic presentation of evolutionary relationships between past and present forms of life. The current dominant paradigm for evolution is based on phylogenetic analysis of genes encoding small subunit rRNA.

tRNA Short-chain RNA molecules present in the cell (in at least 20 varieties, each variety capable of combining with a specific amino acid) that attach the correct amino acid to the protein chain that is being synthesized by the ribosome of the cell.

ultraviolet light (UV) Light waves that have a shorter wavelength than visible light, but longer wavelength than X rays.

vertical gene transfer Vertical gene transfer occurs when an organism receives genetic material from its ancestor (e.g., its parent or a species from which it evolved).

viable plate counts A way to count culturable microorganisms in environmental samples, on solid agar media. The number of microbial colonies that grow on agar plates is inversely proportional to the degree the environmental sample was diluted.

virus A small particle, containing DNA or RNA, that infects cells. Viruses are obligate intracellular parasites; they can reproduce only by invading and taking over other cells as they lack the cellular machinery for self-reproduction.

Index

Page numbers in *italic* refer to figures; page numbers in **bold** refer to tables